Scottish Graduate Series

Neutrinos in Particle Physics, Astrophysics and Cosmology

F. J. P. Soler
University of Glasgow, Scotland, UK

Colin D. Froggatt
University of Glasgow, Scotland, UK

Franz Muheim
University of Edinburgh, Scotland, UK

CRC Press
Taylor & Francis Group
Boca Raton London New York

CRC Press is an imprint of the
Taylor & Francis Group, an **informa** business
A TAYLOR & FRANCIS BOOK

CRC Press
Taylor & Francis Group
6000 Broken Sound Parkway NW, Suite 300
Boca Raton, FL 33487-2742

First issued in paperback 2019

ISBN-13: 978-1-4200-8239-5 (hbk)
ISBN-13: 978-0-367-38649-8 (pbk)

Library of Congress Cataloging-in-Publication Data

Neutrinos in particle physics, astrophysics and cosmology / editors, F.J.P. Soler, Colin D. Froggatt, Franz Muheim.
p. cm. -- (Scottish graduate series ; 61)
Includes bibliographical references and index.
ISBN 978-1-4200-8239-5 (alk. paper)
1. Neutrinos. 2. Neutrino astrophysics. I. Soler, F. J. P. (F. J. Paul), 1962- II. Froggatt, C. D. III. Muheim, Franz, 1960-

QC793.5.N42N55 2008
539.7'215--dc22 2008031914

Visit the Taylor & Francis Web site at
http://www.taylorandfrancis.com

and the CRC Press Web site at
http://www.crcpress.com

SUSSP Proceedings

1	1960	Dispersion Relations
2	1961	Fluctuation, Relaxation and Resonance in Magnetic Systems
3	1962	Polarons and Excitons
4	1963	Strong Interactions and High Energy Physics
5	1964	Nuclear Structure and Electromagnetic Interactions
6	1965	Phonons in Perfect and Imperfect Lattices
7	1966	Particle Interactions at High Energy
8	1967	Methods in Solid State and Superfluid Theory
9	1968	Physics of Hot Plasmas
10	1969	Quantum Optics
11	1970	Hadronic Interactions of Photons and Electrons
12	1971	Atoms and Molecules in Astrophysics
13	1972	Properties of Amorphous Semiconductors
14	1973	Phenomenology of Particles at High Energy
15	1974	The Helium Liquids
16	1975	Non-linear Optics
17	1976	Fundamentals of Quark Models
18	1977	Nuclear Structure Physics
19	1978	Metal Non-metal Transitions in Disordered Solids
20	1979	Laser-Plasma Interactions: 1
21	1980	Gauge Theories and Experiments at High Energy
22	1981	Magnetism in Solids
23	1982	Laser-Plasma Interactions: 2
24	1982	Lasers: Physics, Systems and Techniques
25	1983	Quantitative Electron Microscopy
26	1983	Statistical and Particle Physics
27	1984	Fundamental Forces
28	1985	Superstrings and Supergravity
29	1985	Laser-Plasma Interactions: 3
30	1985	Synchrotron Radiation
31	1986	Localisation and Interaction
32	1987	Computational Physics

/continued

SUSSP Proceedings (continued)

33	1987	Astrophysical Plasma Spectroscopy
34	1988	Optical Computing
35	1988	Laser-Plasma Interactions: 4
36	1989	Physics of the Early Universe
37	1990	Pattern Recognition and Image Processing
38	1991	Physics of Nanostructures
39	1991	High Temperature Superconductivity
40	1992	Quantitative Microbeam Analysis
41	1992	Spatial Complexity in Optical Systems
42	1993	High Energy Phenomenology
43	1994	Determination of Geophysical Parameters from Space
44	1994	Simple Quantum Systems
45	1994	Laser-Plasma Interactions 5: Inertial Confinement Fusion
46	1995	General Relativity
47	1995	Laser Sources and Applications
48	1996	Generation and Application of High Power Microwaves
49	1997	Physical Processes in the Coastal Zone
50	1998	Semiconductor Quantum Optoelectronics
51	1998	Muon Science
52	1998	Advances in Lasers and Applications
53	1999	Soft and Fragile Matter
54	2000	The Restless Universe
55	2001	Heavy Flavour Physics
56	2002	Ultrafast photonics
57	2003	Large Hadron Collider Phenomenology
58	2004	Hadron Physics
59	2004	Soft Condensed Matter Physics in Molecular and Cell Biology
60	2005	Laser Plasma Interactions
61	2006	Neutrinos in Particle Physics, Astrophysics and Cosmology
62	2007	Extra-Solar Planets
63	2008	High pressure Physics
64	2008	Advanced Techniques in Electron Spin Resonance

Lecturers

Chris T Sachrajda	The University of Southampton, Southampton, U.K.
Boris Kayser	Fermi National Accelerator Laboratory, Batavia, Illinois, USA.
Kevin McFarland	University of Rochester, Rochester, New York, USA.
Guido Altarelli	Universita' di Roma Tre, Rome, Italy and CERN, Geneva, Switzerland.
Aldo Serenelli	Institute for Advanced Study, Princeton, New Jersey, USA.
David Wark	Imperial College London and STFC, Rutherford Appleton Laboratory.
Georg G. Raffelt	Max-Planck-Institut für Physik (Werner-Heisenberg-Institut) München, Germany.
Deborah A. Harris	Fermi National Accelerator Laboratory, Batavia, Illinois, USA.
Takaaki Kajita	Institute for Cosmic Ray Research, University of Tokyo, Japan.
Edwin Blucher	The University of Chicago, Chicago, Illinois, USA.
Beate Bornschein	Tritium Laboratory Karlsruhe, Forschungszentrum Karlsruhe, Germany.
Kai Zuber	Department of Physics and Astronomy, University of Sussex, Sussex, U.K.
Yoshitaka Kuno	Osaka University, Osaka, Japan.
Wilfried Buchmüller	Deutsches Elektronen-Synchrotron DESY, Hamburg, Germany.
Sergio Pastor	Instituto de Física Corpuscular (CSIC-Universitat de València), Valencia, Spain.

All participants (alphabetical order)

1 Guido ALTARELLI
2 Erica ANDREOTTI
3 Artur ANKOWSKI
4 Luisa ARRABITO
5 Magali BESNIER
6 Ed BLUCHER
7 Mathieu BONGRAND
8 Beate BORNSCHEIN
9 Gwenaelle BROUDIN-BAY
10 Wilfried BUCHMULLER
11 Nathalie CHON-SEN
12 Lucia CONSIGLIO
13 Pierre Luc DROUIN
14 Stephan EISENHARDT
15 Jesus ESCAMILLA
16 Maria Catalina ESPINOZA HERNANDEZ
17 Nasim FATEMI GHOMI
18 Colin FROGGATT
19 Rashida FROGGATT
20 Nigel GLOVER
21 Felix GONZALEZ CANALES
22 Vannia GONZALEZ
23 Deborah HARRIS
24 Jeffrey John HARTNELL
25 Irene HIDALGO
26 Benjamin JANUTTA
27 Fernando JIMENEZ ALBUQUERQUE
28 Takaaki KAJITA
29 Boris KAYSER
30 Shiva KING
31 Yoshi KUNO
32 Yves Lemiere
33 Jacobo LOPEZ PAVON
34 Rui LUO
35 Victoria MARTIN
36 Kevin MCFARLAND
37 Aaron MISLIVEC
38 Teresa MORRODAN-UNDAGOITIA
39 Franz MUHEIM
40 Alex MURPHY
41 Minja MYYRYLAINEN
42 Irina NASTEVA
43 Leanne O'DONNELL
44 Christopher ORME
45 Sergio PASTOR
46 Alessandra PASTORE
47 Elizabeth PEACH
48 Ken PEACH
49 Aleksandra PIOROWSKA
50 Magdalena Zofia POSIADALA
51 Georg RAFFELT
52 Cipriano RIBEIRO NEI
53 Bernd RIENHOLD
54 Janne Henri Tapani RIITTINEN
55 Daniel ROYTHORNE
56 Chris SACHRAJDA
57 Rikard SANDSTROM
58 John Francois van SCHALKWYK
59 Aldo SERENELLI
60 Alicia Cecylia SMOLIN-JOLIEC
61 Paul SOLER
62 Benjamin STILL
63 Ian TAYLOR
64 Ali Uliv YILMAZER
65 Tomasz WACHALA
66 Kenny WALARON
67 Mark WARD
68 Dave WARK
69 Bjorn Sonke WONSAK
70 Michael WRUM
71 Kai ZUBER

Organising Committee

Professor K. Peach	University of Oxford and Royal Holloway	*Co-Director*
Professor N. Glover	IPPP, University of Durham	*Co-Director*
Dr A. Murphy	University of Edinburgh	*Secretary*
Prof C.D. Froggatt	University of Glasgow	*Treasurer and Co-Editor*
Dr F. Muheim	University of Edinburgh	*Co-Editor*
Dr F.J.P. Soler	University of Glasgow	*Co-Steward and Co-Editor*
Dr V. Martin	University of Edinburgh	*Co-Steward*
Dr S. Eisenhardt	University of Edinburgh	*Co-Steward*
Ms L. O'Donnell	University of Edinburgh	*Administration*

Preface

It is just over 75 years since Pauli's famous letter of 4th December 1930 proposing his "desperate remedy" to save the "exchange theorem of statistics and the law of conservation of energy". He proposed that there "could exist in the nuclei electrically neutral particles, that I wish to call neutrons, which have spin 1/2", which nowadays we call "neutrinos". For more than 20 years, it was thought unlikely that neutrinos could be observed because they interacted so weakly with matter. However, about 50 years ago, Reines and Cowan detected them experimentally at the Savannah River reactor. The rest is, as the famous saying has it, "history".

Except that the history of the neutrino is not yet complete. Initially, the "neutrino questions" concerned the nature of the neutrino - does it exist and what are its properties? In the 1970s and 1980s, the neutrino was used to investigate the nature of the weak interaction, and as a sensitive probe of nucleon structure. Now, in the last decade of the 20th century and the first decade of the 21st century, the focus has shifted back to the nature of the neutrino - what are the parameters describing neutrino oscillations, what are the neutrino masses, what is the character of the neutrino (is it a Dirac or Majorana particle)? Studying these neutrino properties requires more powerful neutrino sources, which will also allow the more precise studies of conventional "neutrino physics" - structure functions, weak currents, etc. Neutrinos are important components of astrophysics and cosmology - the physics of stars and the rate of cooling of the Universe, for example, depend in part upon them.

The lectures in this book cover all aspects of neutrino physics. We hope that this will provide a work of reference for people new to the field and a comprehensive summary for those already involved. The lectures cover the basic experimental data and future prospects for experiments on neutrino oscillations, neutrino interactions and the measurement of the neutrino properties. Neutrino phenomenology is discussed in the context of the Standard Model of particles and their interactions, along with the other relevant "Standard Models" of the solar energy cycle and of big-bang cosmology, and the impact of neutrinos on astrophysical phenomena. Finally, there are lectures on the origins of neutrino masses and the possible role of neutrinos in the baryon asymmetry of the Universe.

The editors would like to thank Kenny Walaron for help in the preparation of this manuscript. We would also like to thank all of the lecturers and participants for their enthusiasm, for physics and for life, which helped make this a truly memorable school.

It is becoming increasingly important that we take our message to the public. Consequently, during the School, there was a public lecture on the neutrino by Professor David Wark (Imperial College and Rutherford Appleton Laboratory), given in the afternoon mainly for schools and in the evening for the general public; both were well attended. We are particularly grateful to the sponsors for their support of this event, which attracted significant local publicity.

But Summer Schools are not just about science - they are about dialogue, discussion, meeting people, and forming lifelong friendships. The School succeeded in this secondary aim, aided by a full social programme, including a memorable whisky tasting provided by Colin Scott and his team from Chivas Bros Ltd, and the friendly environment provided by the staff of the John Burnet Hall. Within this relaxed atmosphere, the scientific discussions and personal interactions flourished.

The Organizing Committee acknowledge the support of the PPARC (the UK Particle Physics and Astronomy Research Council), the Institute for Particle Physics Phenomenology in Durham, the Scottish Universities Summer Schools in Physics, the Institute of Physics, British Energy for supporting the public lectures on neutrino physics held as part of the school, and the Physics and Astronomy Departments of the Universities of Edinburgh, Glasgow and St. Andrews (part of the Scottish Universities Physics Alliance), without which the school would not have been possible.

Neutrinos may be elusive but they are important. Without them, the sun does not shine, supernovae do not happen, the trace elements which make life possible would not exist, and perhaps the Universe itself would be a cold, empty space. We hope that you will find all that you want to know about the neutrino, and more, in this book.

Nigel Glover and Ken Peach

Co-Directors, January 2008

Contents

Section I: Neutrinos in the Standard Model and Beyond

The Standard Model of Particle Physics

Chris T. Sachrajda

School of Physics and Astronomy, The University of Southampton,
Southampton SO17 1BJ, UK

1 Introduction

In this course I will briefly review the construction of the Standard Model of particle physics (SM) and the tools being used to make phenomenological predictions. Of course it is not possible during 5 lectures to present detailed derivations of the results or to do justice to the rich structure of the theory. I will assume that the students have previously attended graduate courses in quantum field theory and particle physics, so my aim will be to tell a coherent story, reminding the audience of the key elements of the standard model and outlining the use of flavour physics to explore its limits.

There are many excellent textbook which cover the elements of the standard model in detail. In preparing these lectures I have not followed the presentation in any single textbook, however ref. (Peskin & Schroeder, 2005) contains both the elementary field theory and the construction of the standard model.

1.1 Elements of the Standard Model

The SM is a gauge theory based on the gauge group

$$SU(2) \times U(1) \times SU(3)\,, \tag{1}$$

where the $SU(3)$ corresponds to Quantum Chromodynamics (QCD), the theory of the Strong Nuclear force and the $SU(2) \times U(1)$ to the Electroweak theory. The number of gauge bosons is $3+1+8=12$ (recall that the number of gauge bosons for an $SU(n)$ gauge group, which necessarily transform under the adjoint representation, is n^2-1). As we will discuss in detail below, these 12 gauge bosons are the $W^{\pm}$, Z^0, the photon γ and the eight gluons.

The fermions come in three generations of quarks and leptons:

1st Generation:	u	d	e^-	ν_e
2nd Generation:	c	s	μ^-	ν_μ
3rd Generation:	t	b	τ^-	ν_τ

Finally there is the Higgs boson, the physical scalar particle which is the consequence of the Higgs mechanism for the generation of the masses of the particles.

In spite of the simplicity of these elements, the SM contains a rich and subtle structure which continues to be tested to great precision at experimental facilities, so far with frustrating success. It does however, leave many unanswered questions and is surely incomplete. The purpose of these lecture is to review the structure and status of the SM and the tools being used to make SM calculations.

1.2 Some Unanswered Questions

Even before we examine the structure of the theory in detail we can raise some important questions:

1. Why are the charges of the electron (Q_e) and proton (Q_P) equal and opposite. From the particle data group (Yao et al, 2006) we know that

$$\left|\frac{Q_e+Q_P}{Q_e}\right| < 10^{-21}\,. \tag{2}$$

This strongly suggests that quarks and leptons are related, and yet in the standard model the leptons do not interact strongly and yet the quarks do. This in turn provides a motivation for the unification of forces and to many of the attempts to construct theories of physics *beyond the standard model* (BSM).

2. Why are there three generations of quarks and leptons? This is a modern variation of Isidor Rabi's famous quotation "Who ordered that?" after the muon was discovered in the 1930s.

3. What is the reason behind the huge variation in fermion masses?

In this course I will introduce the elements of the electroweak theory (lectures 1 and 2) and of QCD (lecture 3). The last two lectures will be devoted to an introductory discussion of *flavour physics*, i.e. to a study of the weak decays of strongly interacting particles and related processes. Flavour physics is one of the central activities in our field being used to explore the limits of the standard model and to search for signatures of new physics. As will be discussed in some detail at this school, the recent confirmation of neutrino masses and oscillations leads to parallel features in the lepton sector, in which fermionic flavour eigenstates are not the mass eigenstates. In the final two lectures, I will discuss flavour physics in the quark sector. The contents of the five lectures in this course are as follows:

Lecture 1	Spontaneous symmetry breaking
Lecture 2	The Electroweak theory
Lecture 3	Quantum Chromodynamics (QCD)
Lecture 4	Flavourdynamics and Non-Perturbative QCD I
Lecture 5	Flavourdynamics and Non-Perturbative QCD II

2 Lecture 1 – Spontaneous Symmetry Breaking

2.1 Weak Decays and Massive Vector Bosons

The short range nature of the weak force ($\sim 10^{-18}$ m) together with detailed experimental studies lead to the construction of the Fermi Model as a first approximation for the description of weak interactions. In this model the Lagrangian density is given by

$$\mathscr{L} = -\frac{G_F}{\sqrt{2}}\, j^{\dagger}_{\mu}\, j^{\mu} \tag{3}$$

where G_F is the Fermi constant and j_μ is the weak $V-A$ current:

$$j^\mu = \bar{\nu}_e\gamma^\mu(1-\gamma^5)e + \bar{\nu}_\mu\gamma^\mu(1-\gamma^5)\mu + \bar{\nu}_\tau\gamma^\mu(1-\gamma^5)\tau + \text{ Hadronic Terms}\,.$$

G_F has dimensions of $[m]^{-2}$, ($G_F \simeq 1.17\times 10^{-5}\,\text{GeV}^{-2}$), which implies that the loop corrections diverge as powers of the cut-off and therefore the 4-Fermion theory is non-renormalizable and inconsistent. Nevertheless, tree-level estimates of weak processes obtained using the lagrangian in eq. (3) generally yield very good first approximations to the rates for weak processes.

As an example of a weak decay process consider μ-decay, $\mu \to e\nu_\mu\bar{\nu}_e$, represented by the following diagram:

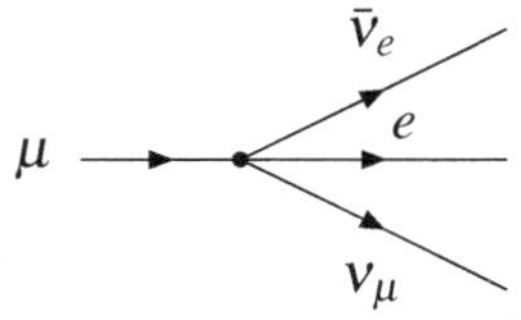

A natural suggestion for the cure of the problem of non-renormalizability is the introduction of an *intermediate vector boson*, which is sufficiently heavy to account for the short-range of the interaction. The point-like four-fermion interaction is replaced by the exchange of a vector boson, represented by the wiggly line in the diagram:

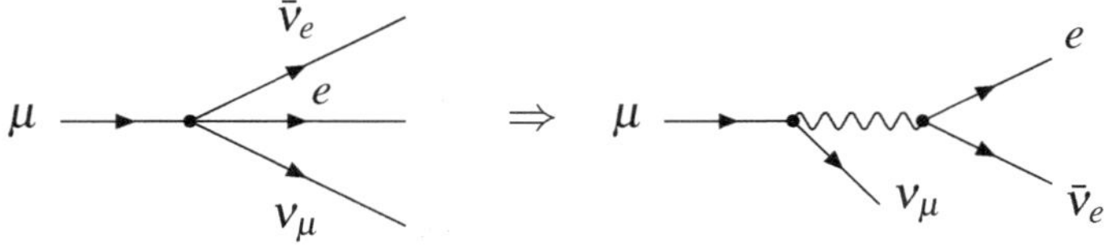

In such a picture, the dimensionful coupling constant G_F is an *effective coupling*, approximating the exchange of the vector boson, specifically

$$G_F \to \frac{g^2}{k^2 - M^2} \tag{4}$$

where g is a dimensionless coupling constant, M is the mass of the boson and k is the momentum flowing through the propagator. If the components of k are much smaller than M (as is the case in the tree-level diagram for μ-decay above), then we can drop the momentum in the denominator of the propagator and we indeed arrive at an effective coupling constant with dimension $[m]^{-2}$ ($G_F \propto g^2/M^2$). At larger momenta, and in particular in loops, we cannot neglect k^2 in the denominator of the propagator.

Looking in more detail however, it turns out that we have not yet solved the problem of renormalizability. The propagator of a massive vector boson is given by

$$\frac{-i}{k^2 - M^2}\left\{g^{\mu\nu} - \frac{k^\mu k^\nu}{M^2}\right\}, \tag{5}$$

where μ, ν are Lorentz indices. For large momenta the propagator $\sim ik^\mu k^\nu/(M^2 k^2)$ instead of being proportional to $1/k^2$ as is the case for the propagators of scalar bosons or those of massless vectors. By power counting we therefore see that the problem of non-renormalizability remains.

The above picture is part of the standard model, but the mass of the vector boson is generated by spontaneous symmetry breaking and the Higgs mechanism to which we now turn.

2.2 Spontaneous Symmetry Breaking and Goldstone Bosons

2.2.1 Spontaneous Breaking of a Discrete Symmetry

I start by briefly describing a model of a real scalar field ϕ with a discrete symmetry and the following potential:

$$V(\phi) = -\frac{\mu^2}{2}\phi^2 + \frac{\lambda}{4!}\phi^4. \tag{6}$$

This is like the scalar field theories which you would have studied in your introductory field theory courses except that the mass term has the *wrong* sign. For a constant field ϕ the potential has the form:

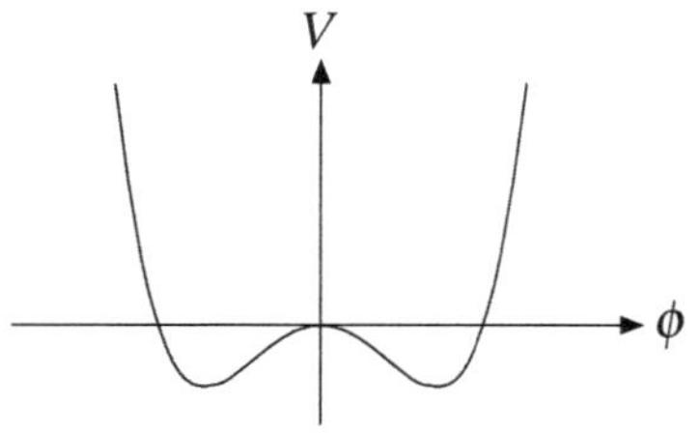

The lowest-energy classical field configuration is the translationally invariant field

$$\phi(x) = \phi_0 = \pm v = \pm\sqrt{\frac{6}{\lambda}}\,\mu. \tag{7}$$

v is called the *Vacuum Expectation Value of* ϕ. To interpret this theory, imagine quantum fluctuations close to one of the minima, $+v$ say. To study the quantum fluctuations it is convenient to write

$$\phi(x) = v + \sigma(x) \tag{8}$$

so that

$$V(\sigma) = \frac{1}{2}(2\mu^2)\,\sigma^2 + \sqrt{\frac{\lambda}{6}}\,\mu\sigma^3 + \frac{\lambda}{4!}\sigma^4 . \tag{9}$$

This is now a standard field theory of a scalar field, with mass $\sqrt{2}\mu$ and with cubic and quartic interactions. The original discrete $\phi \to -\phi$ symmetry is now hidden in the relations between the three coefficients (i.e. the mass and the coupling constants of the cubic and quartic terms in eq.(9)) in terms of two parameters λ and μ.

This is a particularly simple example of *spontaneous symmetry breaking*. The $\phi \to -\phi$ symmetry of the lagrangian is not a symmetry of the vacuum.

Note that in quantum mechanics the situation is qualitatively different. In one dimension for example, if $V(x) = ax^4 - bx^2$, then the ground state energy eigenfunction is a symmetric wavefunction peaked around the two minima at $x = \pm\sqrt{b/2a}$. Tunnelling allows transitions between the two minima of the potential. In field theory for such tunnelling we would require the field to flip values from $\pm v$ to $\mp v$ at every point in space-time. Thus we have an infinite potential barrier and no such tunnelling.

2.2.2 Goldstone Bosons – Continuous Symmetries

More interesting features occur when we have the spontaneous breaking of a continuous symmetry. An instructive example is the linear sigma model, describing the interactions of N real scalar fields:

$$\mathscr{L} = \frac{1}{2}(\partial_\mu\phi^i)^2 + \frac{\mu^2}{2}(\phi^i)^2 - \frac{\lambda}{4}\left[(\phi^i)^2\right]^2 . \tag{10}$$

In eq.(10) there are implicit sums over i in each $(\phi^i)^2$ over $1 \le i \le N$. $\mathscr{L}$ is invariant under $O(N)$ rotational symmetry and the potential is minimized for any $\{\phi_0^i\}$ such that

$$(\phi_0^i)^2 = \frac{\mu^2}{\lambda} . \tag{11}$$

This is a condition only on the length of the vector ϕ_0 and we therefore have a continuous infinity of equivalent vacua.

For definiteness, let us choose

$$\phi_0 = (0, 0, \cdots, 0, v) \quad \text{with} \quad v = \frac{\mu}{\sqrt{\lambda}} . \tag{12}$$

Again we define the shifted fields:

$$\phi^i(x) = (\pi^j(x), v + \sigma(x)) \qquad j = 1, \cdots, N-1 . \tag{13}$$

in terms of which the Langrangian becomes:

$$\begin{aligned} \mathscr{L} &= \frac{1}{2}(\partial_\mu\pi^j)^2 + \frac{1}{2}(\partial_\mu\sigma)^2 - \frac{1}{2}(2\mu^2)\,\sigma^2 - \sqrt{\lambda}\mu\sigma^3 \\ &\quad -\sqrt{\lambda}\mu(\pi^j)^2\sigma - \frac{\lambda}{4}\sigma^4 - \frac{\lambda}{2}(\pi^j)^2\sigma^2 - \frac{\lambda}{4}\left[(\pi^j)^2\right]^2 . \end{aligned} \tag{14}$$

The most striking feature of the Lagrangian is that there is no term proportional to $(\pi^j)^2$. Thus the spectrum of this theory contains $N-1$ massless *Goldstone* Bosons and 1 massive boson (σ) with mass $\sqrt{2}\mu$. The existence of the Goldstone Bosons can be understood in terms of the zero modes.

Goldstone's theorem, which I will not prove here, states that the number of Goldstone bosons is equal to the number of *broken symmetries*, i.e. the number of generators of the potential which are not symmetries of the vacuum. Let us check this theorem for the case of the sigma model. The rotation group $O(N)$ has $N(N-1)/2$ generators. The vacuum field ϕ_0 in eq:(12) has a residual $O(N-1)$ symmetry with $(N-1)(N-2)/2$ generators. The number of broken symmetries is therefore $N(N-1)/2-(N-1)(N-2)/2 = N-1$, which is indeed equal to the number of Goldstone bosons.

There do not appear to be any massless scalar bosons in nature (although the pions, which are so much lighter than other hadrons, are interpreted as being the pseudo-Goldstone bosons of the spontaneous breaking of chiral symmetry). Spontaneous Symmetry Breaking however, when combined with gauge theories, is a central feature in the Higgs Mechanism for mass generation as we will now see.

2.2.3 The Abelian Higgs Model

We now introduce spontaneous symmetry breaking into gauge theories. A particularly instructive example is the theory of a complex scalar field coupled to the electromagnetic field (and itself)

$$\mathscr{L} = -\frac{1}{4}(F_{\mu\nu})^2 + |D_\mu\phi|^2 - V(\phi) \tag{15}$$

where D is the covariant derivative, $D_\mu = \partial_\mu + ieA_\mu$ and

$$V(\phi) = -\mu^2\phi^*\phi + \frac{\lambda}{2}(\phi^*\phi)^2. \tag{16}$$

$\mathscr{L}$ is invariant under the local (Abelian) $U(1)$ gauge transformation:

$$\phi(x) \to e^{i\alpha(x)}\phi(x), \qquad A_\mu(x) \to A_\mu(x) - \frac{1}{e}\partial_\mu\alpha(x). \tag{17}$$

For $\mu^2 < 0$ this is simply the quantum electrodynamics of a charged scalar boson. For $\mu^2 > 0$ however, the $U(1)$ symmetry is spontaneously broken.

For $\mu^2 > 0$, the minimum of this potential is at:

$$|\langle\phi\rangle| = |\phi_0| = \left(\frac{\mu^2}{\lambda}\right)^{\frac{1}{2}}. \tag{18}$$

For the further discussion we choose the minimum to be in the positive real direction (i.e. ϕ_0 to be real and positive) and define the shifted fields $\phi_{1,2}$:

$$\phi(x) = \phi_0 + \frac{1}{\sqrt{2}}(\phi_1(x) + i\phi_2(x)). \tag{19}$$

The potential can be rewritten in terms of the fields $\phi_{1,2}$:

$$V(\phi) = -\frac{1}{2\lambda}\mu^4 + \frac{1}{2}2\mu^2\phi_1^2 + O(\phi_i^3)\,. \tag{20}$$

ϕ_1 is a scalar with mass $\sqrt{2}\mu$ and ϕ_2 is the massless Goldstone Boson. The $O(\phi_i^3)$ terms contain the interactions involving the scalars.

We now consider the interactions with the gauge field:

$$|D_\mu\phi|^2 = \frac{1}{2}(\partial_\mu\phi_1)^2 + \frac{1}{2}(\partial_\mu\phi_2)^2 + \sqrt{2}e\phi_0 A_\mu\partial^\mu\phi_2 + e^2\phi_0^2 A_\mu A^\mu + \cdots\,. \tag{21}$$

The presence of the quadratic term in the vector field, $e^2\phi_0^2 A_\mu A^\mu$ implies that the photon (i.e. the gauge boson) has acquired a mass m_A, where

$$m_A^2 = 2e^2\phi_0^2\,. \tag{22}$$

There is also a peculiar two-point term between the Goldstone Boson ϕ_2 and the photon:

$$\sqrt{2}e\phi_0 A_\mu\partial^\mu\phi_2\,, \tag{23}$$

represented by the diagram:

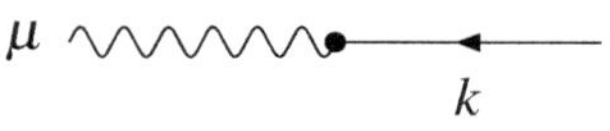

$$= i\sqrt{2}e\phi_0(-ik^\mu) \quad = \quad m_A k^\mu$$

To interpret this term it is convenient to go to the *Unitary Gauge*, i.e. to choose a gauge transformation $\alpha(x)$ at each point such that $\phi_2(x) = 0$. Now we have a theory with a real scalar field ϕ_1 and a massive vector field A_μ (with 3 degrees of freedom) so that the total number of degrees of freedom is 1+3=4. When we originally wrote down the theory we had a massless vector theory (with two degrees of freedom) and two scalars; again we had a total of 4 degrees of freedom. The Goldstone Boson becomes absorbed as the longitudinal degree of freedom of the massive vector boson, or in the vernacular the Goldstone boson has been *eaten* by the vector.

We now come back to the renormalizability of the theory. In the unitary gauge the propagator of the vector boson is:

$$\frac{-i}{k^2 - M_A^2}\left(g^{\mu\nu} - \frac{k^\mu k^\nu}{M_A^2}\right)\,. \tag{24}$$

So in this gauge the spectrum is the physical one, but renormalizability is not manifest.

To discuss renormalizability it is convenient to define the 't Hooft gauge by adding a gauge-fixing term to the Lagrangian:

$$\mathscr{L}_{\mathrm{GF}} = -\frac{1}{2\xi}\left(\partial_\mu A^\mu + \xi M_A\phi_2\right)^2\,. \tag{25}$$

Now the propagator of the vector boson is

$$\frac{-i}{k^2 - M_A^2}\left(g^{\mu\nu} - \frac{k^\mu k^\nu}{k^2}\right) - \frac{i\xi}{k^2 - \xi M_A^2}\frac{k^\mu k^\nu}{k^2}\,, \tag{26}$$

and that of ϕ_2 is

$$\frac{i}{k^2 - \xi M_A^2}. \tag{27}$$

Now the power-counting is manifestly correct for a renormalizable theory, but both the propagators have poles at gauge-dependent positions, so that unitarity is not manifest. In fact the good features of both gauges are correct and 't Hooft and Veltmann won the 1999 Nobel Prize for *elucidating the quantum structure of electroweak interactions in physics* which included the demonstration of the renormalizability and consistency of spontaneously broken field theories, particularly non-Abelian ones (to which we now turn).

2.2.4 Towards $SU(2) \times U(1)$

Textbooks give many examples of the spontaneous breaking of non-abelian gauge symmetries. In general the number of massive vector bosons is equal the number of *broken generators*, i.e. the number of symmetries of the action which are not symmetries of a vacuum state.

An illustrative example is the Georgi-Glashow Model (1972) in which an $SU(2)$ gauge theory with the Higgs in the adjoint representation is broken to $U(1)$. The $U(1)$ could be electromagnetism and the two massive vectors could have been the $W^{\pm}$, the charged intermediate vector bosons of the weak interactions. The theory contains 9 degrees of freedom, which can be viewed as 3 from the scalars and 3×2 from the 3 massless vectors (i.e. as in the theory as originally written) or as 1 from the physical scalar, 2 from the massless vector and 2×3 from the 2 massive vectors. The discovery of the existence of the weak *neutral current* interactions in the early 1970s implied that we also need a neutral massive vector boson (the Z^0) and hence a different theory. The standard model is based on the gauge group $SU(2) \times U(1)$.

2.2.5 $SU(2) \times U(1)$

After considerable amount of experimental and theoretical study, the Standard Model emerged based on the gauge group $SU(2) \times U(1)$. The complex Higgs fields are assigned to the fundamental representation of the $SU(2)$; the complex doublet contains four real fields. The covariant derivative for the theory is:

$$D_\mu \phi = (\partial_\mu - igA_\mu^a \tau^a - \frac{1}{2} g' B_\mu)\,\phi\,, \tag{28}$$

where $\tau^a = \sigma^a/2$ ($a = 1,2,3$) and the σ's are the Pauli spin matrices. g and g' are the $SU(2)$ and $U(1)$ coupling constants respectively; since the gauge group is a direct product of two simple groups the two couplings are different. The general gauge transformation depends on four parameters:

$$\phi \rightarrow e^{i\alpha^a \tau^a} e^{i\beta/2} \phi \tag{29}$$

where a $U(1)$ charge of +1/2 has been assigned to the Higgs fields.

Imagine now the Higgs Potential to be such that a minimum occurs at

$$\langle \phi \rangle = \frac{1}{\sqrt{2}} \begin{pmatrix} 0 \\ v \end{pmatrix}. \tag{30}$$

We can readily see that gauge transformation with $\alpha^1 = \alpha^2 = 0$ and $\alpha^3 = \beta$ leaves $\langle\phi\rangle$ unchanged. There therefore remains an unbroken $U(1)$ symmetry so that the $SU(2)\times U(1)$ symmetry has been broken to $U(1)$, which we identify with electromagnetism. On the basis of the general discussion above, we therefore expect 1 massless vector boson (the photon) and three massive vectors (the $W^\pm$ and Z^0) and one physical Higgs scalar. We will now see that this is indeed the case.

2.2.6 The spectrum of gauge bosons

The mass terms for the vector bosons are derived from the $|D_\mu\phi|^2$ term in the Lagrangian density which we rewrite explicitly as:

$$\frac{1}{2}\begin{pmatrix} 0 & v \end{pmatrix}\left(gA^a_\mu\tau^a + \frac{1}{2}g'B_\mu\right)\left(gA^{\mu\, b}\tau^b + \frac{1}{2}g'B^\mu\right)\begin{pmatrix} 0 \\ v \end{pmatrix} \tag{31}$$

which, using the anti-commutation properties of the σ-matrices, can readily be rewritten as

$$\frac{1}{2}\frac{v^2}{4}\left\{g^2(A^1_\mu)^2 + g^2(A^2_\mu)^2 + (-gA^3_\mu + g'B_\mu)^2\right\}. \tag{32}$$

Thus we arrive at the expected spectrum of vector bosons:

$$W^\pm = \frac{1}{\sqrt{2}}\left(A^1_\mu \mp iA^2_\mu\right) \qquad m_W = g\frac{v}{2} \tag{33}$$

$$Z_\mu = \frac{1}{\sqrt{g^2+g'^2}}\left(gA^3_\mu - g'B_\mu\right) \qquad m_z = \sqrt{g^2+g'^2}\,\frac{v}{2} \tag{34}$$

$$A_\mu = \frac{1}{\sqrt{g^2+g'^2}}\left(g'A^3_\mu + gB_\mu\right) \qquad m_\gamma = 0. \tag{35}$$

It is convenient and conventional to introduce the *weak mixing angle* θ_W:

$$\begin{pmatrix} Z^0 \\ A \end{pmatrix} = \begin{pmatrix} \cos\theta_W & -\sin\theta_W \\ \sin\theta_W & \cos\theta_W \end{pmatrix}\begin{pmatrix} A^3 \\ B \end{pmatrix} \tag{36}$$

so that

$$\cos\theta_W = \frac{g}{\sqrt{g^2+g'^2}}, \quad \text{and} \quad \sin\theta_W = \frac{g'}{\sqrt{g^2+g'^2}}. \tag{37}$$

In terms of θ_W, at tree level the relation between the masses of the W and Z is:

$$m_W = m_Z\cos\theta_W\,.$$

I postpone the inclusion of fermions until the second lecture. I end this lecture by briefly discussing the mass of the Higgs boson.

2.2.7 What is the mass of the Higgs boson?

The best limit on the mass of the standard model Higgs Boson from direct searches comes from LEP-2 $M_h > 114.4\,\text{GeV}$ (Barate *et al.*, 2003). Precision electroweak studies suggest that the mass is actually close to this bound; the preferred values deduced from loop contributions of the Higgs boson give $m_h = 85^{+39}_{-28}\,\text{GeV}$ (results from reference, LEPEWWG). Within the Standard Model, the $\Delta\chi^2$ curve derived from high-Q^2 precision electroweak measurements performed at LEP at CERN, the SLD at SLAC and CDF, and D0 at Fermilab, as a function of the Higgs-boson mass is as follows (results from reference, LEPEWWG):

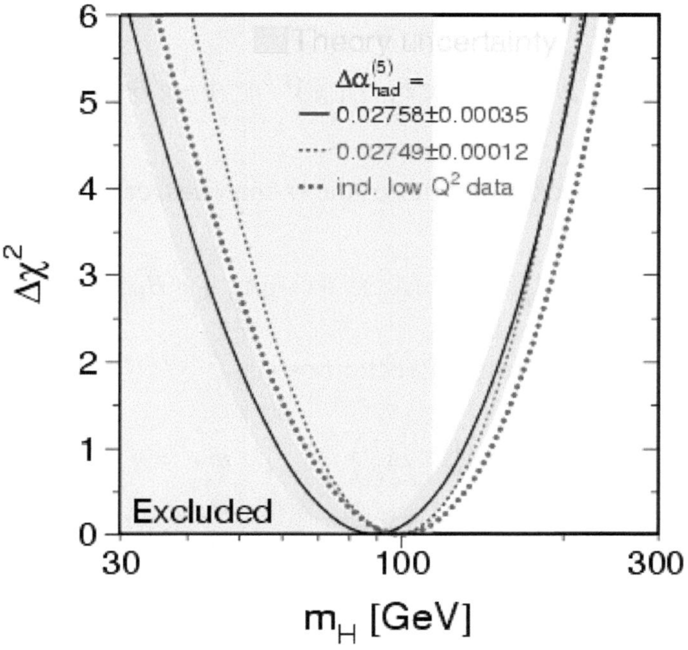

Although the Higgs boson has not been observed directly yet, the gauge bosons have been and they satisfy the relation $m_W = m_Z \cos\theta_W$ (up to calculable radiative corrections). We can therefore ask whether this is evidence for a single complex Higgs doublet and a single physical Higgs boson? The answer is *not necessarily*. The sector of the theory responsible for the symmetry breaking has a global $SU(2) \times U(1)$ symmetry (which is promoted to a local symmetry when couplings to the gauge bosons are introduced). If as the gauge symmetry is broken the global $SU(2)$ symmetry remains unbroken, then the three Goldstone bosons and the corresponding Noether currents transform as triplets under this *custodial* $SU(2)$ *symmetry* and we recover the above relation. A possible example could have been Chiral Symmetry breaking in QCD where the custodial $SU(2)$ is the well known vector isospin symmetry. However, in that case, the value of the decay constant f_π, defined from the matrix element of the weak axial current:

$$\langle 0|J^5_\mu|\pi(p)\rangle = i f_\pi p^\mu \tag{38}$$

is three orders of magnitude too small ($f_\pi = 132\,\text{MeV}$) to play a significant rôle in electroweak symmetry breaking. It might nevertheless be possible to construct a viable model of electroweak symmetry breaking based on a hypothetical new strong interaction analogous to QCD with a chiral symmetry breaking scale $F \sim 175\,\text{GeV}$. This is the motivation for technicolour models (Hill and Simmons, 2003).

3 Lecture 2 – The Electroweak Theory

3.1 Fermions

3.1.1 Chirality

From experimental measurements it has been deduced that only the left-handed components of the fermions participate in charged current weak interactions, i.e. the W's only couple to the left-handed components of the quarks and leptons. The left and right handed projections of fermion fields are defined by:

$$\psi_L = P_L \psi = \frac{1}{2}(1-\gamma^5)\psi \qquad \psi_R = P_R \psi = \frac{1}{2}(1+\gamma^5)\psi . \tag{39}$$

Under the parity transformation $\psi_L(x_0,\vec{x}) \to \psi_R(x_0,-\vec{x})$ and $\psi_R(x_0,\vec{x}) \to \psi_L(x_0,-\vec{x})$. As indicated above P_L and P_R are projection operators

$$P_L^2 = P_L \quad \text{and} \quad P_R^2 = P_R \qquad (P_L P_R = P_R P_L = 0,\ P_L + P_R = I) \tag{40}$$

The vector current and scalar density are given in terms of the left and right-handed fields by:

$$\bar{\psi}\gamma^\mu\psi = \bar{\psi}_L\gamma^\mu\psi_L + \bar{\psi}_R\gamma^\mu\psi_R \quad \text{and} \quad \bar{\psi}\psi = \bar{\psi}_L\psi_R + \bar{\psi}_R\psi_L . \tag{41}$$

(Thus for QCD with N massless fermions we have a $U(N)\times U(N)$ (global) chiral symmetry - I come back to this in later lectures.)

In order to accommodate the observed nature of the parity violation the left and right-handed fermions are assigned to different representations of $SU(2)\times U(1)$, and the left handed ($V-A$, i.e. vector – axial current) nature of the (charged current) weak interactions implies that the right-handed fields are singlets of $SU(2)$.

3.1.2 Including Fermions

For a general representation of fermions the covariant derivative takes the form:

$$D_\mu = \partial_\mu - igA_\mu^a T^a - ig'YB_\mu , \tag{42}$$

where the T^a are the corresponding generators of $SU(2)$ and the Y's are the weak-hypercharges. The covariant derivative can be rewritten in terms of the mass-eigenstates as:

$$D_\mu = \partial_\mu - \frac{ig}{\sqrt{2}}(W_\mu^+T^+ + W_\mu^-T^-) - i\frac{g^2T^3 - g'^2Y}{\sqrt{g^2+g'^2}}Z_\mu - i\frac{gg'}{\sqrt{g^2+g'^2}}(T^3+Y)A_\mu . \tag{43}$$

From eq. (43)we deduce that the electric charge operator is

$$Q = T_3 + Y \quad \text{and} \quad e = \frac{gg'}{\sqrt{g^2 + g'^2}}. \qquad (Q = -1 \text{ for the electron}). \tag{44}$$

In the standard model, the left-handed quarks and leptons are assigned to doublets of $SU(2)$ and the right-handed fermions are singlets.

Consider the assignment of the $SU(2)$ and $U(1)$ quantum numbers for the fermions, remembering that $Q = T_3 + Y$. The left handed leptons are assigned to the doublet:

$$E_L = \begin{pmatrix} \nu_e \\ e \end{pmatrix}_L . \tag{45}$$

In order to have the correct electric charges we must have $Y_{\nu_e} = Y_{e_L} = -1/2$. For the right-handed lepton fields $T_3 = 0$ and hence $Y_{e_R} = -1$. In the standard model we do not have a right-handed neutrino (but as will become clear during the school, the standard model picture of massless neutrinos has to be modified following the observation of neutrino oscillations!). For the left-handed quark fields we have the left-handed doublet:

$$Q_L = \begin{pmatrix} u \\ d \end{pmatrix}_L . \tag{46}$$

with $Y_{Q_L} = 1/6$. The right-handed quark fields therefore have $Y_{u_R} = 2/3$ and $Y_{d_R} = -1/3$. Similar assignments are made for the other two generations.

With the assignments of quantum numbers described above, the terms in the Lagrangian involving the fermions then take the form:

$$\begin{aligned} \mathscr{L} &= \bar{E}_L(i\,\partial\!\!\!/)E_L + \bar{e}_R(i\,\partial\!\!\!/)e_R + \bar{Q}_L(i\,\partial\!\!\!/)Q_L + \bar{u}_R(i\,\partial\!\!\!/)u_R + \bar{d}_R(i\,\partial\!\!\!/)d_R \\ &\quad + g\left(W^+_\mu J^{\mu+}_W + W^-_\mu J^{\mu-}_W + Z^0_\mu J^\mu_Z\right) + eA_\mu J^\mu_{\text{EM}}, \end{aligned} \tag{47}$$

where

$$J^{\mu+}_W = \frac{1}{\sqrt{2}}(\bar{\nu}_L\gamma^\mu e_L + \bar{u}_L\gamma^\mu d_L); \tag{48}$$

$$J^{\mu-}_W = \frac{1}{\sqrt{2}}(\bar{e}_L\gamma^\mu \nu_L + \bar{d}_L\gamma^\mu u_L); \tag{49}$$

$$\begin{aligned} J^\mu_Z &= \frac{1}{\cos\theta_W}\left\{\frac{1}{2}\bar{\nu}_L\gamma^\mu\nu_L + \left(\sin^2\theta_W - \frac{1}{2}\right)\bar{e}_L\gamma^\mu e_L + \sin^2\theta_W\,\bar{e}_r\gamma^\mu e_R \right. \\ &\quad + \left(\frac{1}{2} - \frac{2}{3}\sin^2\theta_W\right)\bar{u}_L\gamma^\mu u_L - \frac{2}{3}\sin^2\theta_W\,\bar{u}_R\gamma^\mu u_R \\ &\quad \left. + \left(\frac{1}{3}\sin^2\theta_W - \frac{1}{2}\right)\bar{d}_L\gamma^\mu d_L + \frac{1}{3}\sin^2\theta_W\,\bar{d}_R\gamma^\mu d_R\right\}; \end{aligned} \tag{50}$$

$$J^\mu_{\text{EM}} = -\bar{e}\gamma^\mu e + \frac{2}{3}\bar{u}\gamma^\mu u - \frac{1}{3}\bar{d}\gamma^\mu d. \tag{51}$$

3.1.3 Fermion Masses and Yukawa Couplings

The standard mass term for the fermions is of the form

$$m\bar{\psi}\psi = m\bar{\psi}_L\psi_R + m\bar{\psi}_R\psi_L\,.$$

This term is manifestly not invariant under the $SU(2)_L$ gauge symmetry and this can be shown to spoil renormalizability.

In the Standard Model, mass terms for the fermions are generated through Yukawa couplings to the Higgs doublet, for example:

$$\Delta\mathscr{L}_e = -\lambda_e\,(\bar{E}^i_L\phi^i)\,e_R + \text{h.c.} \tag{52}$$

where $i = 1,2$ is the $SU(2)$ label. As before, we rewrite the complex doublet ϕ in terms of the fields shifted by $\langle\phi\rangle$, so that

$$\Delta\mathscr{L}_e = -\frac{\lambda_e\,v}{\sqrt{2}}\,\bar{e}_L e_R + \text{h.c.} + \text{ interaction terms} \tag{53}$$

In this picture therefore

$$m_e = \frac{\lambda_e v}{\sqrt{2}} \tag{54}$$

and we have generated a mass-term for the electron in a gauge invariant way. We have traded the mass parameter m_e for the Yukawa coupling λ_e.

λ_e is very small ($v \simeq 250\,\text{GeV}$) and the problem of understanding the pattern of fermion masses becomes the problem of understanding the pattern of Yukawa couplings. At this stage, in spite of considerable theoretical effort, this problem is not solved.

We can choose a gauge such that the scalar field is written in the form

$$\phi(x) = \frac{1}{\sqrt{2}}\begin{pmatrix} 0 \\ v + h(x) \end{pmatrix}, \tag{55}$$

where $h(x)$ is the physical Higgs scalar. The electron's Yukawa term in the Lagrangian now takes the form $\mathscr{L}_e = -m_e\left(1 + \frac{h}{v}\right)\bar{e}e$.

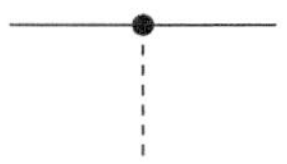
$$= -i\frac{m_e}{v}$$

By construction of the theory, it is a general feature that the couplings of the Higgs boson h are proportional to the masses (or squares of masses) of the particles it is interacting with. *This is a fundamental ingredient in the phenomenology of Higgs searches.*

For the down quark we can introduce a similar Yukawa term to that of the electron. For the up quark, this clearly does not work, but we can exploit the existence of the invariant antisymmetric tensor ε^{ij}.

$$\begin{aligned}
\Delta\mathscr{L}_q &= -\lambda_d\bar{Q}^i_L\phi^i d_R - \lambda_u\varepsilon^{ij}\bar{Q}^i_L\phi^{\dagger j}u_R + h.c. \\
&= -\frac{\lambda_d\,v}{\sqrt{2}}\bar{d}_L d_R - \frac{\lambda_u\,v}{\sqrt{2}}\bar{u}_L u_R + h.c. + \text{ interaction terms} \\
&= -m_d\left(1+\frac{h}{v}\right)\bar{d}d - m_u\left(1+\frac{h}{v}\right)\bar{u}u\,.
\end{aligned} \tag{56}$$

(Note that apart from being singlets under $SU(2)$, the terms in the action also have zero net hypercharge.)

3.2 Quark Mixing

Consider the following two experimental branching ratios:

$$B(K^- \to \pi^0 e^- \bar{\nu}_e) \simeq 5\% \ (K^-_{\ell 3} \text{ Decay}) \quad \text{and} \quad B(K^- \to \pi^- e^+ e^-) < 3 \times 10^{-7}. \tag{57}$$

The quark flow diagrams for these two processes are as follows:

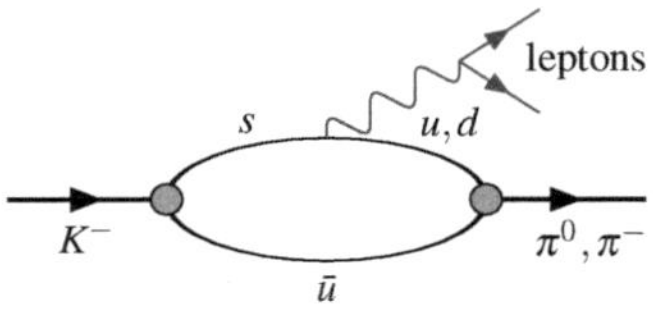

The strange quark in the initial-state K^- is replaced by the u or d quark in the final-state pion.

Measurements like this show that $s \to u$ (charged-current) transitions are not very rare, but that *Flavour Changing Neutral Current* (FCNC) transitions, such as $s \to d$ are. In the picture that we have developed so far, there are no transitions between fermions of different generations, and in view of the experimental observations of processes such as $K^- \to \pi^0$ semileptonic decays mentioned in eq.(57) this has to be modified. The picture which has emerged is the Cabibbo-Kobayashi-Maskawa (CKM) theory of quark mixing which we now consider.

3.2.1 CKM Theory

In the CKM theory the (quark) mass eigenstates are not the same as the weak-interaction eigenstates which we have been considering up to now.

Let

$$U' = \begin{pmatrix} u' \\ c' \\ t' \end{pmatrix} = U_u \begin{pmatrix} u \\ c \\ t \end{pmatrix} = U_u U \quad \text{and} \quad D' = \begin{pmatrix} d' \\ s' \\ b' \end{pmatrix} = U_d \begin{pmatrix} d \\ s \\ b \end{pmatrix} = U_d D, \tag{58}$$

where the primes ($'$s) denote the weak interaction eigenstates and U_u and U_d are unitary matrices.

For neutral currents vertices

$$\bar{U}' \cdots U' = \bar{U} \cdots U \quad \text{and} \quad \bar{D}' \cdots D' = \bar{D} \cdots D \tag{59}$$

and no FCNC are induced. In eq.(59) the ellipses represent Dirac Matrices, but only contain the identity matrix ($I = U_u^\dagger U_u = U_d^\dagger U_d$) in flavour space.

For charged currents vertices

$$J_W^{\mu +} = \frac{1}{\sqrt{2}} \bar{U}'_L \gamma^\mu D'_L = \frac{1}{\sqrt{2}} \bar{U}_L U_u^\dagger \gamma^\mu U_d D_L = \frac{1}{\sqrt{2}} \bar{U}_L \gamma^\mu (U_u^\dagger U_d) D_L \equiv \frac{1}{\sqrt{2}} \bar{U}_L \gamma^\mu V_{\text{CKM}} D_L. \tag{60}$$

Now the flavour structure is non-trivial, transitions between different generations are possible and their relative strength is give by $V_{\rm CKM}$, the Cabibbo, Kobayashi, Maskawa matrix (or the CKM matrix). We can therefore represent the charged-current interactions in the schematic form

$$J^+_\mu = (\bar{u},\bar{c},\bar{t}\,)_L\gamma_\mu V_{\rm CKM}\begin{pmatrix} d \\ s \\ b \end{pmatrix}_L . \tag{61}$$

The 2005 Particle Data Group summary for the magnitudes of the entries in the CKM matrix is (Eidelman *et al.*, 2005):

$$\begin{pmatrix} 0.9739-0.9751 & 0.221-0.227 & 0.0029-0.0045 \\ 0.221-0.227 & 0.9730-0.9744 & 0.039-0.044 \\ 0.0048-0.014 & 0.037-0.043 & 0.9990-0.9992 \end{pmatrix} . \tag{62}$$

An important question is *how many parameters are there?* in the CKM matrix for a theory with N_g generations ($N_g = 3$, of course, in the physical world).

- An $N_g \times N_g$ unitary matrix has N_g^2 real parameters.
- It may seem that we can absorb $2N_g$ parameters into unphysical phases of the quark fields. However, changing the phases of all the quark fields by the same amount leaves the (charged-current) Lagrangian invariant. The number of unphysical parameters is therefore $(2N_g - 1)$.
- Thus the CKM matrix contains $(N_g - 1)^2$ physical parameters to be determined.

Parametrizations of the CKM Matrix For $N_g = 2$ there is only one parameter, which is conventionally chosen to be the Cabibbo angle:

$$V_{\rm CKM} = \begin{pmatrix} \cos\theta_c & \sin\theta_c \\ -\sin\theta_c & \cos\theta_c \end{pmatrix} . \tag{63}$$

For $N_g = 3$, there are 4 real parameters. Three of these can be interpreted as angles of rotation in three dimensions (e.g. the three Euler angles) and the fourth is a phase. The general parametrization recommended by the Particle Data Group is

$$\begin{pmatrix} c_{12}c_{13} & s_{12}c_{13} & s_{13}e^{-i\delta_{13}} \\ -s_{12}c_{23} - c_{12}s_{23}s_{13}e^{i\delta_{13}} & c_{12}c_{23} - s_{12}s_{23}s_{13}e^{i\delta_{13}} & s_{23}c_{13} \\ s_{12}s_{23} - c_{12}c_{23}s_{13}e^{i\delta_{13}} & -c_{12}s_{23} - s_{12}c_{23}s_{13}e^{i\delta_{13}} & c_{23}c_{13} \end{pmatrix} \tag{64}$$

where c_{ij} and s_{ij} represent the cosines and sines respectively of the three angles θ_{ij}, $ij = 12$, 13 and 23. δ_{13} is the phase parameter.

For most purposes it is convenient and conventional to use approximate parametrizations, based on the hierarchy of values in V_{CKM} ($s_{12} \gg s_{23} \gg s_{13}$). These are much simpler that that in eq.(64). The most popular of these is the *Wolfenstein parametrization*:

$$V_{\text{CKM}} = \begin{pmatrix} 1-\frac{\lambda^2}{2} & \lambda & A\lambda^3(\rho - i\eta) \\ -\lambda & 1-\frac{\lambda^2}{2} & A\lambda^2 \\ A\lambda^3(1-\rho-i\eta) & -A\lambda^2 & 1 \end{pmatrix} . \tag{65}$$

$\lambda = s_{12}$ is approximately the Cabibbo angle. A, ρ and η are real numbers that a priori were intended to be of order unity. Corrections to the parametrization are of $O(\lambda^4)$.

The Unitarity Triangle The unitarity of the CKM-matrix gives us a set of relations between the entries and a particularly useful one is:

$$V_{ud}V_{ub}^* + V_{cd}V_{cb}^* + V_{td}V_{tb}^* = 0 . \tag{66}$$

In terms of the Wolfenstein parameters, the components on the left-hand side are given by:

$$\begin{aligned} V_{ud}V_{ub}^* &= A\lambda^3[\bar\rho + i\bar\eta] + O(\lambda^7) \\ V_{cd}V_{cb}^* &= -A\lambda^3 + O(\lambda^7) \\ V_{td}V_{tb}^* &= A\lambda^3[1-(\bar\rho + i\bar\eta)] + O(\lambda^7) , \end{aligned}$$

where $\bar\rho = \rho(1-\lambda^2/2)$ and $\bar\eta = \eta(1-\lambda^2/2)$.

The unitarity relation can be represented schematically by the famous "unitarity triangle" (obtained after scaling out a factor of $A\lambda^3$).

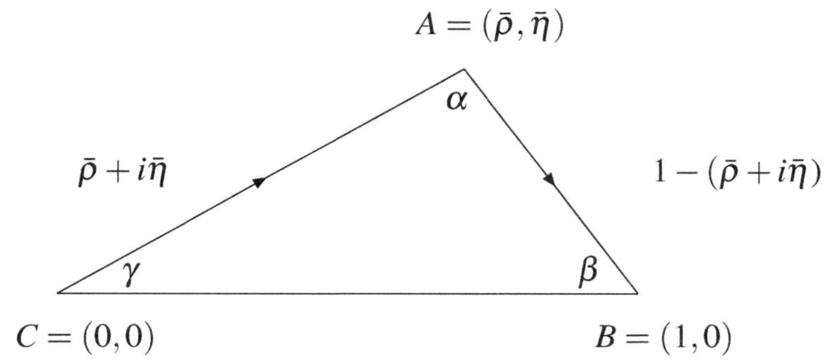

A particularly important approach to *exploring the limits of the SM* is to over-determine the position of the vertex A to check for consistency. A large number of processes are studied, and for each process the locus of possible positions of A is determined. Assuming that all the predictions are given correctly by the Standard Model all the loci will intersect in a consistent way. Of course in practice there are theoretical and experimental uncertainties so that each locus is not a single curve, but a band. I will discuss some of the processes being used in the standard unitarity triangle analyses in lectures 4 and 5. There has been a very significant reduction in the errors during recent years, and frustratingly the measurements are consistent within the uncertainties. The currently allowed region for the vertex A in the $(\bar\rho, \bar\eta)$-plane is shown in fig. 1.

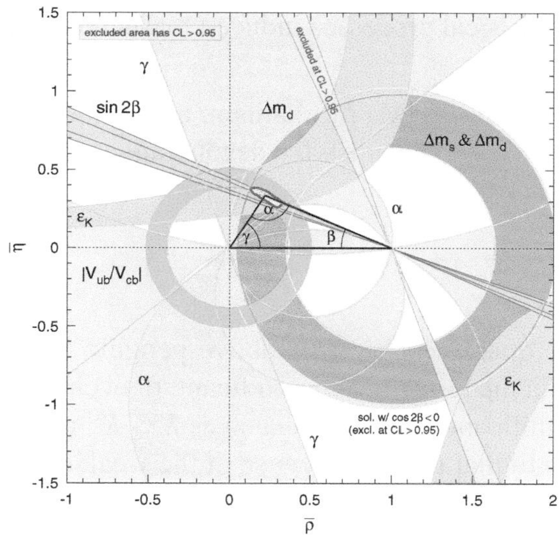

Figure 1. *Currently allowed region for the vertex A of the unitarity triangle in the* $(\bar{\rho}, \bar{\eta})$ *plane (Yao et al, 2006). Copyright (2006) IOP Publishing.*

3.2.2 Flavour Changing Neutral Currents (FCNC)

We have seen that in the SM, unitarity implies that there are no FCNC reactions at tree level, i.e. there are no vertices of the type:

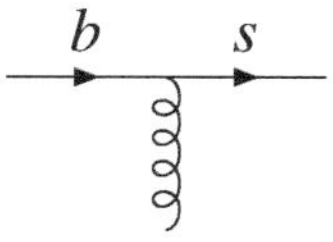

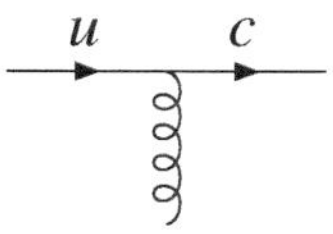

.

Quantum loops, however, can generate FCNC reactions, through *box* diagrams or *penguin* diagrams. For example, for $\bar{B}^0 - B^0$ mixing we have the following box-diagrams:

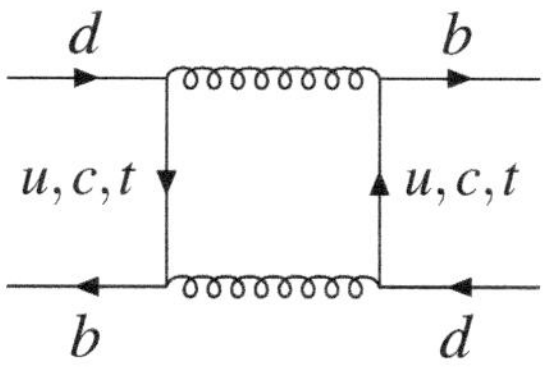

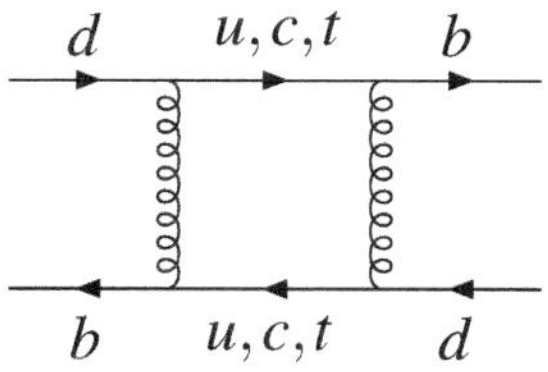

Examples of penguin diagrams relevant for $b \to s$ (FCNC) transitions are:

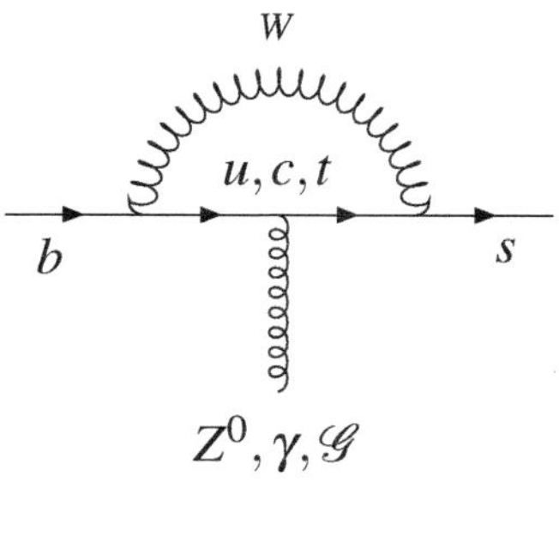

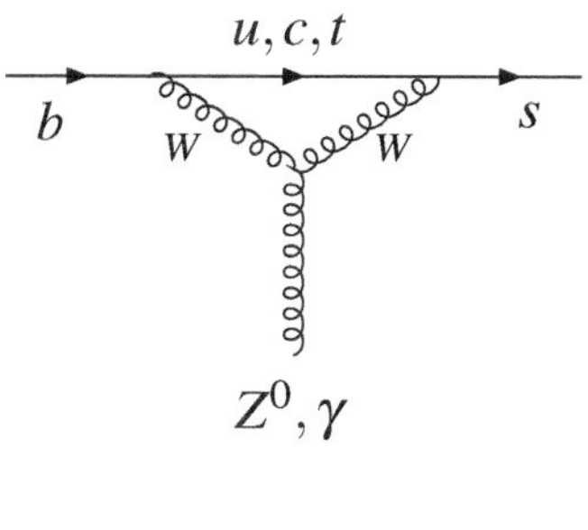

We will discuss several of the physical processes induced by such loop-effects in lectures 4 and 5.

The Glashow-Illiopoulos-Maiani (GIM) mechanism, a consequence of the unitarity of the CKM matrix, implies that FCNC effects vanish for degenerate quarks (i.e. for $m_u = m_c = m_t$). For example unitarity implies

$$V_{ub}V_{us}^* + V_{cb}V_{cs}^* + V_{tb}V_{ts}^* = 0$$

so that for degenerate up-like quarks each of the above penguin vertices would vanish. For some processes the top-quark in the loop gives a significant, or even dominant, contribution and the GIM mechanism is not significant. For others, such as K^0–$\bar{K}^0$ mixing the GIM mechanism provides a further suppression to that present because of the weak interactions.

3.2.3 The Discrete Symmetries *P*, *C* and *CP*

Even after more than 50 years since its discovery, I find parity violation to be an exciting concept. Today there are many detailed studies of the subtle phenomenon of *CP*-violation (see below) to explore the validity of the Standard Model. *CP*-violation is also needed to understand how we arrived at the matter dominated universe in which we live from the Big Bang (although it appears that the CP-violation generated by the phase in the CKM matrix is insufficient to account for the matter-antimatter asymmetry of the universe). In this subsection I present a brief overview of the discrete symmetries *P*, *C* and *CP*.

Parity: The parity transformation is the change of sign of the spatial coordinates:

$$(\vec{x},t) \to (-\vec{x},t).$$

Under this transformation the vector and axial-vector fields transform as:

$$V_\mu(\vec{x},t) \to V^\mu(-\vec{x},t) \;\; \text{and} \;\; A_\mu(\vec{x},t) \to -A^\mu(-\vec{x},t).$$

Left-handed components of fermions $\psi_L = (\frac{1}{2}(1-\gamma^5)\psi)$ transform into right-handed ones $\psi_R = (\frac{1}{2}(1+\gamma^5)\psi)$, and vice-versa. Since charged-current weak interactions in the SM only involve the left-handed components, parity is not a good symmetry of the weak force, whereas QCD and QED are invariant under parity transformations.

Charge Conjugation: Charge conjugation is a transformation which relates each complex field ϕ with $\phi^\dagger$. Under *C* the currents transform as follows:

$$\bar{\psi}_1\gamma_\mu\psi_2 \to -\bar{\psi}_2\gamma_\mu\psi_1 \;\; \text{and} \;\; \bar{\psi}_1\gamma_\mu\gamma_5\psi_2 \to \bar{\psi}_2\gamma_\mu\gamma_5\psi_1,$$

where ψ_i represents a spinor field of type (flavour or lepton species) i.

CP – Under the combined CP-transformation, the currents transform as:

$$\bar{\psi}_1\gamma_\mu\psi_2 \to -\bar{\psi}_2\gamma^\mu\psi_1 \quad \text{and} \quad \bar{\psi}_1\gamma_\mu\gamma_5\psi_2 \to -\bar{\psi}_2\gamma^\mu\gamma_5\psi_1,$$

where the fields on the left (right) hand side are evaluated at $(\vec{x},t)$ ($(-\vec{x},t)$).

Consider now a charged current interaction:

$$(W_\mu^1 - iW_\mu^2)\bar{U}^i\gamma^\mu(1-\gamma^5)V_{ij}D^j + (W_\mu^1 + iW_\mu^2)\bar{D}^j\gamma^\mu(1-\gamma^5)V_{ij}^*U^i, \tag{67}$$

U^i and D^j are up and down type quarks of flavours i and j respectively. Under a CP transformation, the interaction term transforms to:

$$(W_\mu^1 + iW_\mu^2)\bar{D}^j\gamma^\mu(1-\gamma^5)V_{ij}U^i + (W_\mu^1 - iW_\mu^2)\bar{U}^i\gamma^\mu(1-\gamma^5)V_{ij}^*D^j. \tag{68}$$

CP-invariance requires V to be real (or more strictly that any phases must be able to be absorbed into the definition of the quark fields). Since for 3-generations there is a physical phase in the CKM matrix, if the phase is non-zero (as is the case in nature) then CP-violation will be induced.

3.2.4 The Higgs Mass and Interactions

I end this lecture by a brief discussion of the Higgs sector. Imagine that the Higgs potential is given by

$$V = -\mu^2(\phi^\dagger\phi) + \lambda(\phi^\dagger\phi)^2 \quad \text{and write} \quad \phi = \begin{pmatrix} 0 \\ \frac{1}{\sqrt{2}}(v+h(x)) \end{pmatrix} \quad \text{where} \quad v^2 = \frac{\mu^2}{\lambda}. \tag{69}$$

Rewriting the potential in terms of the physical higgs field $h(x)$ we find:

$$V = \mu^2h^2 + \sqrt{\lambda}\mu h^3 + \frac{\lambda}{4}h^4. \tag{70}$$

We know that $v = \mu/\sqrt{\lambda} = 250\,\text{GeV}$ from the known value of M_W and other quantities. The mass of the Higgs is $\sqrt{2}\mu$. Today, we have no direct way of knowing this. The larger that m_h is, the stronger are the Higgs self interactions. Precision studies indicate that the Higgs boson is not very heavy, but if surprisingly it turned out that these indications were wrong then we would have an interesting strongly interacting Higgs sector.

Finally, I have to stress that even if the overall picture presented here is correct, the Higgs sector may be more complicated than that given by a single complex doublet.

4 Lecture 3 – QCD

In this lecture I will briefly review some of the key elements of Quantum Chromodynamics (QCD), the theory of the strong nuclear force.

4.1 The QCD Lagrangian

QCD is a non-abelian gauge theory with an $SU(3)$ gauge group and with the following Lagrangian density:

$$\mathscr{L} = -\frac{1}{4}(F^a_{\mu\nu})^2 + \bar{\psi}(i\not{D} - m)\psi + \mathscr{L}_{\text{GF}} \tag{71}$$

where $a = 1, 2, \cdots, 8$ is an adjoint label. Each flavour of quark transforms under the fundamental representation of the $SU(3)$ colour group and the gluons transform under the adjoint representation (as do all gauge bosons under their gauge groups). The field strength tensor $F^a_{\mu\nu}$ is given by

$$F^a_{\mu\nu} = \partial_\mu A^a_\nu - \partial_\nu A^a_\mu + g f^{abc} A^b_\mu A^c_\nu \tag{72}$$

and the f^{abc} are the structure constants of $SU(3)$ (defined by $[T^a, T^b] = i f^{abc} T^c$, where the T^as are the generators of the group). The covariant derivative D_μ is given by

$$D_\mu = \partial_\mu - i g A^a_\mu T^a \,. \tag{73}$$

$\mathscr{L}_{\text{GF}}$ is the gauge-fixing term which I will not discuss here. In order to perform perturbative calculations we need to make a choice of gauge; physical quantities, of course, are independent of this choice.

The interaction vertices (and the corresponding Feynman rules) can readily be obtained from the QCD Lagrangian (71). Neglecting ghost-fields which generally appear in the gauge-fixing terms, the vertices are:

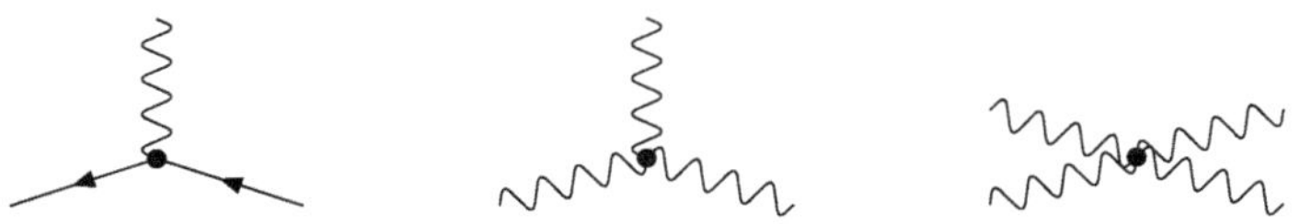

A major difference with quantum electrodynamics is the presence of gluon self-interactions (the three and four gluon vertices above). While the photon is electrically neutral, the gluons transform under the adjoint representation of $SU(3)$ (i.e. they are nonsinglets) and hence self interactions are present.

4.2 Asymptotic Freedom

Asymptotic Freedom, together with infrared safety which I consider in the following subsection, allows us to make perturbative predictions in QCD for a large range of important hard processes. The property of asymptotic theory, unique to non-abelian gauge theories, is the fact that the renormalized coupling decreases to zero as the (momentum) scale at which it is defined increases. The Nobel prize for its discovery was awarded in 2004 to Gross, Politzer and Wilczek (Gross and Wilczek, 1973). Consider the perturbative expansion of the β-function, defined as the logarithmic derivative of the renormalized coupling:

$$\beta(g) \equiv \mu \frac{\partial g}{\partial \mu} = -\beta_0 \frac{g^3}{16\pi^2} - \beta_1 \frac{g^5}{(16\pi^2)^2} - \beta_2 \frac{g^7}{(16\pi^2)^3} - \cdots \tag{74}$$

where μ is the renormalization scale [1]. The first three coefficients (β_0 and β_1 are universal, whereas β_2 is presented here in the $\overline{\text{MS}}$ scheme) are given by:

$$\beta_0 = 11 - \frac{2}{3}n_f, \quad \beta_1 = 102 - \frac{38}{3}n_f, \quad \beta_2 = \frac{2857}{2} - \frac{5033}{18}n_f + \frac{325}{54}n_f^2, \tag{75}$$

where n_f is the number of quarks with mass less than the scale μ. The next term in the expansion, β_3 is also known (Larin S A *et al.*, 1997). The key feature of the expansion (74) is that the first term is negative which implies that the coupling constant decreases with the scale μ. Keeping only the first two terms in the β-function for illustration, the solution of the differential equation in (74) is:

$$\alpha_s(\mu) \equiv \frac{g^2(\mu)}{4\pi} = \frac{4\pi}{\beta_0 \log(\mu^2/\Lambda^2)} \left\{ 1 - \frac{\beta_1}{\beta_0^2} \frac{\log\left[\log(\mu^2/\Lambda^2)\right]}{\log(\mu^2/\Lambda^2)} + \frac{\beta_1^2}{\beta_0^4 \log^2(\mu^2/\Lambda^2)} \times \left(\left(\log[\log(\mu^2/\Lambda^2)] - \frac{1}{2} \right)^2 + \frac{\beta_2\beta_0}{\beta_1^2} - \frac{5}{4} \right) \right\}, \tag{76}$$

where Λ can be considered as a constant of integration. Indeed we can trade the dimensionless coupling constant of QCD (g) for the parameter Λ (which has dimensions of mass). Sidney Coleman referred to this trade-off as *dimensional transmutation* which is a manifestation that, even with massless quarks, the quantum effects in QCD induce a scale.

For illustration the one-loop graphs which need to be evaluated to determine the coefficient β_0 (and, had you done it in time, to win a Nobel prize) are:

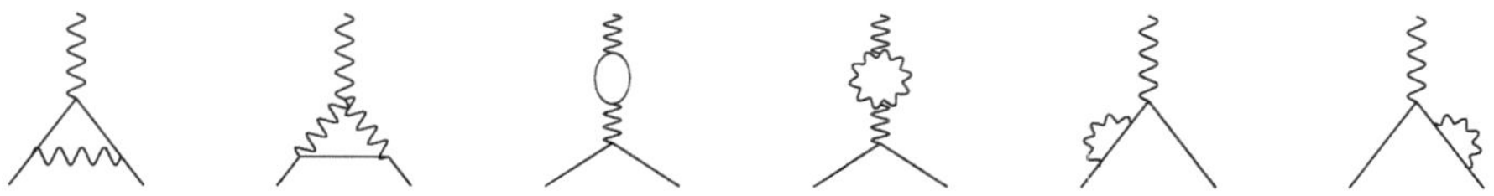

+ ghosts

The *running* of the coupling constant is illustrated in Fig.2, where values obtained from a variety of experimentally measured quantities are presented as a function of the scale (Yao et al, 2006). It has become standard to quote the coupling constant at a renormalization scale of M_Z and the current value is (Yao et al, 2006):

$$\alpha_s(M_Z) = 0.1176 \pm 0.002\,. \tag{77}$$

4.3 Infrared Safety

If in a hard process there is only one large scale, Q say, then it is natural to express the corresponding perturbative series in terms of the coupling renormalized at a scale Q using some appropriate renormalization scheme (such as the $\overline{\text{MS}}$-scheme). However if the predictions also depend on a low momentum, such as quark masses or momenta of the order of the typical scale of QCD (Λ) then we cannot make reliable perturbative predictions. Fortunately for many processes such long-distance effects are not present and for these infrared-safe quantities we can make perturbative predictions.

[1]The derivative in eq.(74) is written as a partial derivative, implying that it is to be taken w.r.t.μ keeping the bare coupling g_0 fixed.

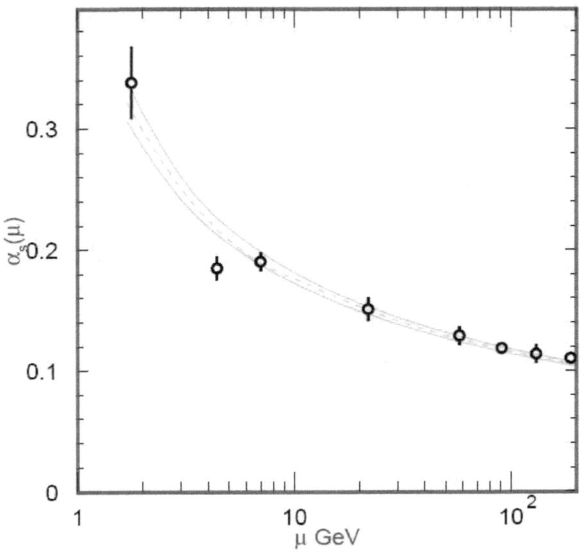

Figure 2. *Results for $\alpha_s(\mu)$ at values of μ where they are measured, (τ-width, Υ-decays, Deep Inelastic Scattering, e^+e^- Event Shapes at 22 and 59 GeV, Z–Width, e^+e^- Event Shapes at 135 and 189 GeV) (Yao et al, 2006). Copyright (2006) IOP Publishing.*

4.3.1 $\sigma(e^+e^- \to$ hadrons)

I illustrate what is meant by infrared safety by considering the inclusive process $e^+e^- \to$ hadrons. Consider the following one-loop diagram contributing to the $e^+e^- \to q\bar{q}$ amplitude:

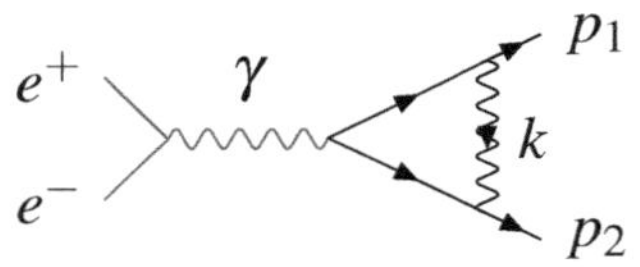

To evaluate the full cross-section, we should also include the contribution of the Z_0 (which couples to the quarks as in eq. (50)) as well as the photon. The present discussion holds also in that case, and so for simplicity I do not mention the Z_0 explicitly.

To illustrate the behaviour at small momenta ($|k_\mu| \ll \sqrt{p_1 \cdot p_2}$) this diagram corresponds to an integral of the form: [2]:

$$I \equiv \int \frac{d^4k}{(2\pi)^4} \frac{1}{(k^2+i\varepsilon)((p_1+k)^2-m^2+i\varepsilon)((p_2-k)^2-m^2+i\varepsilon)}, \tag{78}$$

where the three factors in the denominator of the integrand correspond to the propagators of the two quarks and the gluon in the loop. If the quarks are on-shell $p_1^2 = p_2^2 = m^2$, at small momenta we have

$$I \sim \int \frac{d^4k}{(2\pi)^4} \frac{1}{k^2(2p_1\cdot k)(-2p_2\cdot k)} \quad \text{which is logarithmically divergent.} \tag{79}$$

[2]For small momenta the numerator is independent of the loop-momentum and so we simply set it to 1 here.

The presence of such *infrared divergences* indicates that long-distance/low-momentum contributions are important and therefore there is a danger that asymptotic freedom may not be sufficient to make reliable predictions in perturbation theory. Of course if we had taken the quarks to have been off-shell, there would have been no divergences, but the result would have depended on how much off-shell they were. The off-shellness, which typically is of $O(\Lambda)$, again depends on non-perturbative physics.

For *inclusive* reactions, such as $e^+e^- \to$ hadrons, the infrared divergences cancel between diagrams with virtual and real gluons. This is a generalization of the Bloch-Nordsieck (1937) theorem (Block and Nordsieck, 1937) from QED to QCD. For example, at $O(\alpha_s)$ the following diagrams contribute to $\sigma(e^+e^- \to \text{hadrons})$:

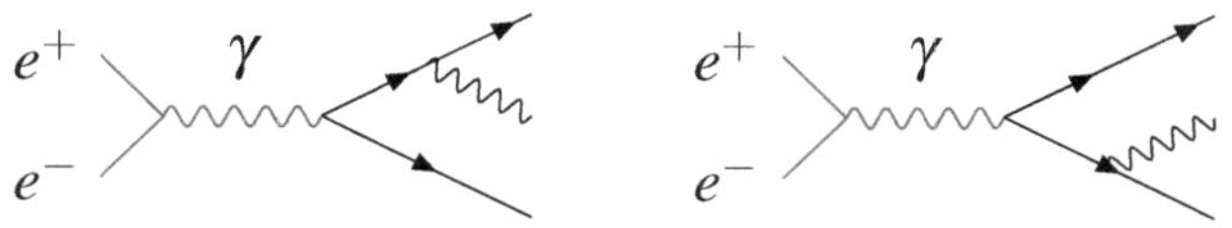

These also contribute infrared divergent terms to $\sigma(e^+e^- \to \text{hadrons})$ at $O(\alpha_s)$, which cancel those of diagrams with virtual loops. $\sigma(e^+e^- \to \text{hadrons})$ is free of infrared divergences to any order of perturbation theory.

The standard physical interpretation in QED is that in any experiment we cannot distinguish the electron e from e together with a number of soft γs, where the γs are too soft to be detected. It is therefore not unreasonable to have to sum over all experimentally indistinguishable contributions.

Infrared divergences are not the only source of *mass singularities*. Consider two massless particles moving parallel to each other (in the z-direction say).

$$q_1 = \omega_1\,(1,0,0,1)\,, \quad q_2 = \omega_2\,(1,0,0,1)\,. \tag{80}$$

The square of the invariant mass of the pair is $(q_1+q_2)^2$ which is zero. Thus when we have two *collinear* massless particles emitted at a vertex, there may be a propagator close to its mass-shell and hence a potential singularity. Indeed, when internal particles are collinear with external ones we get *collinear divergences*. The Kinoshita-Lee-Naunberg theorem implies that collinear divergences cancel when we sum over all degenerate final and initial states. For QCD perturbative corrections to $\sigma(e^+e^- \to \text{hadrons})$ only the sum over final states has to be performed (since the initial state leptons do not emit or absorb gluons) and the collinear divergences also cancel. The standard physical interpretation for the cancellation of collinear divergences is that we cannot distinguish q from q+a collinear gluon (for example), where the collinearity is below the angular resolution of the detector. Again it is therefore not unreasonable to have to sum over all experimentally indistinguishable contributions.

To give you some confidence in the above statements, consider the consequences of unitarity:

$$SS^\dagger = I \quad \Rightarrow \quad (I+iT)(I-iT^\dagger) = I \quad \Rightarrow \quad 2\text{Im}\langle i\,|\,T\,|\,i\rangle = \sum_n |\langle i\,|\,T\,|n\rangle|^2 \tag{81}$$

where S and T are the S and T matrices respectively. This is the *Optical Theorem*. Thus $\sigma(e^+e^- \to \text{hadrons})$ is proportional to the imaginary part of the e^+e^- forward amplitude. But when we look at diagrams such as:

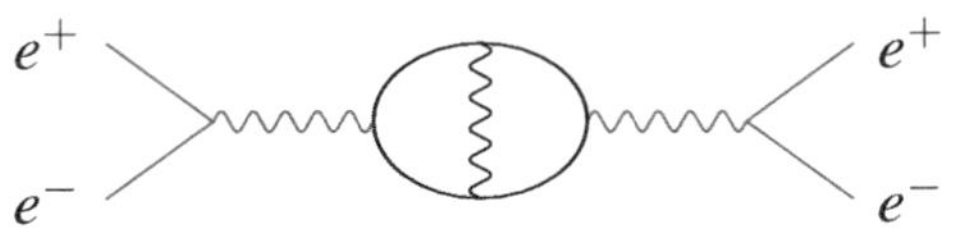

power counting readily shows us that there are no mass singularities, and so the mass singularities must cancel between the separate contributions to the cross section. This is indeed what happens.

Consider now the cross-section for $e^+e^- \longrightarrow \gamma^* \longrightarrow$ hadrons. It takes the form:

$$\sigma = \sigma_0 \left(3\sum_f Q_f^2\right)\left(1+\frac{\alpha_s(\mu)}{\pi}+\frac{\alpha_s^2(\mu)}{(4\pi)^2}\left\{4\beta_0 \log\left(\frac{\mu^2}{Q^2}\right)+c_2\right\}+\cdots\right) \tag{82}$$

where

- σ_0 is the lowest order $e^+e^- \longrightarrow \mu^+\mu^-$ cross section.
- μ is the renormalization scale at which the coupling is defined;
- the form of the logarithms is fixed by the renormalization group (i.e. independence of σ on μ) and the absence of mass-singularities;
- c_2 is a constant.

In order to avoid *Large Logarithms* we should choose $\mu^2 \simeq Q^2$:

$$\sigma = \sigma_0 \left(3\sum_f Q_f^2\right)\left(1+\frac{\alpha_s(Q)}{\pi}+1.411\frac{\alpha_s^2(Q)}{(\pi)^2}-12.8\frac{\alpha_s^3(Q)}{(\pi)^3}+\cdots\right) \tag{83}$$

4.3.2 Event Shape Variables

For about 30 years now we have been trying to get fundamental information about quark and gluon interactions from the observed hadrons in e^+e^- annihilation. An instructive example of a measurable quantity which is not calculable because it is not infrared safe is *Sphericity*, proposed by the SLAC group in 1974:

$$\hat{S} \equiv \frac{3}{2}\min_{\text{axes}} \frac{\sum_i |p_\perp^i|^2}{\sum_i |\vec{p}^{\,i}|^2}\,. \tag{84}$$

The summations are over all the final state hadrons, the $p_\perp^i$'s refer to the components of momentum which are perpendicular to the specified axis and the minimization is over all axes through the interaction point. For events in which there are precisely two back-to-back jets the axis aligned with the jets gives the minimum and $\hat{S} = 0$. For isotropic events the normalization has been chosen to give $\hat{S} = 1$. $\hat{S}$ is experimentally measurable but is not calculable in perturbative QCD. To see this note that

$$(p_\perp^1)^2 + (p_\perp^2)^2 \neq (p_\perp^1 + p_\perp^2)^2,$$

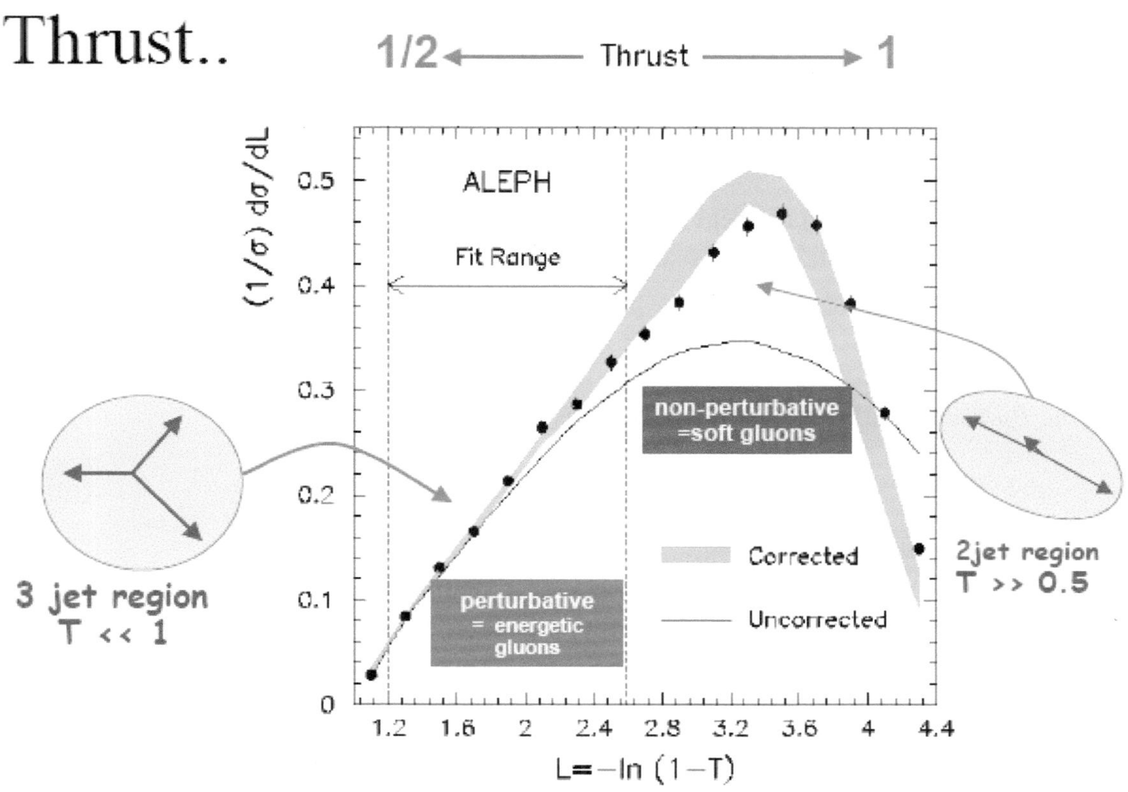

Figure 3. *A Thrust distribution from the LEP QCD Working Group (LEP QCD Working Group, 2003).*

so that, for example, the contribution from a quark will not cancel that from a quark and a gluon with the same total perpendicular component of momentum.

Today many infrared-safe event shape variables are being used, based on summations of terms which are linear in the momenta (rather than quadratic as in eq.(84)). A classic example is *thrust*, defined by

$$T = \max_{\text{axes}} \frac{\sum_i |\vec{p}_i \cdot \vec{n}|}{\sum_i |\vec{p}_i|}, \tag{85}$$

so that $T = 1$ for a two-jet event and $1/2$ for a spherical event. A thrust distribution obtained from LEP is shown in fig. 3

4.4 Deep Inelastic Scattering

In the previous section we considered e^+e^- annihilation, which has no hadrons in the initial state. We now turn to processes with hadrons in the initial state and start by considering *deep inelastic lepton-hadron scattering* with a single hadron in the initial state. To be specific consider the process $ep \to e + X$:

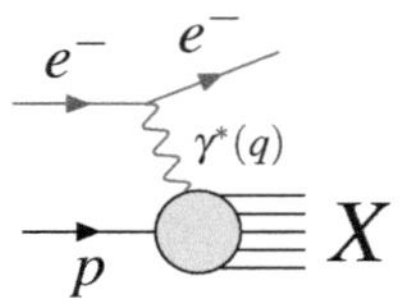

The incoming lepton can also be a μ or a ν. The exchanged boson can also be a Z^0, or in the case of charged-current interactions a W, although in this presentation we will restrict our discussion to the γ. The kinematic region we will be interested in has $-q^2$ and $2p\cdot q$ large (where *large* means w.r.t. Λ) and

$$x \equiv \frac{-q^2}{2p\cdot q} \sim O(1). \tag{86}$$

x is called the *Bjorken* x and is experimentally measurable for each event.

$$(p+q)^2 > 0 \Rightarrow q^2 + 2p\cdot q(+p^2) > 0 \Rightarrow 0 \leq x \leq 1.$$

Therefore x lies between 0 and 1.

In trying to understand the physics of deep-inelastic scattering, much intuition was gained from the Feynman-Bjorken parton picture. Noting that the typical scale of strong-interactions is 1 fm or 200 MeV, consider a frame in which $|\vec{p}|$ is large, and imagine that the struck quark carries a fraction ξ of the proton's momentum.

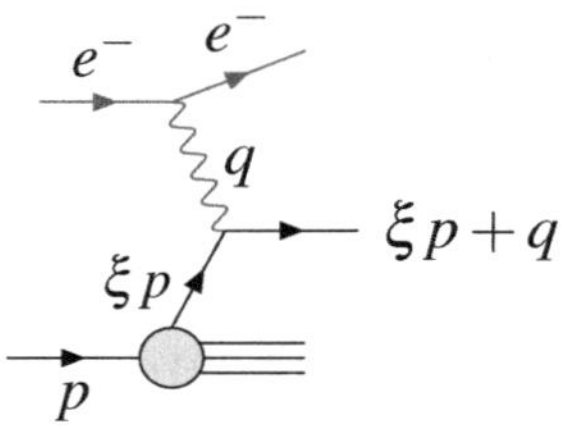

Assuming that the struck quark has virtuality of $O(\Lambda)$ we have:

$$(\xi p+q)^2 \simeq 0 \Rightarrow 2\xi p\cdot q + q^2 \simeq 0 \Rightarrow \xi = x. \tag{87}$$

Thus we arrive at the remarkable fact that the experimentally measurable quantity x gives the fraction of the proton's momentum carried by the struck quark (in the *infinite momentum* frame).

Let the probability density of finding the quark f with longitudinal fraction x of the proton's momentum be $f_{q_f}(x)$. $f_{q_f}(x)$ is called the *parton distribution function*. In the parton model the cross-section for deep inelastic scattering is given by:

$$\sigma(e^-(k)p(p) \to e^-(k')X) = \int_0^1 d\xi \sum_f f_{q_f}(\xi)\, \sigma(e^-(k)q_f(\xi p) \to e^-(k')\, q_f(\xi p+q)). \tag{88}$$

We will consider the QCD corrections later, but for now we stay in the parton model, where

$$\frac{d^2\sigma}{dx\,dy} = \left(\sum_f x f_{q_f}(x)\, Q_f^2\right) \frac{2\pi\alpha^2 s}{q^4}\,[1+(1-y)^2], \tag{89}$$

where $s=(p+k)^2 \simeq 2p\cdot k$ and $y=(2p\cdot q)/s$. In the rest-frame of the proton, y is the fraction of the electron's energy which is transferred to the proton.

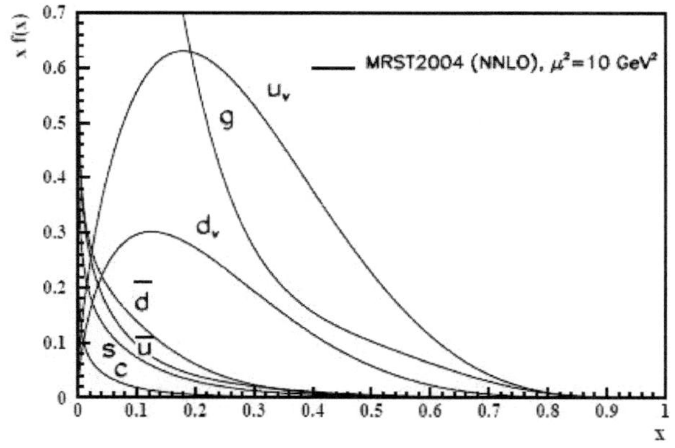

Figure 4. *Parton distribution functions at a scale of 10 GeV^2 (Yao et al., 2006). See this reference for citations to the original literature. Copyright (2006) IOP Publishing.*

Deep Inelastic Scattering gives the momentum distribution of quarks in the proton. With an intermediate photon we get the contributions from the different quarks weighted by the square of the charge of the corresponding flavour as in eq. (89). Information about different linear combinations of the distribution functions can be obtained from ν scattering, consider for example the diagrams:

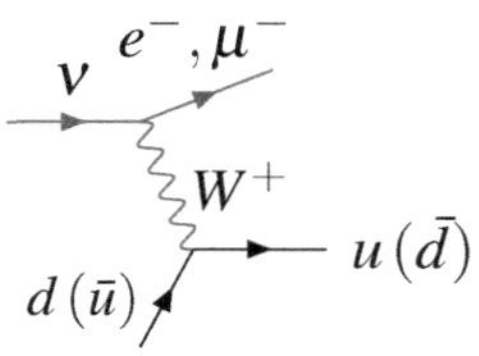

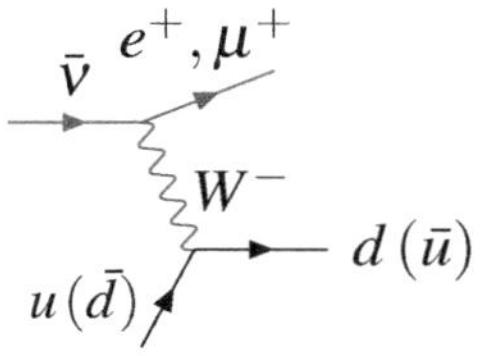

Thus a ν beam probes the distribution of d-quarks (or $\bar{u}$ antiquarks and similarly for the second and even the third generation) whereas a $\bar{\nu}$ beam probes the distribution of u-quarks (or $\bar{d}$ antiquarks etc.):

$$\frac{d^2\sigma(\nu p \to \mu^- X)}{dxdy} = \frac{G_F^2 s}{\pi}\left[x f_d(x) + x(1-y)^2 f_{\bar{u}}(x)\right] \tag{90}$$

$$\frac{d^2\sigma(\bar{\nu} p \to \mu^+ X)}{dxdy} = \frac{G_F^2 s}{\pi}\left[x(1-y)^2 f_u(x) + x f_{\bar{d}}(x)\right] \tag{91}$$

By combining information from e, μ and ν deep inelastic scattering (and more) we get information about each of the distribution functions. In fig. 4 there is a sketch of the distribution functions for the different constituents of the proton.

4.4.1 Hard Scattering Processes in Hadronic Collisions

Before leaving the parton model, let us consider some hard scattering process in hadron-hadron collisions, represented by the following diagram:

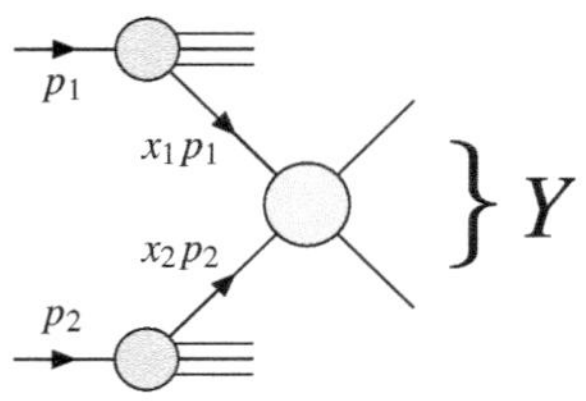

Y is a component of the final state chosen to ensure that the process is indeed hard. For example, Y can be a heavy particle (resonance, Higgs; such processes are called Drell-Yan Processes) or two (or more) jets at large transverse momentum. The cross-section for these processes can be written in the form:

$$\sigma(h_1(p_1)+h_2(p_2)\to Y+X)=\int_0^1 dx_1\int_0^1 dx_2\sum_{f_1,f_2} f_{f_1}(x_1)f_{f_2}(x_2)\sigma(f_1+f_2\to Y)\,. \qquad (92)$$

The f_{f_i} s are "known" from Deep Inelastic Scattering. It is in this way (modified to take QCD corrections into account) that we were able to make predictions for the cross sections for W and Z production at the SPS which led to the discovery of these bosons, or are able to make predictions for Higgs Boson production at the LHC.

4.4.2 Deep Inelastic Scattering and QCD

We start our discussion by recalling the optical theorem, a consequence of unitarity, which relates total cross sections to forward elastic amplitudes. For deep inelastic scattering, this can be represented by the diagram:

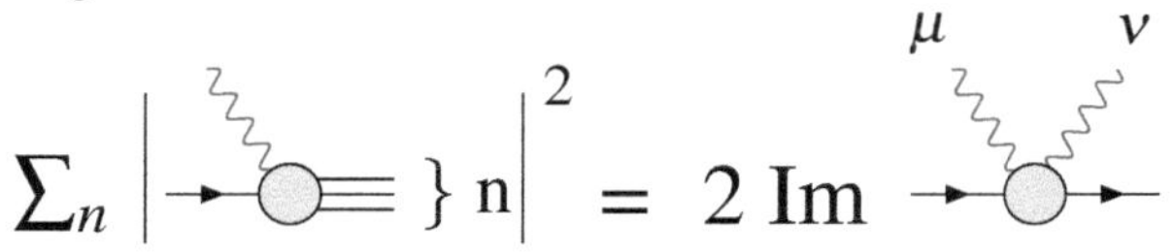

where the sum on the left-hand side is over all intermediate states n and since we are dealing with the QCD effects we have removed the leptons. Instead of calculating the cross section directly we can therefore evaluate the virtual forward Compton amplitude:

$$W^{\mu\nu}(x,q^2)=i\int d^4x\, e^{iq\cdot x}\langle p|T\{J^\mu(x)J^\nu(0)\}|p\rangle \qquad (93)$$

Lorentz and parity invariance and current conservation allow us to write $W^{\mu\nu}$ in terms of scalar functions:

$$W^{\mu\nu}=\left(-g^{\mu\nu}+\frac{q^\mu q^\nu}{q^2}\right)W_1(x,q^2)+\left(p^\mu-q^\mu\frac{p\cdot q}{q^2}\right)\left(p^\nu-q^\nu\frac{p\cdot q}{q^2}\right)W_2(x,q^2)\,, \qquad (94)$$

where $W_{1,2}$ are invariant functions. With Z_0 exchange so that parity is no longer a good symmetry, there is a third *structure function* W_3 multiplying the tensor $\varepsilon^{\mu\nu\alpha\beta}p_\alpha q_\beta$.

In the parton model

$$\mathrm{Im}\,W_1(x)=\pi\sum_f Q_f^2 f_f(x) \quad\text{and}\quad \mathrm{Im}\,W_2(x)=\frac{4\pi}{ys}\sum_f Q_f^2 x f_f(x) \qquad (95)$$

so that

$$\mathrm{Im}\, W_1 = \frac{ys}{4x}\,\mathrm{Im}\, W_2. \tag{96}$$

In a commonly used notation $\nu = p \cdot q$, $F_1 \equiv \mathrm{Im}\, W_1$, $F_2 \equiv \mathrm{Im}\, \nu W_2$ so that in the parton model $F_2 = 2xF_1$.

In QCD there are diagrams such as

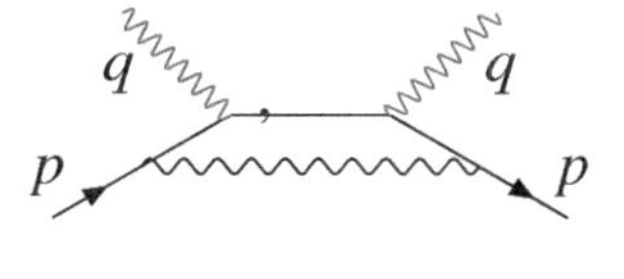

where the external lines with momentum p represent the struck quark. These one-loop diagrams can be evaluated and give a contribution proportional to $\alpha_s \log(q^2/p^2)$. The (collinear) mass singularities do not cancel, in spite of the KLN theorem, because we do not sum over all degenerate initial states. Thus in QCD the structure functions (and parton distribution functions) are functions of q^2 as well as x.

I refer to the standard field theory textbooks, or the original papers of Gross and Wilczek (Gross and Wilczek, 1974) for the use of the Operator Product Expansion (OPE) to determine the q^2 behaviour of the structure functions. The results can also be obtained from the DGLAP equations, which are perhaps more intuitive (let $t = \log(q^2/q_0^2)$):

$$\frac{dq^{\mathrm{NS}}(x,t)}{dt} = \frac{\alpha_s(t)}{2\pi}\int_x^1 \frac{dy}{y} q^{\mathrm{NS}}(y,t)\, P_{q\to q}\left(\frac{x}{y}\right) \tag{97}$$

$$\frac{dq^{\mathrm{S}}(x,t)}{dt} = \frac{\alpha_s(t)}{2\pi}\int_x^1 \frac{dy}{y}\left\{ q^{\mathrm{S}}(y,t)\, P_{q\to q}\left(\frac{x}{y}\right) + g(y,t)\, P_{g\to q}\left(\frac{x}{y}\right)\right\} \tag{98}$$

$$\frac{dg(x,t)}{dt} = \frac{\alpha_s(t)}{2\pi}\int_x^1 \frac{dy}{y}\left\{ q^{\mathrm{S}}(y,t)\, P_{q\to g}\left(\frac{x}{y}\right) + g(y,t)\, P_{g\to g}\left(\frac{x}{y}\right)\right\}, \tag{99}$$

where the superscripts NS and S stand for flavour non-singlet and singlet respectively. The *evolution* equations are different for the singlet (q^{S}) and non-singlet (q^{NS}) quark distributions since the gluon distribution (g) at one value of q^2 can only affect the flavour singlet quark distribution at another value. The *splitting functions* can be represented by the simple diagrams:

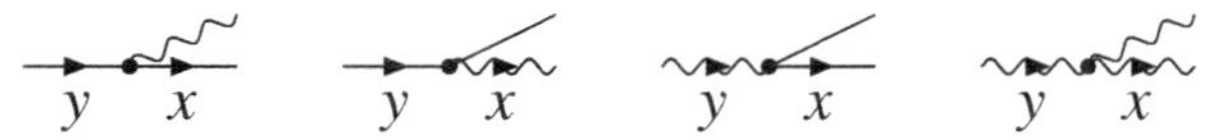

$P_{q\to q}(x/y)$ $P_{q\to g}(x/y)$ $P_{g\to q}(x/y)$ $P_{g\to g}(x/y)$

A nice heuristic physical interpretation emerges from the calculations. Imagine first probing a relativistic proton with a photon of virtuality given by q^2 and observing a quark with fraction y of the proton's momentum. As we improve the resolution (i.e. as we increase q^2) we may be able to resolve that in fact we were observing a quark with fraction x of the proton's momentum and a gluon with fraction $y-x$. From such arguments we may expect that as we increase q^2 we will have more partons with smaller momenta and fewer with large momenta. This is indeed what is obtained from the detailed calculations in QCD and observed experimentally. I illustrate this behaviour by presenting in fig. 5 a compendium of results on the scaling violations in deep inelastic scattering. Results for the structure function F_2 are presented for bins in x and are plotted as a function of q^2. In fig. 5 we indeed observe the fall with q^2 of the structure functions at large x and the rise at small x.

I end this lecture with some concluding comments:

- Our ability to perform calculations in perturbative QCD is limited to short-distance phenomena. For much of the structure of the proton the typical scale is Λ (or 1 fm in coordinate space) and out of reach of perturbation theory. For the structure functions what we can calculate are the (logarithmic) scaling violations at large q^2, i.e. the behaviour of the structure functions with q^2. We cannot calculate the structure functions themselves.

- The photon and weak bosons do not couple to the gluon and hence we do not measure the gluon distribution directly. However the gluon distribution function does affect the scaling violations and therefore the quark distribution functions. By measuring the behaviour of the structure functions with q^2 we are therefore able to determine the gluon distribution in the proton. The gluon distribution functions at a scale of $10\,\mathrm{GeV}^2$ is sketched in fig. 4.

- The *factorization* of hadron-hadron hard-scattering cross sections into a convolution of parton distribution functions (as measured in DIS experiments) and perturbatively calculable parton scattering cross sections is also valid in QCD.

- We can therefore make predictions for specific cross sections (such as that for Higgs production) at the LHC in terms of the measured distribution functions.

5 Flavourdynamics and Non-Perturbative QCD

Theories beyond the standard model of particle physics generally have new particles which are heavier than the energy scales which have been explored to date. Even though these particles may not have been observed directly, they would in general contribute through quantum loops to physical processes. For example, the supersymmetric partners of the known particles would appear in box and penguin diagrams. The size of the contributions depends of course on the masses and couplings, but it may be expected that the effects may be observable in processes which are *rare* in the standard model. An important area of current investigation is to study a large variety of weak processes and check for consistency within the standard model. The framework for doing this is to over-determine the Wolfenstein parameters $\bar{\rho}$ and $\bar{\eta}$ from many different processes. To date no (incontrovertible) inconsistencies have emerged which itself places constraints on the parameter space of new theories. As the precision improves, it is possible that the inconsistencies will appear yielding important clues about the new physics.

In order to understand whether a given rate or CP-asymmetry is consistent with the standard model or not we frequently have to compute non-perturbative QCD effects in the corresponding process. Indeed our limited ability to do so is frequently the source for the largest uncertainty in the determination of fundamental physics from experimental measurements. In the remainder of the course I provide an introduction to the subject of flavour physics and discuss one important approach to the evaluation of non-perturbative QCD effects, lattice QCD. I will illustrate

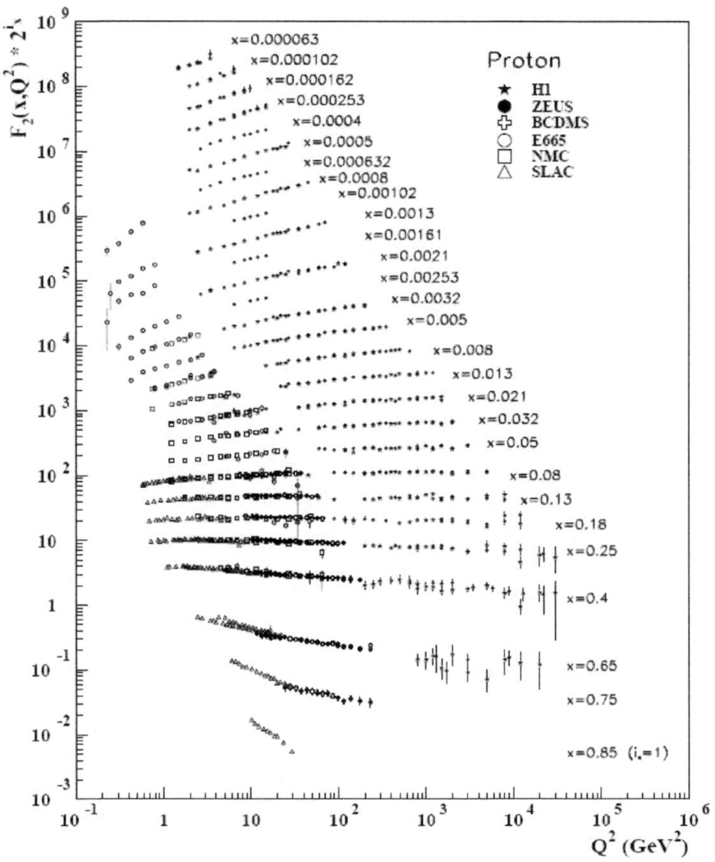

Figure 5. *Scaling violations in the structure function F_2 (Eidelman et al., 2005). See this reference for citations to the original literature. Copyright (2004) Elsevier.*

the determination of the CKM matrix elements using some of the standard physical examples. For a complete source of up-to-date results see the website of the latest CKM workshop (ckm2006).

5.1 Leptonic Decays of Mesons

To illustrate how non-perturbative hadronic effects complicate the extraction of *CKM* matrix elements consider the leptonic decays of pseudoscalar mesons in general and of the *B*-meson in particular. The quark flow diagram for this process is sketched in the following diagram:

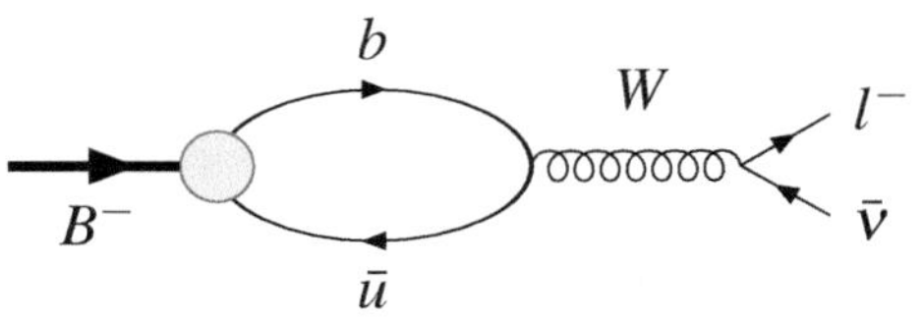

The non-perturbative QCD effects are contained in the matrix element

$$\langle 0|\bar{b}\gamma^{\mu}(1-\gamma^5)u|B(p)\rangle \ . \tag{100}$$

Parity is a good symmetry of the strong interactions which implies that only the axial current contributes to the decay and Lorentz invariance implies that

$$\langle 0|\,\bar{b}\gamma^\mu\gamma^5 u\,|B(p)\rangle = i f_B p^\mu \,, \tag{101}$$

where f_B is a constant (with dimensions of mass). Thus we see that all non-perturbative QCD effects are contained in a single constant, f_B, the B-meson's *(leptonic) decay constant.* (The corresponding constant for the pion is $f_\pi \simeq 132\,\text{MeV}$.)

Earlier this year the Belle collaboration observed this decay for the first time, quoting the branching ratio:

$$B(B\to\tau\nu_\tau) = \left(1.06^{+0.34}_{-0.28}\,(\text{stat.})^{+0.18}_{-0.16}\,(\text{syst.})\right)\times 10^{-4}\,. \tag{102}$$

What does this mean for the CKM matrix elements. Using the definition of f_B in eq.(101), we can write the branching ratio as

$$B(B\to\tau\nu_\tau) = f_B^2\,|V_{ub}|^2\,\frac{G_F^2 m_B m_\tau^2}{8\pi}\left(1-\frac{m_\tau^2}{m_B^2}\right)^2\tau_B\,. \tag{103}$$

Thus the measurement of the branching ratio gives us information about $f_B|V_{ub}|$ and in order to determine V_{ub} we need to know f_B which requires non-perturbative QCD. I now take a brief detour to describe one of the principle techniques for evaluating hadronic matrix elements non-perturbatively, *Lattice QCD*.

5.2 Introduction to Lattice Phenomenology

Lattice phenomenology starts with the evaluation of correlation functions of the form:

$$\langle 0|\,O(x_1,x_2,\cdots,x_n)\,|0\rangle = \frac{1}{Z}\int [dA_\mu]\,[d\psi]\,[d\bar{\psi}]\,e^{iS}\,O(x_1,x_2,\cdots,x_n)\,, \tag{104}$$

where $O(x_1,x_2,\cdots,x_n)$ is a multilocal operator composed of quark and gluon fields and Z is the partition function:

$$Z = \int [dA_\mu]\,[d\psi]\,[d\bar{\psi}]\,e^{iS}\,. \tag{105}$$

These formulae are written in Minkowski space, whereas lattice calculations are performed in Euclidean space ($\exp(iS)\to\exp(-S)$ etc.). The physics which can be studied depends on the choice of the multilocal operator O. The functional integral is performed by discretising space-time and using Monte-Carlo integration.

5.2.1 Two Point Correlation Functions

As a simple example, consider two-point correlation functions of the form:

$$C_2(t) = \int d^3x\, e^{i\vec{p}\cdot\vec{x}}\,\langle 0|\,J(\vec{x},t)\,J^\dagger(\vec{0},0)\,|0\rangle\,, \tag{106}$$

where J and $J^\dagger$ are any interpolating operators for the hadron H which we wish to study and the time t is taken to be positive. We assume that H is the lightest hadron which can be created by $J^\dagger$. (We take $t > 0$, but it should be remembered that lattice simulations are frequently performed on periodic lattices, so that both time-orderings contribute.)

Inserting a complete set of states $\{|n\rangle\}$ between the operators in eq.(106):

$$\begin{aligned} C_2(t) &= \sum_n \int d^3x\, e^{i\vec{p}\cdot\vec{x}}\, \langle 0|\, J(\vec{x},t)\, |n\rangle\, \langle n|J^\dagger(\vec{0},0)\, |0\rangle \\ &= \int d^3x\, e^{i\vec{p}\cdot\vec{x}}\, \langle 0|\, J(\vec{x},t)\, |H\rangle\, \langle H|J^\dagger(\vec{0},0)\, |0\rangle + \cdots \end{aligned} \qquad (107)$$

where the ellipses represent contributions from heavier states with the same quantum numbers as H. Finally using translational invariance we obtain:

$$C_2(t) = \frac{1}{2E}\, e^{-iEt} \left| \langle 0|J(\vec{0},0)|H(p)\rangle \right|^2 + \cdots\,, \qquad (108)$$

where $E = \sqrt{m_H^2 + \vec{p}^2}$. Eq. (108) is illustrated by the diagram:

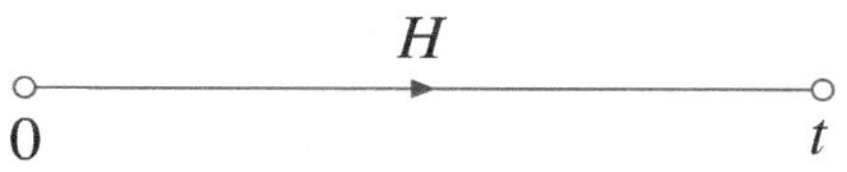

In Euclidean space $\exp(-iEt)$ is replaced by $\exp(-Et)$. By fitting $C_2(t)$ to the form above, both the energy (or, if $\vec{p} = 0$, the mass) and the modulus of the matrix element

$$\left| \langle 0|J(\vec{0},0)|H(p)\rangle \right| \qquad (109)$$

can be evaluated. For example if $J = \bar{u}\gamma^\mu\gamma^5 d$ then the decay constant of the π-meson can be evaluated,

$$\left| \langle 0|\, \bar{u}\gamma^\mu\gamma^5 d\, |\pi^+(p)\rangle \right| = f_\pi\, p^\mu\,, \qquad (110)$$

(the physical value of f_π is 132 MeV). In this way we can evaluate hadronic masses and simple *weak matrix elements* non-perturbatively.

It is useful to introduce the concept of an *effective mass*. At zero momentum ($\vec{p} = 0$)

$$C_2(t) = \text{Constant} \times e^{-mt} + \text{contributions from heavier states} \qquad (111)$$

and the effective mass is defined by:

$$m_{\text{eff}}(t) = \log\left(\frac{C(t)}{C(t+1)}\right)\,. \qquad (112)$$

A typical effective-mass plot for a pseudoscalar meson (taken from a current UKQCD simulation) is shown below:

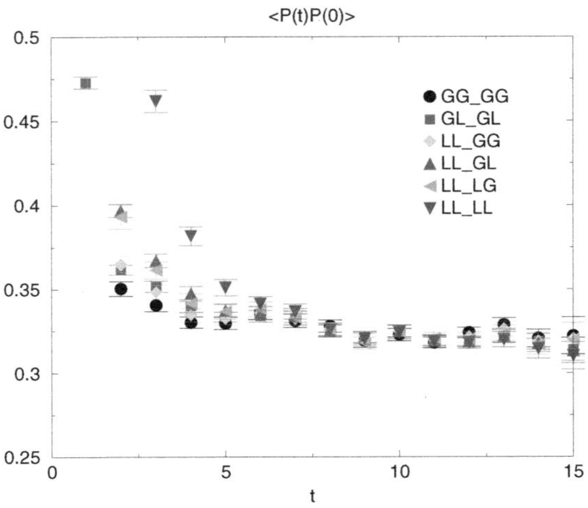

The different symbols correspond to different interpolating operators (J) and we observe the expected behaviour, i.e. that for small values of t there is no plateau because a number of hadronic states contribute, whereas at larger values of t the correlation function is dominated by the ground state with a single effective mass.

5.2.2 Three-Point Correlation Functions

For several of the important phenomenological applications of lattice QCD it is necessary to compute three-point correlation function of the form:

$$C_3(t_x,t_y) = \int d^3x\,d^3y\; e^{i\vec{p}\cdot\vec{x}}\; e^{i\vec{q}\cdot\vec{y}}\; \langle 0|\,J_2(\vec{x},t_x)\,O(\vec{y},t_y)\,J_1^\dagger(\vec{0},0)\,|0\rangle\;, \tag{113}$$

where $J_{1,2}$ are interpolating operators for hadrons 1 and 2 and we assume that $t_x > t_y > 0$. C_3 is illustrated by the diagram:

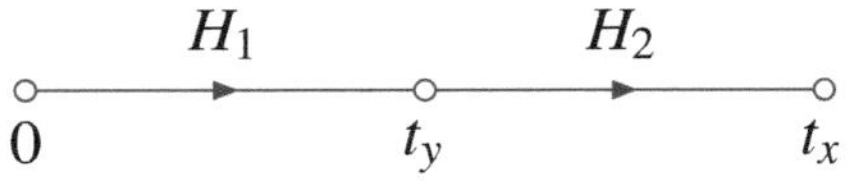

For sufficiently large time intervals t_y and $t_x - t_y$

$$\begin{aligned} C_3(t_x,t_y) &\simeq \frac{e^{-E_1 t_y}}{2E_1}\,\frac{e^{-E_2(t_x-t_y)}}{2E_2}\,\langle 0|J_2(0)|H_2(\vec{p})\rangle \\ &\quad\times\langle H_2(\vec{p})|O(0)|H_1(\vec{p}+\vec{q})\rangle\;\langle H_1(\vec{p}+\vec{q})|J_1^\dagger(0)|0\rangle\;, \end{aligned} \tag{114}$$

where $E_1^2 = m_1^2 + (\vec{p}+\vec{q})^2$ and $E_2^2 = m_1^2 + \vec{p}^{\,2}$.

From the evaluation of two-point functions we know the masses of the two hadrons and the matrix elements of the form $|\langle 0|J|H_{1,2}(\vec{p})\rangle|$. Thus, from the evaluation of three-point functions we obtain matrix elements of the form $|\langle H_2|O|H_1\rangle|$. Important examples of physical quantities which can be obtained in this way include:

- $K^0 - \bar{K}^0$ ($B^0 - \bar{B}^0$) mixing. In this case

$$O = \bar{s}\gamma^\mu(1-\gamma^5)d\;\bar{s}\gamma_\mu(1-\gamma^5)d\,.$$

- Semileptonic and rare radiative decays of hadrons of the form $B \longrightarrow \pi, \rho +$ leptons or $B \longrightarrow K^*\gamma$. Now O is a quark bilinear operator such as $\bar{b}\gamma^\mu(1-\gamma^5)u$ or an *electroweak penguin* operator.

5.2.3 Systematic Uncertainties

Although in principle lattice computations are *ab initio* calculations in QCD, the limitations in available computing resources lead to systematic uncertainties. In particular hadronic properties are computed on a lattice of finite volume and with a non-zero spacing between neighbouring lattice points.

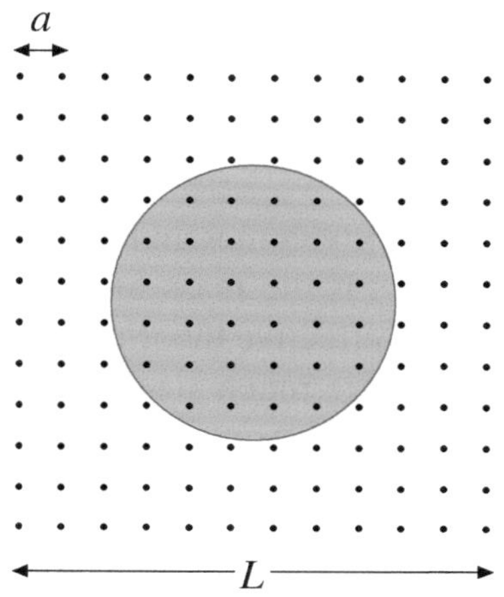

In principle we would like to have a lattice which is much bigger that the size(s) of the hadron(s) and with a lattice spacing which is much smaller that the typical scale of strong interactions ($\sim 1\,\mathrm{fm}$):

$$L \gg 1\,\mathrm{fm} \quad \text{and} \quad a^{-1} \gg \Lambda_{\mathrm{QCD}}\,. \tag{115}$$

Typically in full QCD we can have about $24-32$ points in each spatial direction and so compromises have to be made. A typical choice may be to take the lattice spacing to be about .1 fm and the spatial extent of the lattice to be $2.5-3$ fm. Of course, much effort is being devoted to the control and reduction of the systematic uncertainties, the most significant of which in general is the *chiral extrapolation*, i.e. errors due to the fact that simulations are performed with unphysically heavy u and d quarks and the results then have to be extrapolated to the chiral limit. Wherever possible, we use chiral perturbation theory to guide the extrapolation, but it is still very rare to observe chiral logarithms. It is not appropriate in these introductory lectures to discuss the treatment of the uncertainties in detail; I will just mention that in addition to the errors due to the chiral extrapolation, the finite volume and the non-zero lattice spacing, we have those due to the renormalization of bare lattice quantities and *statistical ones* due to the fact that the functional integral is evaluated by Monte-Carlo sampling.

So what is the lattice result for f_B? Before answering this question we have to note that for heavy quark physics there is an additional complication. If, as mentioned above, a typical lattice spacing is $a \simeq 0.1$ fm, so that $a^{-1} \simeq 2\,\mathrm{GeV}$, such a lattice cannot host a propagating b-quark. Lattice simulations in heavy quark physics therefore require the use of *effective theories* which I describe in the next section. The reviewer at this year's lattice conference presented as his best lattice results for f_{D_s} and f_{B_s} (Onogi, 2006).

$$f_{B_s} = 240(30)\,\text{MeV} \quad \text{and} \quad f_{D_s} = 260(20)\,\text{MeV}\,. \tag{116}$$

The reason for the choice of mesons with a strange quark is that the chiral extrapolation then only involves the sea quarks and is milder. For B and D mesons the decay constants are smaller and two recent reviewers quoted the following for f_B:

$$f_B = 189 \pm 27\,\text{MeV}(\mathit{Hashimoto}, 2005) \quad \text{and} \quad f_B = 216 \pm 22\,\text{MeV}(\mathit{Davies}, 2005)\,. \tag{117}$$

$$f_B = 189 \pm 27\,\text{MeV} \quad \text{and} \quad f_B = 216 \pm 22\,\text{MeV}(\mathit{Davies}, 2005)\,. \tag{118}$$

5.3 The Heavy Quark Effective Theory - HQET

B-physics is playing a central rôle in flavourdynamics and it is useful to exploit the symmetries which arise when $m_Q \gg \Lambda_{\text{QCD}}$, where m_Q is the mass of a heavy quark. The *Heavy Quark Effective Theory* (HQET), as well as other effective theories, is proving invaluable in the study of heavy quark physics. For scales which are much smaller than m_Q the physics in HQET is the same as in QCD. For scales $O(m_Q)$ and greater, the physics is different, but can be *matched* onto QCD using perturbation theory. The non-perturbative physics in the same in the HQET as in QCD.

For a comprehensive review of the foundations of the HQET see (Neubert, 1994). In order to illustrate the features of the theory consider the propagator of a (free) heavy quark:

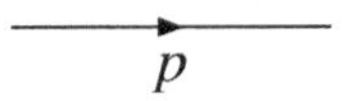

$$= i\frac{\not p + m}{p^2 - m_Q^2 + i\varepsilon}\,.$$

If the momentum of the quark p is not far from its mass shell,

$$p_\mu = m_Q v_\mu + k_\mu\,, \tag{119}$$

where $|k_\mu| \ll m_Q$ and v_μ is the (relativistic) four velocity of the hadron containing the heavy quark ($v^2 = 1$), then

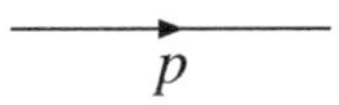

$$= i\frac{1+\not v}{2}\frac{1}{v\cdot k + i\varepsilon} + O\left(\frac{|k_\mu|}{m_Q}\right).$$

$(1+\not v)/2$ is a projection operator, projecting out the *large* components of the spinors.

The heavy quark propagator above can be obtained from the gauge-invariant action

$$\mathscr{L}_{HQET} = \bar{h}(iv\cdot D)\frac{1+\not v}{2}h \tag{120}$$

where h is the spinor field of the heavy quark. $\mathscr{L}_{HQET}$ is independent of m_Q, which implies the existence of symmetries (or *Scaling Laws*) relating physical quantities corresponding to different heavy quarks (in practice the b and c quarks). As will be seen from the following section, the light degrees of freedom are also not sensitive to the spin of the heavy quark, which leads to a spin-symmetry relating physical properties of heavy hadrons of different spins.

5.3.1 Spin Symmetry in the HQET

We now present an example of the spin symmetry. Consider the correlation function:

$$\int d^3x \, \langle 0|J_H(x)\, J_H^\dagger(0)\,|0\rangle\,, \tag{121}$$

$J_H^\dagger$ and J_H are interpolating operators which can create or annihilate a heavy hadron H, where we take H to be a pseudoscalar or vector meson. The hadron is produced at rest, with four velocity $v = (1,\vec{0})$. For example we can take take $J_H = \bar{h}\gamma^5 q$ for the pseudoscalar meson and $J_H = \bar{h}\gamma^i q$ $(i = 1,2,3)$ for the vector meson. From eq. (120) we see that the spin structure of the propagator of the heavy quark [3] with $\vec{v} = \vec{0}$ is $(1+\gamma^0)/2$ so that correlation function is identical in the two cases except for the factor

$$\gamma^5 \frac{1+\gamma^0}{2} \gamma^5 = \frac{1-\gamma^0}{2} \tag{122}$$

when H is a pseudoscalar meson, and

$$\gamma^i \frac{1+\gamma^0}{2} \gamma^i = -3\,\frac{1-\gamma^0}{2} \tag{123}$$

when it is a vector meson. Since the correlation functions are proportional to $\sim \exp(-iM_H t)$, the pseudoscalar and vector mesons are degenerate (up to relative corrections of $O(\Lambda^2_{QCD}/m_Q)$):

$$M_P = M_V + O(\Lambda^2_{QCD}/m_Q)\,. \tag{124}$$

(or $M_V^2 - M_P^2 =$ constant.) Another important heavy quark scaling law which is derived in a similar way is for the pseudoscalar decay constants pseudoscalar mesons and shows that they satisfy $f_P \sim 1/\sqrt{M_P}$.

5.4 Semileptonic Decays

The CKM matrix elements V_{cb} and V_{ub} can be determined from either inclusive or exclusive semileptonic B-decays. I start with a discussion of exclusive decays. Note that the matrix element V_{us} is determined in a similar way from $K \to \pi$ semileptonic decays. The quark flow diagram for exclusive semileptonic B-decays is:

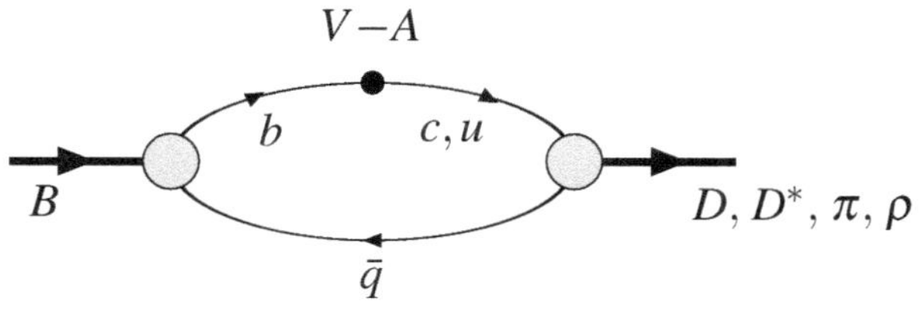

The final state can be a D or D^* meson for $b \to c$ transitions and a π or a ρ-meson for $b \to u$ transitions. As was the case for leptonic decays, in order to determine the CKM matrix elements we must calculate the non-perturbative QCD effects.

[3]When evaluating the correlation functions using eq. (104) we need the quark propagators in the presence of a background gauge-field configuration.

Space-Time symmetries allow us to parametrise the non-perturbative strong interaction effects in terms of invariant form-factors. For example, for decays into a pseudoscalar meson P ($= \pi$ or D for example)

$$\langle P(k)|V^\mu|B(p)\rangle = f^+(q^2)\left[(p+k)^\mu - \frac{m_B^2 - m_P^2}{q^2}\,q^\mu\right] + f^0(q^2)\,\frac{m_B^2 - m_P^2}{q^2}\,q^\mu\,, \tag{125}$$

where $q = p - k$. For decays into a vector V ($= \rho$ or D^* for example), a conventional decomposition is

$$\begin{aligned}\langle V(k,\varepsilon)|V^\mu|B(p)\rangle &= \frac{2V(q^2)}{m_B + m_V}\varepsilon^{\mu\gamma\delta\beta}\varepsilon^*_\beta p_\gamma k_\delta \\ \langle V(k,\varepsilon)|A^\mu|B(p)\rangle &= i(m_B{+}m_V)A_1(q^2)\varepsilon^{*\mu} - i\frac{A_2(q^2)}{m_B{+}m_V}\varepsilon^*{\cdot}p\,(p{+}k)^\mu + i\frac{A(q^2)}{q^2}2m_V\varepsilon^*{\cdot}p\,q^\mu\,,\end{aligned} \tag{126}$$

where ε is the polarization vector of the final-state meson and again $q = p{-}k$.

For $B \to D^*$ decays, HQET implies that

$$\begin{aligned}\frac{d\Gamma}{d\omega} &= \frac{G_F^2}{48\pi^3}(m_B{-}m_{D^*})^2 m_{D^*}^3\sqrt{\omega^2{-}1}\,(\omega{+}1)^2 \times \\ &\quad \left[1 + \frac{4\omega}{\omega{+}1}\frac{m_B^2 - 2\omega m_B m_{D^*} + m_{D^*}^2}{(m_B - m_{D^*})^2}\right]|V_{cb}|^2\,\mathscr{F}^2(\omega)\,,\end{aligned} \tag{127}$$

where $\mathscr{F}(\omega)$ is the Isgur-Wise function combined with perturbative and power corrections. ($\omega = v_B \cdot v_{D^*}$) The key point is that $\mathscr{F}(1) = 1$ up to power corrections and calculable perturbative corrections, so the uncertainties are small.

Why is $\mathscr{F}(1) = 1$? In QCD, form-factors between degenerate states are 1 by current conservation (e.q. the electromagnetic form-factor of pseudoscalar mesons). Although, for example, the B and D mesons are not degenerate the leading order Lagrangian of the HQET is independent of the heavy quark mass. This heavy quark symmetry leads to $\mathscr{F}(1) = 1$.

To determine the difference of $\mathscr{F}(1)$ from 1 using lattice QCD, it has proved effective to use the method of double ratios (Hashimoto, 2000). For example one can evaluate

$$\mathscr{R}_+ = \frac{\langle D|\bar{c}\gamma^4 b|\bar{B}\rangle\,\langle\bar{B}|\bar{b}\gamma^4 c|D\rangle}{\langle D|\bar{c}\gamma^4 c|D\rangle\,\langle\bar{B}|\bar{b}\gamma^4 b|\bar{B}\rangle} = |h_+(1)|^2 \tag{128}$$

with

$$h_+(1) = \eta_V\left\{1 - \ell_P\left(\frac{1}{2m_c} - \frac{1}{2m_b}\right)^2\right\}.$$

By calculating $\mathscr{R}_+$ and similar ratios of $V \leftrightarrow P$ and $V \leftrightarrow V$ matrix elements all three ℓ's required for the evaluation of the $1/m_Q$ corrections to $\mathscr{F}(1)$ can be determined. A recent result from the FNAL/MILC/HPQCD Collaborations (Okamoto, 2004) gives

$$|V_{cb}| = 3.9(1)(3) \times 10^{-2}. \tag{129}$$

The form factors for $D \to \pi, K$ semileptonic decays are also being evaluated.

5.4.1 V_{cb} from Inclusive Decays

V_{cb} can also be obtained from the total semileptonic rate, and from lepton energy and hadronic mass spectra. The main tool is the Operator Product Expansion which gives a series in inverse powers of m_b, m_c .

$$\Gamma = |V_{cb}|^2 \hat{\Gamma}_0 m_b^5(\mu)\,(1 + A_{\rm EW}) A^{\rm pert}(r,\mu) \left\{ z_0 + \frac{z_2(r)}{m_b^2} + \frac{z_3(r)}{m_b^3} + \cdots \right\} \tag{130}$$

where $r = m_c/m_b$ and the z's are known functions depending on non-perturbative parameters which are determined from the spectra (Kowalewski and Mannel). As the labels suggest $A_{\rm EW}$ and $A^{\rm pert}$ are contributions from electromagnetic effects and from QCD perturbation theory respectively. It is difficult however, to quantify any violations of quark-hadron duality.

The 2006 Particle Data Group summary gives:

$$|V_{cb}| = (41.7 \pm 0.7)\,10^{-3} \quad \text{(inclusive)}; \qquad |V_{cb}| = (40.9 \pm 1.8)\,10^{-3} \quad \text{(exclusive)}. \tag{131}$$

5.4.2 $B \to \pi$ Exclusive Semileptonic Decays from the Lattice

Lattice simulations enable the evaluation of the form factors for $B \to \pi$ (and $B \to \rho$) semileptonic decays, illustrated by the quark-flow diagram:

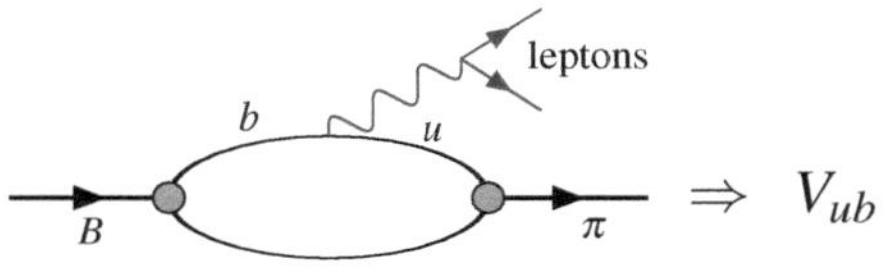

The HQET is of significantly less help here than it was for $B \to D$ decays and so the determination of V_{ub} has a significantly larger uncertainty. In lattice determinations of the form-factors, in order to keep the discretization errors small, the momentum of the pion must be small. We therefore obtain the form factors at large values of q^2. These can be used with theoretical constraints on the decay amplitudes and experimental results for the rates which are now available in q^2 bins to obtain V_{ub} with reasonable precision. Some recent results for $B \to \pi$ form factors are plotted below (Onogi, 2005):

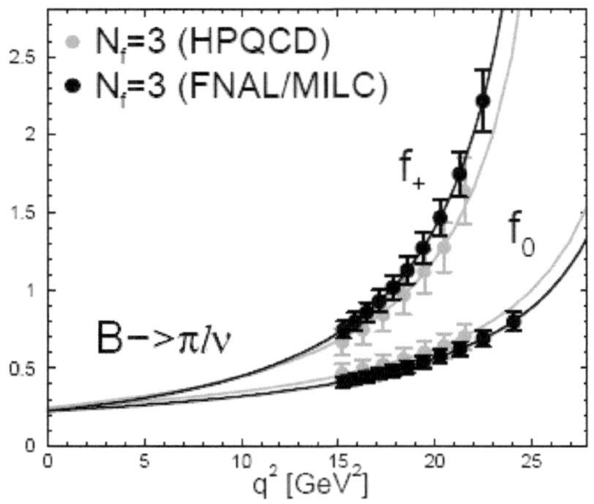

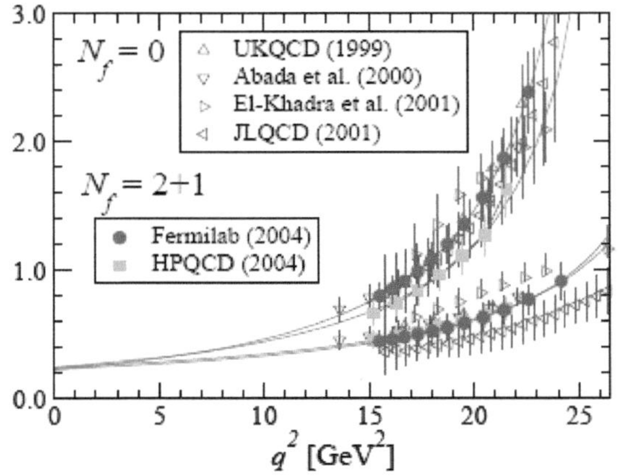

The corresponding values of V_{ub} from the two unquenched simulations are:

HPQCD (Shigemitsu, 2005)	(Staggered Light & NRQCD Heavy)	$\lvert V_{ub}\rvert = 4.04(20)(44)(53)\times 10^{-3}$
FNAL/MILC (Okamoto, 2005)	(Staggered Light & Fermilab Heavy)	$\lvert V_{ub}\rvert = 3.48(29)(38)(47)\times 10^{-3}$

V_{ub} is also determined from inclusive decays $\bar{B}\to X_u\ell\bar{\nu}_\ell$ using the heavy-quark expansion. The difficulty in this approach is to remove the backgrounds from the larger $\bar{B}\to X_c\ell\bar{\nu}_\ell$ decays. If this is done by going towards the end-point so that $b\to c$ decays are not possible, then we need non-perturbative input (the *shape function*) which limits the precision.

The 2006 Particle Data Group summary gives:

$$|V_{ub}| = (4.40\pm 0.2\pm 0.27)\,10^{-3}\quad \text{(inclusive)};\qquad |V_{ub}| = (3.84^{+0.67}_{-0.49})\,10^{-3}\quad \text{(exclusive)}. \tag{132}$$

The theoretical uncertainties in the inclusive and exclusive determinations of V_{cb} and V_{ub} are very different and it is reassuring that the results are consistent (although there is some *tension* for V_{ub}). In terms of the Wolfenstein parameters:

$$|V_{ub}|^2 = A^2\lambda^6(\bar{\rho}^2+\bar{\eta}^2), \tag{133}$$

so that an accurate determination of V_{ub} would give us the circle in the $\bar{\rho}-\bar{\eta}$ plane, with its centre at the origin, on which the vertex A of the unitarity triangle must lie. The current situation is summarized by the dark-green ring in fig. 1.

5.5 Kaon Physics

We now turn to some topics in kaon-physics and in particular to $K^0-\bar{K}^0$ mixing and $\Delta S=1$ decays.

$K^0-\bar{K}^0$ mixing is mediated through the box diagrams:

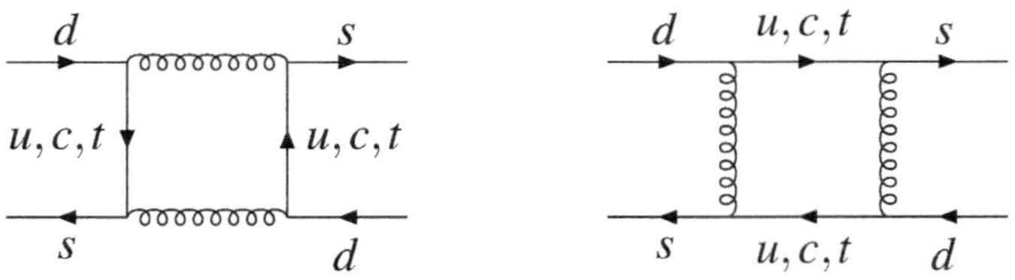

K^0 and $\bar{K}^0$ are not eigenstates of CP, indeed they transform into each other under the CP transformation. The CP-eigenstates (K_1 and K_2) are linear combinations of the two strong-interaction eigenstates:

$$|K_1\rangle = \frac{1}{\sqrt{2}}(|K^0\rangle + |\bar{K}^0\rangle)\quad \text{so that}\quad CP|K_1\rangle = |K_1\rangle \tag{134}$$

and

$$|K_2\rangle = \frac{1}{\sqrt{2}}(|K^0\rangle - |\bar{K}^0\rangle)\quad \text{so that}\quad CP|K_2\rangle = -|K_2\rangle\,. \tag{135}$$

I use the phase convention so that $CP|K^0\rangle = |\bar{K}^0\rangle$.

Because of the complex phase in the CKM-matrix, the physical states (the mass eigenstates), $|K_S\rangle$ and $|K_L\rangle$, differ from $|K_1\rangle$ and $|K_2\rangle$ by a small admixture of the other state:

$$|K_S\rangle = \frac{|K_1\rangle + \bar{\varepsilon}\,|K_2\rangle}{(1+|\bar{\varepsilon}|^2)^{\frac{1}{2}}} \quad \text{and} \quad |K_L\rangle = \frac{|K_2\rangle + \bar{\varepsilon}\,|K_1\rangle}{(1+|\bar{\varepsilon}|^2)^{\frac{1}{2}}}\,, \tag{136}$$

The parameter $\bar{\varepsilon}$ depends on the phase convention chosen for $|K^0\rangle$ and $|\bar{K}^0\rangle$.

For $K \to \pi\pi$ and $K \to \pi\pi\pi$ decays, the two pion states are CP-even and the three-pion states are CP-odd, which implies that the dominant decays are:

$$K_S \to \pi\pi \quad \text{and} \quad K_L \to 3\pi\,. \tag{137}$$

There is little phase-space for the decay into three pions and this is the reason why K_L is much longer lived than K_S. K_L and K_S are not precisely CP-eigenstates however, and so $K_L \to 2\pi$ and $K_S \to 3\pi$ decays may occur at a low level. CP-violating decays which occur due to the fact that the mass eigenstates are not CP-eigenstates are called *indirect CP-violating decays*.

A measure of the strength of indirect CP-violation is given by the physical parameter ε_K defined by the ratio:

$$\varepsilon_K \equiv \frac{A\,(K_L \to (\pi\pi)_{I=0})}{A\,(K_S \to (\pi\pi)_{I=0})} = (2.280 \pm 0.013)\,10^{-3}\,e^{i\frac{\pi}{4}}\,. \tag{138}$$

Direct CP-violating decays are those in which a CP-even (-odd) state decays into a CP-odd (-even) one. For K_L decays this is illustrated in the following schematic diagram:

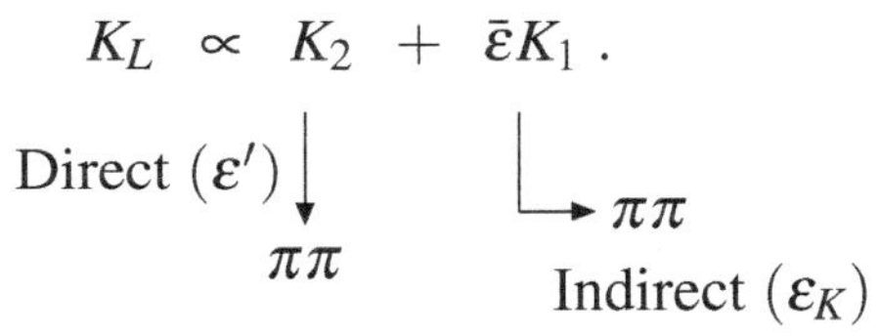

To get some insight into the origin of ε' consider the following contributions to $K \to \pi\pi$ decays:

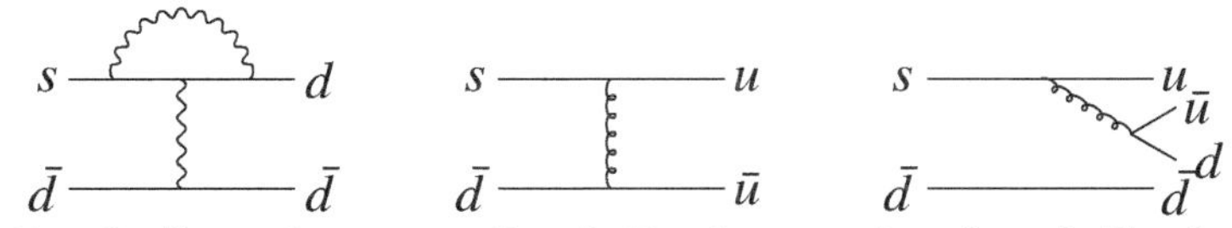

$I = 0$, Complex (a) — $I = 0$, Real (b) — $I = 0$ or 2, Real (c)

Thus direct CP-violation in kaon decays manifests itself as a non-zero relative phase between the $I = 0$ and $I = 2$ amplitudes. We also have *strong phases*, δ_0 and δ_2, which are independent of the form of the weak Hamiltonian. The amplitudes for $K^0 \to \pi\pi$ decays can therefore be written in the form:

$$A(K^0 \to \pi^+\pi^-) = \sqrt{\frac{2}{3}}A_0\,e^{i\delta_0} + \sqrt{\frac{1}{3}}A_2\,e^{i\delta_2} \tag{139}$$

$$A(K^0 \to \pi^0\pi^0) = \sqrt{\frac{2}{3}}A_0\,e^{i\delta_0} - 2\sqrt{\frac{1}{3}}A_2\,e^{i\delta_2}\,. \tag{140}$$

The parameter ε', which is used as a measure of CP-violation is defined by:

$$\varepsilon' = \frac{\omega}{\sqrt{2}} e^{i\phi} \left(\frac{\mathrm{Im} A_2}{\mathrm{Re} A_2} - \frac{\mathrm{Im} A_0}{\mathrm{Re} A_0} \right) , \tag{141}$$

where

$$\omega \equiv \frac{\mathrm{Re} A_2}{\mathrm{Re} A_0} \quad \text{and} \quad \phi = \frac{\pi}{2} + \delta_2 - \delta_0 \simeq \frac{\pi}{4} . \tag{142}$$

ε' is manifestly zero if the phases of the $I = 0$ and $I = 2$ weak amplitudes are the same.

At this stage it is worth raising the $\Delta I = 1/2$ rule, a puzzle which has been around for more than 50 years. In terms of ω it can be posed as *Why is* ω^{-1} *so large?* Experimentally we know that $\omega^{-1} \simeq 22$ and yet we still do not understand well this enhancement of the $I = 0$ amplitude relative to the $I = 2$ one.

Experimentally the two parameters ε_K (which, following standard conventions I rename from now on as ε, $\varepsilon \equiv \varepsilon_K$) and ε' can be determined by measuring the ratios:

$$\eta_{00} \equiv \frac{A(K_L \to \pi^0\pi^0)}{A(K_S \to \pi^0\pi^0)} \simeq \varepsilon - 2\varepsilon' \tag{143}$$

$$\eta_{+-} \equiv \frac{A(K_L \to \pi^+\pi^-)}{A(K_S \to \pi^+\pi^-)} \simeq \varepsilon + \varepsilon' . \tag{144}$$

Direct CP-violation is found to be considerably smaller than indirect violation. By measuring the decays and using

$$\left| \frac{\eta_{00}}{\eta_{+-}} \right|^2 \simeq 1 - 6\,\mathrm{Re}\left(\frac{\varepsilon'}{\varepsilon} \right) + \cdots , \tag{145}$$

the NA31 and E371 experiments have measured ε'/ε, and the combined result is:

$$\varepsilon'/\varepsilon = (17.2 \pm 1.8)\, 10^{-4} . \tag{146}$$

ε and the Unitarity Triangle: In order to determine the CKM matrix elements present in $K^0 - \bar{K}^0$ mixing, we need to know the following hadronic matrix element:

$$\langle \bar{K}^0 \,|\, \mathcal{H}_{\mathrm{eff}}^{\Delta S=2} \,|\, K^0 \rangle . \tag{147}$$

The form of the effective Hamiltonian is

$$\mathcal{H}_{\mathrm{eff}}^{\Delta S=2} = \frac{G_F^2}{16\pi^2} M_W^2 \, \mathscr{X} \, O^{\Delta S=2}(\mu) \tag{148}$$

where $\mathscr{X}$ is a function of the CKM-matrix elements, with coefficients which can be calculated perturbatively and which depend on the $(u,)c$ and t masses. The non-perturbative QCD corrections are contained in the matrix element:

$$\langle \bar{K}^0 \,|\, \bar{s}\gamma^\mu (1-\gamma^5) d \; \bar{s}\gamma_\mu (1-\gamma^5) d \,|\, K^0 \rangle \equiv \frac{8}{3} m_K^2 f_K^2 B_K(\mu) . \tag{149}$$

The uncertainty in B_K is a major restriction on the Unitarity Triangle analysis.

There have been many lattice determinations of B_K. Until recently these have been largely in the quenched approximation (in which vacuum polarization effects are neglected) and two recent reviewers at major conferences gave the following averages:

$$B_K^{\overline{\text{MS}}}(2\,\text{GeV}) = 0.58(4) \qquad \text{S.Hashimoto (ICHEP 2004)}(Hashimoto, 2005) \qquad (150)$$

$$B_K^{\overline{\text{MS}}}(2\,\text{GeV}) = 0.58(3) \qquad \text{C.Dawson (Lattice 2005)}(Dawson, 2006)\,. \qquad (151)$$

Since, by today's standards quenched simulations are relatively undemanding on computing resources, the remaining systematic uncertainties in eqs.(150) and (151) are relatively small. Dynamical computations (i.e. ones in full QCD, with vacuum polarization effects included) of B_K are underway by a number of collaborations, but so far the results are very preliminary. In particular we still do not have precise extrapolations to the continuum limit. Chris Dawson, by comparing unquenched and quenched results at similar masses and lattice spacings, "guesstimated" the physical result to be

$$B_K^{\overline{\text{MS}}}(2\,\text{GeV}) = 0.58(3)(6) \qquad \text{C.Dawson (Lattice 2005)}\,, \qquad (152)$$

but it is only a guesstimate. We need to wait until reliable dynamical results are available in the next year or two for a more reliable unquenched result.

A precise determination of ε would fix the vertex A to lie on a hyperbola in the $(\bar{\rho}, \bar{\eta})$ plane:

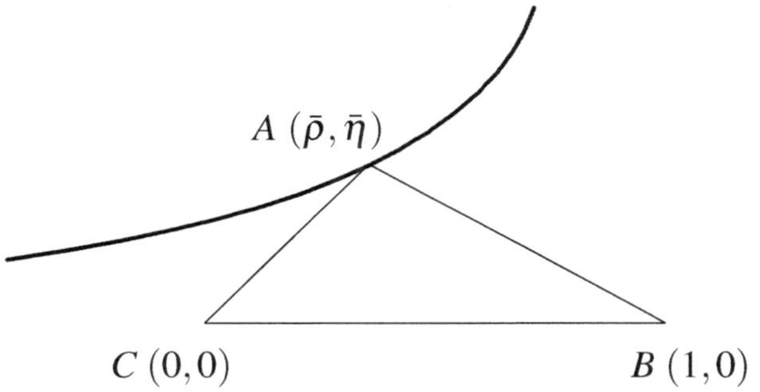

Of course in practice there are both experimental and theoretical uncertainties (the theoretical errors are the dominant ones) so that the hyperbola is replaced by a band. The current status is summarized by the light green band in fig. 1.

5.6 *B*-Physics

IN this section we consider a number of important processes in B-physics.

5.6.1 $B^0 - \bar{B}^0$ Mixing

The first topic which we consider is neutral B-meson mixing, mediated by the box diagrams:

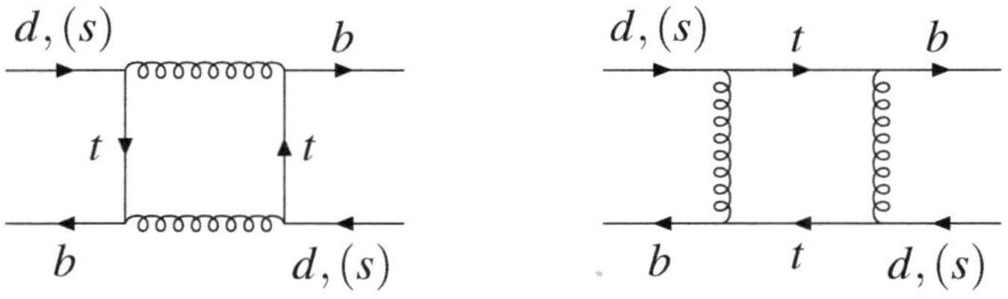

The top quark dominates and hence from the measured mass differences we obtain V_{td} and V_{ts}. Similarly to $K^0 - \bar{K}^0$ mixing, the non-perturbative QCD effects are contained in the matrix element of the $\Delta B = 2$ operator:

$$O^{\Delta B=2} = \bar{b}\gamma^{\mu}(1-\gamma^5)d\,\bar{b}\gamma_{\mu}(1-\gamma^5)d \equiv \frac{8}{3}m_B^2 f_B^2 B_B(\mu). \tag{153}$$

The uncertainty in this matrix element dominates that in the final answer for $|V_{td}|$.

The Particle Data Group (2006) use $\Delta m_d = 0.507 \pm 0.004$ as the experimental value and take the lattice result $f_{B_d}\sqrt{\hat{B}_{B_d}} = (244 \pm 11 \pm 24)$ MeV to obtain

$$|V_{td}| = (7.4 \pm 0.8) \times 10^{-3}. \tag{154}$$

The uncertainties are reduced in the lattice calculation of the ratio

$$\xi = \frac{f_{B_s}\sqrt{B_{B_s}}}{f_{B_d}\sqrt{B_{B_d}}} = 1.21 \pm 0.04^{+0.04}_{-0.01} \quad \Rightarrow \left|\frac{V_{td}}{V_{ts}}\right| = 0.208^{+0.008}_{-0.006}, \tag{155}$$

where the new Tevatron result of $\Delta m_s = (17.31^{+0.33}_{-0.18} \pm 0.07)\ \mathrm{ps}^{-1}$ has been used.

5.6.2 The Golden Mode - $B \to K_S J/\Psi$

As we have seen, the difficulty in extracting fundamental information from experimental measurements in nonleptonic B-decays is largely due to our limited ability to quantify the non-perturbative QCD effects. There are a handful of cases where these hadronic effects cancel, and the best example is the CP-asymmetry in the decay $B \to K_S J/\Psi$ which gives us $\sin(2\beta)$ with remarkable precision. I will now review the theoretical background behind this statement.

Mixing Induced CP-Violating Decays In order to study CP-violation we need to be sensitive to the weak phase and so we have to use *interference* in some way. In general there may be several weak contributions to the amplitude, each with a different weak phase. Although in such a case we do have interference of course, it is difficult to control the hadronic effects. For the golden mode, there is a single weak phase (to an excellent approximation) and the interference comes from the mixing of B^0 and $\bar{B}^0$ as I now explain.

The two neutral mass-eigenstates are linear combinations of the flavour states B^0 and $\bar{B}^0$:

$$|B_L\rangle = \frac{1}{\sqrt{p^2+q^2}}\left(p\,|B^0\rangle + q\,|\bar{B}^0\rangle\right) \tag{156}$$

$$|B_H\rangle = \frac{1}{\sqrt{p^2+q^2}}\left(p\,|B^0\rangle - q\,|\bar{B}^0\rangle\right), \tag{157}$$

where p and q are complex parameters.

Starting with a B^0 meson at time $t = 0$, its subsequent evolution is governed by the Schrödinger equation with a 2×2 mass-matrix of the form (required by CPT invariance)

$$M - \frac{i\Gamma}{2} = \begin{pmatrix} A & p^2 \\ q^2 & A \end{pmatrix}.$$

where A, p and q are complex parameters. The state evolves at time t into

$$|B^0_{\rm phys}(t)\rangle = g_+(t)\,|B^0\rangle + \left(\frac{q}{p}\right) g_-(t)|\bar{B}^0\rangle\,, \tag{158}$$

where

$$\begin{aligned} g_+(t) &= \exp\left[-\frac{\Gamma t}{2}\right]\exp[-iMt]\cos\left(\frac{\Delta M t}{2}\right), \\ g_-(t) &= \exp\left[-\frac{\Gamma t}{2}\right]\exp[-iMt]\,i\sin\left(\frac{\Delta M t}{2}\right), \end{aligned}$$

and $M = (M_H + M_L)/2$. Similarly, starting with a $\bar{B}^0$ meson at $t = 0$, the time evolution is given by

$$|\bar{B}^0_{\rm phys}(t)\rangle = (p/q)\,g_-(t)|\bar{B}^0\rangle + g_+(t)\,|\bar{B}^0\rangle. \tag{159}$$

Decays of Neutral B-Mesons into CP-Eigenstates Let f_{CP} be a CP-eigenstate and $A,\bar{A}$ be the amplitudes

$$A \equiv \langle f_{CP}|\mathscr{H}|B^0\rangle \quad \text{and} \quad \bar{A} \equiv \langle f_{CP}|\mathscr{H}|\bar{B}^0\rangle. \tag{160}$$

Defining

$$\lambda \equiv \frac{q}{p}\frac{\bar{A}}{A} \tag{161}$$

we have

$$\langle f_{CP}\,|\,\mathscr{H}\,|\,B^0_{\rm phys}\rangle = A\,[g_+(t) + \lambda\, g_-(t)] \quad \text{and} \quad \langle f_{CP}\,|\,\mathscr{H}\,|\,\bar{B}^0_{\rm phys}\rangle = A\,\frac{p}{q}\,[g_-(t) + \lambda\, g_+(t)]. \tag{162}$$

The time-dependent rates for initially pure B^0 or $\bar{B}^0$ states to decay into the CP-eigenstate f_{CP} at time t are given by:

$$\begin{aligned} \Gamma(B^0_{\rm phys}(t) \to f_{CP}) &= |A|^2 e^{-\Gamma t}\left[\frac{1+|\lambda|^2}{2} + \frac{1-|\lambda|^2}{2}\cos(\Delta M t) - {\rm Im}\,\lambda\,\sin(\Delta M t)\right] \quad (163) \\ \Gamma(\bar{B}^0_{\rm phys}(t) \to f_{CP}) &= |A|^2 e^{-\Gamma t}\left[\frac{1+|\lambda|^2}{2} - \frac{1-|\lambda|^2}{2}\cos(\Delta M t) + {\rm Im}\,\lambda\,\sin(\Delta M t)\right] \quad (164) \end{aligned}$$

The time-dependent CP-asymmetry is defined as:

$$\begin{aligned} \mathscr{A}_{f_{CP}}(t) &\equiv \frac{\Gamma(B^0_{\rm phys}(t) \to f_{CP}) - \Gamma(\bar{B}^0_{\rm phys}(t) \to f_{CP})}{\Gamma(B^0_{\rm phys}(t) \to f_{CP}) + \Gamma(\bar{B}^0_{\rm phys}(t) \to f_{CP})} \quad (165) \\ &= \frac{(1-|\lambda|^2)\cos(\Delta M t) - 2{\rm Im}\,\lambda\,\sin(\Delta M t)}{1+|\lambda|^2}. \quad (166) \end{aligned}$$

If $|q/p| = 1$ (which is the case if $\Delta\Gamma \ll \Delta M$) and $|\bar{A}/A| = 1$ (examples of this will be presented below) then $|\lambda| = 1$ and the first term on the right-hand side above vanishes.

The generic form of the amplitudes A and $\bar{A}$ is:

$$A = \sum_i A_i e^{i\delta_i} e^{i\phi_i} \quad \text{and} \quad \bar{A} = \sum_i A_i e^{i\delta_i} e^{-i\phi_i} \tag{167}$$

where the sum is over all the contributions to the process; the A_i are real; the δ_i are the strong phases corresponding to final state rescattering and the ϕ_i are the phases from the CKM matrix.

In the most favourable situation, all the contributions to the decay amplitudes have a single CKM phase (ϕ_D say) so that

$$\frac{\bar{A}}{A} = \exp(-2i\phi_D). \tag{168}$$

Since $\Gamma_{12} << M_{12}$, $q/p = \sqrt{M_{12}^*/M_{12}} \equiv \exp(-2i\phi_M)$, and

$$\lambda = \exp(-2i(\phi_D + \phi_M)) \quad \text{so that} \quad \text{Im}\,\lambda = -\sin(2(\phi_D + \phi_M)). \tag{169}$$

From the box diagrams we have:

$$\left(\frac{q}{p}\right)_{B_d} = \frac{V_{td}V_{tb}^*}{V_{td}^*V_{tb}} \quad \text{and} \quad \left(\frac{q}{p}\right)_{B_s} = \frac{V_{ts}V_{tb}^*}{V_{ts}^*V_{tb}}. \tag{170}$$

Consider now those processes in which the b-quark decays through the subprocesses of the form $b \rightarrow d_j u_i \bar{u}_i$, where the suffices i, j in u_i and d_j label the flavours of the up and down-like quarks respectively. The corresponding tree-level diagram is

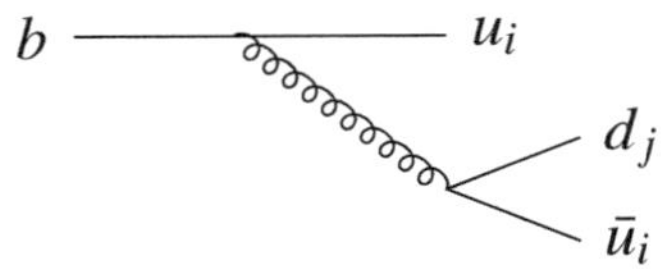

for which

$$\frac{\bar{A}}{A} = \frac{V_{ib}V_{ij}^*}{V_{ib}^*V_{ij}}. \tag{171}$$

Now let us focus on the specific process $B_d \rightarrow J/\Psi K_S$. In this case

$$\lambda(B \rightarrow J/\Psi K_S) = \frac{V_{td}V_{tb}^*}{V_{td}^*V_{tb}}\,\frac{V_{cs}V_{cd}^*}{V_{cs}^*V_{cd}}\,\frac{V_{cb}V_{cs}^*}{V_{cb}^*V_{cs}} = -\sin(2\beta) \tag{172}$$

where we recall that

$$\beta = \arg\left(-\frac{V_{cd}V_{cb}^*}{V_{td}V_{tb}^*}\right), \tag{173}$$

and

- the first factor is $(q/p)_{B_d}$;
- the second factor is the analogous one for the final state kaon;
- the third factor is $\bar{A}/A$, with $u_i = c$ and $d_j = s$.

Thus from a measurement of the asymmetry we obtain $\sin(2\beta)$ with (almost) no hadronic uncertainties. The "almost" is inserted in the last sentence because, as always, there are other weak contributions, but unusually they are very small in this case. For example, in addition to the tree contribution above there is also a small penguin contribution to this process:

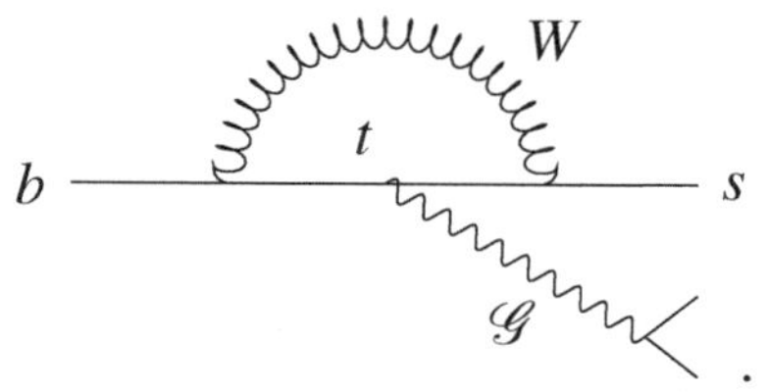

The important feature for this process is that the phase of the penguin contribution is that of $V_{tb}V_{ts}^*$, which is equal (to an excellent approximation) to that of $V_{cb}V_{cs}^*$. We therefore have a single weak phase and hence hadronic uncertainties are negligible in the determination of the $\sin(2\beta)$ from this process (hence the label *golden mode*). Unfortunately such an (almost) ideal situation is very rare.

PDG 2006 average the results from BaBar and Belle and obtain

$$\sin(2\beta) = 0.687 \pm 0.032 . \tag{174}$$

The uncertainty is still decreasing: for example in the 2000 edition of PDG book, $\sin(2\beta) = 0.78 \pm 0.08$.

6 Summary and Conclusions

In these lectures I have tried to remind you of the main elements of the Standard Model of Particle Physics and to describe some of the attempts to explore its limits in the quark sector. We know from the observation of ν oscillations that there is new physics to understand and Flavour Physics will continue to be a powerful tool with which to unravel the structure of physics beyond the standard model. (I did not have time to discuss the recent developments in the determination of α and γ from two-body B-decays.) Since these lectures have been prepared the Particle Data Group has published the 2006 edition of the Review of Particle Properties (Yao et al, 2006), updating some of the results quoted here.

Finally I would like to register my warm thanks to the organisers for inviting me to such an enjoyable school, to the students for the stimulating questions and to everyone for your excellent company.

References

Barate R *et al.* LEP Working Group for Higgs boson searches Phys. Lett. B **565** (2003) 61 [arXiv:hep-ex/0306033]

Block F and Nordsieck A, Phys. Rev. **52** (1937) 54.

Ckm2006, `http://ckm2006.hepl.phys.nagoya-u.ac.jp/homepage.html`

Davies C H, `http://www.lip.pt/events/2005/hep2005/talks/hep2005_talk_ChristineDavies.pdf`

Dawson C, 2006 PoS **LAT2005** (2006) 007.

Eidelman S *et al.*[Particle Data Group], Phys. Lett. B **592** (2004) 1 and 2005 partial update, `http://pdg.lbl.gov/2005/reviews/contents_sports.html`

Gross D J and Wilczek, F, 1973 Phys. Rev. Lett. **30** (1973) 1343; H. D. Politzer, Phys. Rev. Lett. **30** (1973) 1346.

Gross D J and Wilczek F, 1974 Phys. Rev. D **8** (1973) 3633; Phys. Rev. D **9** (1974) 980.
Hashimoto S *et al.*, 2000, Phys. Rev. D **61** (2000) 014502 [arXiv:hep-ph/9906376].
Hashimoto S, 2005, Int. J. Mod. Phys. A **20** (2005) 5133 [arXiv:hep-ph/0411126].
Hill C T and Simmons E H, Phys. Rept. **381** (2003) 235 [arXiv:hep-ph/0203079]
Kowalewski R and Mannel T, *Determination of* V_{cb} *and* V_{ub} *in (Yao, 2005)*
Larin S A *et al.* Phys. Lett. B **400**, 379 (1997) [arXiv:hep-ph/9701390]
LEPEWWG *The LEP Electroweak Working Group's home page*,
`http://lepewwg.web.cern.ch/LEPEWWG/`
LEP QCD Working Group, presented by Roger Jones,
`http://lepqcd.web.cern.ch/LEPQCD/annihilations/LEPQCDWGReportMarch03.pdf`
Neubert M, Phys. Rept. **245** (1994) 259 [arXiv:hep-ph/9306320].
Okamoto M, 2004 [Fermilab Lattice Collaboration], arXiv:hep-lat/0412044
Okamoto M et al., 2005 (Fermilab/MILC Collaboration), Nucl. Phys. (Proc. Suppl.) B **140** (2005) 461.
Onogi T, 2005, Chamonix Flavour Dynamics Workshop, October 2005
Onogi T, 2006 arXiv:hep-lat/0610115
Peskin M E and Schroeder D V, *An Introduction To Quantum Field Theory, Addison-Wesley (1995)*
Shigemitsu J *et al.*, 2005 (HPQCD collaboration) Nucl. Phys. (Proc. Suppl.) B **140** (2005) 464.
Yao W M et al. (Particle Data Group),2006, *J. Phys.* **G33** 1

Neutrino Oscillation Phenomenology

Boris Kayser

Fermi National Accelerator Laboratory, USA

1 Introduction

Progress on our understanding of the neutrinos continues to be exhilarating. This progress is due mainly to experiments on neutrino oscillation. Here, we explain the physics of oscillation in vacuum and in matter.

2 The physics of neutrino oscillation

Treatments of the physics of neutrino oscillation may be found in, for example, [1, 2]. Here, we give a slightly modified treatment, and explain some points that have caused puzzlement, such as the fact that, even if neutrinos are their own antiparticles, their interaction with matter can still cause a difference between neutrino and so-called "antineutrino" oscillations.

We assume that the couplings of the neutrinos and charged leptons to the W boson are correctly described by the Standard Model, extended to take leptonic mixing into account. These couplings are then summarized by the Lagrangian

$$\mathscr{L}_W = -\frac{g}{\sqrt{2}} \sum_{\substack{\alpha=e,\mu,\tau \\ i=1,2,3}} (\overline{\ell_{L\alpha}} \gamma^\lambda U_{\alpha i} \nu_{Li} W_\lambda^- + \overline{\nu_{Li}} \gamma^\lambda U_{\alpha i}^* \ell_{L\alpha} W_\lambda^+) \ . \tag{1}$$

Here, L denotes left-handed chiral projection, ℓ_α is the charged-lepton mass eigenstate of flavor α (ℓ_e is the electron, ℓ_μ the muon, and ℓ_τ the tau), and ν_i is a neutrino mass eigenstate. The constant g is the semiweak coupling constant, and U is the leptonic mixing matrix [3]. Supposing, as assumed by Eq. (1), that there are only three charged-lepton mass eigenstates, and three neutrino mass eigenstates, U is 3×3, and may be written as

$$U = \begin{bmatrix} U_{e1} & U_{e2} & U_{e3} \\ U_{\mu 1} & U_{\mu 2} & U_{\mu 3} \\ U_{\tau 1} & U_{\tau 2} & U_{\tau 3} \end{bmatrix} \ . \tag{2}$$

In the extended Standard Model, the 3×3 mixing matrix U is unitary, and we shall assume that this is also true in nature. However, we note that if there are "sterile" neutrinos (neutrinos that do not couple to the W or Z boson), then there are $N > 3$ neutrino mass eigenstates, and the leptonic mixing matrix U that is unitary is $N \times N$, rather than 3×3. The 3×3 matrix of Eq. (2) is then just a submatrix, and is not unitary [4].

Supposing that the unitary mixing matrix is $N \times N$, not because of the existence of sterile neutrinos but because there are N conventional lepton generations, how many physically-significant parameters does U contain? To see how many, we note first that an $N \times N$ complex matrix contains N^2 entries, each of which may have a real and an imaginary part. Thus, the matrix can be fully specified by $2N^2$ real parameters. If the matrix is unitary, then each of its columns must be a vector of unit length: $\sum_\alpha |U_{\alpha i}|^2 = 1$; $i = 1, N$. Together, these conditions are N constraints. In addition, each pair of columns in U must be orthogonal vectors: $\sum_\alpha U^*_{\alpha i} U_{\alpha j} = 0$; $i, j = 1, N$ with $i \neq j$. Taking into account that each of these $N(N-1)/2$ orthogonality conditions has both a real and an imaginary part, we see that these conditions impose $N(N-1)$ constraints. Thus, the number of independent parameters in a general $N \times N$ unitary matrix is $2N^2 - N - N(N-1) = N^2$. However, in the case of our unitary matrix, U, some of these parameters may be removed. From Eq. (1), $\langle \ell_\alpha | \mathscr{L}_W | \nu_i W^- \rangle \propto U_{\alpha i}$. Now, without affecting the physics, we are always free to redefine the state $\langle \ell_\alpha |$ by multiplying it by a phase factor: $\langle \ell_\alpha | \rightarrow \langle \ell'_\alpha | = \langle \ell_\alpha | e^{-i\varphi_\alpha}$. Clearly, this has the effect of multiplying the $U_{\alpha i}$, for all i, by the same factor: $U_{\alpha i} \rightarrow U'_{\alpha i} = e^{-i\varphi_\alpha} U_{\alpha i}$. If there are N ℓ_α, this phase redefinition of them may be used to remove N phases from U. It might be thought that analogous phase redefinition of the neutrinos ν_i could be used to remove additional phases. However, unlike the quarks and charged leptons, the neutrino mass eigenstates ν_i may be their own antiparticles: $\overline{\nu_i} = \nu_i$. This possibility motivates the search for neutrinoless nuclear double beta decay, as discussed at this school by K. Zuber. If $\overline{\nu_i} = \nu_i$, then physically significant phases cannot be eliminated by phase redefinition of the ν_i [5]. To allow for the possibility that $\overline{\nu_i} = \nu_i$, we shall retain the phases that can be eliminated only when $\overline{\nu_i} \neq \nu_i$. Then U is left with $N^2 - N$ physically significant parameters. These are commonly chosen to be "mixing angles"—parameters that would be present even if U were real—and complex phase factors. To see how many of the parameters are mixing angles, and how many are phases, let us imagine for a moment that U is real. Then it can be fully specified by its N^2 real entries. These are subject to the unitarity requirement that the N columns of U all have unit length: $\sum_\alpha U^2_{\alpha i} = 1$, $i = 1, N$, and the requirement that all $N(N-1)/2$ pairs of columns be orthogonal: $\sum_\alpha U_{\alpha i} U_{\alpha j} = 0$, $i, j = 1, N$ with $i \neq j$. Hence, a real mixing matrix U for N generations has $N^2 - N - N(N-1)/2 = N(N-1)/2$ physically significant parameters, and a complex one has this number of mixing angles. Since a complex U has $N(N-1)$ physically significant parameters in all, the fact that $N(N-1)/2$ of them are mixing angles means that the remaining $N(N-1)/2$ must be complex phase factors.

In summary, a complex $N \times N$ unitary mixing matrix U for N lepton generations may contain—

N(N-1)/2	mixing angles
N(N-1)/2	complex phase factors
N(N-1)	physically significant parameters in all

Throughout most of these lecture notes, we will assume that $N = 3$. Then the mixing matrix contains three mixing angles and three complex phase factors. It can be shown that this matrix can be written in the form

$$U = \begin{bmatrix} 1 & 0 & 0 \\ 0 & c_{23} & s_{23} \\ 0 & -s_{23} & c_{23} \end{bmatrix} \times \begin{bmatrix} c_{13} & 0 & s_{13}e^{-i\delta} \\ 0 & 1 & 0 \\ -s_{13}e^{i\delta} & 0 & c_{13} \end{bmatrix} \times \begin{bmatrix} c_{12} & s_{12} & 0 \\ -s_{12} & c_{12} & 0 \\ 0 & 0 & 1 \end{bmatrix}$$
$$\times \begin{bmatrix} e^{i\xi_1/2} & 0 & 0 \\ 0 & e^{i\xi_2/2} & 0 \\ 0 & 0 & 1 \end{bmatrix} . \quad (3)$$

Here, $c_{ij} \equiv \cos\theta_{ij}$ and $s_{ij} \equiv \sin\theta_{ij}$, where the θ_{ij} are the three mixing angles. The quantities δ, ξ_1, and ξ_2 are the three complex phases.

From Eq. (1), we observe that the amplitude for the decay $W^+ \to \overline{\ell_\alpha} + \nu_i$ to yield the particular charged-lepton mass eigenstate $\overline{\ell_\alpha}$ in combination with the particular neutrino mass eigenstate ν_i is proportional to $U^*_{\alpha i}$. Thus, if we define the "neutrino state of flavor α", $|\nu_\alpha\rangle$, with $\alpha = e, \mu$, or τ, to be the neutrino state that accompanies the particular charged lepton $\overline{\ell_\alpha}$ in leptonic W^+ decay, then we must have

$$|\nu_\alpha\rangle = \sum_{i=1}^{3} U^*_{\alpha i} |\nu_i\rangle \ . \quad (4)$$

From Eq. (1), the amplitude for this ν_α to interact and produce the particular charged-lepton ℓ_β is proportional to

$$\sum_{i=1}^{3} U_{\beta i} U^*_{\alpha i} = \delta_{\beta\alpha} \ , \quad (5)$$

where we have invoked the unitarity of U. We see that when a ν_e, the neutrino born in a W^+ decay that produced an $\bar{e}$, interacts and produces a second charged lepton, the latter can only be an e. Similarly for ν_μ and ν_τ.

We may invert Eq. (4) to obtain

$$|\nu_i\rangle = \sum_{\alpha=e,\mu,\tau} U_{\alpha i} |\nu_\alpha\rangle \ . \quad (6)$$

This expresses the mass eigenstate $|\nu_i\rangle$ in terms of the states of definite flavor, $|\nu_\alpha\rangle$. We see that the flavor-α fraction of $|\nu_i\rangle$ is simply $|U_{\alpha i}|^2$.

2.1 Neutrino oscillation in vacuum

Consider the vacuum neutrino oscillation experiment depicted schematically in the upper part of Figure 1. A neutrino source produces, via W exchange, the charged lepton $\overline{\ell_\alpha}$ of flavor α, plus an accompanying neutrino that, by definition, must be a ν_α. The neutrino then propagates, in vacuum, a distance L to a target/detector. There, it interacts via W exchange and produces a second charged lepton ℓ_β of flavor β. Thus, at the moment of its interaction in the detector, the neutrino is a ν_β. If the flavors α and β are different, then, during the neutrino's trip to the detector, it has changed, or "oscillated", from a ν_α into a ν_β.

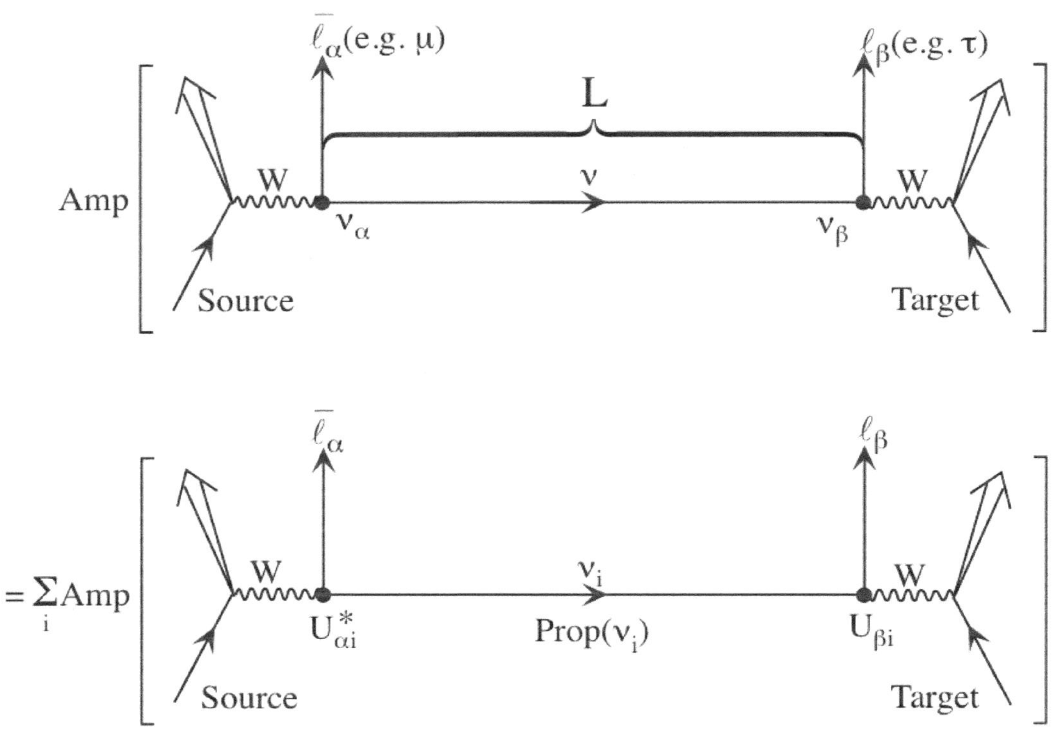

Figure 1. *Neutrino flavor change (oscillation) in vacuum. "Amp" denotes an amplitude.*

In the neutrino mass eigenstate basis, the particle that travels from the neutrino source to the detector is one or another of the mass eigenstates ν_i. In a given event, we will not know which ν_i was actually involved. Hence, the amplitude for the oscillation $\nu_\alpha \rightarrow \nu_\beta$, Amp $(\nu_\alpha \rightarrow \nu_\beta)$, is a coherent sum over the contributions of all the ν_i, as shown in the lower part of Figure 1. The contribution of an individual ν_i is a product of three factors. The first is the amplitude for the neutrino produced together with the charged lepton $\overline{\ell_\alpha}$ to be, in particular, a ν_i. From Eq. (1), this amplitude is $U^*_{\alpha i}$, as indicated in Figure 1. The second factor, Prop (ν_i), is the amplitude for the mass eigenstate ν_i to propagate from the source to the detector. The final factor is the amplitude for the charged lepton created when the ν_i interacts in the detector to be, in particular, an ℓ_β. From Eq. (1), this amplitude is $U_{\beta i}$.

From elementary quantum mechanics, the propagation amplitude Prop(ν_i) is simply exp $[-im_i\tau_i]$, where m_i is the mass of ν_i, and τ_i is the proper time that elapses in the ν_i rest frame during its propagation. By Lorentz invariance, $m_i\tau_i = E_i t - p_i L$, where L is the lab-frame distance between the neutrino source and the detector, t is the lab-frame time taken for the beam to traverse this distance, and E_i and p_i are, respectively, the lab-frame energy and momentum of the ν_i component of the beam.

Once the absolute square $|\text{Amp}\,(\nu_\alpha \rightarrow \nu_\beta)|^2$ is taken to compute the probability for the oscillation $\nu_\alpha \rightarrow \nu_\beta$, only the *relative* phases of the propagation amplitudes Prop (ν_i) for different mass eigenstates will have physical consequences. From the discussion above, the relative phase of Prop (ν_1) and Prop (ν_2), $\delta\phi(12)$ is given by

$$\begin{aligned}\delta\phi(12) &= (E_2t - p_2L) - (E_1t - p_1L) \\ &= (p_1 - p_2)L - (E_1 - E_2)t \ . \end{aligned} \tag{7}$$

In practice, experiments do not measure the transit time t. However, Lipkin has shown [6] that, to an excellent approximation, t may be taken to be $L/\bar{v}$, where

$$\bar{v} \equiv \frac{p_1 + p_2}{E_1 + E_2} \tag{8}$$

is an approximation to the average of the velocities of the ν_1 and ν_2 components of the beam. We then have

$$\begin{aligned}\delta\phi(12) &\cong \frac{p_1^2 - p_2^2}{p_1 + p_2}L - \frac{E_1^2 - E_2^2}{p_1 + p_2}L \\ &= (m_2^2 - m_1^2)\frac{L}{p_1 + p_2} \cong (m_2^2 - m_1^2)\frac{L}{2E} \ , \end{aligned} \tag{9}$$

where, in the last step, we have used the fact that for highly relativistic neutrinos, p_1 and p_2 are both approximately equal to the beam energy E. We conclude that all the relative phases in Amp$(\nu_\alpha \to \nu_\beta)$ will be correct if we take

$$\text{Prop}\,(\nu_i) = e^{-im_i^2 L/2E} \ . \tag{10}$$

Combining the factors that appear in the lower part of Figure 1, we have

$$\text{Amp}\,(\nu_\alpha \to \nu_\beta) = \sum_i U_{\alpha i}^* e^{-im_i^2 L/2E} U_{\beta i} \ . \tag{11}$$

Squaring, and making judicious use of the unitarity of U, we find that the probability of $\nu_\alpha \to \nu_\beta$, $\text{P}\,(\nu_\alpha \to \nu_\beta)$, is given by

$$\begin{aligned}\text{P}(\nu_\alpha \to \nu_\beta) &= |\text{Amp}\,(\nu_\alpha \to \nu_\beta)|^2 \\ &= \delta_{\alpha\beta} - 4\sum_{i>j} \text{Re}\,(U_{\alpha i}^* U_{\beta i} U_{\alpha j} U_{\beta j}^*) \sin^2(\Delta m_{ij}^2 L/4E) \\ &\quad + 2\sum_{i>j} \text{Im}\,(U_{\alpha i}^* U_{\beta i} U_{\alpha j} U_{\beta j}^*) \sin(\Delta m_{ij}^2 L/2E) \ . \end{aligned} \tag{12}$$

Here, $\Delta m_{ij}^2 \equiv m_i^2 - m_j^2$ is the splitting between the squared masses of ν_i and ν_j. It is clear from the derivation of Eq. (12) that this expression would hold for any number of flavors and equal number of mass eigenstates.

Given that the particles described by the oscillation probability of Eq. (12) are born with an $\overline{\ell_\alpha}$ and convert into an ℓ_β in the detector, they are *neutrinos*, rather than *antineutrinos* (should there be a difference). To obtain the corresponding oscillation probability for antineutrinos, we observe that $\overline{\nu_\alpha} \to \overline{\nu_\beta}$ is the CPT-mirror image of $\nu_\beta \to \nu_\alpha$. Thus, if CPT invariance holds,

$$\text{P}(\overline{\nu_\alpha} \to \overline{\nu_\beta}) = \text{P}(\nu_\beta \to \nu_\alpha) \ . \tag{13}$$

Now, from Eq. (12), we see that

$$P(\nu_\beta \rightarrow \nu_\alpha;\, U) = P(\nu_\alpha \rightarrow \nu_\beta;\, U^*) \ . \tag{14}$$

Hence, assuming CPT invariance holds,

$$P(\overline{\nu_\alpha} \rightarrow \overline{\nu_\beta};\, U) = P(\nu_\alpha \rightarrow \nu_\beta;\, U^*) \ . \tag{15}$$

That is, the probability for oscillation of an antineutrino is the same as that for a neutrino, except that the mixing matrix U is replaced by its complex conjugate. Thus, from Eq. (12),

$$\begin{aligned} P(\overset{(-)}{\nu_\alpha} \rightarrow \overset{(-)}{\nu_\beta}) &= \delta_{\alpha\beta} - 4\sum_{i>j} \mathrm{Re}\,(U^*_{\alpha i}U_{\beta i}U_{\alpha j}U^*_{\beta j})\sin^2(\Delta m^2_{ij}L/4E) \\ &\quad \overset{+}{\underset{(-)}{}} 2\sum_{i>j} \mathrm{Im}\,(U^*_{\alpha i}U_{\beta i}U_{\alpha j}U^*_{\beta j})\sin(\Delta m^2_{ij}L/2E) \ . \end{aligned} \tag{16}$$

We see that if U is not real, the probabilities for $\nu_\alpha \rightarrow \nu_\beta$ and for the corresponding antineutrino oscillation, $\overline{\nu_\alpha} \rightarrow \overline{\nu_\beta}$, will in general differ. Since $\nu_\alpha \rightarrow \nu_\beta$ and $\overline{\nu_\alpha} \rightarrow \overline{\nu_\beta}$ are CP-mirror-image processes, this difference will be a violation of CP invariance.

As Eq. (16) makes clear, neutrino oscillation in vacuum from one flavor α into a different one β implies nonzero mass splittings Δm^2_{ij}, hence nonzero neutrino masses. It also implies nontrivial leptonic mixing. That is, the mixing matrix U cannot be diagonal.

Including the so-far omitted factors of $\hbar$ and c, we have

$$\Delta m^2_{ij}\frac{L}{4E} = 1.27\,\Delta m^2_{ij}(\mathrm{eV}^2)\frac{L(\mathrm{km})}{E(\mathrm{GeV})} \ . \tag{17}$$

From Eq. (16), if the U matrix cooperates, the probability for $\nu_\alpha \rightarrow \nu_\beta$, $\beta \neq \alpha$, will be appreciable if the kinematical phase difference in Eq. (17) is $\mathcal{O}(1)$ or larger. This requires only that for some ij,

$$\Delta m^2_{ij}(\mathrm{eV}^2) \gtrsim \frac{E(\mathrm{GeV})}{L(\mathrm{km})} \ . \tag{18}$$

Thus, for example, an experiment that studies 1 GeV neutrinos that travel a distance $L \sim 10^4$km, the diameter of the earth, will be sensitive to neutrino (mass)^2 splittings Δm^2_{ij} as small as 10^{-4}eV^2. Through quantum interference between neutrino mass eigenstates of different masses, neutrino oscillation gives us sensitivity to very tiny (mass)^2 splittings. However, as Eq. (16) underscores, oscillation cannot determine the masses m_i of the individual mass eigenstates. To learn those will require another approach.

There are basically two kinds of neutrino oscillation experiments. In the first, an *appearance* experiment, one starts with a beam of neutrinos that initially are purely of flavor α, and looks for the appearance in this beam of neutrinos of a new flavor β, $\beta \neq \alpha$, that were not originally present in the beam. In the second kind of experiment, a *disappearance* experiment, one starts with a known flux of ν_α, and looks to see whether some of the initial ν_α flux disappears as the beam travels.

By the definition of "probability", the probability that a neutrino changes flavor, plus the probability that it does not change flavor, must equal unity. That is, we must have

$$\sum_\beta P(\nu_\alpha \rightarrow \nu_\beta) = \sum_\beta P(\overline{\nu_\alpha} \rightarrow \overline{\nu_\beta}) = 1 \ , \tag{19}$$

where the sum is over all final flavors β, including the initial flavor α. From the unitarity of U, which implies that $\sum_\beta U_{\beta i} U^*_{\beta j} = \delta_{ij}$, it immediately follows that the oscillation probabilities of Eq. (16) do obey this constraint.

Neutrino flavor oscillation does not change the total flux in a neutrino beam. It merely redistributes it among the flavors. However, if we create a beam of neutrinos that at birth are of some active (i.e., weakly interacting) flavor, (muon neutrinos, for example), and some of these neutrinos oscillate into sterile (i.e., non-interacting) flavors, then some of the total *active* neutrino flux will have disappeared.

The combination of the CPT-invariance constraint of Eq. (13) and the probability constraint of Eq. (19) has powerful consequences for CP violation. To see this, consider the CP-violating differences

$$\Delta_{\alpha\beta} \equiv \mathrm{P}(\nu_\alpha \to \nu_\beta) - \mathrm{P}(\overline{\nu_\alpha} \to \overline{\nu_\beta}) \ . \tag{20}$$

If CPT invariance holds, then from Eq. (13)

$$\Delta_{\beta\alpha} = -\Delta_{\alpha\beta} \ . \tag{21}$$

In particular,

$$\Delta_{\alpha\alpha} = 0 \ . \tag{22}$$

That is, there can be no CP-violating difference between the survival probabilities $\mathrm{P}(\nu_\alpha \to \nu_\alpha)$ and $\mathrm{P}(\overline{\nu_\alpha} \to \overline{\nu_\alpha})$. Hence, there can be no observable CP violation in a disappearance experiment. Now, from Eq. (19), it follows that

$$\sum_\beta \Delta_{\alpha\beta} = 0 \ , \tag{23}$$

where the sum runs over all flavors, including $\beta = \alpha$. However, in view of Eq. (22), Eq. (23) implies that

$$\sum_{\beta \neq \alpha} \Delta_{\alpha\beta} = 0 \ . \tag{24}$$

If there are only three neutrino flavors, ν_e, ν_μ, and ν_τ, then this constraint implies that, in particular,

$$\Delta_{e\mu} + \Delta_{e\tau} = 0 \qquad \text{and} \qquad \Delta_{\mu e} + \Delta_{\mu\tau} = 0 \ . \tag{25}$$

From these relations and Eq. (21), we see that

$$\Delta_{e\mu} = \Delta_{\mu\tau} = \Delta_{\tau e} = -\Delta_{\mu e} = -\Delta_{\tau\mu} = -\Delta_{e\tau} \equiv \Delta \ . \tag{26}$$

In summary, if CPT holds, then the CP-violating difference $\Delta_{\alpha\beta} = \mathrm{P}(\nu_\alpha \to \nu_\beta) - \mathrm{P}(\overline{\nu_\alpha} \to \overline{\nu_\beta})$ can be nonvanishing only for $\beta \neq \alpha$. If, in addition, there are only three flavors, then the six possibly-nonvanishing $\Delta_{\alpha\beta}$, shown in Eq. (26), must all be equal, apart from a predicted minus sign [7]. (If there are more than three flavors, then Eq. (26) need not hold.)

Counter to intuition, the CP-violating difference $\Delta_{\alpha\beta} \equiv \mathrm{P}(\nu_\alpha \to \nu_\beta) - \mathrm{P}(\overline{\nu_\alpha} \to \overline{\nu_\beta})$ between neutrino and what we conventionally call "antineutrino" oscillation probabilities can still be nonvanishing even when the ν_i are identical to their antiparticles. Indeed, $\Delta_{\alpha\beta}$ is actually completely independent of whether the ν_i are their own antiparticles or not. We illustrate this by comparing the processes $\nu_\mu \to \nu_e$ and "$\overline{\nu_\mu} \to \overline{\nu_e}$", depicted in Figure 2. In $\nu_\mu \to \nu_e$,

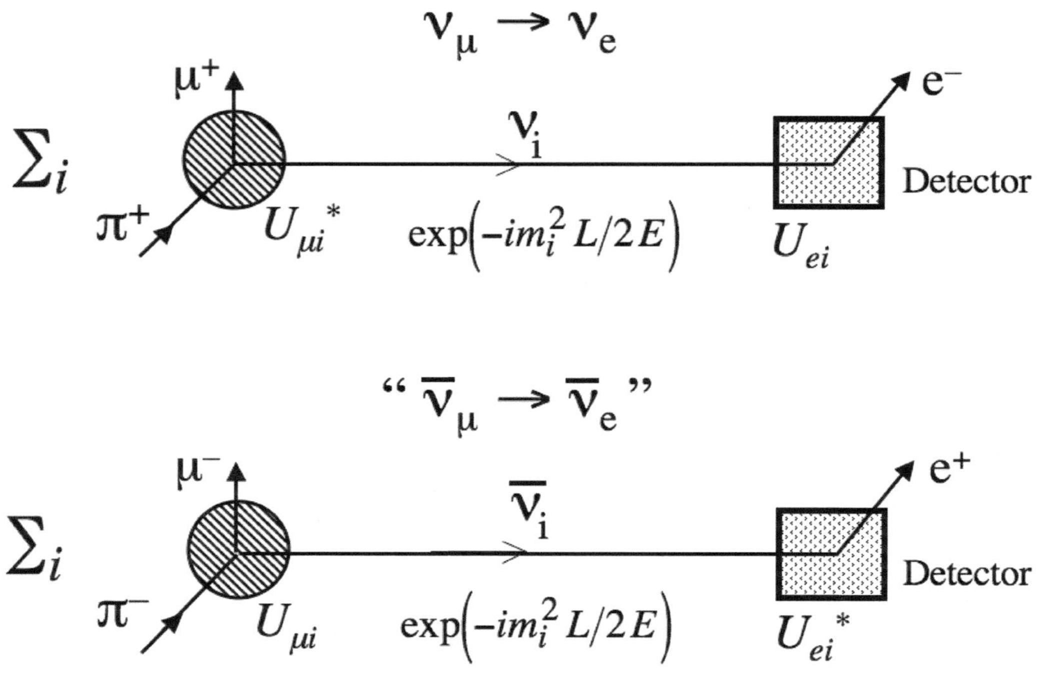

Figure 2. *The CP-mirror-image oscillations $\nu_\mu \to \nu_e$ and "$\overline{\nu_\mu} \to \overline{\nu_e}$". In each process, the particle that travels down the beamline is one or another of the mass eigenstates, and the amplitude is a coherent sum over the contributions of these eigenstates, as indicated. In "$\overline{\nu_\mu} \to \overline{\nu_e}$", the mass eigenstate $\overline{\nu_i}$ may or may not be identical, apart from its polarization, to the corresponding ν_i in $\nu_\mu \to \nu_e$. The propagator for this $\overline{\nu_i}$,* $\exp(-im_i^2 L/2E)$, *is identical to that for the corresponding ν_i in either case. The elements of the U matrix that, according to Eq. (1), appear at the beam-particle production and detection vertices are shown.*

the neutrino is created together with a μ^+ in π^+ decay. After traveling down a beamline to a detector, it is detected via its production of an e^-. In the corresponding "antineutrino" oscillation, "$\overline{\nu_\mu} \to \overline{\nu_e}$", the particle that travels down the beamline is created together with a μ^- in π^- decay, and is detected via its production in the detector of an e^+. One never directly observes the particle that travels down the beamline; it is an intermediate state. In terms of the charged leptons that one does (or at least can) observe, $\nu_\mu \to \nu_e$ and "$\overline{\nu_\mu} \to \overline{\nu_e}$" are clearly different, CP-mirror-image processes: the first involves a μ^+ and e^-, while the second involves a μ^- and e^+. Thus, even if $\overline{\nu_i} = \nu_i$, $\nu_\mu \to \nu_e$ and "$\overline{\nu_\mu} \to \overline{\nu_e}$" can have different probabilities, and if they do, the difference is a violation of CP invariance.

Even if $\overline{\nu_i} = \nu_i$, the beam particle will create an e^- in $\nu_\mu \to \nu_e$, but an e^+ in "$\overline{\nu_\mu} \to \overline{\nu_e}$", because it is oppositely polarized in the two processes. Due to the chirally left-handed structure of the weak interaction, reflected in the Lagrangian of Eq. (1), the beam particle will have helicity $h = -1/2$ in the first process, but $h = +1/2$ in the second. Due to this same parity-violating left-handed structure, the $h = -1/2$ beam particle will create an e^- (via the first term in Eq. (1)) in $\nu_\mu \to \nu_e$, while the $h = +1/2$ beam particle will create an e^+ (via the second term in Eq. (1)) in "$\overline{\nu_\mu} \to \overline{\nu_e}$".

From the amplitude factors displayed in Figure 2, we see that while

$$\text{Amp}(\nu_\mu \to \nu_e) = \sum_i U^*_{\mu i} e^{-im_i^2 \frac{L}{2E}} U_{ei} \ , \tag{27}$$

$$\text{Amp}(\overline{\nu_\mu} \to \overline{\nu_e}) = \sum_i U_{\mu i} e^{-im_i^2 \frac{L}{2E}} U^*_{ei} \ . \tag{28}$$

These expressions hold whether $\overline{\nu_i} = \nu_i$ or not. Thus, in either case, if the CP-violating phase δ in Eq. (3) is not zero or π, so that U is complex, the interference terms in $\text{P}(\nu_\mu \to \nu_e)$ and $\text{P}(\overline{\nu_\mu} \to \overline{\nu_e})$ will differ. As a result, the CP-violating difference $\text{P}(\nu_\mu \to \nu_e) - \text{P}(\overline{\nu_\mu} \to \overline{\nu_e})$ will be nonzero. Furthermore, the value of this difference will not depend on whether $\overline{\nu_i} = \nu_i$, and this value will be correctly implied by Eq. (16), which holds regardless of whether $\overline{\nu_i} = \nu_i$.

The general expression for $\text{P}(\overset{(-)}{\nu_\alpha} \to \overset{(-)}{\nu_\beta})$, Eq. (16), simplifies considerably in some important special cases. One such case is the simplified world in which there are only two charged leptons, say e and μ, two corresponding neutrinos of definite flavor, ν_e and ν_μ, and two neutrino mass eigenstates, ν_1 and ν_2, that make up ν_e and ν_μ. From our earlier analysis of the number of parameters in a mixing matrix, we know that the 2×2 unitary mixing matrix U for this two-flavor world may contain one mixing angle and one complex phase factor. It may easily be shown that U may be written in the form

$$U \equiv \begin{bmatrix} U_{e1} & U_{e2} \\ U_{\mu 1} & U_{\mu 2} \end{bmatrix} = \begin{bmatrix} \cos\theta & \sin\theta \\ -\sin\theta & \cos\theta \end{bmatrix} \times \begin{bmatrix} e^{i\xi/2} & 0 \\ 0 & 1 \end{bmatrix} \ , \tag{29}$$

where θ is the mixing angle and ξ is the phase. With $\Delta m^2_{21} \equiv \Delta m^2$ the sole (mass)^2 splitting in the problem, we find from the U of Eq. (29) and the general expression of Eq. (16) that

$$\text{P}(\overset{(-)}{\nu_e} \to \overset{(-)}{\nu_\mu}) = \text{P}(\overset{(-)}{\nu_\mu} \to \overset{(-)}{\nu_e}) = \sin^2 2\theta \sin^2(\Delta m^2 L/4E) \ , \tag{30}$$

and that

$$\text{P}(\overset{(-)}{\nu_e} \to \overset{(-)}{\nu_e}) = \text{P}(\overset{(-)}{\nu_\mu} \to \overset{(-)}{\nu_\mu}) = 1 - \sin^2 2\theta \sin^2(\Delta m^2 L/4E) \ . \tag{31}$$

As we know, the real world contains (at least) three charged leptons ℓ_α, three corresponding neutrinos of definite flavor ν_α, and three underlying neutrino mass eigenstates ν_i that make up the ν_α. Thus, the two-neutrino oscillation formulae of Eqs. (30) and (31) do not apply. However, if there are only three flavors, then under certain circumstances, rather similar simple formulae do apply. To see this, we note that the three-neutrino (mass)^2 spectrum has been observed to have the form shown in Figure 3 [2]. The splitting Δm^2_{21}, which drives the behavior of solar neutrinos, is roughly 30 times smaller than $\Delta m^2_{32} \cong \Delta m^2_{31}$, which drives the behavior of atmospheric neutrinos. (It is not known whether the closely-spaced pair ν_1-ν_2 is at the bottom or the top of the spectrum.) If an experiment is performed with L/E such that $\Delta m^2_{32} L/E = \mathcal{O}(1)$, then $\Delta m^2_{21} L/E \ll 1$, and in first approximation, this experiment cannot "see" the small splitting Δm^2_{21}. Neglecting this small splitting in Eq. (16), this equation and the unitarity of U imply that, for $\beta \neq \alpha$,

$$\text{P}(\overset{(-)}{\nu_\alpha} \to \overset{(-)}{\nu_\beta}) \cong 4|U_{\alpha 3} U_{\beta 3}|^2 \sin^2(\Delta m^2_{32} L/4E) \ . \tag{32}$$

Similarly, they imply that, for $\beta = \alpha$,

$$\text{P}(\overset{(-)}{\nu_\alpha} \to \overset{(-)}{\nu_\alpha}) \cong 1 - 4|U_{\alpha 3}|^2(1 - |U_{\alpha 3}|^2) \sin^2(\Delta m^2_{32} L/4E) \ . \tag{33}$$

We see that, by measuring these simple oscillation probabilities, experiments with $\Delta m^2_{32} L/4E = \mathcal{O}(1)$ can determine the flavor content of the isolated member of the spectrum, ν_3.

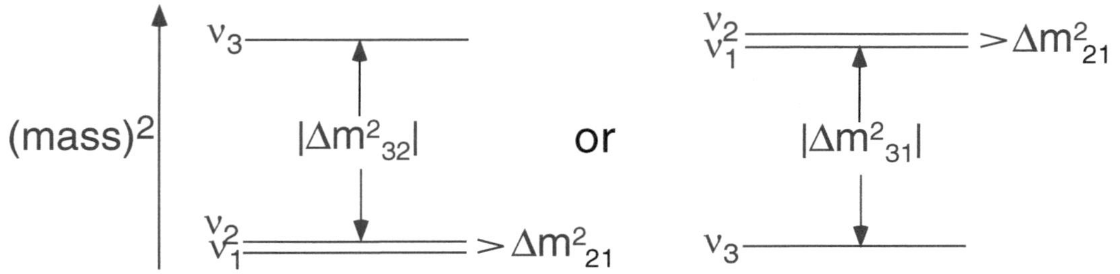

Figure 3. *The three-neutrino (mass)^2 spectrum.*

2.2 Neutrino oscillation in matter

Inside matter, the coherent forward scattering of neutrinos from the electrons, protons, and neutrons that make up the matter leads to neutrino effective masses and mixing angles that differ from their vacuum counterparts. As a result, inside matter, the probabilities for neutrino oscillations differ from their vacuum counterparts.

The Standard-Model interactions between neutrinos and other particles do not change flavor. Thus, barring hypothetical non-Standard-Model flavor-changing interactions, the observation of neutrino flavor change implies neutrino mass and leptonic mixing, even if the observation involves neutrinos passing through matter.

Neutrino propagation in matter may be conveniently treated via the laboratory-frame Schrödinger time-evolution equation

$$i\frac{\partial}{\partial t}\Psi(t) = \mathscr{H}\Psi(t) \ . \tag{34}$$

Here, t is the time, and $\Psi(t)$ is a multi-component neutrino wave function. Its α component, $\Psi_\alpha(t)$, is the amplitude for the neutrino to have flavor α at time t. If there are N flavors, the Hamiltonian $\mathscr{H}$ is an $N \times N$ matrix in flavor space. In matter, this matrix includes interaction energies arising from neutrino-matter interactions mediated by W or Z exchange. According to the Standard Model, the Z-mediated interactions neither change neutrino flavor nor depend on the flavor. Thus, they add to $\mathscr{H}$ a term proportional to the identity matrix. Such a term shifts all the eigenvalues of $\mathscr{H}$ by a common amount, leaving the splittings between the eigenvalues unchanged. Now, as we have seen when discussing neutrino flavor oscillation in vacuum, the amplitude for oscillation depends only on the *relative* phases of the different neutrino eigenstates. This means that it depends only on the splittings between the eigenvalues, and will not be affected by an interaction that merely shifts all the eigenvalues by the same amount. Thus, if our purpose is to treat neutrino flavor oscillation, we may omit the Z-exchange contribution to $\mathscr{H}$.

The W-exchange contribution is another matter. From the Standard Model, it follows that coherent forward ν_e-electron scattering via the W-exchange diagram of Figure 4 adds to the ν_e-ν_e element of $\mathscr{H}$, $\mathscr{H}_{\nu_e\nu_e}$, an interaction energy

$$V = \sqrt{2}G_F N_e \ . \tag{35}$$

Here, G_F is the Fermi coupling constant, and N_e is the number of electrons per unit volume in the matter through which the neutrinos are passing. The Fermi constant appears in V because it is a measure of the amplitude for the diagram in Figure 4, and the density N_e appears because

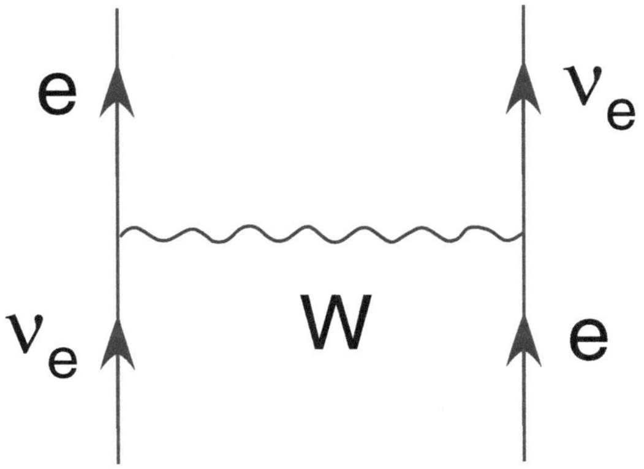

Figure 4. *The W-exchange interaction that modifies neutrino flavor oscillation in matter.*

the coherent scattering amplitude will obviously depend on how many electrons are present to contribute. The Standard Model tells us that for antineutrinos in matter, V is replaced by $-V$.

Since ν_e is the only neutrino flavor that couples to an electron and a W, W-mediated $\nu - e$ scattering affects *only* the ν_e-ν_e element of $\mathscr{H}$. Thus, its contribution to $\mathscr{H}$ is not proportional to the identity matrix, and it does affect neutrino flavor oscillation.

Neutrino flavor change in matter is illustrated by the case where there are only two significant flavors, say ν_e and ν_μ, and, correspondingly, two significant mass eigenstates. The Hamiltonian $\mathscr{H}$ in the Schrödinger equation, Eq. (34), is then a 2×2 matrix in ν_e-ν_μ space. Taking into account ν_e-e scattering via W exchange, but omitting some irrelevant contributions that are proportional to the identity matrix, we readily find [1] that

$$\mathscr{H} = \frac{\Delta m_M^2}{4E} \begin{bmatrix} -\cos 2\theta_M & \sin 2\theta_M \\ \sin 2\theta_M & \cos 2\theta_M \end{bmatrix} . \tag{36}$$

Here, E is the energy of the neutrinos, and Δm_M^2 and θ_M are, respectively, the effective (mass)^2 splitting and the effective mixing angle in matter. These effective quantities are related to their vacuum counterparts, Δm^2 and θ, by

$$\Delta m_M^2 = \Delta m^2 \sqrt{\sin^2 2\theta + (\cos 2\theta - x_\nu)^2} \tag{37}$$

and

$$\sin^2 2\theta_M = \frac{\sin^2 2\theta}{\sin^2 2\theta + (\cos 2\theta - x_\nu)^2} . \tag{38}$$

In these expressions,

$$x_\nu \equiv \frac{2\sqrt{2} G_F N_e E}{\Delta m^2} \tag{39}$$

is a measure of the importance of the effects of matter. In vacuum, N_e and consequently x_ν vanishes, and, as confirmed by Eqs. (37) and (38), Δm_M^2 and θ_M revert to the vacuum values, Δm^2 and θ, respectively.

Imagine an accelerator-generated neutrino beam that travels a distance $L \sim 1000$ km through the earth to a detector. The electron density N_e encountered by this beam will be that of the earth's mantle, and approximately constant. Then x_ν, Δm^2_M, θ_M, and $\mathscr{H}$ will all be ~ position-independent. From Eqs. (36) and (34) and straightforward quantum mechanics, it follows that

$$P(\nu_e \to \nu_\mu) = P(\nu_\mu \to \nu_e) = \sin^2 2\theta_M \sin^2(\Delta m^2_M L/4E) \ . \quad (40)$$

This is the usual two-neutrino oscillation result, Eq. (30), except that the vacuum parameters θ and Δm^2 are replaced by their counterparts in matter, θ_M and Δm^2_M. If $N_e \to 0$, so that $x_\nu \to 0$, the oscillation probabilities in matter of Eq. (40) become the vacuum probabilities of Eq. (30), as they must.

The size of the effect of matter may be judged by the size of x_ν. For the illustrative beam that we are considering, the actual three-neutrino vacuum (mass)^2 splitting Δm^2 that will most strongly influence flavor oscillation is probably the large one, $\Delta m^2_{31} \cong \Delta m^2_{32}$. Experimentally, $\Delta m^2_{31} \simeq 2.4 \times 10^{-3}\,\text{eV}^2$ [8]. For this Δm^2, we find from Eq. (39) that

$$|x_\nu| \simeq E/12\,\text{GeV} \ . \quad (41)$$

Thus, for $E = 0.5\,\text{GeV}$, the matter effect is quite small, for $E = 2\,\text{GeV}$ it is modest, and for $E = 20\,\text{GeV}$ it is large.

As already mentioned, when antineutrinos, rather than neutrinos, propagate through matter, the interaction energy V is replaced by $-V$. It follows readily that, as a result, x_ν, Eq. (39), is replaced in Eqs. (37)-(38) by

$$x_{\bar{\nu}} \equiv -x_\nu \ . \quad (42)$$

We see that this change has the consequence that, within matter, the effective (mass)^2 splitting and the effective mixing angle for antineutrinos are different than they are for neutrinos. As a result, within matter the flavor oscillation of an antineutrino beam will differ from that of a neutrino beam. The two-flavor Hamiltonian $\mathscr{H}$ is still given by Eq. (36), and the two-flavor oscillation probability in matter of constant density is still given by Eq. (40), but the quantities Δm^2_M and θ_M have different values than they did in the neutrino case.

Earlier, we raised the possibility that neutrinos are their own antiparticles. That is, we imagined that, *for a given momentum $\vec{p}$ and helicity h,* perhaps each neutrino mass eigenstate ν_i is identical to its antiparticle: $\nu_i(\vec{p},h) = \overline{\nu_i}(\vec{p},h)$. Suppose that this is indeed the case. Will the interaction with matter still cause the flavor oscillation of an "antineutrino" beam to differ from that of a "neutrino" beam within matter? In practical terms the answer is "yes". The reason is that, in practice, the "neutrino" and "antineutrino" beams that we study are never of the same helicity. A "neutrino" is the particle "ν" produced, for example, in the decay $W^+ \to e^+ + \nu$. As already noted, owing to the chirally left-handed structure of the weak interaction, this "ν" will be of left-handed helicity: $h = -1/2$. In contrast, an "antineutrino" is the particle "$\bar{\nu}$" produced in $W^- \to e^- + \bar{\nu}$. As already noted, owing to the same structure of the weak interaction, this "$\bar{\nu}$" will be of right-handed helicity: $h = +1/2$. Since the weak interaction is not invariant under parity, the interaction in matter of the left-handed "ν" and the right-handed "$\bar{\nu}$" will be quite different, even if helicity is the only difference between the "ν" and the "$\bar{\nu}$". Only the first term on the right-hand side of Eq. (1) can couple an incoming left-handed beam particle to an electron, while only the second term can couple an incoming right-handed beam particle. These two terms lead to different scattering amplitudes. These amplitudes do not depend on whether the "ν" and "$\bar{\nu}$" beams differ only in helicity, or in some other way as well.

Future accelerator neutrino experiments hope to study $\nu_\mu \to \nu_e$ and $\overline{\nu_\mu} \to \overline{\nu_e}$ in matter under conditions where all three of the known neutrino mass eigenstates $\nu_{1,2,3}$, or equivalently both of the known splittings Δm^2_{31} and Δm^2_{21}, play significant roles. The oscillation probabilities are then more complicated than the expression of Eq. (40). However, since $\alpha \equiv \Delta m^2_{21}/\Delta m^2_{31} \sim 1/30$ [2] and $\sin^2 2\theta_{13} < 0.2$ [9], the probability for $\nu_\mu \to \nu_e$ in matter is well approximated by [10]

$$P(\nu_\mu \to \nu_e) \cong \sin^2 2\theta_{13} T_1 - \alpha \sin 2\theta_{13} T_2 + \alpha \sin 2\theta_{13} T_3 + \alpha^2 T_4 \ , \tag{43}$$

where

$$T_1 = \sin^2\theta_{23} \frac{\sin^2[(1-x_\nu)\Delta]}{(1-x_\nu)^2} \ , \tag{44}$$

$$T_2 = \sin\delta \sin 2\theta_{12} \sin 2\theta_{23} \sin\Delta \frac{\sin(x_\nu\Delta)}{x_\nu} \frac{\sin[(1-x_\nu)\Delta]}{(1-x_\nu)} \ , \tag{45}$$

$$T_3 = \cos\delta \sin 2\theta_{12} \sin 2\theta_{23} \cos\Delta \frac{\sin(x_\nu\Delta)}{x_\nu} \frac{\sin[(1-x_\nu)\Delta]}{(1-x_\nu)} \ , \tag{46}$$

and

$$T_4 = \cos^2\theta_{23} \sin^2 2\theta_{12} \frac{\sin^2(x_\nu\Delta)}{x_\nu^2} \ . \tag{47}$$

In these expressions, $\Delta \equiv \Delta m^2_{31} L/4E$ is the kinematical phase of the oscillation. and x_ν is the matter-effect quantity defined by Eq. (39), with Δm^2 now taken to be Δm^2_{31}. In the appearance probability $P(\nu_\mu \to \nu_e)$, the T_1 term represents the oscillation due to the splitting Δm^2_{31}, the T_4 term represents the oscillation due to the splitting Δm^2_{21}, and the T_2 and T_3 terms are the CP-violating and CP-conserving interference terms, respectively.

The probability for the corresponding antineutrino oscillation, $P(\overline{\nu_\mu} \to \overline{\nu_e})$, is the same as the probability $P(\nu_\mu \to \nu_e)$ given by Eqs. (43)-(47), but with x_ν replaced by $x_{\bar{\nu}} = -x_\nu$ and $\sin\delta$ by $-\sin\delta$: both the matter effect and CP violation lead to a difference between the $\nu_\mu \to \nu_e$ and $\overline{\nu_\mu} \to \overline{\nu_e}$ oscillation probabilities. In view of the dependence of x_ν on Δm^2_{31}, and in particular on the sign of Δm^2_{31}, the matter effect can reveal whether the neutrino mass spectrum has the closely-spaced ν_1-ν_2 pair at the bottom or the top (see Figure 3). However, to determine the nature of the spectrum, and to establish the presence of CP violation, it obviously will be necessary to disentangle the matter effect from CP violation in the neutrino-antineutrino oscillation probability difference that is actually observed. To this end, complementary measurements will be extremely important. These can take advantage of the differing dependences on the matter effect and on CP violation in $P(\nu_\mu \to \nu_e)$ and $P(\overline{\nu_\mu} \to \overline{\nu_e})$.

Acknowledgments

It is a pleasure to thank H. Lipkin, S. Parke, and L. Stodolsky for useful conversations relevant to the physics of these lectures. I am grateful to Susan Kayser for her crucial role in the preparation of the manuscript.

References

[1] Kayser, B., "Neutrino Physics", in the*Proceedings of the SLAC Summer Institute of 2004*, eConf **C040802**, L004 (2004): hep-ph/0506165.

[2] Kayser, B., "Neutrino Mass, Mixing, and Flavor Change", to appear in the 2008 edition of the *Review of Particle Physics*, by The Particle Data Group. This reference includes the phenomenology of neutrino oscillation and a summary of what we have learned about the neutrinos so far from experiment.

[3] This matrix is sometimes referred to as the Maki-Nakagawa-Sakata matrix, or as the Pontecorvo-Maki-Nakagawa-Sakata matrix, in recognition of the pioneering contributions of these scientists to the physics of mixing and oscillation.
See Maki, Z., Nakagawa, M., and Sakata, S., *Prog. Theor. Phys.* **28**, 870 (1962);
Pontecorvo, B., *Zh. Eksp. Teor. Fiz.* **53**, 1717 (1967) [*Sov. Phys. JETP* **26**, 984 (1968)].

[4] For a discussion of the possibility of a nonunitary leptonic mixing matrix, see Antusch, S. *et al.*, *JHEP* **0610**, 084, (2006).

[5] Kayser, B., "CP Effects When Neutrinos Are Their Own Antiparticles", in *CP Violation*, ed. C. Jarlskog (World Scientific, Singapore, 1989) p. 334.

[6] Lipkin, H., *Phys. Lett.* **B642**, 366 (2006).

[7] We thank S. Petcov for a long-ago conversation on how to obtain this result in a simple way.

[8] The MINOS Collaboration (Michael, D. *et al.*), *Phys. Rev. Lett.* **97**, 191801 (2006), and talks by MINOS collaboration members updating their results.

[9] The CHOOZ Collaboration (Apollonio, M. *et al.*),*Eur. Phys. J.* **C27**, 331 (2003);
Fogli, G. *et al.*, *Prog. Part. Nucl. Phys.* **57**, 742 (2006).

[10] Cervera, A. *et al.*, *Nucl. Phys.* **B579**, 17 (2000);
Freund, M., *Phys. Rev.* **D64**, 053003 (2001).

Neutrino Interactions

Kevin McFarland

University of Rochester, Rochester, NY, USA 14627

1 Introduction and Motivations

The study of neutrino interaction physics played an important role in establishing the validity of the theory of weak interactions and electroweak unification. Today, however, the study of interactions of neutrinos takes a secondary role to studies of the properties of neutrinos, such as masses and mixings. This brief introduction describes the historical role that the understanding of neutrino interactions has played in neutrino physics and what we need to understand about neutrino interactions to proceed in future experiments aimed at learning more about neutrinos.

The original application of neutrino interactions was the discovery of the neutrino itself. For most physicists today, who came of age professionally well after the first observation of neutrinos, it takes a bit of thought to understand the perspective of the experimenters seeking to discover the neutrino. A close analogy today might be the search for interactions of weakly interacting massive (WIMP) dark matter particles. In order to sensibly design an experiment to search for a new particle and to interpret the results, an experimenter needs guidance about the probable type and rate of interactions to be observed. For the case of WIMP dark matter, information about the strength of interactions comes from the standard cosmological model which relates modern day abundance of dark matter to production and annihilation cross sections.

In the case of neutrinos in the early 1950s, the guiding principle was the Fermi "four fermion" theory of the weak interaction (Fermi 1934). This theory introduced a four-fermion vertex connecting a neutron n, a proton p, an electron e^- and an anti-neutrino $\bar{\nu}$ to explain neutron decay, $n \to pe^-\bar{\nu}$ in terms of a single unknown coupling constant, G_F. Because that single constant governed the strength of all weak interactions among these particles, the Fermi theory led to definite prediction for neutrino interactions involving these particles. The prediction for the cross section of $\bar{\nu}p \to e^+n$ was first derived by Bethe and Peierls shortly after the Fermi theory was published (Bethe and Peierls 1934). For neutrinos with energies of a few MeV from a reactor, a typical cross section in this theory was predicted to be $\sigma_{\bar{\nu}p} \sim 5 \times 10^{-44}$ cm^2. Interestingly, this prediction for reactor neutrino cross sections is still accurate today, up to a factor of two required to account for the then unknown phenomenon of maximal parity violation in the weak interaction! This small cross section is, as we all recognize today, the primary challenge in performing experiments with neutrinos. By contrast, the cross section for the corresponding electromagnetic process with a photon γ at similar energies is $\sigma_{\gamma p} \sim 10^{-25}$ cm^2. The tiny neutrino cross section means that the mean free path of reactor neutrinos with energies of a few MeV in steel is approximately ten light years.

With these predictions in place, the stage was set for the two critical measurements establishing the existence and nature of the neutrinos from nuclear reactors: the Davis *et al* null measurement of the reaction $\bar{\nu} + {}^{37}Cl \rightarrow {}^{37}Ar + X$ and Reines and Cowan's observation of $\bar{\nu} p \rightarrow e^+ n$ in 1955-56. In modern language, the latter measurement establishes the existence of the neutrino and validates the universality of the Fermi theory and the former non-measurement shows that the neutrino and anti-neutrino carry an opposite conserved lepton number which forbids $\bar{\nu} n \rightarrow e^- p$ (Reines 1996).

1.1 A Cautionary Tale: Discovery of the Weak Neutrino Current

A more sobering story involving knowledge of neutrino cross sections involves the discovery of the weak neutral current in neutrino interactions. No textbook would be complete without the requisite picture of the famous single electron event in the Gargamelle bubble chamber, attributed to $\bar{\nu}_e e^- \rightarrow \bar{\nu}_e e^-$. While this event is a wonderful illustration of a weak neutrino process, it was not the discovery channel. As we will see, the cross section for this reaction is exceedingly small, and concerns about backgrounds and the lack of corroborating information in such a reaction make it a difficult channel in which to claim a discovery. The discovery measurement for the weak neutral current involves processes where neutrinos scatter off of the nuclei in the target allowing the experimenters to measure a quantity such as

$$R^{\nu} = \frac{\sigma(\nu_\mu N \rightarrow \nu_\mu X)}{\sigma(\nu_\mu N \rightarrow \mu^- X)} \tag{1}$$

or its analog with an anti-neutrino beam. Figure 1 shows these two measurements compared with the prediction of the electroweak standard model as a function of its single parameter not constrained by low energy data, $\sin^2\theta_W$, which is the weak mixing angle or Weinberg angle.

This major triumph for the standard model of electroweak unification was sadly complicated by an involved saga which ultimately boiled down to uncertainties in translating observed events to the measurement of R^{ν}. Experimentally, the measurement consists of identifying events as either containing or not containing of final state muon and using this distinguishing feature to separate charged and neutral current interactions. Very low energy muons are difficult to separate from other particles, primarily charged mesons, produced in inelastic scattering from nuclei, and so these events constitute a background to the neutral current sample. Equally problematic for this measurement are neutral current events which produce charged hadrons in the final state that are confused with energetic muons. Without a good model for the production of these charged mesons or a good understanding of the probability of confusing charged mesons with muons in the detector, the experimental problem of isolating sufficiently clean samples with high statistics hobbled efforts to produce a convincing observations by both of the competing collaborations, Gargamelle at CERN and HWPF at Fermilab (Galison 1983). It is notable that this important discovery was never honored with a Nobel prize, despite its critical role in validating the electroweak theory.

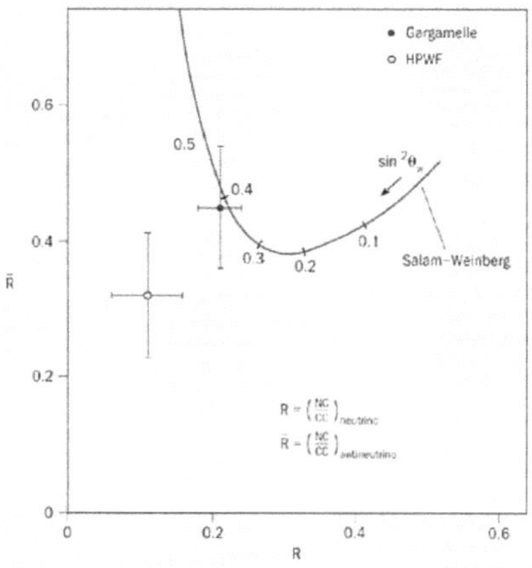

Figure 1. *Measurement of $R^{\bar{\nu}}$ vs. R^{ν} from the Gargamelle and HWPF collaborations compared with the prediction of the electroweak standard model.*

1.2 Cross Section Knowledge and Next Generation Oscillation Experiments

The current and next generation of accelerator neutrino oscillation experiments are again facing limitations arising from knowledge of neutrino cross sections. The physics roadmap of precisely measuring the "atmospheric" oscillation parameters, measuring θ_{13}, determining the neutrino mass hierarchy and measuring the CP violating phase, δ, has driven an experimental program to be realized in several steps. Currently this program is the measurement of $\nu_\mu \to \nu_\mu$ transition probabilities in wide band beams with baseline L and mean energies E near $L/E \sim 400$ km/GeV (K2K and MINOS), and the measurement of $\nu_\mu \to \nu_\tau$ near τ production threshold (OPERA). In the near future, it includes narrowband (off-axis) beam experiments again near $L/E \sim 400$ km/GeV to precisely measure $\nu_\mu \to \nu_e$ transitions in neutrino and anti-neutrino beams (T2K and NOvA). Most likely, completion of this program will require a new generation of experiments to study these transitions at the second oscillation maximum as well, $L/E \sim 1200$ km/GeV, either in narrow band beams (T2KK) or wideband beams (discussed in FNAL to DUSEL proposals). Practical considerations limit the range of possible baselines to $L \lesssim 2000$ km because of available sites and achievable event rates and $E \gtrsim 0.5$ GeV because of the roughly quadratic drop in the signal cross section and because of significant nuclear effects with neutrinos energies below this limit. This implies that the neutrinos to be studied will have $0.5 < E_\nu < 5$ GeV. As we will see, this region is at the threshold for inelastic interactions on nucleons, which is a particularly difficult energy region to model and is lacking in data to contribute to understanding the relevant effects governing the details of cross sections.

Knowledge of cross sections impacts a $\nu_\mu \to \nu_\mu$ disappearance measurement in this energy regime because, regardless of experimental techniques, the details of the final state will impact the separation of signal from background and the measurement of neutrino energy in a given event. Figure 2 illustrates the effect of backgrounds on the measurement of the maximum oscillation probability on the T2K experiment. If the background to the signal, in this case primarily from inelastic charged-current events, cannot be accurately estimated, then it

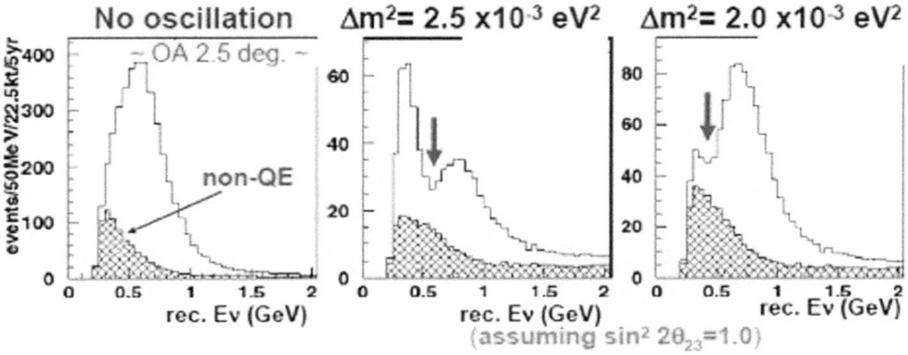

Figure 2. *The expected far detector ν_μ candidate spectrum in the T2K experiment for $\theta_{23} = \pi/4$. The hatched area in each plot shows expected backgrounds.*

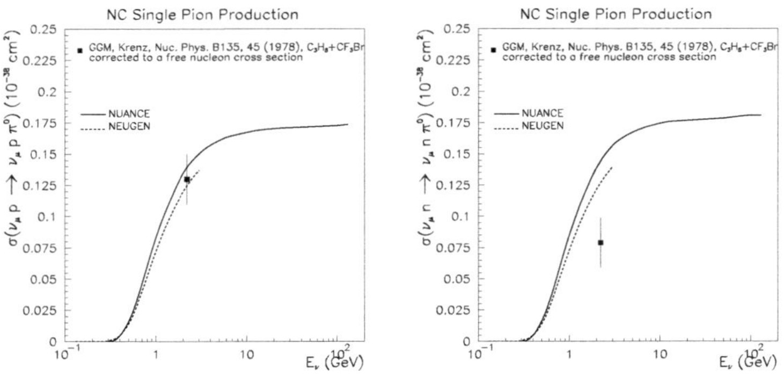

Figure 3. *Knowledge of single π^0 neutrino production cross sections as a function of energy before K2K or MiniBooNE results (Zeller 2003).*

becomes difficult to measure the depth of the oscillation "dip" which is used to measure θ_{23}. In a broadband beam like that of the MINOS experiment where the neutrinos at the energy of maximal oscillation have an energy near 2 GeV, the differences in energy response between baryons, charged pions and neutral pions in the final state lead to a significant uncertainty in reconstructed energy. This uncertainty in turn impacts the measurement of the energy of the oscillation "dip" which determines δm^2_{23}.

Because the $\nu_\mu \to \nu_e$ oscillation probability is so low, the major impact on these appearance experiments, such as T2K and NOvA, is expected to be from backgrounds to electron appearance. The major such background is the production of neutral pions which decay into photons that shower and mimic electrons, either because of a merging or loss of γ rings in a Cerenkov detector or because of the merging or loss of one γ in a calorimetric detector. Unfortunately, this background is poorly constrained by existing data (Figure 3). The challenge becomes apparent when looking at the precision needed for the physics goals of these experiments. Ultimately, as illustrated in Figure 4, the transition probabilities will need to be measured with sub-percent precision to measure the effect of CP violation in neutrinos. This places strict requirements on the understanding of ν_e backgrounds in both neutrino and anti-neutrino beams.

This interest in neutrino interactions in the 0.5 to 5 GeV energy region has led to the proposal and construction of a number of dedicated neutrino cross section experiments designed

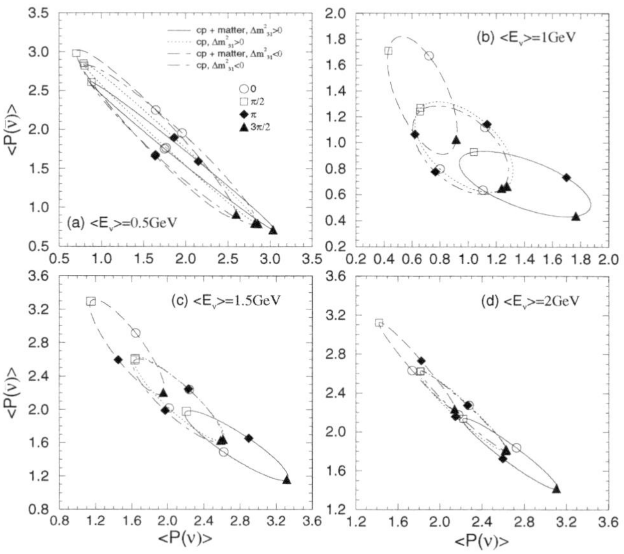

Figure 4. *The anti-neutrino vs. neutrino $\nu_\mu \to \nu_e$ transition probability in percent a baseline of 700 km at different energies for different mass hierarchies and values of the CP violating phase δ (Minakata and Nunokata 2001). Copyright (2001) IOP Publishing.*

to make these measurements. The K2K experiment recently built a near detector, "SciBar", designed for such measurements which is currently running as "SciBooNE" in the Fermilab Booster neutrino beam. The MINERvA experiment is currently under construction for future operation in the Fermilab NuMI beamline.

2 Pointlike Interactions

For a pedagogical explanation of neutrino cross section phenomenology, it is helpful to start with the scattering of neutrinos from effectively massless pointlike fermions, such as neutrino-electron scattering. Although this interaction is of limited practical interest for accelerator oscillation experiments, the calculation of pointlike scattering will serve multiple purposes as we begin to explore more complicated cross section phenomenology. First, in the high energy limit of neutrino-nucleon scattering, a good approximation to deep inelastic scattering is to consider the scattering of neutrinos on point-like quark constituents of the nucleus. Second, the study of scattering from pointlike particles will make a good point of departure from which to study effects such as initial and final state masses and the effect of structure in a target fermion. Therefore, please suspend skepticism of the usefulness of this particular exercise, and let us begin to consider neutrino scattering on electrons.

The style of the lectures was to present examples illustrating the phenomena of neutrino interactions. Accordingly, what follows below uses heuristic arguments and does not follow a style of rigorous proof. To paraphrase the humorist Michael Feldman, readers who are sticklers for the whole truth should write their own lectures.

Table 1. *Weak neutral-current couplings g_L and g_R.*

Z coupling	g_L	g_R
ν	$1/2$	0
e,μ,τ	$-1/2+\sin^2\theta_W$	$\sin^2\theta_W$
u,c,t	$1/2-(2/3)\sin^2\theta_W$	$-(2/3)\sin^2\theta_W$
d,s,b	$-1/2+(1/3)\sin^2\theta_W$	$(1/3)\sin^2\theta_W$

2.1 Weak Interactions and Neutrinos

The modern view of the weak interaction is not the four fermion interaction of Fermi's theory, but rather an interaction mediated by the exchange of massive W and Z bosons. In the low momentum limit, where the mediating boson is far off shell, the weak interaction Hamiltonian governing the process $\nu f \to \ell/\nu + f'$ is

$$\mathscr{H}_{weak} = \frac{4G_F}{\sqrt{2}}\left[\bar{\ell}/\bar{\nu}\gamma_\mu\frac{(1-\gamma_5)}{2}\nu\right]\left[\bar{f}'\gamma^\mu\left(g_L\frac{1-\gamma_5}{2}+g_R\frac{1+\gamma_5}{2}\right)f\right]+\text{h.c.} \tag{2}$$

where f, f', l and ν stand for an initial and final state fermion, lepton and neutrino, respectively, g_L and g_R are the weak neutral-current couplings, γ_μ are the standard Dirac matrices and $\gamma_5 \equiv i\gamma_0\gamma_1\gamma_2\gamma_3$. Note that, like the Fermi theory, this form makes reference to a single coupling constant, G_F, to which we shall return later. It does include an important component not recognized in the Fermi theory, namely parity non-conservation. The factor $(1-\gamma_5)/2$ is a projection operator onto left-handed states for fermions and right-handed states for anti-fermions.

The Hamiltonian above also has provision for a neutral-current interaction, mediated by the Z, in which the neutrino remains a neutrino, and a charged-current interaction, mediated by the W in which the neutrino becomes a charged lepton. A neutrino, weak or flavor, eigenstate, ν_e, ν_μ or ν_τ, is associated with the production of a charged lepton of the same generation in the charged-current weak interaction. The weak interaction is maximally parity-violating in the charged-current interaction, selecting only left-handed fermions, and therefore the right handed charged-current couplings are zero. However, in the case of the neutral weak interaction, these couplings are given in terms of the electromagnetic and weak couplings by the electroweak unification theory and their values for each species of fermion are given in Table 1. Note the right-handed neutrino has no weak couplings, neither in the neutral nor the charged current, which makes it unique among the fermions.

The rigorous definition of this "handedness", or chirality, is equivalent to the definition of the left-handed (right-handed) projection operator, $(1 \mp \gamma_5)/2$. If a particle is massless, this chirality is equivalent to its helicity, i.e. the projection of its spin σ along the direction of the particle, $\sigma \cdot \hat{p}$. The Hamiltonian above indicates that neutrinos produced or participating in weak interactions will be entirely left-handed. Since neutrinos do have mass, m_ν, this implies that while the neutrino will primarily be negative helicity, there will be a small positive helicity component, frame-dependent, and proportional to m_ν/E_ν where E_ν is the neutrino energy. For most practical purposes, this positive helicity component can be entirely neglected.

The final aspect of this form of the weak interaction to be explained is the Fermi constant itself, $G_F \approx 1.166 \times 10^{-5}$ GeV^{-2}. The dimensions and size of the Fermi constant, which make

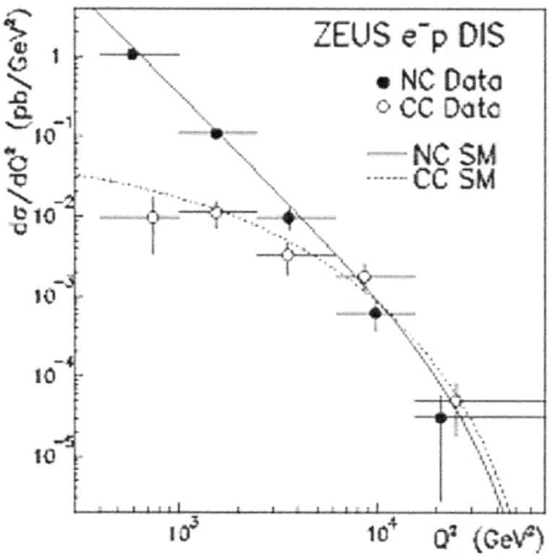

Figure 5. *Similarity of the strength of weak and electromagnetic interactions at high momentum transfer as illustrated by measurements of neutral and charged-current ep scattering measured by the ZEUS experiment at HERA.*

the weak interaction "weak" at low energies, have their origin in the propagator associated with the exchange of the W boson. For a two body massless weak scattering process,

$$\frac{d\sigma}{dq^2} \propto \frac{1}{(q^2 - M_W^2)^2}, \tag{3}$$

where M_W and q are the mass of and the four-momentum carried by the W boson[1]. For $|q^2| \ll M_W^2$, this propagator term gives a factor of M_W^{-4}. In the case of the electromagnetic interaction, this same term becomes q^{-4} since the mass of the exchanged boson, the photon, is zero. Figure 5 shows cross sections of the neutral current process, $e^-p \to e^-p$, which has contributions from both γ and Z^0 exchange, and the charged current process, $e^-p \to \nu X$, which is purely weak. In these processes, $q^2 < 0$, and we usually write $Q^2 = -q^2$ by convention. We see that when $Q^2 < M_W^2$, the neutral current cross section is rapidly falling with Q^2, while the charged-current cross section is roughly constant. However, beginning at $Q^2 \sim M_W^2$, both cross sections are roughly comparable and falling steeply with Q^2. In the electroweak theory, G_F can be expressed in terms of M_W and an overall weak coupling constant, g_W, as

$$G_F = \frac{\sqrt{2}}{8}\left(\frac{g_W}{M_W}\right)^2; \tag{4}$$

therefore, g_W is a coupling of $\mathcal{O}(1)$ and roughly the same size as the electromagnetic coupling constant in this unified theory of the two interactions.

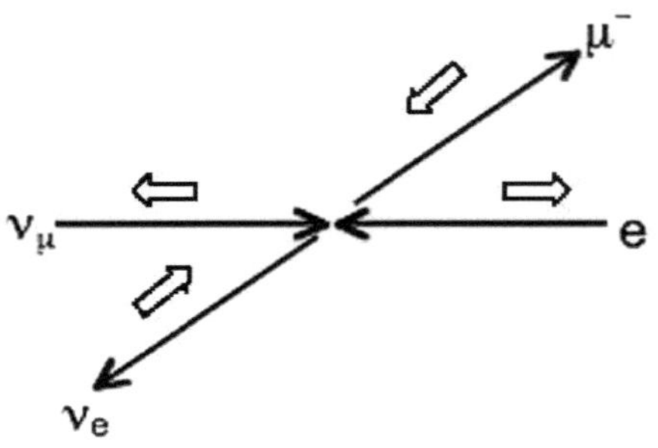

Figure 6. *Helicity in the massless limit of neutrino-electron scattering.*

2.2 Neutrino Electron Scattering

With this background, we are ready to calculate a neutrino-electron scattering cross section. For pedagogical simplicity, first consider $\nu_\mu e^- \to \mu^- \nu_e$ at sufficiently high energies so that we may neglect all masses in the problem, including the mass of the final state muon, but not at such high energies that we need worry about the effect of Q^2 on the propagator $1/(Q^2+M_W^2)^2$. In this limit, chirality is equivalent to helicity. In the center-of-mass frame we can easily see that the two left-handed and negative helicity particles in the final state have a total spin along the interaction axis of $J_z = 0$ (Figure 6), and therefore there is no preferred center-of-mass scattering angle. Thus,

$$\sigma = \int_0^{Q^2_{MAX}} dQ^2 \frac{d\sigma}{dQ^2} \propto \int_0^{Q^2_{MAX}} dQ^2 \frac{1}{(Q^2+M_W^2)^2} = \frac{Q^2_{MAX}}{M_W^4}. \tag{5}$$

That's it! To complete the evaluation of the cross section, we need only find the constant of proportionality, which turns out to be $g_W^4/32\pi = M_W^4 \times G_F^2/\pi$, and the maximum Q^2 that can be exchanged. Here Q^2, the negative of the square of the four-momentum carried by the W boson, is $-(\underline{e} - \underline{\nu}_e)^2$, where the underlined terms represent four-vectors. It is simple to show that, in terms of E_ν^* and θ^* the center-of-mass energy and scattering angle, $Q^2 = 2E_\nu^{*2}(1-\cos\theta^*)$. This means that Q^2 ranges between 0 and $4E_\nu^{*2} = s$ where $\sqrt{s}$ is the total available center-of-mass energy. The cross section is therefore

$$\sigma = \frac{G_F^2 s}{\pi}. \tag{6}$$

Numerically, this turns out to be $\sigma = 17.2 \times 10^{-42}$ cm$^2 \times E_\nu$/GeV. The proportionality to the neutrino energy in the lab frame comes about computationally because, if the target electron is at rest, $s = m_e^2 + 2m_e E_\nu$, and the m_e^2 term can be neglected for neutrino beam energies of interest. More fundamentally, this proportionality to energy is a generic feature of pointlike neutrino scattering at Q^2 below M_W^2 squared, since $d\sigma/dQ^2$ is constant.

We now consider another process, this time the neutral current elastic scattering $\nu_\mu e^- \to \nu_\mu e^-$. How is this process different from our previous example under the same energy approximations? First, the process is a neutral current one, and therefore, as can be seen in Table 1,

[1]Suffice it to say that some details are glossed over in this statement. It is rigorously true for a neutrino impinging on a target at rest in the lab that $d\sigma/dq^2 = |\mathscr{M}|^2/(64\pi p_\nu^2 M_T^2)$ where M_T is the mass of the target and $\mathscr{M}$ is the matrix element of the scattering process. There are many steps between this statement and the assertion above that the propagator "factors" out as shown above.

the interaction couples to both the left-handed and right-handed electron. In the massless left-handed case, as before, the total spin along the interaction axis is 0, but in the right-handed case, the total spin along this axis is 1. The right-handed case therefore differs from the case shown in Figure 6 because if the target electron and the outgoing lepton spins are flipped, there is a preference for forward scattering as opposed to backward scattering due to conservation of angular momentum along the interaction axis. Therefore, while $d\sigma/d\theta^*$ is constant for the left-handed target lepton,

$$\frac{d\sigma_{J_z=1}}{d\theta^*} \propto \left(\frac{1+\cos\theta^*}{2}\right)^2 \tag{7}$$

for the right-handed target lepton. Integrating this over all solid angles leads to the conclusion that $\sigma_{J_z=1} = \sigma_{J_z=0}/3$, where the reduced cross section can be understood from the suppression of non-forward scattering due to the projection of spin from the initial to the final state axes.

The couplings enter linearly into the matrix element and, therefore, are squared in the cross sections. With the effect of the initial state spin accounted for, we can write

$$\begin{aligned}
\sigma_{J_z=0} &= \frac{G_F^2 s}{\pi}\left(-\frac{1}{2}+\sin^2\theta_W\right)^2 \\
\sigma_{J_z=1} &= \frac{1}{3}\frac{G_F^2 s}{\pi}\left(\sin^2\theta_W\right)^2 \\
\sigma(\nu_\mu e^- \to \nu_\mu e^-)_{TOT} &= \frac{G_F^2 s}{\pi}\left(\frac{1}{4}-\sin^2\theta_W+\frac{4}{3}\sin^4\theta_W\right).
\end{aligned} \tag{8}$$

Generalizing from the examples here, it's possible to derive all the neutrino-electron elastic scattering processes in the massless limit. Some, such as $\nu_e e^- \to \nu_e e^-$ have the added complexity of both neutral and charged-current contributions. In the case of this reaction, the analysis is the same as that for $\nu_\mu e^- \to \nu_\mu e^-$ scattering above with one exception. The charged-current gives an additional process contributing to the scattering from left-handed electrons. Because these processes have identical initial and final states, they interfere, and therefore are correctly computed by adding amplitudes rather than cross sections. This leads to an effective left-handed coupling for the process of $-1/2+g_L = -1+\sin^2\theta_W$, where the $-1/2$ term represents the coupling of the charged-current to the left-handed electron. This results in a cross section of

$$\sigma(\nu_e e^- \to \nu_e e^-)_{TOT} = \frac{G_F^2 s}{\pi}\left(1-2\sin^2\theta_W+\frac{4}{3}\sin^4\theta_W\right), \tag{9}$$

much larger than that of the neutral-current only process.

2.3 The Effect of Initial and Final State Masses

To this point, we have neglected the effect of massive particles. This is not always a reasonable approximation, as it is simple to illustrate with our initial example, $\nu_\mu e^- \to \mu^-\nu_e$. In the lab frame with a stationary target electron, the total center-of-mass energy squared, $s = m_e^2 + 2m_e E_\nu$. However, in order to produce a muon in the final state at all, $s \geq m_\mu^2$. Solving, we find that this reaction only occurs at all when the neutrino energy

$$E_\nu > \frac{m_\mu^2 - m_e^2}{2m_e}, \tag{10}$$

which is approximately 11 GeV. Therefore, for practical cases such as this one, we need a way to account for the effect of the final state mass.

Recall that in our original derivation of this cross section in Equation 5, we noted that we had to integrate the roughly constant differential cross section $d\sigma/dQ^2$ over the range of available Q^2 from zero up to a maximum value. In the massless limit, the range of Q^2 is, in fact, 0 to s as asserted above. In the presence of initial and final state masses, these limits are more complicated:

$$\begin{aligned} Q^2_{MAX,MIN} &= (p^*_\nu \pm p^*_\mu)^2 - \frac{(m^2_\mu - m^2_e)^2}{4s} \\ \Rightarrow Q^2_{MAX} - Q^2_{MIN} &= 4p^*_\nu p^*_\mu \\ &\approx s\left(1 - \frac{m^2_\mu}{s}\right)\left(1 + \mathcal{O}\left(\frac{m^2_e}{m^2_\mu}\right)\right). \end{aligned} \quad (11)$$

In summary, the process is suppressed relative to its massless cross section by a factor of $1 - m^2_\mu/s$. This suppression is a factor that, while not general, recurs often in calculations of mass suppression due to a single massless particle in the final state.

Now consider a more complicated case of the inverse beta-decay reaction in which reactor neutrinos were discovered, $\bar{\nu}_e p \to e^+ n$. Here *both* particles in the final state are heavier than their initial state counterparts: $m_e \approx 0.5$ MeV and $m_n - m_p \approx 1.3$ MeV. We can calculate the threshold energy, E^{MIN}_ν, of the reaction by observing that the heavy nucleon in the final state will have zero kinetic energy to zeroth order in m_e/m_n. Equating the initial and final state s under this condition, we find

$$E^{MIN}_\nu \approx \frac{(m_n + m_e)^2 - m^2_p}{2m_p}, \quad (12)$$

which is approximately 1.8 MeV. If we define $\delta E \equiv E_\nu - E^{MIN}_\nu$, we can then write

$$\begin{aligned} s &= (\underline{\nu} + \underline{p}) \\ &= m^2_p + 2m_p(\delta E + E^{MIN}_\nu) \\ &= 2m_p \times \delta E + (m_n + m_e)^2. \end{aligned} \quad (13)$$

Then the mass suppression factor is

$$\begin{aligned} \xi_{mass} \equiv 1 - \frac{m^2_{final}}{s} &= \frac{2m_p \times \delta E}{(m_n + m_e)^2 + 2m_p \times \delta E} \\ &\approx \begin{cases} \delta E \times \frac{2m_p}{(m_n+m_e)^2} & \text{if } \delta E \ll m_p \\ 1 - \frac{(m_n+m_e)^2}{2m_p \times \delta E} & \text{if } \delta E \gg m_p \end{cases} \end{aligned} \quad (14)$$

Note that for $\delta E \ll m_p$, the mass suppression ξ_{mass} is linear in δE, and therefore near threshold the cross section will increase quadratically: one power from the δE dependence in Equation 14 and one power from the linear increase in cross section with energy from pointlike scattering.

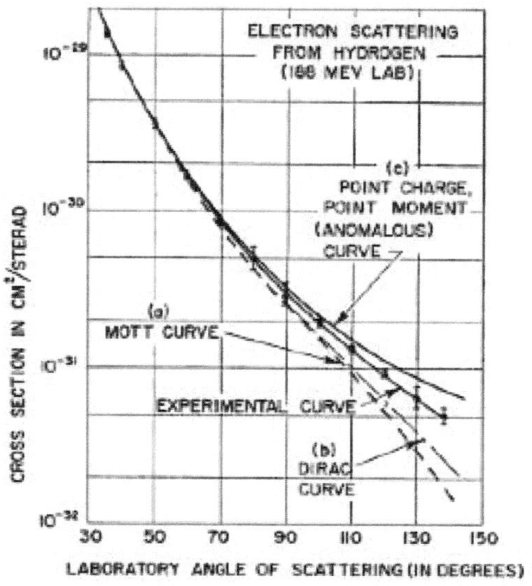

Figure 7. *Angular dependence of cross sections for scattering of* 188 *MeV electrons. The data measure the proton charge radius to be* $(0.7 \pm 0.2) \times 10^{-15}$ *m. (McAlister and Hofstadter 1956). Copyright (1956) American Physical Society.*

3 Beyond Pointlike Scattering

The astute reader will note, however, that I have yet to write down a cross section for inverse beta decay because we are still missing a key ingredient to do so. Although electrons are pointlike, the protons of inverse beta decay most certainly are not. In the next section of this lecture, we will consider the effect of the structure of the target on neutrino interactions.

3.1 Target Structure in νN Elastic Scattering

To begin our exploration of scattering from pointlike scattering, we will continue our investigation of inverse beta decay, $\bar{\nu}_e p \to e^+ n$. This reaction is termed "quasi-elastic"[2] in the sense that the target nucleon remains a single nucleon in the final state and only changes its charge in the charged-current weak interaction.

The target proton differs from an electron in several important respects. The couplings of composite particles like the proton are not predicted by the electroweak theory, nor is the anomalous magnetic moment, $(g-2)/2$, necessarily small for a composite particle. Finally, the weak couplings may have a dependence on Q^2 which reflects the finite size of the particle. Figure 7 shows data from some of the original measurements of proton structure in ~ 200 MeV electron scattering. The increase in angle corresponds to an increase in the Q^2 of the electromagnetic interaction. As can be seen, the experimental data do not agree with the prediction of the proton as a Dirac particle, but require not only anomalous magnetic moment but also finite-sized charge distribution to explain the suppression at high Q^2 relative to a point charge.

[2]Beware of nuclear physicists using the term "quasi-elastic". It is also used to indicate nuclear dissociation in electromagnetic interactions, such as $e^- d \to e^- pn$.

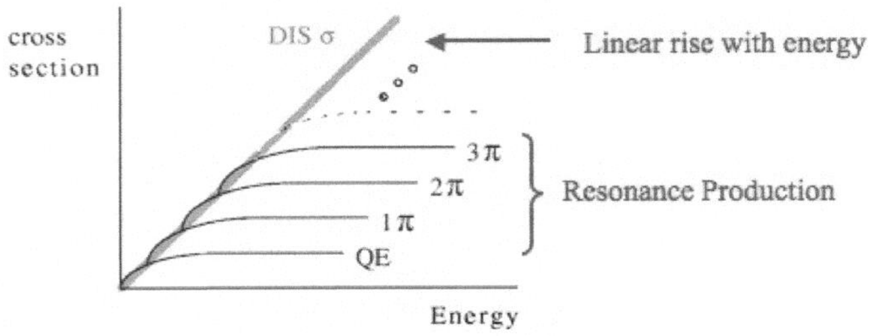

Figure 8. *A schematic diagram representing the rise of the cross section with energy as inelastic channels open up with energy.*

The full cross section for inverse beta decay is

$$\sigma(\bar{\nu}_e p \to e^+ n) = \frac{G_F s}{\pi} \times \cos^2\theta_C \times \xi_{mass} \times \left(g_V^2 + 3g_A^2\right), \tag{15}$$

where the first term is the point-like scattering cross section result we derived for neutrino-electron scattering, the θ_C term takes into account the charged-current quark mixing transition from a u quark to a d quark, ξ_{mass} is defined in Equation 14, and g_V and g_A are the proton form factors. As mentioned above, the proton form factors and the relevant (small) momentum transfer for this process at low energy are not predicted by the electroweak theory, and must be experimentally determined. g_V at low momentum transfer is the electric charge of the proton, +1, and g_A is determined by the neutron lifetime to be -1.26.

3.2 Deep Inelastic Scattering

Of course, another difference between a strongly bound target, such as a nucleus, and an electron is that the strongly bound target can be broken apart in the final state to create different particles. In such a case, what do we qualitatively expect to happen to the cross section?

Consider first the elastic scattering process of neutrinos on nucleons. This total cross section will rise linearly with energy when the energy is sufficiently low. However, if the Q^2 of the reaction is high enough, the differential elastic cross section, $d\sigma/dQ^2$ will start to fall with Q^2 because the nucleon will break apart when too much Q^2 is transferred. At some point, the cross section no longer rises with energy because the elastic process only occurs up to a finite Q^2, and the s at this high energy exceeds that Q^2. However, at the same point, new inelastic processes, such as the production of a single pion will become energetically possible. These will rise with energy, initially quadratically and then linearly until they too reach their Q^2 limit, at which point their cross section stops rising with energy. As illustrated in Figure 8, this process repeats itself, resulting in a linear rise of the total cross section with energy.

Of course, a linear rise with energy is exactly what is expected in the case of point like scattering. The picture above, while possibly helpful in the region of transition between elastic and inelastic to be discussed in Section 4.1, is awkward for understanding the high energy behavior of inelastic scattering. Instead, we model this process as the deep inelastic scattering of neutrinos from *quarks* inside the strongly bound system. These quarks are fundamental particles, and therefore the cross section of neutrino quark scattering will rise linearly with energy.

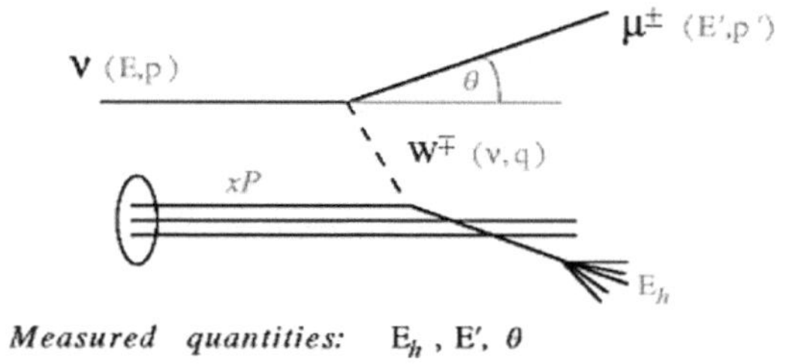

Figure 9. *Kinematic quantities in deep inelastic scattering.*

We first need a common language of kinematics that is relevant for an inelastic process such as $\nu N \to \ell X$ or its neutral current counterpart, $\nu N \to \nu X$. As shown in Figure 9, we define the energy and four-momentum in the lab frame of the incoming neutrino, the outgoing lepton and the weak boson, respectively, to be $p = (E, p)$, $p' = (E', p')$ and $q = (\nu, q)$, and we also define the lab scattering angle of the outgoing lepton as θ, the four-momentum of the target as P, and the energy of the hadronic recoil in the lab frame as E_h. As before we define the negative of the W four-momentum squared

$$Q^2 \equiv -q^2 = -(p' - p)^2 \approx 4EE' \sin^2(\theta/2). \tag{16}$$

Note that this definition is given purely in terms of variables on the well-defined leptonic side of the event. We will follow this convention as much as possible, expressing quantities in the lab in terms of leptonic variables and the initial target mass, M_T. We may define other invariants, such as the lab energy transfer, ν, the inelasticity, y, and the Feynman scaling variable, x:

$$\begin{aligned} \nu &\equiv \frac{q \cdot P}{\sqrt{P \cdot P}} = E - E', \\ y &\equiv \frac{q \cdot P}{p \cdot P} = \frac{\nu}{E}, \\ x &\equiv \frac{-q \cdot q}{2(p \cdot q)} = \frac{Q^2}{2M_T \nu}. \end{aligned} \tag{17}$$

The center-of-mass scattering energy, $\sqrt{s}$, and the mass of the hadronic recoil system, W, can also be written in term of leptonic variables x, y and ν and the target mass M_T:

$$\begin{aligned} s &\equiv (p + P)^2 = M_T^2 + \frac{Q^2}{xy}, \\ W^2 &\equiv (q + P)^2 = M_T^2 + 2M_T \nu - Q^2. \end{aligned} \tag{18}$$

In the picture of neutrinos scattering from constituents of strongly bound systems, the "parton" interpretation of deep inelastic scattering, the variable x has a special interpretation as the fractional momentum of the target nucleon carried by the parton in a frame where the target momentum is very large. The common picture of this frame is that the nucleon, as seen by the incoming lepton, is flat and static because of length contraction and time dilation, and the incoming lepton interactions with a single one of these frozen partons, carrying a momentum fraction x. In this picture, we can define effective masses for the initial and final state partons,

$$\begin{aligned} m_q^2 &= x^2 P^2 = x^2 M_T^2, \\ m_{q'}^2 &= (xP + q)^2 \end{aligned} \tag{19}$$

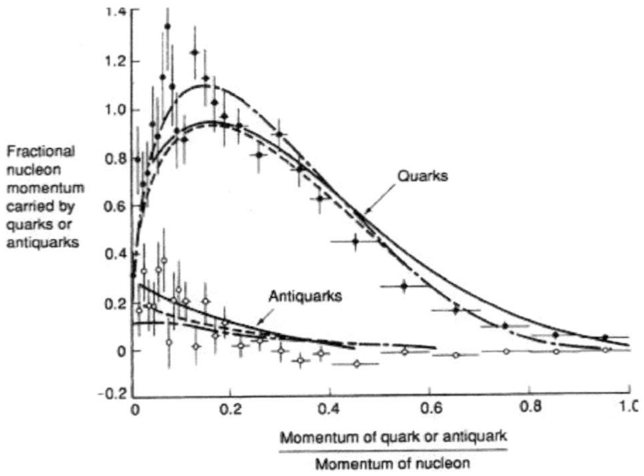

Figure 10. *Distribution of quark and anti-quark momentum density in the nucleon as a function of x.*

To make sense of deep inelastic scattering, we cannot merely consider the hard process of neutrinos scattering from quarks; we must also place those quarks inside the target hadron. This is made possible by the Factorization Theorems of QCD which allow us to write a scattering cross section for a hadronic process in terms of cross sections for scattering from partons convoluted with a parton distribution function $q_h(x)$:

$$\sigma(\nu+h \to \ell+X) = \sum_q \int dx\, \sigma(\nu+q(x) \to \ell+X) q_h(x). \tag{20}$$

The parton distributions, while not (yet) something we can calculate from principles of QCD, are universal. Therefore, they can be determined in one process and applied to another process.

Figure 10 shows an illustration of typical quark and anti-quark distributions in a nucleon at moderate Q^2. The parton distribution function (PDF), $q(x)$ gives the number density of quarks of a given x. If quarks, carried all the momentum of the nucleon, $\int xq(x)dx = 1$; however, in reality this integral is significantly less than one. This momentum sum also turns out to be logarithmically dependent on Q^2, as are the PDFs more generally. These slow changes with Q^2 are called "scaling violation", in reference to the Feynman scaling variable, x. They result from the strong interactions of the quarks themselves in the nucleon. There is a duality between Q^2 and distance scales, with higher Q^2 interactions probing features at small distance scales. At these small scales, the strong interactions among partons in the nucleon will cause quarks to radiate gluons and gluons to split into quarks and anti-quarks. The net results, whose effects have been calculated quantitatively in perturbative QCD, are that quarks and anti-quarks increase in number at higher Q^2, but their average fractional momentum decreases[3].

3.2.1 Deep Inelastic Scattering as Elastic Neutrino-Quark Scattering

Now that we have established the link between neutrino deep inelastic scattering and elastic neutrino-quark scattering, we can apply what we have learned about elastic scattering to deep

[3]This and other topics in perturbative QCD make for fascinating exploration in detail, but are well beyond the scope of these lectures. I highly recommend the CTEQ Collaboration Handbook of Perturbative QCD (Sterman et al 1995) for a pedagogical introduction to these topics.

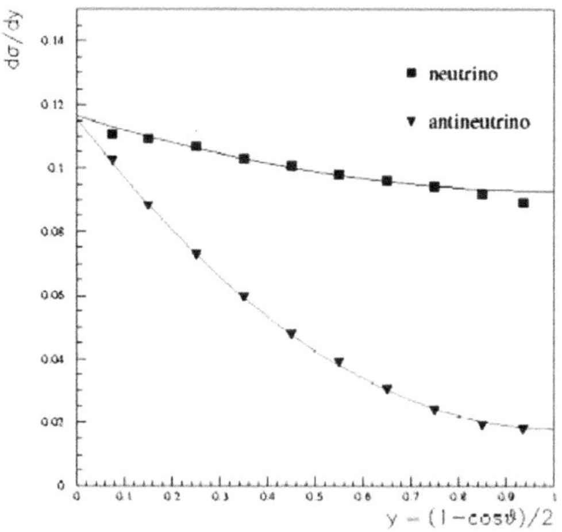

Figure 11. *Neutrino and anti-neutrino deep inelastic scattering cross sections as a function of x.*

inelastic scattering. Recall that for the charged current $\nu q \to \ell q'$ process, we expect a cross section of $G_F s/\pi$, up to a possible angular suppression accounting for total spin in the initial and final state. But $s = M_T^2 + 2M_T E_\nu$ in the lab frame. We have just learned that for each quark, the initial state target mass is xm_N, where m_N is the nucleon mass, so the total effective target mass is of the same order of magnitude as the nucleon mass. Compared with the case of neutrino-electron scattering where the target mass was m_e, the cross section for deep inelastic scattering will be approximately three orders of magnitude larger!

We can also look at chirality and total spin in the reactions of neutrinos and quarks. Again, in the high energy limit where helicity and chirality are equivalent, consider the center-of-mass frame as we did in Figure 6. For the case of a quark, the charged-current weak interaction will pick out left-handed quarks just as it did left-handed electrons, and there will be no net spin along the interaction axis. By contrast, for the case of neutrino scattering from anti-quarks, the target will be right-handed in the center-of-mass frame, and there will be a net spin of 1 along the interaction axis. As we argued in Equation 7, for the case of a right-handed target, the back-scattering is suppressed, and the overall cross section is reduced by a factor of three. A convenient kinematic relationship exists between the center-of-mass scattering angle, θ^* and the inelasticity, y,

$$\left(\frac{1+\cos\theta^*}{2}\right) = 1 - y, \tag{21}$$

and therefore, $d\sigma_{J_z=1}/d\theta^* \propto (1-y)^2$.

The same argument that leads to a back-scattering or high y suppression of the cross section for neutrino-antiquark scattering also holds for antineutrino-quark scattering, and similarly antineutrino-antiquark scattering has no suppression. Therefore

$$\begin{aligned} \frac{d\sigma(\nu q)}{dxdy} &= \frac{d\sigma(\bar{\nu}\bar{q})}{dxdy} \propto 1, \\ \frac{d\sigma(\bar{\nu} q)}{dxdy} &= \frac{d\sigma(\nu\bar{q})}{dxdy} \propto (1-y)^2. \end{aligned} \tag{22}$$

This fact, combined with the smaller momentum fraction carried by antiquarks than is carried by quarks (Figure 10), means that the total anti-neutrino cross section is approximation factor of two smaller than the neutrino cross section on nucleons. Differential cross sections of each are shown in Figure 11.

3.2.2 Structure Functions in Deep Inelastic Scattering

We have approached deep inelastic scattering both from its interpretation as neutrino-quark elastic scattering, and also by purely considering the kinematics. Beyond kinematic constraints, conservation laws and Lorentz invariance also provide model independent constraints on the possible forms of inelastic scattering cross section, and in this picture, information about the structure of the target is contained in a number of general "structure functions". If we consider the case of zero lepton mass, there are three structure functions that can be used to describe the scattering, $2xF_1$, F_2 and xF_3:

$$\frac{d\sigma^{\nu,\bar{\nu}}}{dxdy} \propto \left[y^2 2xF_1(x,Q^2) + \left(2-2y-\frac{M_T xy}{E}\right) F_2(x,Q^2) \pm y(2-y)xF_3(x,Q^2) \right]. \tag{23}$$

Note that xF_3 is a structure function that is not present in electromagnetic interactions, and is only allowed because of the parity violation of the weak interaction.

There is an approximate simplification with a model of massless, free spin-$1/2$ partons, first derived by Callan and Gross, $2xF_1 = F_2$. The Callan-Gross relation implies that the intermediate boson is completely transverse, and so violations of Callan-Gross are often parameterized by R_L, defined so that

$$R_L \equiv \frac{\sigma_L}{\sigma_T} = \frac{F_2}{2xF_1}\left(1+\frac{4M_T x^2}{Q^2}\right). \tag{24}$$

Contributions to R_L arise because of processes internal to the target, like gluon splitting $g \longrightarrow q\bar{q}$ which are calculable in perturbative QCD, and because of the target mass, M_T

Continuing with the assumptions of the validity of the Callan-Gross relation and of massless targets, we can match the y dependence of the structure functions with the y dependence of elastic scattering from quarks and anti-quarks to make assignments of structure functions with parton distributions. In this limit, the coefficient in front of xF_3 simplifies to $1-(1-y)^2$, and the coefficient multiplying $2xF_1 = F_2$ is $1+(1-y)^2$. From Equation 22, the former would be associated with the non-singlet contribution of $q-\bar{q}$ and the later with the sum $q+\bar{q}$. Furthermore, for the charged-current, there is a charge selection, namely, a neutrino cannot produce a quark or anti-quark by sending its W^+ to a target quark unless that target quark has negative charge; otherwise, the resulting final state would have to have charge greater than $+1$ and would not be a quark. Putting all these constraints together, we find:

$$\begin{aligned} 2xF_1^{\nu p,\,\mathrm{CC}} &= x\left[d_p(x)+\bar{u}_p(x)+s_p(x)+\bar{c}_p(x)\right], \\ xF_3^{\nu p,\,\mathrm{CC}} &= x\left[d_p(x)-\bar{u}_p(x)+s_p(x)-\bar{c}_p(x)\right], \end{aligned} \tag{25}$$

where $q_p(x)$ refers to the PDF of a given quark flavor in the proton and where the contribution from third generation quarks, which have very small PDFs, is neglected.

Just as with the neutrino-electron scattering, the neutral current case is more complicated because the neutral current couples to quarks of both helicities with a non-trivial coupling constant. However, unlike the charged-current case, there is no selection based on quark charge. The neutral current structure functions under the same assumptions are:

$$\begin{aligned}
2xF_1^{\nu p,\,\mathrm{NC}} &= x\left[\left(u_L^2+u_R^2\right)\left(u_p(x)+\bar{u}_p(x)+c_p(x)+\bar{c}_p(x)\right)\right.\\
&\quad \left.+\left(d_L^2+d_R^2\right)\left(d_p(x)+\bar{d}_p(x)+s_p(x)+\bar{s}_p(x)\right)\right],\\
xF_3^{\nu p,\,\mathrm{NC}} &= x\left[\left(u_L^2+u_R^2\right)\left(u_p(x)-\bar{u}_p(x)+c_p(x)-\bar{c}_p(x)\right)\right.\\
&\quad \left.+\left(d_L^2+d_R^2\right)\left(d_p(x)-\bar{d}_p(x)+s_p(x)-\bar{s}_p(x)\right)\right],
\end{aligned} \tag{26}$$

where the new notation here, e.g., $u_{L,R}$, refers to the left and right-handed neutral current couplings of up or down type quarks.

Some simplification in the case of the charged-current can be obtained for the practical case where the target material consists of an isoscalar nucleus with equal numbers of neutrons and protons. The light PDFs of the neutron are standardly assumed to be related to the PDFs of the proton by isospin symmetry,

$$\begin{aligned}
u_p(x) &= d_n(x),\\
d_p(x) &= u_n(x),
\end{aligned} \tag{27}$$

and the PDFs of the heavy quarks, $s(x)$ and $c(x)$ are assumed to be identical in neutrons and protons and identical with their anti-quark distributions since they result from gluon splitting and not the valence quark content of the nucleon. Under these assumptions,

$$\begin{aligned}
2xF_1^{\nu N,\,\mathrm{CC}} &= x\left[u(x)+d(x)+\bar{u}(x)+\bar{d}(x)+2s(x)+2c(x)\right]\\
&= x(q(x)+\bar{q}(x)),\\
xF_3^{\nu N,\,\mathrm{CC}} &= x\left[u(x)+d(x)-\bar{u}(x)-\bar{d}(x)+2s(x)-2c(x)\right]\\
&= x(q_{val}(x)+2s(x)-2c(x))
\end{aligned} \tag{28}$$

where the PDFs written are those of the proton and where $q_{val}(x) \equiv q(x)-\bar{q}(x)$. Note the particularly simple forms, these structure functions have, at least in the limit of neglecting the heavy quarks.

3.2.3 ν_τ Charged Current Interactions

A challenging endeavor to apply our theory of deep inelastic scattering is ν_τ appearance experiments such as OPERA. The full calculation of lepton mass effects is beyond the scope of these lectures. Note that all that has preceded this, including the definitions of the structure functions, assumed massless leptons. But again, we can apply our mass suppression formalism to get an approximation of the effect.

As we argued in Equation 11, the generic form of the mass suppression is $(1-m_{final}^2/s)$. Since deep inelastic scattering is neutrino-quark elastic scattering, the relevant quantity for s here is the s of the neutrino-quark system, which is $\hat{s}=xs$. The form of the mass suppression for τ production from a given parton x is then $1-m_\tau^2/(xs)$. This implies that at low x the mass suppression will be large at much higher energies than at high x, and thus qualitatively, the rise of the cross section relative to muon neutrino charged current scattering will be very slow. This can be seen in the full calculation illustrated in Figure 12.

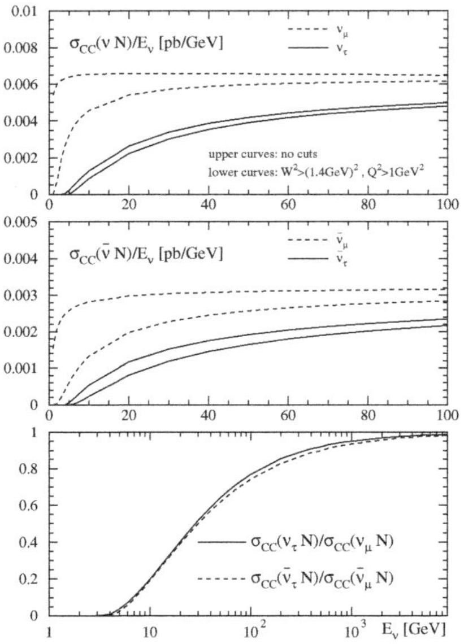

Figure 12. *Mass suppression of ν_τ deep inelastic scattering cross sections (Kretzer and Reno 2002). Copyright (2002) American Physical Society.*

3.2.4 Charm Production by Neutrinos

Another way that massive final state corrections can enter into deep inelastic scattering is the production of charm quarks in neutrino deep inelastic scattering. Although there are few charm quarks to be found in the proton itself (after all, $m_c > m_p$), the sea has a large number of strange quarks, roughly half as many as either of the light quark species. Since the Cabibbo-favored charged-current process turns these strange quarks into charm quarks, production of charm quarks is a significant fraction of the charged-current cross section.

Let's return to the kinematic variables of Equation 17 to study the effect of the final state quark mass. Production of a charm quark in the final state implies that $m_c^2 = (q + \xi P)$, where ξ represents the fractional momentum of the initiating quark instead of the usual Feynman scaling variable x. If $\xi \ll 1$, then

$$\xi \approx \frac{-q^2 + m_c^2}{2P \cdot q} = \frac{Q^2 + m_c^2}{2M_T \nu} = x\left(1 + \frac{m_c^2}{Q^2}\right). \tag{29}$$

The reason introducing ξ as distinct from x now becomes obvious. The x variable as defined in terms of leptonic side variables is no longer the same as the fractional momentum carried by the target quark, but is in fact smaller. Therefore, for a given set of scattering kinematics, the initiating quark must carry a higher fractional momentum and thus will be less common than in the case where a light quark is produced in the final state. This formalism for treating the production of massive quarks is referred to as "slow rescaling".

One of the best ways to actually measure the strange quark content of the nucleon is to measure charged-current charm quark production tagged by the semi-muonic decay of charm

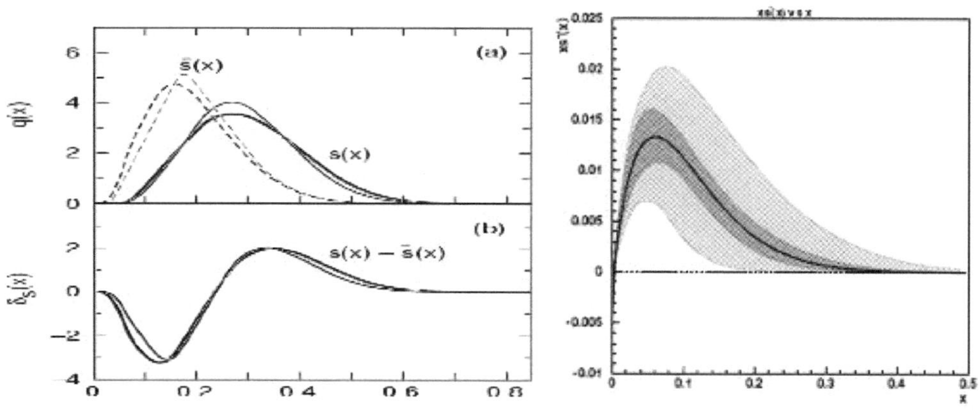

Figure 13. *Non-perturbative model of Brodsky and Ma for $s(x) - \bar{s}(x)$ (left) compared with data from NuTeV (right).*

in a neutrino experiment. In other words, determine $s(x)$ by measuring $\nu_\mu + s \rightarrow \mu^- + c + X$, $c \rightarrow \mu^+ + X$ and its anti-neutrino analog, each of which give two high momentum muons in the final state and are commonly referred to as "dimuon" events.

There is currently some debate about parton distributions regarding whether the strange and anti-strange seas carry equal momentum. Strange quarks and anti-quarks generated by perturbative processes should be nearly symmetric in momentum, but there are non-perturbative effects that can lead to differences in their momentum. The NuTeV experiment has recently completed an analysis of dimuon events induced by neutrino beams and anti-neutrino beams which therefore separately measure the strange and anti-strange quark distributions in the nucleus. Figure 13 shows a comparison of one theoretical prediction with the measurement of this momentum asymmetry from NuTeV.

4 Transitions between Elastic and Inelastic Scattering

To this point, we have explored elastic scattering of neutrinos from pointlike particles and a high energy limit of neutrinos scattering inelastically from nucleons where the neutrino effectively scatters from free quarks in the target. If we look at the cross sections shown in Figure 14, we see both the elastic and deeply inelastic cross sections co-existing over a broad region with a significant component over nearly an order of magnitude in energy being the "barely inelastic" process of single pion production. This transition occurs at these energy values because the "binding energy" of the the target nucleon is approximately λ_{QCD}, which is the scale of a typical momentum exchange for scattering of a neutrino with 1 GeV energy.

This section of the lectures will explore a few features of regions of transition. Because it is of the most interest for oscillations, we will largely focus on the transition between nucleon elastic and inelastic at neutrino energies near a GeV, but we will conclude with comments on other transition regions of interest.

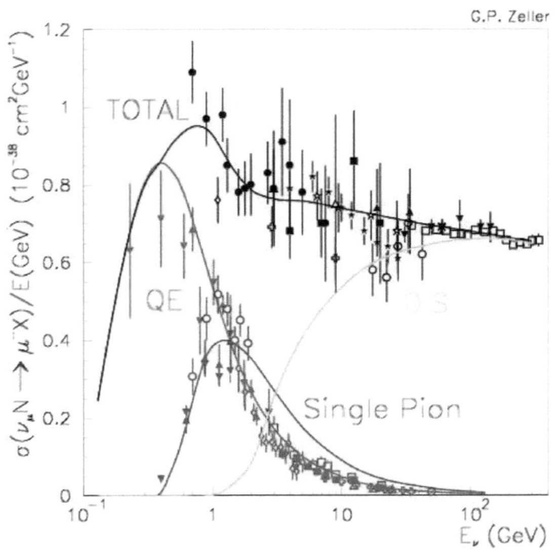

Figure 14. *A compilation of neutrino cross sections, shown as σ/E_ν, in the GeV region with quasi-elastic, deep inelastic and single pion cross sections shown separately. (Figure courtesy G.P. Zeller).*

4.1 The GeV Region

We have exhaustively described the deep inelastic scattering limit of Figure 14, but have not spent much time describing the quasi-elastic cross section, e.g. $\nu_\mu n \to \mu^- p$. At low energies, we expect by the same arguments given in other cases of lepton mass, a suppression due to the muon mass going roughly as

$$\begin{aligned} \frac{\sigma}{\sigma_{\text{massless}}} &\sim 1 - \frac{(M_T + m_\mu)^2}{s} \\ &= 1 - \frac{M_T^2 + 2M_T m_\mu + m_\mu^2}{M_T^2 + 2E_\nu M_T} \\ &\approx \frac{2E_\nu}{M_T} \quad \text{if } E_\nu << M_T. \end{aligned} \tag{30}$$

Thus the cross section at low energies will be quadratic until $E_\nu \sim m_T/2$, as shown in Figure 14. However, we also see that the cross section stops growing above $E_\nu \sim 1$ GeV. As we observed in Section 3.2, above a sufficiently high Q^2, $d\sigma/dQ^2$ begins to fall because interactions at higher Q^2 tend to break apart the target nucleon and therefore are not quasi-elastic.

Nucleon structure also plays a significant role in quasi-elastic scattering. As with deep inelastic scattering, it is relatively straightforward to write a cross section formula for quasi-elastic scattering; however, in the end there are unknown form factors that enter the calculation which must be determined experimentally. The cross section is usually parameterized in terms of vector, F_V, and axial vector form factors, F_A,

$$F_{V,A} \approx \frac{F_{V,A}(0)}{\left(1 + Q^2/M_{V,A}^2\right)^2}, \tag{31}$$

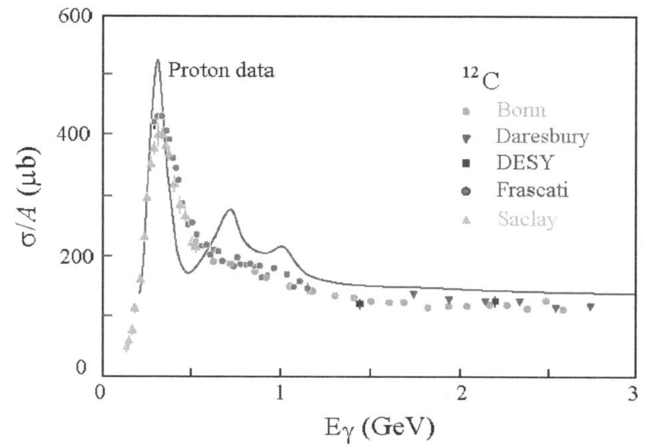

Figure 15. *Photo-absorption data on protons (line) and nuclei (data points) as a function of energy illustrating the effect of Fermi smearing on resonance production.*

in the so-called "dipole approximation" (Llewellyn Smith 1972). These are only phenomenological approximations to the true form factors, and precise measurements of the vector form factor F_V in electron scattering show significant deviations from the dipole form at $Q^2 \gg M_V^2$ where $M_V \approx 0.84$ GeV. The axial vector form factor parameters are well determined only at $Q^2 \approx 0$ (recall the discussion following Equation 15), and the best current estimates from data of the axial mass give $M_A \approx 1.1$ GeV with significant theoretical and experimental uncertainties.

As the cross section becomes barely inelastic, this region is often called the "resonance region" because it is dominated by the production of discrete baryon resonances in the final state. Recall that the mass of the hadronic system, W, is given by $W^2 = M_T^2 + 2M_T\nu(1-x)$. In the barely inelastic regime, this cannot take any arbitrary value because there must be a baryonic state available at that mass. As the solid line in Figure 15 illustrates in a different process, photo-nuclear absorption, there are discrete excitation lines corresponding to specific broad baryon resonances. The lowest mass excited baryonic state is the $\Delta(1232)$ resonance which is very visible and separated as the first peak in Figure 15. Above the $\Delta(1232)$, resonances tend to overlap one another and approach a continuum.

How is it possible to understand such a complicated set of overlapping final states? One way to gain a good qualitative and the beginning of a quantitative understanding is through Bloom-Gilman duality. The ideal of duality is that, on average, one can model the behavior of discrete hadronic states through the behavior of their underlying quark content. This emperically successful idea straddles the border between asymptotically free and confined states in QCD.

The most famous example of Bloom-Gilman duality is illustrated in the quantity $R = \sigma(e^+e^- \rightarrow \text{hadrons})/\sigma(e^+e^- \rightarrow \mu^+\mu^-)$. Shown in Figure 16 is R compared against a prediction from the quark model that $R = N_C \sum Q_q^2$, where Q_q and N_C are the charge and number of color states for the quark, respectively. The sum runs over all quark states that can be produced at a given s. This method works well to describe the cross-section over a complicated mix of final states that can be found well above the $s\bar{s}$ (ϕ) threshold, and also in the region above the $c\bar{c}$ and $b\bar{b}$ narrow resonances.

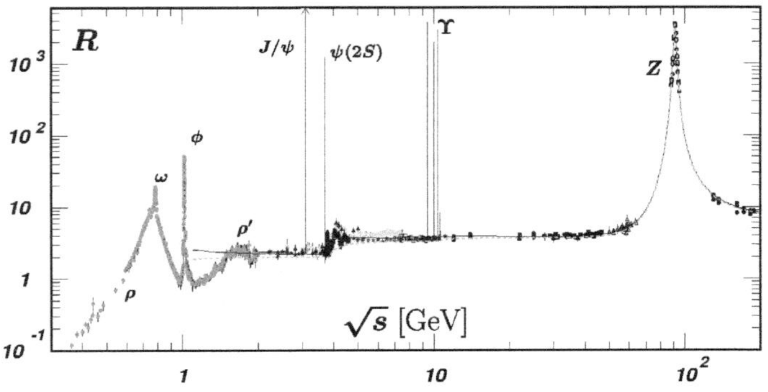

Figure 16. *The ratio of e^+e^- annihilation cross section into hadrons divided by that into muons*

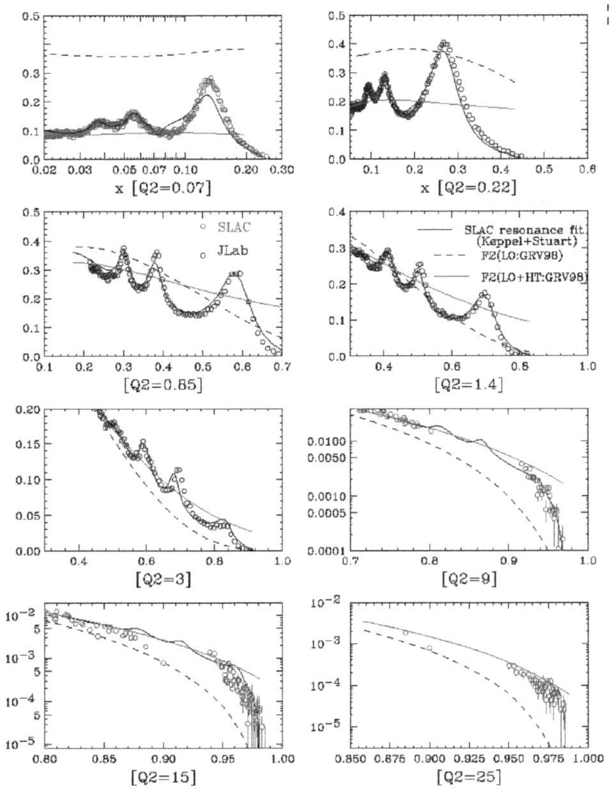

Figure 17. *Electron scattering data in the baryon resonance region compared to a quark-model calculation (Bodek and Yang 2002). Copyright (2002) Elsevier.*

Duality has also been applied successfully to describe the resonance region in electron scattering data (Bodek and Yang 2002), and a comparison of a quark model with actual electron scattering data is shown in Figure 17. The resonance structure essentially appears as modulations on top of the quark model prediction. This approach is now being used in most modern neutrino generators attempting to interpolate the region of hadronic mass squared, W^2, best treated as production of discrete resonances and the higher energy region where parton model calculations of deep inelastic scattering are good approximations.

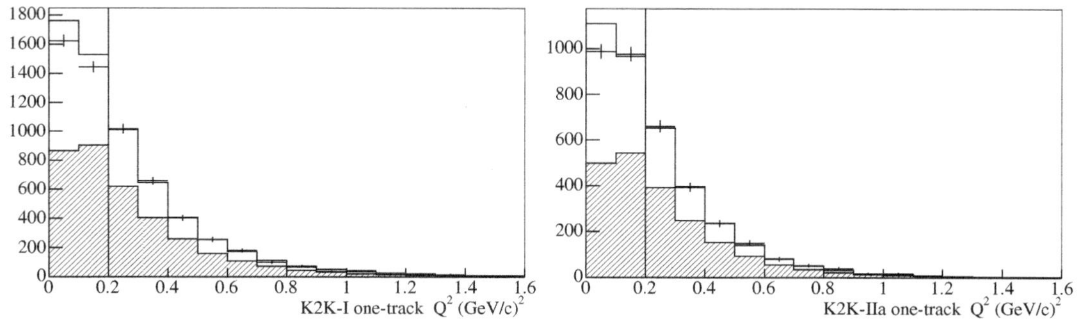

Figure 18. *Q^2 distribution in a quasi-elastic enriched sample in the SciBar detector at KEK (Gran 2006) shown for two different running periods. The estimation of the number of events due to quasi-elastic processes in each bin is shown by the shaded area. Copyright (2006) American Physical Society.*

4.2 The Effect of the Nucleus

A significant complication for the understanding of neutrino interactions in future experiments is the need to model cross sections on a variety of nuclei. Exclusive neutrino interactions in the GeV energy range of the type that must be understood in future experiments are particularly sensitive to modification in the nuclear medium, and there are few definitive models and little data currently available to help to understand the effects. In this section, we will survey some of the relevant phenomenology.

One effect that is relevant for all energy regimes is the motion of the target nucleon to the nucleus. This is often called "Fermi smearing", and it can have a dramatic effect. Figure 15 illustrates how significant this effect is on the production of resonances in photo-nuclear absorption. The proton data represented by the solid line clearly shows multiple resonance peaks, but in ^{12}C, the same resonances become indistinct due to Fermi smearing except for the well separated $\Delta(1232)$ resonance. Similar dramatic effects can be seen in reconstruction of quasi-elastic events, and in scattering from high x partons in deep inelastic scattering.

Quasi-elastic scattering at low Q^2 has a unique nuclear effect. Because the final state nucleon will not necessarily be energetic enough to leave the nucleus, its creation may be suppressed if there is no free nuclear state available due to the Pauli exclusion principle. This effect is often called Pauli blocking. The effect can be modeled, although it is unclear how well these models work or how universal they are. Even with a model for Pauli blocking in their predictions, both the SciBar and MiniBooNE detectors see significant deficits of events from scattering off of carbon at low Q^2 (Figure 18).

Another significant effect is the rescattering in the nuclear medium of hadrons which are sufficiently energetic to escape the target nucleus. Because the nucleus is so dense, the material traversed when a produced particle escapes the target nucleus is significant compared to the amount of nuclear matter it sees when traveling through macroscopic amounts of detector material afterward. Therefore, it is not surprising that the probability of a reinteraction in the nucleus is significant. Such a reinteraction may be particularly difficult for an experiment relying on knowledge of exclusive final states, such use of two-body kinematic constraints in reconstructing quasi-elastic events, or when concerned about backgrounds to ν_e appearance from π^0s. Again, these effects are not well studied, although there is some promise in the use

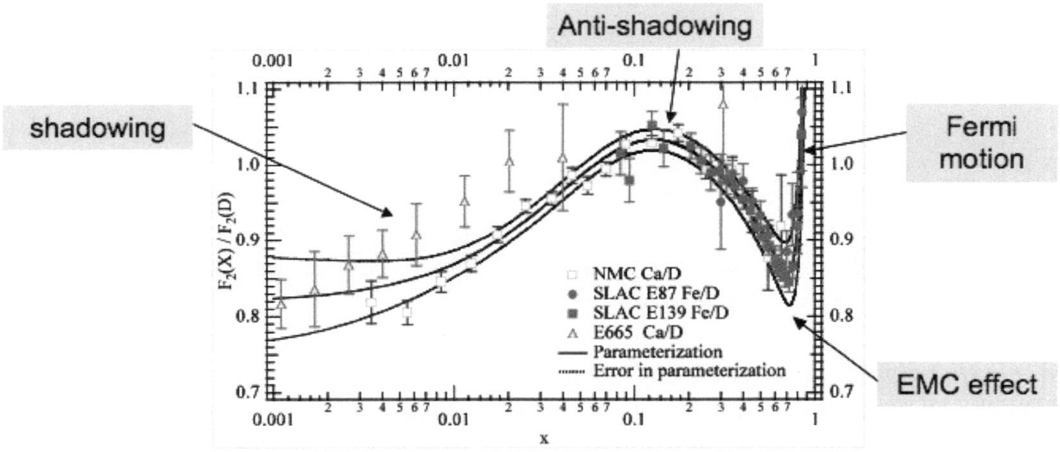

Figure 19. *Nuclear effects on parton distributions.*

of electron scattering data to constrain models of such final state reinteractions.

In the deep inelastic scattering region, nuclear effects are well measured in charged lepton scattering and are often parameterized in terms of their effect on parton distribution functions as shown in Figure 19. At high x, the same Fermi smearing described above leads to a dramatic increase in the rare partons carrying very high x. The region of moderate x, in the "valence quark" region is suppressed through an effect generally named the "EMC" effect after the first experiment to observe it. At $x \sim 0.1$, there is a small enhancement of the PDFs sometimes referred to as "anti-shadowing" and PDFs at low x appear to be dramatically suppressed due to "shadowing". Because these effects have only been measured in charged-lepton scattering and because, with the exception of the Fermi smearing and perhaps shadowing, there are plausible but not definitive theoretical interpretations of the effect, it is not clear whether the modifications to PDFs are in fact universal, or whether the effects in neutrino neutral and charged-current scattering will be different. Most likely, data from neutrino scattering experiments on a variety of nuclei, including light nuclei, will be required to resolve this question.

4.3 Other Regimes of Transition

There are other regions of transition, usually associated with binding thresholds. Binding energies of electrons in atoms are $\lesssim Z^2 m_e c^2 \alpha_E M$ which can cover a broad range in energy from a few eV to 10^5 eV, and certainly at very low energies these bindings can affect neutrino scattering from atomic electrons. However, this is not in an energy range that has effected neutrino oscillation experiments to date. There is also a transition region associated with the binding energy of nucleons inside the nucleus that ranges from 0.1–10 MeV. This binding energy most definitely has had an impact on oscillation physics in a number of experiments. The SNO experiment uses charged and neutral-current reactions $\nu_e d \to ppe^-$ and $\nu d \to pn\nu$ on deuterons and elastic scattering from atomic electrons as its oscillation signatures. For the few MeV neutrinos from the sun, the thresholds of atomic electrons, < 1 keV even for oxygen, are irrelevant. However, the 2.2 MeV binding energy of the deuteron and the characteristic sharp quadratic turn-on with neutrino energy at the threshold is a significant theoretical uncertainty in reaction rates for low energy neutrinos.

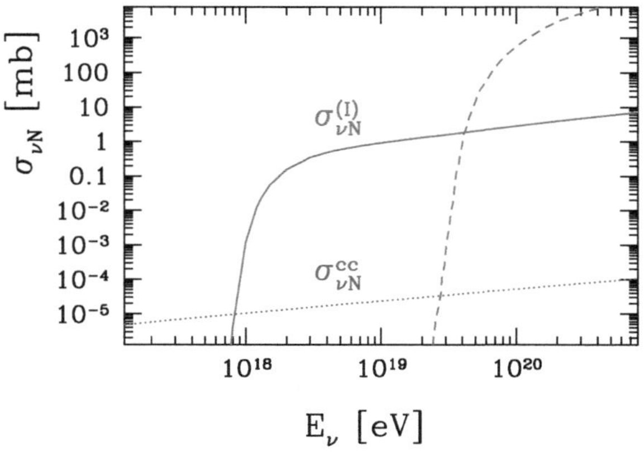

Figure 20. *Possible modifications of ultra-high neutrino energy cross sections could be quite dramatic if new degrees of freedom can be excited at high energies. This model (Fodor 2003) shows an example using electroweak instantons. Copyright (2002) Elsevier.*

A more interesting possibility might be realized in cosmic ray physics involving neutrinos. Although our current knowledge from energy frontier colliders limits the energy scale at which quarks and leptons might be bound states to $\gtrsim$ 10 TeV, beyond this it is possible that some new process turns on as illustrated in Figure 20. It is interesting to remember from our discussions that the "background" deep inelastic cross section will continue to grow with energy approximately linearly until $Q^2 > M_W^2$, when the propagator term (Equation 3) will begin to drop with increasing Q^2. This effect begins to be noticeable at neutrino energies of $\sim$ 10 TeV, and is so significant at high energies that baseline "QCD" cross section in Figure 20 is barely increasing with energy.

5 Conclusions

By way of conclusions, I will offer in compact form what I think are the most important points for the student to take away from these lecture notes.

The understanding of neutrino interactions is one of the keys to precision measurements of neutrino oscillations at accelerators. It will soon limit precision in the current program of ν_μ disappearance at atmospheric baselines. In the future cross section uncertainties, if not addressed with new data, will play a significant role in the ultimate precision of the $\nu_\mu \to \nu_e$ and $\bar{\nu}_\mu \to \bar{\nu}_e$ measurements needed to untangle the neutrino mass hierarchy and to search for leptonic CP violation.

The neutrino scattering rate is generally proportional to energy. Specifically this is true for scattering from pointlike particles when the $Q^2 \ll M_W^2$. When it is not true for an exclusive process below this Q^2 threshold, then there is some physics limiting the maximum momentum transfer, such as a threshold above which the target breaks up.

Neutrino target structure (atom, nucleus, nucleon) is a significant complication to precise theoretical calculation of cross section on neutrinos, particularly near inelastic thresholds. Tools like quark-hadron duality are helpful for modeling the major features, but detailed predictive models require additional data we do not currently have in hand, particularly when knowledge of the nuclear environment is needed to make a firm prediction.

Acknowledgments

I am grateful to the organizers of summer school for providing a stimulating environment and program of study and for their hard work in recruiting students to the school. I thank Dave Casper, Rik Gran, Debbie Harris, Jorge Morfin, Tsuyoshi Nakaya and Sam Zeller for providing material for or useful comments on these lectures.

References

Bethe, H. and Peierls, R. (1934), *Nature* **133**, 532.
Bodek, A. and Yang, U.-K. (2002), *Nucl. Phys. Proc. Suppl.* **112** 70.
Fermi, E. (1934), *Z. Physik* **88**, 161.
Fodor, Z. *et al.* (2003), *Phys. Lett.* **B561** 191.
Galison, P. (1983), *Rev. Mod. Physics* **55**, 477.
Gran, R. *et al.* (2006), *Phys. Rev.* **D74**, 052002.
Kretzer, S. and Reno, M.H. (2002), *Phys. Rev.* **D66** 113007.
Llewellyn Smith, C.H. (1972), *Phys. Rep.* **3C**, 261.
McAlister, R. and Hofstadter, R. (1955), *Phys. Rev.* **102** 851.
Minakata, H. and Nunokawa, H. (2001), *Jour. HEP* **0110** 001.
Reines, F. (1996), *Rev. Mod. Physics* **68** 317.
Sterman, G. *et al.* (1995), *Rev. Mod. Physics* **67**, 157.
Zeller, G.P. (2003), *arXiv:hep-ex* 0312061.

Models of Neutrino Masses and Mixings

Guido Altarelli

Universita' di Roma Tre, Rome, Italy and CERN, Geneva, Switzerland

1 Introduction

At the School I gave three lectures on neutrino masses and mixings. Much of the material covered in my first two lectures is written down in a review on the subject that I published not long ago with F. Feruglio [1]. Moreover, there has been some (necessary and useful) overlap with other particularly related courses at this School, e.g. with [2], [3], [4] and with the experimental talks. Here, I make a relatively short summary (with updates) of the content of my first two lectures, referring to our review for a more detailed presentation, and then I expand on the content of the third lecture which was dedicated to recent work on A4 models of tri-bimaximal neutrino mixing which were not covered in the review.

By now there is convincing evidence for solar and atmospheric neutrino oscillations. The Δm^2 values and mixing angles are known with fair accuracy. A summary of the results, taken from Ref. [5] is shown in Table 1. For the Δm^2 we have: $\Delta m^2_{atm} \sim 2.4\ 10^{-3}$ eV^2 and $\Delta m^2_{sol} \sim 7.9\ 10^{-5}$ eV^2. As for the mixing angles, two are large and one is small. The atmospheric angle θ_{23} is large, actually compatible with maximal but not necessarily so: at 3σ: $0.29 \lesssim \sin^2\theta_{23} \lesssim 0.71$ with central value around 0.44. The solar angle θ_{12}, the most precisely measured, is large, $\sin^2\theta_{12} \sim 0.31$, but certainly not maximal (by about 6 σ now). The third angle θ_{13}, strongly limited mainly by the CHOOZ experiment, has at present a 3σ upper limit given by about $\sin^2\theta_{13} \lesssim 0.04$.

In spite of this experimental progress there are still many alternative routes in constructing models of neutrino masses. This variety is mostly due to the considerable ambiguities that remain. First of all, it is essential to know whether the LSND signal, which has not been confirmed by KARMEN and is currently being double-checked by MiniBoone, will be confirmed or will be excluded. If LSND is right we probably need at least four light neutrinos; if not we can do with only the three known ones, as we assume here in the following. Then, as neutrino oscillations only determine mass squared differences, a crucial missing input is the absolute scale of neutrino masses. Also the pattern of the neutrino mass spectrum is not known: it could be approximately degenerate with $m^2 >> \Delta m^2_{ij}$ or of the inverse hierarchy type (with the 1,2 solar doublet on top) or of the normal hierarchy type (with the solar doublet below).

Table 1. *Best fit values of squared mass differences and mixing angles [5]*

	lower limit (2σ)	best value	upper limit (2σ)
$(\Delta m^2_{sun})_{\rm LA}$ (10^{-5} eV^2)	7.2	7.9	8.6
Δm^2_{atm} (10^{-3} eV^2)	1.8	2.4	2.9
$\sin^2\theta_{12}$	0.27	0.31	0.37
$\sin^2\theta_{23}$	0.34	0.44	0.62
$\sin^2\theta_{13}$	0	0.009	0.032

The following experimental information on the absolute scale of neutrino masses is available. From the endpoint of tritium beta decay spectrum we have an absolute upper limit of 2 eV (at 95% C.L.) on the mass of "$\bar{\nu}_e$" [6], [7], which, combined with the observed oscillation frequencies under the assumption of three CPT-invariant light neutrinos, represents also an upper bound on the masses of all active neutrinos. Less direct information on the mass scale is obtained from neutrinoless double beta decay ($0\nu\beta\beta$) [8]. The discovery of $0\nu\beta\beta$ decay would be very important because it would directly establish lepton number violation and the Majorana nature of ν's. The present limit from $0\nu\beta\beta$ is affected by a relatively large uncertainty due to ambiguities on nuclear matrix elements. We quote here two recent limits (90%c.l.): $|m_{ee}| < 0.33-1.35$ eV [IGEX(^{76}Ge) [9]] or $|m_{ee}| < (0.2-1.1)$ eV [Cuoricino(^{130}Te) [10]], where $m_{ee} = \sum U^2_{ei} m_i$ in terms of the mixing matrix and the mass eigenvalues (see eq.(5)). Complementary information on the sum of neutrino masses is also provided by measurements in cosmology [3], where an extraordinary progress has been made in the last years, in particular data on the cosmic microwave background (CMB) anisotropies (WMAP), on the large scale structure of the mass distribution in the Universe (SDSS, 2dFGRS) and from the Lyman alpha forest [11]. WMAP by itself is not very restrictive: $\sum_i |m_i| < 2.11$ eV (at 95% C.L.). Combining CMB data with those on the large scale structure one obtains $\sum_i |m_i| < 0.68$ eV. Adding also the data from the Lyman alpha forest one has $\sum_i |m_i| < 0.17$ eV [12]. But this last combination is questionable because of some tension (at $\sim 2\sigma$'s) between the Lyman alpha forest data and those on the large scale structure. In any case, the cosmological bounds depend on a number of assumptions (or, in fashionable terms, priors) on the cosmological model. In summary, from cosmology for 3 degenerate neutrinos of mass m, depending on which data sets we include and on our degree of confidence in cosmological models, we can conclude that $m \lesssim 0.06-0.23-0.7$ eV.

Given that neutrino masses are certainly extremely small, it is really difficult from the theory point of view to avoid the conclusion that L conservation is probably violated. In fact, in terms of lepton number violation the smallness of neutrino masses can be naturally explained as inversely proportional to the very large scale where lepton number L is violated, of order the grand unification scale M_{GUT} or even the Planck scale M_{Pl}. If neutrinos are Majorana particles, their masses arise from the generic dimension-five non renormalizable operator of the form:

$$O_5 = \frac{(Hl)_i^T \lambda_{ij} (Hl)_j}{M} + h.c. \quad , \tag{1}$$

with H being the ordinary Higgs doublet, l_i the SU(2) lepton doublets, λ a matrix in flavour space, M a large scale of mass and a charge conjugation matrix C between the lepton fields is understood. Neutrino masses generated by O_5 are of the order $m_\nu \approx v^2/M$ for $\lambda_{ij} \approx \mathrm{O}(1)$, where $v \sim \mathrm{O}(100\ \mathrm{GeV})$ is the vacuum expectation value of the ordinary Higgs. A particular realization leading to comparable masses is the see-saw mechanism, where M derives from the exchange of heavy right-handed ν_R's and becomes their Majorana mass matrix: the resulting effective light Majorana neutrino mass matrix reads:

$$m_\nu = m_D^T M^{-1} m_D \quad . \tag{2}$$

that is, the light neutrino masses are quadratic in the Dirac masses and inversely proportional to the large Majorana mass. For $m_\nu \approx \sqrt{\Delta m^2_{atm}} \approx 0.05$ eV and $m_\nu \approx m_D^2/M$ with $m_D \approx v \approx$ 200 GeV we find $M \approx 10^{15}$ GeV which indeed is an impressive indication for M_{GUT}. Thus probably neutrino masses are a probe into the physics at M_{GUT}.

2 The ν-Mixing Matrix

If we take maximal s_{23} ($s_{ij} = \sin\theta_{ij}$) and keep only linear terms in $u = s_{13}e^{i\varphi}$, from experiment we find the following structure of the mixing matrix U_{fi} ($f = e,\mu,\tau$, $i = 1,2,3$), apart from sign convention redefinitions:

$$U_{fi} = \begin{pmatrix} c_{12} & s_{12} & u \\ -(s_{12}+c_{12}u^*)/\sqrt{2} & (c_{12}-s_{12}u^*)/\sqrt{2} & 1/\sqrt{2} \\ (s_{12}-c_{12}u^*)/\sqrt{2} & -(c_{12}+s_{12}u^*)/\sqrt{2} & 1/\sqrt{2} \end{pmatrix} \tag{3}$$

If s_{13} would be exactly zero there would be no CP violations in ν oscillations. A main target of the new planned oscillation experiments is to measure the actual size of s_{13}. In the next decade the upper limit on $\sin^2 2\theta_{13}$ will possibly go down by at least an order of magnitude (T2K, NoνA, DoubleCHOOZ.....) [13]. Even for three neutrinos the pattern of the neutrino mass spectrum is still undetermined: it can be approximately degenerate, or of the inverse hierarchy type or normally hierarchical. Given the observed frequencies and the notation $\Delta m^2_{sun} \equiv \Delta m^2_{12}$, $\Delta m^2_{atm} \equiv |\Delta m^2_{23}|$ with $\Delta m^2_{12} = |m_2|^2 - |m_1|^2 > 0$ and $\Delta m^2_{23} = m_3^2 - |m_2|^2$, the three possible patterns of mass eigenvalues are:

$$\begin{aligned} &\texttt{Degenerate}: |m_1| \sim |m_2| \sim |m_3| \gg |m_i - m_j| \\ &\texttt{Inverted hierarchy}: |m_1| \sim |m_2| \gg |m_3| \\ &\texttt{Normal hierarchy}: |m_3| \gg |m_{2,1}| \end{aligned} \tag{4}$$

The sign of Δm^2_{23} can be measured in the future through matter effects in long baseline experiments [13]. Models based on all these patterns have been proposed and studied and all are in fact viable at present.

The detection of neutrino-less double beta decay, besides its enormous intrinsic importance as direct evidence of L non conservation, would also offer a way to possibly disentangle the 3 cases. The quantity which is bound by experiments is the 11 entry of the ν mass matrix, which in general, from $m_\nu = U^* m_{diag} U^\dagger$, is given by :

$$|m_{ee}| \ = |(1 - s^2_{13})\ (m_1 c^2_{12} \ + \ m_2 s^2_{12}) + m_3 e^{2i\phi} s^2_{13}| \tag{5}$$

Starting from this general formula it is simple to derive the following bounds for degenerate, inverse hierarchy or normal hierarchy mass patterns.

a) Degenerate case. If $|m|$ is the common mass and we set $s_{13} = 0$, which is a safe approximation in this case, because $|m_3|$ cannot compensate for the smallness of s_{13}, we have $m_{ee} \sim |m|(c^2_{12} \pm s^2_{12})$. Here the phase ambiguity has been reduced to a sign ambiguity which is sufficient for deriving bounds. So, depending on the sign we have $m_{ee} = |m|$ or $m_{ee} = |m| \cos 2\theta_{12}$. We conclude that in this case m_{ee} could be as large as the present experimental limit but should be at least of order $O(\sqrt{\Delta m^2_{atm}}) \ \sim \ O(10^{-2}\ \text{eV})$ given that the solar angle cannot be too close to maximal (in which case the minus sign option could be arbitrarily small). The experimental 2-σ range of the solar angle does not favour a cancellation by more than a factor of about 3.

b) Inverse hierarchy case. In this case the same approximate formula $m_{ee} = |m|(c^2_{12} \pm s^2_{12})$ holds because m_3 is small and the s_{13} term in eq.(5) can be neglected. The difference is that here we know that $|m| \approx \sqrt{\Delta m^2_{atm}}$ so that $|m_{ee}| < \sqrt{\Delta m^2_{atm}} \ \sim$ 0.05 eV. At the same time, since a full cancellation between the two contributions cannot take place, we expect $|m_{ee}| > 0.01$ eV.

c) Normal hierarchy case. Here we cannot in general neglect the m_3 term. However in this case $|m_{ee}| \ \sim \ |\sqrt{\Delta m^2_{sun}}\, s^2_{12} \ \pm \ \sqrt{\Delta m^2_{atm}}\, s^2_{13}|$ and we have the bound $|m_{ee}| <$ a few 10^{-3} eV.

Recently some evidence for $0\nu\beta\beta$ was claimed [14] corresponding to $|m_{ee}| \sim (0.2 \div 0.6)$ eV ($(0.1 \div 0.9)$ eV in a more conservative estimate of the involved nuclear matrix elements). This result is not supported by the IGEX and Cuoricino measurements of a comparable sensitivity, but if confirmed it would rule out cases b) and c) and point to case a) or to models with more than 3 neutrinos. In the next few years a new generation of experiments will reach a larger sensitivity on $0\nu\beta\beta$ by about an order of magnitude [8]. If these experiments will observe a signal this would indicate that the inverse hierarchy is realized, if not, then the normal hierarchy case remains a possibility.

3 "Normal" versus "Exceptional" Models

After KamLAND, SNO and WMAP not too much hierarchy in neutrino masses is indicated by experiments:

$$r = \Delta m^2_{sol}/\Delta m^2_{atm} \sim 1/30. \tag{6}$$

Precisely at 2σ: $0.025 \lesssim r \lesssim 0.049$ [5]. Thus, for a hierarchical spectrum, $m_2/m_3 \sim \sqrt{r} \sim 0.2$, which is comparable to the Cabibbo angle $\lambda_C \sim 0.22$ or $\sqrt{m_\mu/m_\tau} \sim 0.24$. This suggests that the same hierarchy parameter (raised to powers with o(1) exponents) applies for quark, charged lepton and neutrino mass matrices. This in turn indicates that, in absence of some special dynamical reason, we do not expect a quantity like θ_{13} to be too small. Indeed it would be very important to know how small the mixing angle θ_{13} is and how close to maximal is θ_{23}. Actually one can make a distinction between "normal" and "exceptional" models. For normal models θ_{23} is not too close to maximal and θ_{13} is not too small, typically a small power of the self-suggesting order parameter $\sqrt{r}$, with $r = \Delta m^2_{sol}/\Delta m^2_{atm} \sim 1/30$. Exceptional models are those where some symmetry or dynamical feature assures in a natural way the near vanishing of θ_{13} and/or of $\theta_{23} - \pi/4$. Normal models are conceptually more economical and much simpler to construct. Typical categories of normal models are:

a) Anarchy. These are models with an approximately degenerate mass spectrum and no ordering principle, no approximate symmetry assumed in the neutrino mass sector [15] [1]. The small value of r is accidental, due to random fluctuations of matrix elements in the Dirac and Majorana neutrino mass matrices. Starting from a random input for each matrix element, the see-saw formula, being a product of 3 matrices, generates a broad distribution of r values. All mixing angles are generically large: so in this case one does not expect θ_{23} to be maximal and θ_{13} must probably be found near its upper bound.

b) Semianarchy. We have seen that anarchy is the absence of structure in the neutrino sector. Here we consider an attenuation of anarchy where the absence of structure is limited to the 23 sector. The typical structure is in this case [16] [1]:

$$m_\nu \approx m \begin{pmatrix} \delta & \varepsilon & \varepsilon \\ \varepsilon & 1 & 1 \\ \varepsilon & 1 & 1 \end{pmatrix} , \tag{7}$$

where δ and ε are small and by 1 we mean entries of $o(1)$ and also the 23 determinant is of $o(1)$. This texture can be realized, for example, without see-saw from a suitable set of $U(1)_F$ charges for (l_1, l_2, l_3), eg $(a, 0, 0)$ appearing in the dim. 5 operator of eq.(1). Clearly, in general we would expect two mass eigenvalues of order 1, in units of m, and one small, of order δ or ε^2. This typical pattern would not fit the observed solar and atmospheric observed frequencies. However, given that $\sqrt{r}$ is not too small, we can assume that its small value is generated accidentally, as for anarchy. We see that, if by chance the second eigenvalue $\eta \sim \sqrt{r} \sim \delta + \varepsilon^2$, we can then obtain the correct value of r together with large but in general non maximal θ_{23} and θ_{12} and small $\theta_{13} \sim \varepsilon$. The guaranteed smallness of θ_{13} is the main advantage over anarchy, and the relation with $\sqrt{r}$ normally keeps θ_{13} not too small. For example, $\delta \sim \varepsilon^2$ in typical $U(1)_F$ models that provide a very economical but effective realization of this scheme .

c) Inverse hierarchy. One obtains inverted hierarchy, for example, in the limit of exact $L_e - L_\mu - L_\tau$ symmetry with $r = 0$ and bi-maximal mixing (both θ_{12} and θ_{23} are maximal) [1]. Simple forms of symmetry breaking cannot sufficiently displace θ_{12} from the maximal value because typically $\tan^2\theta_{12} \sim 1 + o(r)$. Viable normal models can be obtained by arranging large contributions to θ_{23} and θ_{12} from the charged lepton mass diagonalization. But then, it turns out that, in order to obtain the measured value of θ_{12}, the size of θ_{13} must be close to its present upper bound [17]. If indeed the shift from maximal θ_{12} is due to the charged lepton diagonalization, this could offer a possible track to explain the empirical Raidal relation $\theta_{12} + \theta_C = \pi/4$ [18](with present data $\theta_{12} + \theta_C = (47.0 + 1.7 - 1.6)^0$). While it would not be difficult in this case to arrange that the shift from maximal is of the order of θ_C, it is not clear how to guarantee that it is precisely equal to θ_C [19]. Besides the effect of the charged lepton diagonalization, in a see-saw context, one can assume a strong additional breaking of $L_e - L_\mu - L_\tau$ from soft terms in the right-handed M_{RR} Majorana mass matrix [20]. Since ν_R's are gauge singlets and thus essentially uncoupled, a large breaking in M_{RR} does not feedback in other sectors of the lagrangian. In this way one can obtain realistic values for θ_{12} and for all other masses and mixings, in particular also with a small θ_{13}.

d) Normal hierarchy. Particularly interesting are models with the 23 determinant suppressed by the see-saw mechanism [1]: in the 23 sector one needs relatively large mass splittings to fit the small value of r but nearly maximal mixing. This can be obtained if the 23 subdeterminant is suppressed by some dynamical trick. Typical examples are lopsided models with large off diagonal term in the Dirac matrices of charged leptons and/or neutrinos (in minimal SU(5) the d-quark and charged lepton mass matrices are each the transposed of the other, so that large left-handed mixings for charged leptons correspond to large unobservable right-handed mixings for d-quarks). Another class of typical examples is the dominance in the see-saw formula of a small eigenvalue in M_{RR}, the right-handed Majorana neutrino mass matrix. When the 23 determinant suppression is implemented in a 33 context, normally θ_{13} is not protected from contributions that vanish with the 23 determinant, hence with r.

The fact that some neutrino mixing angles are large and even nearly maximal, while surprising at the start, was eventually found to be well compatible with a unified picture of quark and lepton masses within GUTs. The symmetry group at M_{GUT} could be either (SUSY) SU(5) or SO(10) or a larger group. For example, normal models based on anarchy, semianarchy, inverted hierarchy or normal hierarchy can all be naturally implemented by simple assignments of $U(1)_F$ horizontal charges in a semiquantitative unified description of all quark and lepton masses in SUSY SU(5)$\times$ $U(1)_F$. Actually, in this context, if one adopts a statistical criterion, hierarchical models appear to be preferred over anarchy and among them normal hierarchy with see-saw ends up as being the most likely [21].

In conclusion we expect that experiment will eventually find that θ_{13} is not too small and that θ_{23} is sizably not maximal. But if, on the contrary, either θ_{13} very small or θ_{23} very close to maximal will emerge from experiment or both, then theory will need to cope with this fact. Normal models have been extensively discussed in the literature [1], so we concentrate here on examples of exceptional models.

4 Tri-bimaximal Mixing

Here we want to discuss some particular exceptional models where both θ_{13} and $\theta_{23} - \pi/4$ exactly vanish (more precisely, they vanish in a suitable limit, with correction terms that can be made negligibly small) and, in addition, $s_{12} \sim 1/\sqrt{3}$, a value which is in very good agreement with present data. This is the so-called tri-bimaximal or Harrison-Perkins-Scott mixing pattern (HPS) [22], with the entries in the second column all equal to $1/\sqrt{3}$ in absolute value. Here we adopt the following phase convention:

$$U_{HPS} = \begin{pmatrix} \sqrt{\frac{2}{3}} & \frac{1}{\sqrt{3}} & 0 \\ -\frac{1}{\sqrt{6}} & \frac{1}{\sqrt{3}} & -\frac{1}{\sqrt{2}} \\ -\frac{1}{\sqrt{6}} & \frac{1}{\sqrt{3}} & \frac{1}{\sqrt{2}} \end{pmatrix} . \quad (8)$$

In the HPS scheme $\tan^2\theta_{12} = 0.5$, to be compared with the latest experimental determination in Table 1: $\tan^2\theta_{12} = 0.45^{+0.07}_{-0.04}$ (at 1σ). The challenge is to find natural and appealing schemes that lead to this matrix with good accuracy. Clearly, in a natural realization of this model, there must be a very constraining and predictive underlying dynamics. It is interesting to explore particular structures giving rise to this very special set of models in a natural way. In this case we have a maximum of "order" implying special values for all mixing angles. Interesting ideas on how to obtain the HPS mixing matrix have been discussed in refs. [22], [23], [24]. Some attractive models are based on the discrete symmetry A4, which appears as particularly suitable for the purpose, and were presented in ref. [25], [26], [27], [28], [29], [30].

The HPS mixing matrix suggests that mixing angles are independent of mass ratios (while for quark mixings relations like $\lambda_C^2 \sim m_d/m_s$ are typical). In fact in the basis where charged lepton masses are diagonal, the effective neutrino mass matrix in the HPS case is given by $m_\nu = U_{HPS}\mathrm{diag}(m_1, m_2, m_3)U^T_{HPS}$:

$$m_\nu = \left[\frac{m_3}{2}M_3 + \frac{m_2}{3}M_2 + \frac{m_1}{6}M_1\right] . \quad (9)$$

where:

$$M_3 = \begin{pmatrix} 0 & 0 & 0 \\ 0 & 1 & -1 \\ 0 & -1 & 1 \end{pmatrix}, \quad M_2 = \begin{pmatrix} 1 & 1 & 1 \\ 1 & 1 & 1 \\ 1 & 1 & 1 \end{pmatrix}, \quad M_1 = \begin{pmatrix} 4 & -2 & -2 \\ -2 & 1 & 1 \\ -2 & 1 & 1 \end{pmatrix}. \quad (10)$$

The eigenvalues of m_ν are m_1, m_2, m_3 with eigenvectors $(-2,1,1)/\sqrt{6}$, $(1,1,1)/\sqrt{3}$ and $(0,1,-1)/\sqrt{2}$, respectively. In general, disregarding possible Majorana phases, there are six parameters in a real symmetric matrix like m_ν: here only three are left after the values of the three mixing angles have been fixed à la HPS. For a hierarchical spectrum $m_3 >> m_2 >> m_1$, $m_3^2 \sim \Delta m^2_{atm}$, $m_2^2/m_3^2 \sim \Delta m^2_{sol}/\Delta m^2_{atm}$ and m_1 could be negligible. But also degenerate masses and inverse hierarchy can be reproduced: for example, by taking $m_3 = -m_2 = m_1$ we have a degenerate model, while for $m_1 = -m_2$ and $m_3 = 0$ an inverse hierarchy case is realized (stability under renormalization group running strongly prefers opposite signs for the first and the second eigenvalue which are related to solar oscillations and have the smallest mass squared splitting). From the general expression of the eigenvectors one immediately sees that this mass matrix, independent of the values of m_i, leads to the HPS mixing matrix.

It is interesting to recall that the most general mass matrix, in the basis where charged leptons are diagonal, that corresponds to $\theta_{13} = 0$ and θ_{23} maximal is of the form [31]:

$$m = \begin{pmatrix} x & y & y \\ y & z & w \\ y & w & z \end{pmatrix}, \tag{11}$$

Note that this matrix is symmetric under 2-3 or $\mu - \tau$ exchange. It is however not easy to make a model where the $\mu - \tau$ symmetry applies to the whole lepton sector [32]. Imposing the symmetry on $l^T m_\nu l$ does not work because the Dirac mass term $l^c m_D l$ then produces a charged lepton mixing that completely spoils θ_{23} maximal. For example, in the model [33], the $\mu - \tau$ symmetry is badly broken in the charged lepton mass sector and, as a result, for parameter choices that fit the masses, θ_{23} is not necessarily close to maximal and θ_{13} is not too small: finally the model looks like a "normal" model! Similarly a symmetry $\nu_{\mu R} - \nu_{\tau R}$ in the right-handed (RH) neutrino sector does not lead to a $\mu - \tau$ symmetric neutrino mass matrix after the see-saw. A more elaborate broken symmetry is needed, like a set of discrete broken symmetries that make the charged lepton mass matrix diagonal and, at the same time, the Dirac neutrino mass matrix diagonal and $\mu - \tau$ symmetric and finally the permutational $2-3$ symmetry is in the RR Majorana mass matrix. Thus the idea of a "simple" $\mu - \tau$ symmetry ends up with leading to complicated models.

For $\theta_{13} = 0$ there is no CP violation, so that, disregarding Majorana phases, we can restrict our consideration to real parameters. There are four of them in eq.(11) which correspond to three mass eigenvalues and one remaining mixing angle, θ_{12}. In particular, θ_{12} is given by:

$$\sin^2 2\theta_{12} = \frac{8y^2}{(x-w-z)^2 + 8y^2} \tag{12}$$

In the HPS case θ_{12} is also fixed and an additional parameter, for example x, can be eliminated, leading to:

$$m = \begin{pmatrix} z+w-y & y & y \\ y & z & w \\ y & w & z \end{pmatrix}, \tag{13}$$

It is easy to see that the HPS mass matrix in eqs.(9-10) is indeed of the form in eq.(13).

In the next sections we will present models of tri-bimaximal mixing based on the A4 group. We first introduce A4 and its representations and then we show that this group is particularly suited to the problem.

5 The A4 Group

A4 is the group of the even permutations of 4 objects. It has 4!/2=12 elements. Geometrically, it can be seen as the invariance group of a tethraedron (the odd permutations, for example the exchange of two vertices, cannot be obtained by moving a rigid solid). Let us denote a generic permutation $(1,2,3,4) \to (n_1, n_2, n_3, n_4)$ simply by $(n_1 n_2 n_3 n_4)$. A4 can be generated by two

Table 2. *Characters of A4*

Class	χ^1	$\chi^{1'}$	$\chi^{1''}$	χ^3
C_1	1	1	1	3
C_2	1	ω	ω^2	0
C_3	1	ω^2	ω	0
C_4	1	1	1	-1

basic permutations S and T given by $S = (4321)$ and $T = (2314)$. One checks immediately that:

$$S^2 = T^3 = (ST)^3 = 1 \tag{14}$$

This is called a "presentation" of the group. The 12 even permutations belong to 4 equivalence classes (h and k belong to the same class if there is a g in the group such that $ghg^{-1} = k$) and are generated from S and T as follows:

$$\begin{aligned}
C_1 &: I = (1234) \\
C_2 &: T = (2314), ST = (4132), TS = (3241), STS = (1423) \\
C_3 &: T^2 = (3124), ST^2 = (4213), T^2S = (2431), TST = (1342) \\
C_4 &: S = (4321), T^2ST = (3412), TST^2 = (2143)
\end{aligned} \tag{15}$$

Note that, except for the identity I which always forms an equivalence class in itself, the other classes are according to the powers of T (in C4 S could as well be seen as ST^3).

In a finite group the squared dimensions of the inequivalent irreducible representations add up to N, the number of transformations in the group ($N = 12$ in $A4$). $A4$ has four inequivalent representations: three of dimension one, 1, $1'$ and $1''$ and one of dimension 3. It is immediate to see that the one-dimensional unitary representations are obtained by:

$$\begin{array}{lll}
1 & S = 1 & T = 1 \\
1' & S = 1 & T = e^{i2\pi/3} \equiv \omega \\
1'' & S = 1 & T = e^{i4\pi/3} \equiv \omega^2
\end{array} \tag{16}$$

Note that $\omega = -1/2 + \sqrt{3}i/2$ is the cubic root of 1 and satisfies $\omega^2 = \omega^*$, $1 + \omega + \omega^2 = 0$.

The three-dimensional unitary representation, in a basis where the element S is diagonal, is built up from:

$$S = \begin{pmatrix} 1 & 0 & 0 \\ 0 & -1 & 0 \\ 0 & 0 & -1 \end{pmatrix}, \quad T = \begin{pmatrix} 0 & 1 & 0 \\ 0 & 0 & 1 \\ 1 & 0 & 0 \end{pmatrix}. \tag{17}$$

The characters of a group χ_g^R are defined, for each element g, as the trace of the matrix that maps the element in a given representation R. It is easy to see that equivalent representations have the same characters and that characters have the same value for all elements in an equivalence class. Characters satisfy $\sum_g \chi_g^R \chi_g^{S*} = N\delta^{RS}$. Also, for each element h, the character of h in a direct product of representations is the product of the characters: $\chi_h^{R\otimes S} = \chi_h^R \chi_h^S$ and also is equal to the sum of the characters in each representation that appears in the decomposition of $R \otimes S$. The character table of A4 is given in Table 2 [25]. From this Table one derives that indeed there are no more inequivalent irreducible representations other than 1, $1'$, $1''$ and 3. Also, the multiplication rules are clear: the product of two 3 gives $3 \times 3 = 1 + 1' + 1'' + 3 + 3$ and $1' \times 1' = 1''$, $1' \times 1'' = 1$, $1'' \times 1'' = 1'$ etc. If $3 \sim (a_1, a_2, a_3)$ is a triplet transforming by the matrices in eq.(17) we have that under S: $S(a_1, a_2, a_3)^t = (a_1, -a_2, -a_3)^t$ (here the upper index t indicates transposition) and under T: $T(a_1, a_2, a_3)^t = (a_2, a_3, a_1)^t$. Then, from two such triplets $3_a \sim (a_1, a_2, a_3)$, $3_b \sim (b_1, b_2, b_3)$ the irreducible representations obtained from their product are:

$$1 = a_1 b_1 + a_2 b_2 + a_3 b_3 \tag{18}$$

$$1' = a_1 b_1 + \omega^2 a_2 b_2 + \omega a_3 b_3 \tag{19}$$

$$1'' = a_1 b_1 + \omega a_2 b_2 + \omega^2 a_3 b_3 \tag{20}$$

$$3 \sim (a_2 b_3, a_3 b_1, a_1 b_2) \tag{21}$$

$$3 \sim (a_3 b_2, a_1 b_3, a_2 b_1) \tag{22}$$

In fact, take for example the expression for $1'' = a_1 b_1 + \omega a_2 b_2 + \omega^2 a_3 b_3$. Under S it is invariant and under T it goes into $a_2 b_2 + \omega a_3 b_3 + \omega^2 a_1 b_1 = \omega^2 [a_1 b_1 + \omega a_2 b_2 + \omega^2 a_3 b_3]$ which is exactly the transformation corresponding to $1''$.

In eq.(17) we have the representation 3 in a basis where S is diagonal. It is interesting to go to a basis where instead it is T which is diagonal. This is obtained through the unitary transformation:

$$T' = VTV^\dagger = \begin{pmatrix} 1 & 0 & 0 \\ 0 & \omega & 0 \\ 0 & 0 & \omega^2 \end{pmatrix}, \tag{23}$$

$$S' = VSV^\dagger = \frac{1}{3} \begin{pmatrix} -1 & 2 & 2 \\ 2 & -1 & 2 \\ 2 & 2 & -1 \end{pmatrix}. \tag{24}$$

where:

$$V = \frac{1}{\sqrt{3}} \begin{pmatrix} 1 & 1 & 1 \\ 1 & \omega^2 & \omega \\ 1 & \omega & \omega^2 \end{pmatrix}. \tag{25}$$

The matrix $\sqrt{3}V$ is special in that it is a 3x3 unitary matrix with all entries of unit absolute value. It is interesting that this matrix V was proposed long ago as a possible mixing matrix for neutrinos [34]. We shall see in the following that the matrix V appears in $A4$ models as the unitary transformation that diagonalizes the charged lepton mass matrix.

An obvious representation of $A4$ is obtained by considering the 4x4 matrices that directly realize each permutation. For $S = (4321)$ and $T = (2314)$ we have:

$$S_4 = \begin{pmatrix} 0 & 0 & 0 & 1 \\ 0 & 0 & 1 & 0 \\ 0 & 1 & 0 & 0 \\ 1 & 0 & 0 & 0 \end{pmatrix} , \qquad T_4 = \begin{pmatrix} 0 & 1 & 0 & 0 \\ 0 & 0 & 1 & 0 \\ 1 & 0 & 0 & 0 \\ 0 & 0 & 0 & 1 \end{pmatrix} . \tag{26}$$

The matrices S_4 and T_4 satisfy the relations (14), thus providing a representation of A4. Since the only irreducible representations of A4 are a triplet and three singlets, the 4x4 representation described by S_4 and T_4 is not irreducible. It decomposes into the sum of the invariant singlet plus the triplet representation. This decomposition is realized by the unitary matrix [30] U given by:

$$U = \frac{1}{2} \begin{pmatrix} +1 & +1 & +1 & +1 \\ -1 & +1 & +1 & -1 \\ +1 & -1 & +1 & -1 \\ +1 & +1 & -1 & -1 \end{pmatrix} . \tag{27}$$

This matrix maps S_4 and T_4 into matrices that are block-diagonal:

$$U S_4 U^\dagger = \left(\begin{array}{c|c} 1 & 0 \\ \hline 0 & S \end{array} \right) , \qquad U T_4 U^\dagger = \left(\begin{array}{c|c} 1 & 0 \\ \hline 0 & T \end{array} \right) , \tag{28}$$

where S and T are the generators of the three-dimensional representation in eq.(17).

There is an interesting relation [29] between the A_4 model considered so far and the modular group. This relation could possibly be relevant to understand the origin of the A4 symmetry from a more fundamental layer of the theory. The modular group Γ is the group of linear fractional transformations acting on a complex variable z:

$$z \to \frac{az+b}{cz+d} , \qquad ad - bc = 1 , \tag{29}$$

where a, b, c, d are integers. There are infinite elements in Γ, but all of them can be generated by the two transformations:

$$s: \quad z \to -\frac{1}{z} , \qquad t: \quad z \to z+1 , \tag{30}$$

The transformations s and t in (30) satisfy the relations

$$s^2 = (st)^3 = 1 \tag{31}$$

and, conversely, these relations provide an abstract characterization of the modular group. Since the relations (14) are a particular case of the more general constraint (31), it is clear that A4 is a very small subgroup of the modular group and that the A4 representations discussed above are also representations of the modular group. In string theory the transformations (30) operate in many different contexts. For instance the role of the complex variable z can be played by a field, whose VEV can be related to a physical quantity like a compactification radius or a coupling constant. In that case s in eq. (30) represents a duality transformation and t in eq. (30) represents the transformation associated with an "axionic" symmetry.

A different way to understand the dynamical origin of A_4 was recently presented in ref. [30] where it is shown that the A_4 symmetry can be simply obtained by orbifolding starting from a model in 6 dimensions (6D). In this approach A_4 appears as the remnant of the reduction from 6D to 4D space-time symmetry induced by the special orbifolding adopted. There are 4D branes at the four fixed points of the orbifolding and the tetrahedral symmetry of A_4 connects these branes. The standard model fields have components on the fixed point branes while the scalar fields necessary for the A_4 breaking are in the bulk. Each brane field, either a triplet or a singlet, has components on all of the four fixed points (in particular all components are equal for a singlet) but the interactions are local, i.e. all vertices involve products of field components at the same space-time point. This approach suggests a deep relation between flavour symmetry in 4D and space-time symmetry in extra dimensions.

The orbifolding is defined as follows. We consider a quantum field theory in 6 dimensions, with two extra dimensions compactified on an orbifold T^2/Z_2. We denote by $z = x_5 + ix_6$ the complex coordinate describing the extra space. The torus T^2 is defined by identifying in the complex plane the points related by

$$\begin{aligned} &z \to z+1 \\ &z \to z+\gamma \qquad\qquad \gamma = e^{i\frac{\pi}{3}} \quad , \end{aligned} \tag{32}$$

where our length unit, $2\pi R$, has been set to 1 for the time being. The parity Z_2 is defined by

$$z \to -z \tag{33}$$

and the orbifold T^2/Z_2 can be represented by the fundamental region given by the triangle with vertices $0, 1, \gamma$, see Fig. 1. The orbifold has four fixed points, $(z_1, z_2, z_3, z_4) = (1/2, (1+\gamma)/2, \gamma/2, 0)$. The fixed point z_4 is also represented by the vertices 1 and γ. In the orbifold, the segments labelled by a in Fig. 1, $(0, 1/2)$ and $(1, 1/2)$, are identified and similarly for those labelled by b, $(1, (1+\gamma)/2)$ and $(\gamma, (1+\gamma)/2)$, and those labelled by c, $(0, \gamma/2)$, $(\gamma, \gamma/2)$. Therefore the orbifold is a regular tetrahedron with vertices at the four fixed points.

The symmetry of the uncompactified 6D space time is broken by compactification. Here we assume that, before compactification, the space-time symmetry coincides with the product of 6D translations and 6D proper Lorentz transformations. The compactification breaks part of this symmetry. However, due to the special geometry of our orbifold, a discrete subgroup of rotations and translations in the extra space is left unbroken. This group can be generated by two transformations:

$$\begin{aligned} &\mathscr{S}: \;\; z \to z + \tfrac{1}{2} \\ &\mathscr{T}: \;\; z \to \omega z \qquad\qquad \omega \equiv \gamma^2 \quad . \end{aligned} \tag{34}$$

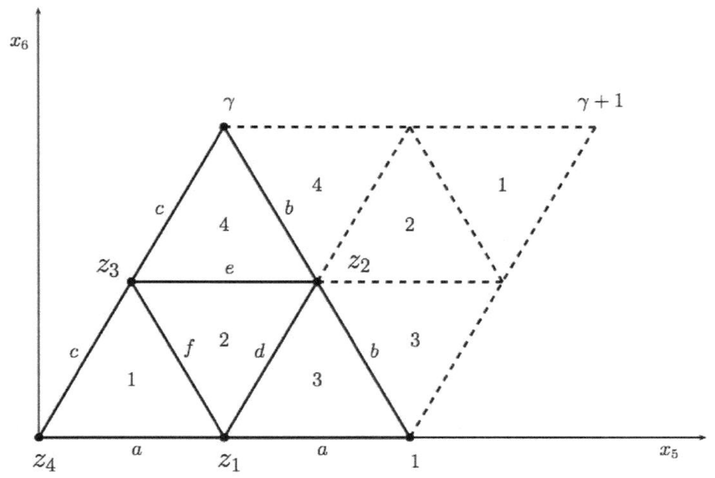

Figure 1. *Orbifold T_2/Z_2. The regions with the same numbers are identified with each other. The four triangles bounded by solid lines form the fundamental region, where also the edges with the same letters are identified. The orbifold T_2/Z_2 is exactly a regular tetrahedron with 6 edges a,b,c,d,e,f and four vertices z_1, z_2, z_3, z_4, corresponding to the four fixed points of the orbifold.*

Indeed $\mathscr{S}$ and $\mathscr{T}$ induce even permutations of the four fixed points:

$$\begin{array}{ll} \mathscr{S}: & (z_1,z_2,z_3,z_4) \to (z_4,z_3,z_2,z_1) \\ \mathscr{T}: & (z_1,z_2,z_3,z_4) \to (z_2,z_3,z_1,z_4) \end{array} , \tag{35}$$

thus generating the group A_4. From the previous equations we immediately verify that $\mathscr{S}$ and $\mathscr{T}$ satisfy the characteristic relations obeyed by the generators of A_4: $\mathscr{S}^2 = \mathscr{T}^3 = (\mathscr{S}\mathscr{T})^3 = 1$. These relations are actually satisfied not only at the fixed points, but on the whole orbifold, as can be easily checked from the general definitions of $\mathscr{S}$ and $\mathscr{T}$ in eq. (34), with the help of the orbifold defining rules in eqs. (32) and (33).

6 Applying A4 to Lepton Masses and Mixings

A typical A4 model works as follows [28], [29]. One assigns leptons to the four inequivalent representations of A4: left-handed lepton doublets l transform as a triplet 3, while the right-handed charged leptons e^c, μ^c and τ^c transform as 1, $1'$ and $1''$, respectively. At this stage we do not introduce RH neutrinos, but later we will discuss a see-saw realization. The flavour symmetry is broken by two real triplets φ and φ' and by a real singlet ξ. These flavon fields are all gauge singlets. We also need one or two ordinary SM Higgs doublets $h_{u,d}$, which we take invariant under A4. The Yukawa interactions in the lepton sector read:

$$\begin{aligned} \mathscr{L}_Y &= y_e e^c(\varphi l) + y_\mu \mu^c(\varphi l)'' + y_\tau \tau^c(\varphi l)' \\ &+ x_a \xi(ll) + x_d(\varphi' ll) + h.c. + \ldots \end{aligned} \tag{36}$$

In our notation, (33) transforms as 1, $(33)'$ transforms as $1'$ and $(33)''$ transforms as $1''$. Also, to keep our notation compact, we use a two-component notation for the fermion fields and we set to 1 the Higgs fields $h_{u,d}$ and the cut-off scale Λ. For instance $y_e e^c(\varphi l)$ stands for $y_e e^c(\varphi l)h_d/\Lambda$, $x_a\xi(ll)$ stands for $x_a\xi(lh_u lh_u)/\Lambda^2$ and so on. The Lagrangian $\mathscr{L}_Y$ contains the lowest order operators in an expansion in powers of $1/\Lambda$. Dots stand for higher dimensional operators that will be discussed later. Some terms allowed by the flavour symmetry, such as the terms obtained by the exchange $\varphi' \leftrightarrow \varphi$, or the term (ll) are missing in $\mathscr{L}_Y$. Their absence is crucial and, in each version of A4 models, is motivated by additional symmetries. For example (ll), being of lower dimension than $(\varphi' ll)$, would be the dominant component, proportional to the identity, of the neutrino mass matrix. In addition to that, the presence of the singlet flavon ξ plays an important role in making the VEV directions of φ and φ' different.

For the model to work it is essential that the fields φ', φ and ξ develop VEVs along the directions:

$$\begin{aligned}\langle\varphi'\rangle &= (v',0,0)\\ \langle\varphi\rangle &= (v,v,v)\\ \langle\xi\rangle &= u \quad .\end{aligned} \tag{37}$$

A crucial part of all serious A4 models is the dynamical generation of this alignment in a natural way. If the alignment is realized, at the leading order of the $1/\Lambda$ expansion, the mass matrices m_l and m_ν for charged leptons and neutrinos are given by:

$$m_l = v_d\frac{v}{\Lambda}\begin{pmatrix} y_e & y_e & y_e \\ y_\mu & y_\mu\omega^2 & y_\mu\omega \\ y_\tau & y_\tau\omega & y_\tau\omega^2 \end{pmatrix} \quad , \tag{38}$$

$$m_\nu = \frac{v_u^2}{\Lambda}\begin{pmatrix} a & 0 & 0 \\ 0 & a & d \\ 0 & d & a \end{pmatrix} \quad , \tag{39}$$

where

$$a \equiv x_a\frac{u}{\Lambda} \quad , \qquad d \equiv x_d\frac{v'}{\Lambda} \quad . \tag{40}$$

Charged leptons are diagonalized by the matrix

$$l \rightarrow Vl = \frac{1}{\sqrt{3}}\begin{pmatrix} 1 & 1 & 1 \\ 1 & \omega^2 & \omega \\ 1 & \omega & \omega^2 \end{pmatrix} l \quad , \tag{41}$$

This matrix was already introduced in eq.(25) as the unitary transformation between the S-diagonal to the T-diagonal 3x3 representation of $A4$. In fact, in this model, the S-diagonal basis is the Lagrangian basis and the T diagonal basis is that of diagonal charged leptons. The great virtue of $A4$ is to immediately produce the special unitary matrix V as the diagonalizing matrix of charged leptons and also to allow a singlet made up of three triplets, $(\phi' ll) = \phi'_1 l_2 l_3 + \phi'_2 l_3 l_1 + \phi'_3 l_1 l_2$ which leads, for the alignment in eq. (37), to the right neutrino mass matrix to finally obtain the HPS mixing matrix.

The charged fermion masses are given by:

$$m_e = \sqrt{3} y_e v_d \frac{v}{\Lambda} \quad , \qquad m_\mu = \sqrt{3} y_\mu v_d \frac{v}{\Lambda} \quad , \qquad m_\tau = \sqrt{3} y_\tau v_d \frac{v}{\Lambda} \quad . \tag{42}$$

We can easily obtain in a a natural way the observed hierarchy among m_e, m_μ and m_τ by introducing an additional $\mathrm{U}(1)_F$ flavour symmetry under which only the right-handed lepton sector is charged. We assign F-charges 0, 2 and $3 \div 4$ to τ^c, μ^c and e^c, respectively. By assuming that a flavon θ, carrying a negative unit of F, acquires a VEV $\langle\theta\rangle/\Lambda \equiv \lambda < 1$, the Yukawa couplings become field dependent quantities $y_{e,\mu,\tau} = y_{e,\mu,\tau}(\theta)$ and we have

$$y_\tau \approx O(1) \quad , \qquad y_\mu \approx O(\lambda^2) \quad , \qquad y_e \approx O(\lambda^{3\div4}) \quad . \tag{43}$$

In the flavour basis the neutrino mass matrix reads [notice that the change of basis induced by V, because of the Majorana nature of neutrinos, will in general change the relative phases of the eigenvalues of m_ν (compare eq.(39) with eq.(44))]:

$$m_\nu = \frac{v_u^2}{\Lambda} \left(\begin{array}{ccc} a+2d/3 & -d/3 & -d/3 \\ -d/3 & 2d/3 & a-d/3 \\ -d/3 & a-d/3 & 2d/3 \end{array} \right) \quad , \tag{44}$$

and is diagonalized by the transformation:

$$U^T m_\nu U = \frac{v_u^2}{\Lambda} \mathtt{diag}(a+d, a, -a+d) \quad , \tag{45}$$

with

$$U = \left(\begin{array}{ccc} \sqrt{2/3} & 1/\sqrt{3} & 0 \\ -1/\sqrt{6} & 1/\sqrt{3} & -1/\sqrt{2} \\ -1/\sqrt{6} & 1/\sqrt{3} & +1/\sqrt{2} \end{array} \right) \quad . \tag{46}$$

The leading order predictions are $\tan^2\theta_{23} = 1$, $\tan^2\theta_{12} = 0.5$ and $\theta_{13} = 0$. The neutrino masses are $m_1 = a+d$, $m_2 = a$ and $m_3 = -a+d$, in units of v_u^2/Λ. We can express $|a|$, $|d|$ in terms of $r \equiv \Delta m^2_{sol}/\Delta m^2_{atm} \equiv (|m_2|^2 - |m_1|^2)/|m_3|^2 - |m_1|^2)$, $\Delta m^2_{atm} \equiv |m_3|^2 - |m_1|^2$ and $\cos\Delta$, Δ being the phase difference between the complex numbers a and d:

$$\begin{aligned} \sqrt{2}|a|\frac{v_u^2}{\Lambda} &= \frac{-\sqrt{\Delta m^2_{atm}}}{2\cos\Delta\sqrt{1-2r}} \\ \sqrt{2}|d|\frac{v_u^2}{\Lambda} &= \sqrt{1-2r}\sqrt{\Delta m^2_{atm}} \quad . \end{aligned} \tag{47}$$

To satisfy these relations a moderate tuning is needed in this model. Due to the absence of (ll) in eq. (36) which we will motivate in the next section, a and d are of the same order in $1/\Lambda$, see eq. (40). Therefore we expect that $|a|$ and $|d|$ are close to each other and, to satisfy eqs. (47), $\cos\Delta$ should be negative and of order one. We obtain:

$$\begin{aligned} |m_1|^2 &= \left[-r + \frac{1}{8\cos^2\Delta(1-2r)}\right]\Delta m^2_{atm} \\ |m_2|^2 &= \frac{1}{8\cos^2\Delta(1-2r)}\Delta m^2_{atm} \\ |m_3|^2 &= \left[1 - r + \frac{1}{8\cos^2\Delta(1-2r)}\right]\Delta m^2_{atm} \end{aligned} \tag{48}$$

If $\cos\Delta = -1$, we have a neutrino spectrum close to hierarchical:

$$|m_3| \approx 0.053 \text{ eV} \quad , \qquad |m_1| \approx |m_2| \approx 0.017 \text{ eV} \quad . \tag{49}$$

In this case the sum of neutrino masses is about 0.087 eV. If $\cos\Delta$ is accidentally small, the neutrino spectrum becomes degenerate. The value of $|m_{ee}|$, the parameter characterizing the violation of total lepton number in neutrinoless double beta decay, is given by:

$$|m_{ee}|^2 = \left[-\frac{1+4r}{9} + \frac{1}{8\cos^2\Delta(1-2r)}\right] \Delta m^2_{atm} \quad . \tag{50}$$

For $\cos\Delta = -1$ we get $|m_{ee}| \approx 0.005$ eV, at the upper edge of the range allowed for normal hierarchy, but unfortunately too small to be detected in a near future. Independently from the value of the unknown phase Δ we get the relation:

$$|m_3|^2 = |m_{ee}|^2 + \frac{10}{9}\Delta m^2_{atm}\left(1 - \frac{r}{2}\right) \quad , \tag{51}$$

which is a prediction of this model.

7 A4 model with an extra dimension

One of the problems we should solve in the quest for the correct alignment is that of keeping neutrino and charged lepton sectors separate, allowing φ and φ' to take different VEVs and also forbidding the exchange of one with the other in interaction terms. One possibility is that this separation is achieved by means of an extra spatial dimension. The space-time is assumed to be five-dimensional, the product of the four-dimensional Minkowski space-time times an interval going from $y = 0$ to $y = L$. At $y = 0$ and $y = L$ the space-time has two four-dimensional boundaries, called "branes". The idea is that matter SU(2) singlets such as e^c, μ^c, τ^c are localized at $y = 0$, while SU(2) doublets, such as l are localized at $y = L$ (see Fig. 2). Neutrino masses arise from local operators at $y = L$. Charged lepton masses are produced by non-local effects involving both branes. The simplest possibility is to introduce a bulk fermion, depending on all space-time coordinates, that interacts with e^c, μ^c, τ^c at $y = 0$ and with l at $y = L$. The exchange of such a fermion can provide the desired non-local coupling between right-handed and left-handed ordinary fermions. Finally, assuming that φ and (φ', ξ) are localized respectively at $y = 0$ and $y = L$, one obtains a natural separation between the two sectors.

Such a separation also greatly simplifies the vacuum alignment problem. One can determine the minima of two scalar potentials V_0 and V_L, depending only, respectively, on φ and (φ', ξ). Indeed, it is shown that there are whole regions of the parameter space where $V_0(\varphi)$ and $V_L(\varphi', \xi)$ have the minima given in eq. (37). Notice that in the present setup dealing with a discrete symmetry such as A4 provides a great advantage as far as the alignment problem is concerned. A continuous flavour symmetry such as, for instance, SO(3) would need some extra structure to achieve the desired alignment. Indeed the potential energy $\int d^4x[V_0(\varphi) + V_L(\varphi', \xi)]$ would be invariant under a much bigger symmetry, $SO(3)_0\times$ $SO(3)_L$, with the $SO(3)_0$ acting on φ and leaving (φ', ξ) invariant and vice-versa for $SO(3)_L$. This symmetry would remove any alignment between the VEVs of φ and those of (φ', ξ). If, for instance, (37) is a minimum of the potential energy, then any other configuration obtained by acting on (37) with $SO(3)_0\times$

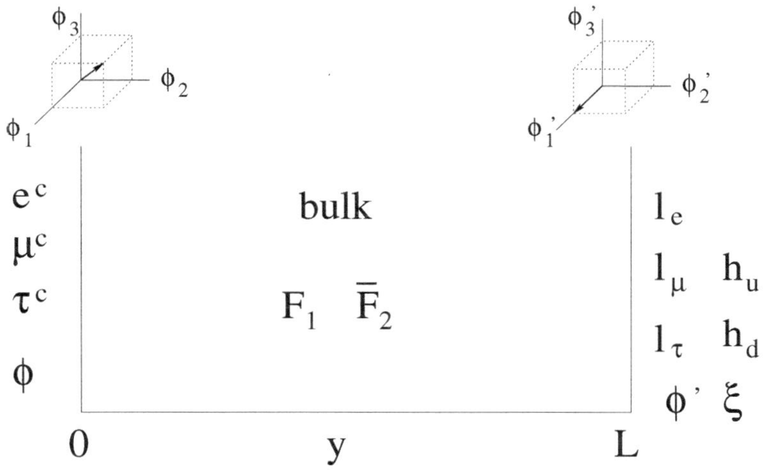

Figure 2. *Fifth dimension and localization of scalar and fermion fields. The symmetry breaking sector includes the A4 triplets φ and φ', localized at the opposite ends of the interval. Their VEVs are dynamically aligned along the directions shown at the top of the figure.*

$SO(3)_L$ would also be a minimum and the relative orientation between the two sets of VEVs would be completely undetermined. A discrete symmetry such as A4 does not have this problem, because applying separate A4 transformations to the minimum solutions on each brane a finite number of degenerate vacua is obtained which can be shown to correspond to the same physics apart from redefinitions of fields and parameters.

The Lagrangian in 5 dimensions includes a bulk fermion field $F(x,y) = (F_1, \overline{F_2})$, singlet under SU(2) with hypercharge $Y = -1$ and transforming as a triplet of A4. One also imposes a discrete Z_4 symmetry under which $(f^c, l, F, \varphi, \varphi', \xi)$ transform into $(-if^c, il, iF, \varphi, -\varphi', -\xi)$. The complete action is

$$\begin{aligned} S &= \int d^4x dy \left\{ \left[iF_1\sigma^\mu\partial_\mu\overline{F}_1 + iF_2\sigma^\mu\partial_\mu\overline{F}_2 + \frac{1}{2}(F_2\partial_yF_1 - \partial_yF_2F_1 + h.c.) \right] \right. \\ &- M(F_1F_2 + \overline{F}_1\overline{F}_2) \\ &+ V_0(\varphi)\delta(y) + V_L(\varphi', \xi)\delta(y-L) \\ &+ \left[Y_e e^c(\varphi F_1) + Y_\mu \mu^c(\varphi F_1)'' + Y_\tau \tau^c(\varphi F_1)' + h.c. \right] \delta(y) \\ &+ \left. \left[\frac{x_a}{\Lambda^2}\xi(ll)h_uh_u + \frac{x_d}{\Lambda^2}(\varphi' ll)h_uh_u + Y_L(F_2 l)h_d + h.c. \right] \delta(y-L) \right\} + \ldots \quad , \end{aligned} \tag{52}$$

where the constants Y have mass dimension -1/2. The first two lines represent the five-dimensional kinetic and mass terms of the bulk field F. The third line is the scalar potential and the remaining terms are the lowest order invariant operators localized at the two branes. Dots stand for the kinetic terms of $f^c, l, \varphi, \varphi', \xi$ and for higher-dimensional operators.

The potential energy is given, at lowest order by:

$$U = \int d^4x \left[V_0(\varphi) + V_L(\varphi', \xi) \right] \quad , \tag{53}$$

and, under the conditions discussed above, is minimized by eqs. (37) [28]. It is clear that at lowest order φ and (φ', ξ) are strictly separated.

We now discuss the effects of the tree-level exchange of F. To this purpose we consider the equations of motion for (F_1, F_2):

$$\begin{aligned} i\sigma^\mu \partial_\mu \overline{F}_2 + \partial_y F_1 - M F_1 &= 0 \\ i\sigma^\mu \partial_\mu \overline{F}_1 - \partial_y F_2 - M F_2 &= 0 \end{aligned} \quad (54)$$

If M is large and positive, we can prove that all the modes contained in (F_1, F_2) become heavy, at a scale greater than or comparable to $1/L$, which we assume to be much higher than the electroweak scale. If we are only interested in energies much lower than $1/L$, we can solve the equations of motion in the static approximation, by neglecting the four-dimensional kinetic term:

$$\begin{aligned} F_1(y) &= F_1(L) e^{M(y-L)} \\ F_2(y) &= F_2(0) e^{-My} \quad . \end{aligned} \quad (55)$$

These equations must be supplemented with appropriate boundary conditions, which can be identified by varying the action S with respect the fields (F_1, F_2). As a final result, as shown in detail in ref. [28], in lowest order approximation the Lagrangian $\mathscr{L}_Y$ of eq. (36) is reproduced and the general discussion applies.

We also recall that, to account for the observed hierarchy of the charged lepton masses, one has to include an additional U(1) flavour symmetry. Therefore, in the present picture, the quantities $Y_{e,\mu,\tau}$ stand for:

$$Y_e = \tilde{Y}_e \left(\frac{\theta}{\Lambda}\right)^4 \quad , \quad Y_\mu = \tilde{Y}_\mu \left(\frac{\theta}{\Lambda}\right)^2 \quad , \quad Y_\tau = \tilde{Y}_\tau \quad , \quad (56)$$

where $\tilde{Y}_{e,\mu,\tau}$ are field-independent constants having similar values. After spontaneous breaking of $U(1)$, the Yukawa couplings y_f possess the desired hierarchy.

8 A4 model with SUSY in 4 Dimensions

We now discuss an alternative supersymmetric solution to the vacuum alignment problem [29]. In a SUSY context, the right-hand side of eq. (36) should be interpreted as the superpotential w_l of the theory, in the lepton sector:

$$\begin{aligned} w_l &= y_e e^c (\varphi l) + y_\mu \mu^c (\varphi l)'' + y_\tau \tau^c (\varphi l)' + \\ &+ (x_a \xi + \tilde{x}_a \tilde{\xi})(ll) + x_b (\varphi' ll) + h.c. + \ldots \end{aligned} \quad (57)$$

where dots stand for higher dimensional operators and where we have also added an additional A4-invariant singlet $\tilde{\xi}$. Such a singlet does not modify the structure of the mass matrices discussed previously, but plays an important role in the vacuum alignment mechanism. A key observation is that the superpotential w_l is invariant not only with respect to the gauge symmetry SU(2)$\times$ U(1) and the flavour symmetry $U(1)_F \times A_4$, but also under a discrete Z_3 symmetry and a continuous $U(1)_R$ symmetry under which the fields transform as shown in the following table.

Field	l	e^c	μ^c	τ^c	$h_{u,d}$	φ	φ'	ξ	$\tilde{\xi}$	φ_0	φ_0'	ξ_0
A4	3	1	1'	1''	1	3	3	1	1	3	3	1
Z_3	ω	ω^2	ω^2	ω^2	1	1	ω	ω	ω	1	ω	ω
$U(1)_R$	1	1	1	1	0	0	0	0	0	2	2	2

We see that the Z_3 symmetry explains the absence of the term (ll) in w_l: such a term transforms as ω^2 under Z_3 and need to be compensated by the field ξ in our construction. At the same time Z_3 does not allow the interchange between φ and φ', which transform differently under Z_3. The singlets ξ and $\tilde{\xi}$ have the same transformation properties under all symmetries and, as we shall see, in a finite range of parameters, the VEV of $\tilde{\xi}$ vanishes and does not contribute to neutrino masses. Charged leptons and neutrinos acquire masses from two independent sets of fields. If the two sets of fields develop VEVs according to the alignment described in eq. (37), then the desired mass matrices follow.

Finally, there is a continuous $U(1)_R$ symmetry that contains the usual R-parity as a subgroup. Suitably extended to the quark sector, this symmetry forbids the unwanted dimension two and three terms in the superpotential that violate baryon and lepton number at the renormalizable level. The $U(1)_R$ symmetry allows us to classify fields into three sectors. There are "matter fields" such as the leptons l, e^c, μ^c and τ^c, which occur in the superpotential through bilinear combinations. There is a "symmetry breaking sector" including the higgs doublets $h_{u,d}$ and the flavons φ, φ', $(\xi, \tilde{\xi})$. Finally, there are "driving fields" such as φ_0, φ_0' and ξ_0 that allow the construction of a non-trivial scalar potential in the symmetry breaking sector. Since driving fields have R-charge equal to two, the superpotential is linear in these fields.

The full superpotential of the model is

$$w = w_l + w_d \tag{58}$$

where, at leading order in a $1/\Lambda$ expansion, w_l is given by eq. (57) and the "driving" term w_d reads:

$$\begin{aligned} w_d &= M(\varphi_0\varphi) + g(\varphi_0\varphi\varphi) + g_1(\varphi_0'\varphi'\varphi') + g_2\tilde{\xi}(\varphi_0'\varphi') + g_3\xi_0(\varphi'\varphi') \\ &+ g_4\xi_0\xi^2 + g_5\xi_0\xi\tilde{\xi} + g_6\xi_0\tilde{\xi}^2 \quad . \end{aligned} \tag{59}$$

At this level there is no fundamental distinction between the singlets ξ and $\tilde{\xi}$. Thus we are free to define $\tilde{\xi}$ as the combination that couples to $(\varphi_0'\varphi')$ in the superpotential w_d. We notice that at the leading order there are no terms involving the Higgs fields $h_{u,d}$. We assume that the electroweak symmetry is broken by some mechanism, such as radiative effects when SUSY is broken. It is interesting that at the leading order the electroweak scale does not mix with the potentially large scales u, v and v'. The scalar potential is given by:

$$V = \sum_i \left|\frac{\partial w}{\partial \phi_i}\right|^2 + m_i^2|\phi_i|^2 + \ldots \tag{60}$$

where ϕ_i denote collectively all the scalar fields of the theory, m_i^2 are soft masses and dots stand for D-terms for the fields charged under the gauge group and possible additional soft breaking terms. Since m_i are expected to be much smaller than the mass scales involved in w_d, it makes sense to minimize V in the supersymmetric limit and to account for soft breaking effects subsequently. A detailed minimization analysis, presented in ref. [29], shows that the desired alignment solution is indeed realized. In ref. [30] we have shown that it is straightforward to reformulate this SUSY model in the approach where the A4 symmetry is derived from orbifolding.

9 Corrections to the Lowest Approximation

The results of the previous sections hold to first approximation. Higher-dimensional operators, suppressed by additional powers of the cut-off Λ, can be added to the leading terms in the lagrangian. These corrections have been classified and discussed in detail in refs. [28], [29]. They are completely under control in our models and can be made negligibly small without any fine-tuning: one only needs to assume that the VEV's are sufficiently smaller than the cutoff Λ. Higher-order operators contribute corrections to the charged lepton masses, to the neutrino mass matrix and to the vacuum alignment. These corrections, suppressed by powers of VEVs/Λ, with different exponents in different versions of A4 models, affect all the relevant observables with terms of the same order: s_{13}, s_{12}, s_{23}, r. If we require that the subleading terms do not spoil the leading order picture, these deviations should not be larger than about 0.05. This can be inferred by the agreement of the HPS value of $\tan^2\theta_{12}$ with the experimental value, from the present bound on θ_{13} or from requiring that the corrections do not exceed the measured value of r. In the SUSY model, where the largest corrections are linear in VEVs/Λ [29], this implies the bound

$$\frac{v}{\Lambda} \approx \frac{v'}{\Lambda} \approx \frac{u}{\Lambda} < 0.05 \tag{61}$$

which does not look unreasonable, for example if VEVs $\sim M_{GUT}$ and $\Lambda \sim M_{Pl}$.

10 See-saw Realization

We can easily modify the previous model to implement the see-saw mechanism. We introduce conjugate right-handed neutrino fields ν^c transforming as a triplet of A4 and we modify the transformation law of the other fields according to the following table:

Field	ν^c	φ'	ξ	$\tilde{\xi}$	φ'_0	ξ_0
A4	3	3	1	1	3	1
Z_3	ω^2	ω^2	ω^2	ω^2	ω^2	ω^2
$U(1)_R$	1	0	0	0	2	2

The superpotential becomes

$$w = w_l + w_d \tag{62}$$

where the 'driving' part is unchanged, whereas w_l is now given by:

$$\begin{aligned} w_l &= y_e e^c(\varphi l) + y_\mu \mu^c(\varphi l)'' + y_\tau \tau^c(\varphi l)' + y(\nu^c l) + (x_A \xi + \tilde{x}_A \tilde{\xi})(\nu^c \nu^c) \\ &+ x_B(\varphi' \nu^c \nu^c) + h.c. + \ldots \end{aligned} \tag{63}$$

dots denoting higher-order contributions. The vacuum alignment proceeds exactly as discussed in section 8 and also the charged lepton sector is unaffected by the modifications. In the neutrino sector, after electroweak and A4 symmetry breaking we have Dirac and Majorana masses:

$$m_\nu^D = y v_u \mathbf{1}, \;\; M = \begin{pmatrix} A & 0 & 0 \\ 0 & A & B \\ 0 & B & A \end{pmatrix} u \;\; , \tag{64}$$

where $\mathbf{1}$ is the unit 3×3 matrix and

$$A \equiv 2x_A \quad , \qquad B \equiv 2x_B \frac{v}{u} \quad . \tag{65}$$

The mass matrix for light neutrinos is $m_\nu = (m_\nu^D)^T M^{-1} m_\nu^D$ with eigenvalues

$$m_1 = \frac{y^2}{A+B} \frac{v_u^2}{u} \quad , \qquad m_2 = \frac{y^2}{A} \frac{v_u^2}{u} \quad , \qquad m_3 = \frac{y^2}{A-B} \frac{v_u^2}{u} \quad . \tag{66}$$

The mixing matrix is the HPS one, eq. (8). In the presence of a see-saw mechanism both normal and inverted hierarchies in the neutrino mass spectrum can be realized. If we call Φ the relative phase between the complex number A and B, then $\cos\Phi > -|B|/2|A|$ is required to have $|m_2| > |m_1|$. In the interval $-|B|/2|A| < \cos\Phi \leq 0$, the spectrum is of inverted hierarchy type, whereas in $|B|/2|A| \leq \cos\Phi \leq 1$ the neutrino hierachy is of normal type. It is interesting that this model is an example of a model with inverse hierarchy, realistic θ_{12} and θ_{23} and, at least in a first approximation, $\theta_{13} = 0$. The quantity $|B|/2|A|$ cannot be too large, otherwise the ratio r cannot be reproduced. When $|B| \ll |A|$ the spectrum is quasi degenerate. When $|B| \approx |A|$ we obtain the strongest hierarchy. For instance, if $B = -2A + z$ ($|z| \ll |A|, |B|$), we find the following spectrum:

$$\begin{aligned} |m_1|^2 &\approx \Delta m^2_{atm}(\frac{9}{8} + \frac{1}{12} r), \\ |m_2|^2 &\approx \Delta m^2_{atm}(\frac{9}{8} + \frac{13}{12} r), \\ |m_3|^2 &\approx \Delta m^2_{atm}(\frac{1}{8} + \frac{1}{12} r). \end{aligned} \tag{67}$$

When $B = A + z$ ($|z| \ll |A|, |B|$), we obtain:

$$\begin{aligned} |m_1|^2 &\approx \Delta m^2_{atm}(\frac{1}{3} r), \\ |m_2|^2 &\approx \Delta m^2_{atm}(\frac{4}{3} r), \\ |m_3|^2 &\approx \Delta m^2_{atm}(1 - \frac{1}{3} r). \end{aligned} \tag{68}$$

These results are affected by higher-order corrections induced by non renormalizable operators with similar results as in the version with no see-saw. In conclusion, the symmetry structure of the model is fully compatible with the see-saw mechanism.

11 Quarks

There are several possibilities to include quarks. At first sight the most appealing one is to adopt for quarks the same classification scheme under A4 that we have used for leptons. Thus we tentatively assume that left-handed quark doublets q transform as a triplet 3, while the right-handed quarks (u^c, d^c), (c^c, s^c) and (t^c, b^c) transform as 1, 1′ and 1”, respectively. We can similarly extend to quarks the transformations of Z_3 and $U(1)_R$ given for leptons in the table of section 8. The superpotential for quarks reads:

$$\begin{aligned} w_q &= y_d d^c(\varphi q) + y_s s^c(\varphi q)'' + y_b b^c(\varphi q)' \\ &+ y_u u^c(\varphi q) + y_c c^c(\varphi q)'' + y_t t^c(\varphi q)' + h.c. + \ldots \end{aligned} \tag{69}$$

It is interesting to note that such an extrapolation to quarks leads to a diagonal CKM mixing matrix in first approximation [25, 26, 29, 35]. In fact, starting from eq. (69) and proceeding as described in detail for the lepton sector, we see that the up quark and down quark mass matrices are separately diagonal with mass eigenvalues which are left unspecified by A4 and with a hierarchy that could be accomodated by a suitable $U(1)_F$ set of charge assignments for quarks. Thus the V_{CKM} matrix is the identity in leading order, providing a good first order approximation.

The problems come when we discuss non-leading corrections. As seen in section 9, first-order corrections to the lepton sector should be typically below 0.05, approximately the square of the Cabibbo angle. Also, by inspecting these corrections more closely, we see that, up to very small terms of order $y^2_{u(d)}/y^2_{t(b)}$ and $y^2_{c(s)}/y^2_{t(b)}$, all corrections are the same in the up and down sectors and therefore they almost exactly cancel in the mixing matrix V_{CKM}. We conclude that, if one insists in adopting for quarks the same flavour properties as for leptons, than new sources of A4 breaking are needed in order to produce an acceptable V_{CKM}.

The A4 classification for quarks and leptons discussed in this section, which leads to an appealing first approximation with $V_{CKM} \sim 1$ for quark mixing and to U_{HPS} for neutrino mixings, is not compatible with A4 commuting with SU(5) or SO(10). In fact for this to be true all particles in a representation of SU(5) should have the same A4 classification. But, for example, both the $Q = (u,d)_L$ LH quark doublet and the RH charged leptons l^c belong to the 10 of SU(5), yet they have different A4 transformation properties. In a recent paper [36] the possibility of classifying all fermion multiplets as triplets was advanced. But the crucial issues of the correct alignment and of reproducing in a natural way the observed hierarchy of, for example, the charged leptons were not addressed and are difficult to realize in this case.

12 Conclusion

In the last decade we have learnt a lot about neutrino masses and mixings. A list of important conclusions have been reached. Neutrinos are not all massless but their masses are very small. Probably masses are small because neutrinos are Majorana particleswith masses inversely proportional to the large scale M of lepton number violation. It is quite remarkable that M is empirically close to $10^{14-15} GeV$ not far from M_{GUT}, so thatneutrino masses fit well in the SUSY GUT picture. Also out of equilibrium decays with CP and L violation of heavy RH neutrinos can produce a B-L asymmetry, then converted near the weak scale by instantons into an amount of B asymmetry compatible with observations (baryogenesis via leptogenesis) [4], [37]. It has been established that neutrinos are not a significant component of dark matter in the Universe. We have also understood there is no contradiction between large neutrino mixings and small quark mixings, even in the context of GUTs.

This is a very impressive list of achievements. Coming to a detailed analysis of neutrino masses and mixings a very long collection of models have been formulated over the years. With a continuous improvement of the data and a progressive narrowing of the values of the mixing angles most of the models have been discarded by experiment. Still the missing elements in the picture like, for example, the scale of the average neutrino m^2, the pattern of the spectrum (degenerate or inverse or normal hierarchy) and the value of θ_{13} have left many different viable alternatives for models. It certainly is a reason of satisfaction that so much has been learnt recently from experiments on neutrino mixings. By now, besides the detailed knowledge of the entries of the V_{CKM} matrix we also have a reasonable determination of the neutrino mixing matrix. It is remarkable that neutrino and quark mixings have such a different qualitative pattern. One could have imagined that neutrinos would bring a decisive boost towards the formulation of a comprehensive understanding of fermion masses and mixings. In reality it is frustrating that no real illumination was sparked on the problem of flavour. We can reproduce in many different ways the observations but we have not yet been able to single out a unique and convincing baseline for the understanding of fermion masses and mixings. In spite of many interesting ideas and the formulation of many elegant models, some of them presented in these lectures, the mysteries of the flavour structure of the three generations of fermions have not been much unveiled.

Acknowledgments

It is a very pleasant duty for me to most warmly thank the Organizers of the School for their kind invitation and for the great hospitality offered to me in St. Andrews. This work has been partly supported by by the European Commission under contract MRTN-CT-2004-503369.

References

[1] G. Altarelli and F. Feruglio, New J. Phys. **6** (2004) 106, [arXiv:hep-ph/0405048].

[2] B. Kayser, these Proceedings.

[3] S. Pastor, these Proceedings.

[4] W. Buchmuller, these Proceedings.

[5] G. L. Fogli et al, hep-ph/0506083, hep-ph/0506307.

[6] W.-M. Yao et al. (Particle Data Book), J. Phys. G 33, 1 (2006).

[7] B. Bornschein, these Proceedings.

[8] K. Zuber, these Proceedings.

[9] C. E. Aalseth at al, Phys. Rev. D **65**, 092007 (2002).

[10] C. Arnaboldi et al, Phys. Rev. Lett. **95**, 142501 (2005).

[11] For a recent review see, for example, J. Lesgourgues and S. Pastor, Phys. Rept. **429** (2006) 307, [arXiv:astro-ph/0603494]

[12] U. Seljak, A. Slosar and P. McDonald, astro-ph/0604335 and refs. therein.

[13] See the lectures by D. Harris, Y. Kuno and E. Blucher, these Proceedings.

[14] H. V. Klapdor-Kleingrothaus, A. Dietz, I. V. Krivosheina and O. Chkvorets, Phys. Lett. B **586** (2004) 198, [arXiv:hep-ph/0404088].

[15] L. J. Hall, H. Murayama and N. Weiner, Phys. Rev. Lett. **84**, 2572 (2000), [arXiv:hep-ph/9911341];

[16] N. Irges, S. Lavignac and P. Ramond, Phys. Rev. D **58**, 035003 (1998);

[17] P.H. Frampton, S.T. Petcov and W. Rodejohann, Nucl. Phys. B687 (2004) 31, [arXiv:hep-ph0401206]; G. Altarelli, F. Feruglio and I. Masina, Nucl. Phys. B689 (2004) 157, [arXiv:hep-ph0402155]; A. Romanino, Phys. Rev. D70 (2004) 013003,[arXiv:hep-ph0402258].

[18] M. Raidal, Phys.Rev.Lett.93,161801,2004, [arXiv: hep-ph/0404046].

[19] H. Minakata and A. Smirnov, Phys.Rev.D70:073009,2004, [arXiv: hep-ph/0405088].

[20] W. Grimus and L. Lavoura, hep-ph/0410279; G. Altarelli and R. Franceschini, hep-ph/0512202.

[21] G. Altarelli, F. Feruglio and I. Masina, JHEP 0301:035,2003,[ArXiv: hep-ph/0210342].

[22] P. F. Harrison, D. H. Perkins and W. G. Scott, Phys. Lett. B **530** (2002) 167, [arXiv:hep-ph/0202074]; P. F. Harrison and W. G. Scott, Phys. Lett. B **535** (2002) 163, [arXiv:hep-ph/0203209]; Z. z. Xing, Phys. Lett. B **533** (2002) 85, [arXiv:hep-ph/0204049]; P. F. Harrison and W. G. Scott, Phys. Lett. B **547** (2002) 219, [arXiv:hep-ph/0210197]. P. F. Harrison and W. G. Scott, Phys. Lett. B **557** (2003) 76, [arXiv:hep-ph/0302025]. P. F. Harrison and W. G. Scott, hep-ph/0402006; P. F. Harrison and W. G. Scott,hep-ph/0403278.

[23] S. F. King, JHEP **0508** (2005) 105, [arXiv:hep-ph/0506297]; I. de Medeiros Varzielas and G. G. Ross, arXiv:hep-ph/0507176. S. F. King and M. Malinsky, hep-ph/0608021.

[24] J. Matias and C. P. Burgess, JHEP **0509** (2005) 052, [arXiv:hep-ph/0508156]; S. Luo and Z. z. Xing, hep-ph/0509065; W. Grimus and L. Lavoura, hep-ph/0509239; F. Caravaglios and S. Morisi, hep-ph/0510321; I . de Medeiros Varzielas, S. F. King and G. G. Ross, hep-ph/0512313; hep-ph/0607045; C. Hagedorn, M. Lindner and R. N. Mohapatra, JHEP **0606** (2006) 042, [arXiv:hep-ph/0602244]; P. Kovtun and A. Zee, Phys. Lett. B **640** (2006) 37, [arXiv:hep-ph/0604169]; R. N. Mohapatra, S. Nasri and H. B. Yu, Phys. Lett. B **639** (2006) 318 [arXiv:hep-ph/0605020]; Z. z. Xing, H. Zhang and S. Zhou, Phys. Lett. B **641** (2006) 189 [arXiv:hep-ph/0607091]; N. Haba, A. Watanabe and K. Yoshioka, hep-ph/0603116.

[25] E. Ma and G. Rajasekaran, Phys. Rev. D **64** (2001) 113012, [arXiv:hep-ph/0106291];

[26] E. Ma, Mod. Phys. Lett. A **17** (2002) 627, [arXiv:hep-ph/0203238].

[27] K. S. Babu, E. Ma and J. W. F. Valle, Phys. Lett. B **552** (2003) 207, [arXiv:hep-ph/0206292]. M. Hirsch, J. C. Romao, S. Skadhauge, J. W. F. Valle and A. Villanova del Moral, hep-ph/0312244, hep-ph/0312265; E. Ma, hep-ph/0404199. E. Ma, Phys. Rev. D **70** (2004) 031901; E. Ma hep-ph/0409075; E. Ma, New J. Phys. **6** (2004) 104; S. L. Chen, M. Frigerio and E. Ma, hep-ph/0504181; E. Ma, Phys. Rev. D **72** (2005) 037301, [arXiv:hep-ph/0505209]; K. S. Babu and

X. G. He, hep-ph/0507217; A. Zee, Phys. Lett. B **630** (2005) 58, [arXiv:hep-ph/0508278]; E. Ma, Mod. Phys. Lett. A **20** (2005) 2601, [arXiv:hep-ph/0508099]; E. Ma, hep-ph/0511133; S. K. Kang, Z. z. Xing and S. Zhou, hep-ph/0511157. X. G. He, Y. Y. Keum and R. R. Volkas, JHEP **0604** (2006) 039, [arXiv:hep-ph/0601001]; B. Adhikary et al, Phys. Lett. B **638** (2006) 345, [arXiv:hep-ph/0603059]; L. Lavoura and H. Kuhbock, hep-ph/0610050.

[28] G. Altarelli and F. Feruglio, Nucl. Phys. B **720** (2005) 64, [arXiv:hep-ph/0504165].

[29] G. Altarelli and F. Feruglio, Nucl. Phys. B **741** (2006) 215, [arXiv:hep-ph/0512103].

[30] G. Altarelli, F. Feruglio and Y. Lin, hep-ph/0610165.

[31] See, for example, W. Grimus and L. Lavoura, hep-ph/0305046.

[32] T. Fukuyama and H. Nishiura, hep-ph/9702253; R. N. Mohapatra and S. Nussinov, Phys. Rev. **D 60**, 013002 (1999); E. Ma and M. Raidal, Phys. Rev. Lett. **87**, 011802 (2001); C. S. Lam, hep-ph/0104116; T. Kitabayashi and M. Yasue, Phys.Rev. **D67** 015006 (2003); W. Grimus and L. Lavoura, hep-ph/0309050; Y. Koide, Phys.Rev. **D69**, 093001 (2004); A. Ghosal, hep-ph/0304090; W. Grimus et al, hep-ph/0408123; A. de Gouvea, Phys.Rev. **D69**, 093007 (2004); R. N. Mohapatra and W. Rodejohann, Phys. Rev. **D 72**, 053001 (2005); T. Kitabayashi and M. Yasue, Phys. Lett., **B 621**, 133 (2005); R. N. Mohapatra and S. Nasri, Phys. Rev. D **71**, 033001 (2005); R. N. Mohapatra, S. Nasri and H. B. Yu,Phys. Lett. B **615**, 231(2005), Phys. Rev. D **72**, 033007 (2005); Y. H. Ahn, Sin Kyu Kang, C. S. Kim, Jake Lee, hep-ph/0602160.

[33] R. N. Mohapatra, S. Nasri and H. B. Yu, hep-ph/0603020.

[34] N. Cabibbo, Phys. Lett. B **72** (1978) 333; L. Wolfenstein, Phys. Rev. D **18** (1978) 958.

[35] E. Ma, H. Sawanaka and M. Tanimoto, hep-ph/0606103.

[36] E. Ma, hep-ph/0607190.

[37] For a recent review, see, for example: W. Buchmuller, R.D. Peccei and T. Yanagida, Ann.Rev.Nucl.Part.Sci.55:311,2005. [arXiv:hep-ph/0502169]

Section II: Neutrinos in Astrophysics

Standard Solar Models

Aldo Serenelli

Institute for Advanced Study, Princeton, USA

To the memory of John N. Bahcall (1934-2005)

1 Introduction

The development of neutrino physics is intimately linked to theoretical studies of the Sun. The Sun is a star going through the Main Sequence phase, the most stable and long evolutionary phase, where stars consume the hydrogen in their cores. Hydrogen burning operates through a series of nuclear reactions that create electron neutrinos ν_e which escape freely from the solar interior and reach the Earth. Detecting solar neutrinos seemed, in the 1960s, a good astrophysics experiment where to test stellar structure and evolution theories. Ray Davis led the team that built the first chlorine neutrino detector and John Bahcall computed the first Standard Solar Models (SSMs) and estimated the theoretical solar neutrino fluxes. In 1968, when results from the neutrino experiment and the solar model were put together the "solar neutrino problem" was born (a brief account of the situation at that moment can be found in Bahcall 1971). The solar neutrino problem lived for almost 35 years, until 2002, when the Sudbury Neutrino Observatory announced the results of the neutral current experiment (Ahmad et al. 2002). Astrophysics had shown the way to new physics. Solar neutrino predictions from SSM calculations have been extensively used to help constraining neutrino parameters such as mixing angles and mass splitting. The solar neutrino problem is solved, but only the rare ^{8}B solar neutrino flux has been directly measured so far. It remains for neutrino experiments to detect the lower energy neutrinos generated in nuclear reactions more relevant to the solar, and stellar, energetics. We hope to learn stellar astrophysics from these neutrino experiments, going back in this way to the original proposal by Ray Davis and John Bahcall.

We have a clear picture of the solar interior. Helioseismology has been the main observational tool in this direction. The Sun is a resonant cavity and more than 10^5 oscillation frequencies have been measured from space-borne missions like SOHO[1] and ground based experiments like BiSON[2]. Inversion techniques (Dziembowski et al. 1990) allow the use of the measured frequencies to determine the internal structure of the Sun: the sound speed and

[1] http://sohowww.nascom.nasa.gov
[2] http://bison.ph.bham.ac.uk

density profiles, constrains on the solar composition and extension of the convective envelope are some of the important quantities that have been determined by helioseismology and that constrain, and support, the SSM predictions. We review some of them in the following sections and give appropriate references there.

In Section 2 we review briefly the basic equations that determine the stellar structure and evolution, give some detail about the nuclear reactions that are relevant in the context of the SSM, and define the constraints an SSM has to satisfy. In Section 3 we present and discuss the predictions of SSM calculations and compare them to observable quantities when possible. Emphasis is put on the neutrino production and propagation properties inside the Sun. A discussion of the "solar abundance problem" is presented towards the end of that section. In Section 4 we discuss briefly the theoretical uncertainties in SSM calculations, focusing on uncertainties in neutrino fluxes.

2 The Standard Solar Model. The Basics

The Standard Solar Model is a working framework within which we try to understand the Sun. It involves a series of assumptions, each of which render the model less approximate to the real Sun, that are necessary to make the SSM calculation possible at all. The validity of the approximations is better judged from the results. The most important assumption is probably that of spherical symmetry. Consistency with this assumption implies that rotation and magnetic fields are not considered in the SSM. There are non-standard models in the literature that try to account for some of the missing physics. They involve, however, adding free parameters to the model that have to be somehow fixed. This is usually done without solid physical arguments, but just trying to match a given observational constraint. We do not take this approach in these lectures.

2.1 Basic Equations

We introduce the basic equations that describe the evolution of stars in general and the Sun in particular. It is a very brief review, not a short course on stellar evolution. Details can be found in classical books like Cox & Giuli (1968), Clayton (1983), Kippenhahn & Weigert (1990) and many others.

2.1.1 Momentum and Mass Conservation

The first equation we consider is the condition for hydrostatic equilibrium, derived from the conservation of momentum by neglecting inertia terms. Let P and ρ be the pressure and density at a distance r from the stellar center and let M be the enclosed mass at that point and G the gravitational constant, then the balance between the gravitational and pressure forces is

$$\frac{dP}{dr} = -\frac{GM\rho}{r^2}. \tag{1}$$

Hydrodynamic effects play a negligible role in determining the structure of the Sun and neglecting them is an excellent approximation.

With spherical symmetry and assuming hydrostatic equilibrium, the continuity equation is simply

$$\frac{\partial M}{\partial r} = 4\pi r^2 \rho. \tag{2}$$

2.1.2 Energy Conservation

Let us consider a spherical shell of thickness dr at a distance r from the stellar center and let L be the energy flux (luminosity) through the shell. Matter inside the shell can exchange heat with the surroundings due to both mechanical work and changes in its internal energy. If dq/dt is the rate of heat exchange per unit mass, then by applying the first law of thermodynamics we get

$$\frac{\partial L}{\partial r} = -4\pi r^2 \rho \frac{dq}{dt}.$$

In the stellar interior, however, there may be additional sources (or sinks) of energy. In the case of the Sun, the best examples are the release of energy by nuclear fusion and the energy losses due to neutrino production. We can account for these kind of sources by simply adding appropriate source terms in the previous equation. We then have

$$\frac{\partial L}{\partial r} = 4\pi r^2 \rho \left(\varepsilon_{\rm nuc} - \varepsilon_\nu - T\frac{ds}{dt} \right). \tag{3}$$

Here $\varepsilon_{\rm nuc}$ and ε_ν represent, respectively, the nuclear energy release and the energy losses by neutrino emission, per unit mass, per unit time produced within the shell of thickness dr, T is the temperature of such shell and s the specific entropy of the material and have made use of $dq = Tds$. It should be noted that the observed luminosity of the Sun is

$$\mathrm{L}_\odot = \int_0^{R_\odot} \left(\frac{dL}{dr} \right) dr.$$

2.1.3 Energy Transport

The presence of a temperature gradient gives rise to a net flux of energy from the central, hotter regions of the Sun. In regions that are dynamically stable, energy is transported by radiation. A detailed derivation of the transport equation is straightforward but rather lengthy and for this reason we omit it here (see Kippenhahn & Weigert 1990, Chap. 5). The resulting equation is

$$\frac{\partial T}{\partial r} = -\frac{GMT\rho}{r^2 P} \nabla_{\rm rad}, \tag{4}$$

where

$$\nabla_{\rm rad} = \frac{3}{16\pi acG} \frac{\kappa LP}{MT^4} \tag{5}$$

is known as the radiative temperature gradient, a is the radiation-density constant, c the speed of light and κ the radiative opacity of stellar matter (typical values for κ in the solar core are of order 1 $\mathrm{cm^2\ g^{-1}}$). When $\nabla_{\rm rad}$ becomes too large (for example in regions where κ is large because of partial ionization of atoms, or regions where L/M is large as it happens in the cores of massive stars), radiative transport becomes inefficient and convective instabilities develop.

In the case of the Sun this happens in the envelope, where κ becomes very large, and there the bulk of energy is transported by convection. Under such a situation we need a theory of convection to determine the temperature gradient. This is not a solved problem in astrophysics and one has to use simplified prescriptions to compute the convective temperature gradient. The most common approach is the Mixing Length Theory (Cox & Giuli 1968). In the case of the solar envelope, convection is very efficient in transporting energy and the temperature gradient in the solar envelope is very close to the adiabatic one (this is not true close to the solar surface, but we ignore this fact in these lectures). We can then safely assume that in the envelope the temperature gradient is in fact adiabatic ($\nabla_{\rm ad}$). The great advantage is that $\nabla_{\rm ad}$ only depends on the equation of state, which is well known, and not on a convection theory. Summarizing, the transport equation can be written in general form as

$$\frac{\partial T}{\partial r} = -\frac{GMT\rho}{r^2 P}\nabla, \tag{6}$$

where $\nabla = \nabla_{\rm rad}$ in the radiative solar interior and $\nabla = \nabla_{\rm ad}$ in the convective envelope. The transition between radiative and convective transport occurs where $\nabla_{\rm rad} = \nabla_{\rm ad}$.

2.1.4 Composition Changes

Changes in the chemical composition of a star occur through nuclear reactions and mixing processes like convection and element diffusion.

We first consider nuclear reactions. Let n_i be the number density of ions of type i that can be created and destroyed in nuclear reactions, and let $\langle\sigma v\rangle$ represent the thermally averaged cross section of a given reaction, then the rate of change of n_i due to nuclear reactions can be written, schematically, as

$$\left(\frac{\partial n_i}{\partial t}\right)_{\rm nuc} = -\sum_j \frac{1}{1+\delta_{ij}} n_i n_j \langle\sigma v\rangle_{ij} + \sum_{k,l} \frac{1}{1+\delta_{kl}} n_k n_l \langle\sigma v\rangle_{kl} \tag{7}$$

where the first term accounts for the destruction of nuclei of type i by fusion in binary reactions with nuclei of type j, and the second one for the creation of nuclei of type i by binary reactions between nuclei of types k and l. The factor containing the Kronecker's delta function avoids double counting over pair of nuclei. A very complete discussion about nuclear cross section calculations and equations describing nuclear networks can be found in Clayton (1983, Chap. 4 and 5).

Mixing processes in stars have different origins but can be modelled as diffusive processes. As a result, the rate of change of n_i can be written generically as

$$\left(\frac{\partial n_i}{\partial t}\right)_{\rm mix} = \frac{1}{r^2}\frac{\partial}{\partial r}\left(D_i r^2 \left(\frac{\partial n_i}{\partial r}\right)\right) \tag{8}$$

where D_i is a diffusion coefficient that can be different for different ions.

Convection is the most important mixing process in stars and is of macroscopic nature. Convection is turbulent in stars and a very efficient mixing process. Because of its macroscopic nature, the convective diffusion coefficient $D_{\rm conv}$ is the same for the different atomic species present in the convective region.

On the other hand, microscopic effects can also affect the chemical composition. Microscopic effects arise when gradients are present: gradients in the abundances of chemical elements, temperature and pressure give rise to, respectively, *concentration diffusion*, *temperature diffusion* and *pressure diffusion*. We only discuss qualitatively the effects of pressure diffusion or *gravitational settling* here because it is the most relevant one for the SSM. Under its influence helium and metals migrate inwards while hydrogen tends to float on top on them. It is a second order but relevant effect in solar modelling, as will be shown in later sections. All the dirty physics in the diffusion equation is hidden in the calculation of the diffusion coefficient. The detailed statistical theory of diffusion is derived in Chapman & Cowling (1970), a simplified discussion is given in Kippenhahn & Weigert (1990), and a detailed calculation of the coefficients in the context of solar models can be found in Thoul et al. (1994).

The diffusion equation is linear in D and the total effect of different mixing processes can be accounted for by defining: $D = D_{\rm conv} + D_{\rm concen} + D_{\rm therm} + D_{\rm grav}$. In convective regions, $D_{\rm conv}$ is more than 15 orders of magnitude larger than any other mixing process. In the solar envelope, convective mixing occurs on timescales of the order of a month, i.e. convective mixing is a very fast process. Because $D_{\rm conv}$ is the same for all atomic species the convection in the Sun ensures a chemically homogeneous convective envelope. In the radiative interior there are no macroscopic mixing processes but[3], as we mentioned before, gravitational settling plays a relevant role. Its importance is strengthened because it "connects" the convective envelope with the radiative interior and heavy elements and helium leak from the convective envelope into the radiative interior, while hydrogen is transported in the opposite direction. Timescales for gravitational settling are of order $10^{10} - 10^{11}$ years.

Summarizing, changes in the composition of a star arise from nuclear fusion and mixing processes. In addition to the equations of stellar structure, we have to solve the set of coupled equations that determine the time evolution of all the atomic species involved in the calculation:

$$\frac{dn_i}{dt} = \left(\frac{\partial n_i}{\partial t}\right)_{\rm nuc} + \left(\frac{\partial n_i}{\partial t}\right)_{\rm mix} \qquad i = 1, \ldots, N \tag{9}$$

where $N = 15$ is the total number of atomic species considered in the solar models discussed in the present lectures.

2.1.5 Summary of equations and input physics

The basic differential equations that determine the structure and evolution of a spherical star are the non-linear partial differential equations (1, 2, 3, 6) and (9) for the N atomic species considered. These equations contain functions which describe the properties of stellar matter such as ρ, $\varepsilon_{\rm nuc}$, $\varepsilon_{\rm nu}$, s, $\nabla_{\rm ad}$, κ, the cross sections $\langle \sigma v \rangle$, and the diffusion coefficients. All these quantities have to be known as a function of P and T and the composition described by $\{n_i\}_{i=1,\ldots,N}$. The most commonly used equation of state (EOS) used in Standard Solar Model calculations is known as the OPAL EOS (Rogers et al. 1996). Radiative opacities are either from the OPAL (Rogers & Iglesias 1996) or OP (Seaton 2005; Badnell et al. 2005) groups. Nuclear reaction cross sections come from a variety of theoretical calculations and experiments; an updated list

[3]There is probably some macroscopic mixing induced by the solar rotation, but this is not accounted for in the SSM. On the other hand, the Sun is a slow rotator, so this is likely to be a minor effect.

of references can be found in Bahcall et al. (2006). Coefficients for microscopic diffusion are obtained from Thoul et al. (1994).

2.2 Nuclear Reactions

The temperature in the solar core is $\sim 10 - 15 \times 10^6$K and the nuclear energy generation occurs by fusion of hydrogen into helium. Hydrogen burning operates by the so-called *pp-chains* and *CNO bi-cycle*, the former being responsible for about 99% of the nuclear energy released in the Sun. The detailed set of reactions of both the pp-chains and CNO bi-cycle is presented in Table 1 together with the energy released per reaction. Additionally, we identify the eight nuclear neutrino sources and give the average neutrino energy for each of these reactions. Note that the pp-chains involve three branches, i.e. three different ways in which the ^{4}He nuclei are produced, the ppI, ppII and ppIII branches. The branching between ppI on one side and the ppII and ppIII branches on the other depends of the fate of the ^{3}He nuclei. The branching between ppII and ppIII depends on whether a ^{7}Be nucleus captures an electron or a proton. For the CNO-bi-cycle there exists one branching point given by the two channels the proton capture on ^{15}N nuclei has. A discussion of the contribution to the solar energetics of the different chains and cycles, as well as results concerning the individual neutrino fluxes , is deferred to Section 3.

We cannot discuss in these notes how the nuclear cross sections can be calculated. Suffice it to say here that, in general, the cross sections can be expressed as a product of a known function that carries the temperature dependence, and a constant factor, called "low energy cross section" or "astrophysical factor" and generally denoted by S. The S factor includes the nuclear physics part of the reaction, and depends only on the nuclei involved in the reaction. The reader interested in the nuclear physics involved behind the calculation of nuclear cross sections is strongly encouraged to read the book by Clayton (1983; Chap. 4). A more simplified discussion in the context of solar models can be found in Bahcall (1989; Chap. 3). A technical discussion of theoretical and experimental results applied to solar fusion cross sections is given in Adelberger et al. (1998).

2.3 SSM: Constraints and Free Parameters

The SSM is a model of the Sun at its present age and, as such, has to reproduce the present-day conditions of the Sun. The measured quantities an SSM has to match are the present-day solar luminosity ($L_\odot$), radius ($R_\odot$), and surface composition represented by the metal to hydrogen mass fraction ratio $(Z/X)_\odot$[4]. Here Z represents the total mass fraction, i.e. it is the added contribution from all elements heavier than helium. The solar structure does not depend just on Z, but actually on the individual abundances of elements like carbon, nitrogen, oxygen, neon, silicon, iron and a few other. The relative abundances for all the elements in the Sun are determined observationally. Compilations of solar abundances relevant in these lectures are Grevesse & Sauval (1998) and Asplund et al. (2005) and we refer to them as the GS98 and AGS05 solar abundances, respectively.

The concept behind an SSM is not just to have a model that satisfies the present-day con-

[4]Element abundance measurements are not absolute but relative to a reference element, usually hydrogen. The absolute values for the abundances can be determined from the constraint $X + Y + Z = 1$.

Chain	Reaction	Q (Mev)	$\langle Q_\nu \rangle$(Mev)	No.
ppI	$p+p \longrightarrow {}^2H+e^+ + \nu_e + \gamma$	1.442	0.265 (pp)	1
	$p+p+e^- \longrightarrow {}^2H + \nu_e$	1.442	1.442 (pep)	2
	${}^2H+p \longrightarrow {}^3He+\gamma$	5.49	–	3
	${}^3He+{}^3He \longrightarrow {}^4He+2p+\gamma$	12.86	–	4
ppII	${}^3He+{}^4He \longrightarrow {}^7Be+\gamma$	1.586	–	5
	${}^7Be+e^- \longrightarrow {}^7Li+\nu_e$	0.862(90%)	0.862(^{7}Be)	6
		0.384(10%)	0.384(^{7}Be)	
	${}^7Li+p \longrightarrow {}^4He+{}^4He+\gamma$	17.347	–	7
ppIII	${}^7Be+p \longrightarrow {}^8B+\gamma$	0.137	–	8
	${}^8B \longrightarrow {}^4He+{}^4He+\nu_e+\gamma$	17.98	6.710 (^{8}B)	9
	${}^3He+p \longrightarrow {}^4He+e^++\nu_e+\gamma$	19.795	9.625 (hep)	10
Cycle	Reaction	Q (Mev)	$\langle Q_\nu \rangle$(Mev)	
CN	${}^{12}C+p \longrightarrow {}^{13}N+\gamma$	1.943	–	11
	${}^{13}N \longrightarrow {}^{13}C+e^++\nu_e$	2.221	0.7067 (^{13}N)	12
	${}^{13}C+p \longrightarrow {}^{14}N+\gamma$	7.551	–	13
	${}^{14}N+p \longrightarrow {}^{15}O+\gamma$	7.297	–	14
	${}^{15}O \longrightarrow {}^{15}N+e^++\nu_e$	2.754	0.9965 (^{15}O)	15
	${}^{15}N+p \longrightarrow {}^{12}C+\alpha$	4.966	–	16
NO	${}^{15}N+p \longrightarrow {}^{16}O+\gamma$	12.128	–	17
	${}^{16}O+p \longrightarrow {}^{17}F+\gamma$	0.600	–	18
	${}^{17}F \longrightarrow {}^{17}O+e^++\nu_e$	2.762	0.9994 (^{17}F)	19

Table 1. *Nuclear reactions in the Sun. The chain or cycle for each reaction is identified in the first column, and the reaction in the second column. Third column gives the total energy released by the reaction and the fourth the average energy carried away by neutrinos when produced. Also identified is the usual name given to the neutrino fluxes. The effective energy that powers the solar luminosity, from each reaction, is* $Q_\gamma = Q - \langle Q_\nu \rangle$.

ditions, but to do so as the result of consistent stellar evolution calculations, i.e. by computing the evolution of a 1 $M_\odot$ star from early evolutionary phases up to the Sun's present age. The considerations that determine the initial model, i.e. the model that represents the initial phase of the solar evolution, are: 1) the total mass lost from the Sun throughout the solar lifetime is negligible compared to the solar mass, i.e. the solar mass can be taken as constant; 2) solar mass stars are initially fully convective and, consequently, chemically homogeneous. The composition of the initial model, i.e. the primordial composition of the Sun, is specified by the initial mass fractions of hydrogen ($X_{\rm ini}$), helium ($Y_{\rm ini}$), and metals ($Z_{\rm ini}$). By definition, they are linked by $X_{\rm ini}+Y_{\rm ini}+Z_{\rm ini}=1$ leaving only two independent unknown quantities. These are two of the degrees of freedom we have to construct an SSM. An SSM has to match $L_\odot$, $R_\odot$ and $(Z/X)_\odot$. We need a third free parameter and it is related to the convection prescription. In the Mixing Length Theory it is the mixing length parameter, $\alpha_{\rm MLT}$, a constant that determines the entropy stratification in the upper (non-adiabatic) layers of the convective envelope of the Sun.

Calculation of an SSM involves the following steps: 1) construct a 1 $M_\odot$ initial model with homogeneous composition specified by $X_{\rm ini}$, $Y_{\rm ini}$ and $Z_{\rm ini}$ and fix the mixing length parameter $\alpha_{\rm MLT}$; 2) evolve this model from $\tau = 0$ to $\tau = \tau_\odot$ by integrating the equations derived in Sec-

Quantity	Value	Method of determination
Mass	$M_\odot = 1.989 \times 10^{33}$g	Kepler's 3rd law
Age	$\tau_\odot = 4.57 \times 10^9$yr	Radioactive dating from meteorites
Luminosity	$L_\odot = 3.842 \times 10^{33}$erg/s	Solar constant
Radius	$R_\odot = 6.9598 \times 10^{10}$cm	Angular diameter
Metals/Hydrogen ratio	$(Z/X)_\odot$=0.0229/0.0165	Photosphere and meteorites

Table 2. *Solar quantities used to calculate an SSM. Mass and age are fixed quantities in the model while the other three are the parameters to be matched by adjusting the initial composition and mixing length parameter. The careful reader will note that two values are given for $(Z/X)_\odot$; this will be discussed in Section 3.3.*

tion 2.1; and 3) check if $L_{model} = L_\odot$, $R_{model} = R_\odot$ and $(Z/X)_{model} = (Z/X)_\odot$ to the required accuracy (usually one part in 10^6). If step 3) is not successful, initial conditions have to be improved (an iterative scheme is easily implemented) and the steps repeated until convergence of the model properties to the present-day conditions is achieved.

3 The Standard Solar Model. Results

The Sun is an ordinary star. As such, it is interesting to understand its evolution, past and future. We start our discussion of the SSM in Section 3.1 by describing the temporal evolution of some relevant quantities. The current epoch SSM is discussed in more detail in Section 8. In Section 3.3 we comment of the SSM results arising from recent solar composition measurements.

3.1 Temporal evolution

We limit our discussion to the Main Sequence (MS) phase, i.e. where the nuclear energy source is hydrogen burning in the stellar core. Figure 1 presents the temporal evolution of many relevant characteristics of an SSM and summarizes the most important results of this Section. In general, quantities are normalized to present-day values; these are given in Tables 2, 3 and 4.

Panel *(a)* shows the evolution of the solar luminosity L and radius R, which are related by $L = 4\pi\sigma R^2 T_{\rm eff}^4$ (here σ is the Stefan-Boltzmann constant and $T_{\rm eff}$ the effective temperature). $T_{\rm eff}$ varies very little during the first 10 Gyr of evolution for a 1 $M_\odot$ star and thus L changes approximately as R^2. Despite the fact the R increases with time, the solar core shrinks. In panel *(b)* we show the evolution of the central density (ρ_c), pressure (P_c), temperature (T_c) and the central mass fraction of hydrogen (X_c). All quantities increase slowly during the first 5 Gyr but while ρ_c and P_c increase by about a factor of 2 during this time, T_c only increases by 20%. During the subsequent 5 Gyr, T_c continues to increase at a similar rate but the increase of ρ_c and P_c becomes steeper and at $\tau = 10$ Gyr they are about 6 and 5 times, respectively, their present-day values. The luminosity of the Sun is derived from hydrogen burning. The central hydrogen mass fraction X_c at the present day is about half the initial hydrogen abundance and its depletion will continue at a similar rate, the Sun is approximately half-way its MS evolution.

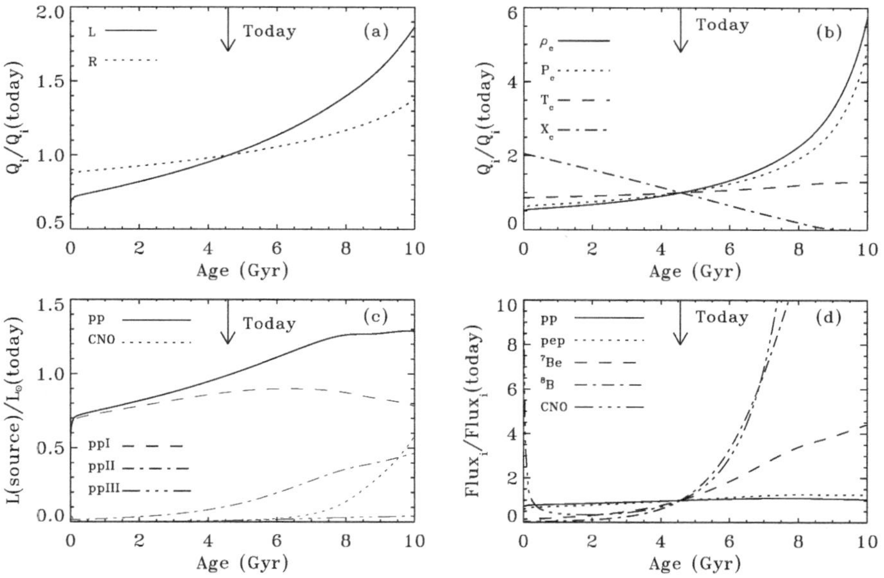

Figure 1. *Evolution of various quantities as a function of age for the Standard Solar Model BS05. Age is given in* 10^9 *yr. Panel (a) shows the luminosity* (L) *and radius* (R)*, and panel (b) the central density* (ρ_c)*, pressure* (P_c)*, temperature* (T_c)*, and hydrogen mass fraction* (X_c)*. Panel (c) shows the fraction of nuclear energy released by the pp-chains (pp) and CNO-cycles (CNO) and the fractions corresponding to the different pp-chains (ppI, ppII and ppIII). Finally, panel (d) shows the evolution of the most relevant neutrino fluxes, pp, pep,* ^{7}*Be,* ^{8}*B and CNO* $\equiv$ $^{13}\mathrm{N}+^{15}\mathrm{O}+^{17}\mathrm{F}$. *All quantities are normalized to present-day values.*

About 99.5% of the nuclear energy in the present-day Sun originates in the pp-chains (see Table 1) leaving CNO cycles as just a marginal contribution. In panel *(c)* we see that the total energy released in both pp-chains and CNO cycles increase with time (their added contribution is almost equivalent to the total solar luminosity at any time). However, because nuclear reactions relevant to the CNO cycles depend very strongly on temperature, their relative contribution increases as T_c does, and is about 1/3 of the total nuclear energy released at 10 Gyr. As an additional comment we note that after the MS, when the Sun will become a Red Giant Star, the nuclear energy release will be completely dominated by the CNO reactions and the pp-chains (now dominant) will represent just a small contribution. It is also instructive to examine the relative contributions of the different pp-chains, in particular those from the ppI- and ppII-chains. ppI is always dominant and relatively constant; during the 10 Gyr of solar evolution it varies by no more than 20%. During the first 6 Gyr its contribution increases slowly because of the relatively weak dependence on temperature (about $\sim T^4$) of the rate of the key reaction $\mathrm{p}(\mathrm{p},\mathrm{e}^+\gamma)\mathrm{d}$. After 7 Gyr it starts to decline because even if the temperature continues to increase, this can not compensate for the decrease in the ^{3}He abundance in the solar core, on which the energy released by the ppI-chain depends quadratically. The ppII-chain, on the other hand, has a similar temperature dependence as ppI but depends only linearly on the ^{3}He abundance and so its contribution to the energy budget steadily increases with time. The relative contribution of the ppIII-chain, to which the ^{8}B neutrino flux is linked, increases continuously along the solar evolution, but in absolute terms is always a negligible contribution to the energetics of the star.

Panel *(d)* shows the evolution of some neutrino fluxes. In particular the pp, pep, ^{7}Be and ^{8}B fluxes, coming from the pp-chains, are all shown. Additionally, the three fluxes from the CNO cycles are added together into the artificial *CNO* flux. Note that the plot is normalized to the present-day flux values. The pp and pep fluxes vary little during the evolution of the Sun. We could have inferred this behavior by looking at the evolution of the energy coming from the ppI-chain in panel *(c)*. The pp flux, in particular, increases slowly from the beginning of the solar evolution when it is about 75% its present-day value up to about 110% its present-day value at about 7 Gyr of age, and then declines slowly. The behavior of the pep flux is qualitatively similar albeit somewhat steeper, going from an initial minimum of 60% its present-day value up to about 125% its present value at 8 Gyr. ^{7}Be and ^{8}B increase continuously along the solar evolution. Again, this behavior can be qualitatively inferred from the temporal evolution of the energy contribution of the ppII and ppIII chains. The ^{7}Be flux begins with only a 13% its present-day value and increases up to a factor of 4 larger than today's value at 10 Gyr. The ^{8}B flux is the one that depends most strongly on temperature and this is manifested in the steep increase that can be seen in the plot. The initial ^{8}B flux is only 3% its value today, while at 10 Gyr it is 25 times its present-day value. The last flux shown, "CNO", is actually the total flux from ^{13}N, ^{15}O and the negligible ^{17}F. The first two contribute comparable amounts during most of the solar evolution and their time evolution is very similar, except for an initial transient phase we discuss later. The reason for the similar behavior and comparable contributions is that these two fluxes are part of the CN cycle (see Table 1) and the rate at which this cycle operates is regulated by the rate of the reaction $^{14}\text{N}(\text{p},\gamma)^{15}\text{O}$. This is the slowest reaction in the cycle by at least a couple of orders of magnitude, meaning that when this reaction occurs, all the others follow almost instantaneously and in this way one ^{13}N neutrino and one ^{15}O are produced together. This picture does not apply during the first evolutionary stages, i.e. the first few hundred millon years or so. The "CNO" flux is initially very large but decreases very rapidly. All these neutrinos are ^{13}N neutrinos with almost no contribution from ^{15}O. This is a consequence of the initial composition of the Sun, representative of the cosmic abundances, that has a ^{12}C abundance which is about two orders of magnitude larger than the equilibrium abundance resulting from the CN cycle. As soon as the temperature in the young Sun reaches 10^7 K the excess ^{12}C is burnt by proton captures producing the large initial ^{13}N (first two reactions in the CN cycle) flux seen on the plot. After this initial transient phase, the evolution of the "CNO" flux is very similar to the evolution of the ^{8}B flux due to the similar temperature dependence.

3.2 Current epoch

In this Section we discuss in detail SSM results corresponding to the present-day Sun. All results presented in this section refer, unless otherwise specified, to the BS05(OP) model from Bahcall et al. (2005) that adopts the Grevesse & Sauval (1998) solar composition. In these notes we call this model BS05(GS98). Discussion of the model uncertainties is omitted in this section and postponed to Sec. 4.

Qnt.	Value	Qnt.	Value	Flux	Value	Flux	Value
R_{CZ}	0.713(0.728)	Cl(SNU)	8.1(6.6)	pp	5.99(6.06)	^{8}B	5.69(4.51)
Y_{surf}	0.243(0.229)	Ga(SNU)	126(119)	pep	1.42(1.45)	^{13}N	3.05(2.00)
$\langle\delta c\rangle$	0.001(0.005)	pp	99.2%(99.5%)	hep	7.93(8.25)	^{15}O	2.31(1.44)
$\langle\delta\rho\rangle$	0.011(0.044)	CNO	0.8%(0.5%)	^{7}Be	4.84(4.34)	^{17}F	5.83(3.25)

Table 3. *Standard Solar Model Predictions: Measurable quantities. First column gives helioseismology quantities: location of the base of the convective zone (in solar radius), surface helium mass fraction, averaged fractional rms differences of the sound speed and density profiles. Second column: expected neutrino rates in the chlorine and gallium experiments assuming no neutrino oscillations and percentage of nuclear fusion energy that is generated in the pp-chains and in the CNO cycles. Last two columns give the expected solar neutrino fluxes on Earth in units of* $10^{10}(pp)$, $10^9(^7\mathrm{Be})$, $10^8(pep,^{13}N,^{15}O)$, $10^6(^8B,^{17}F)$ *and* $10^3(hep)$ $cm^{-2}\ s^{-1}$. *Values without parenthesis refer to the BS05(GS98) model and values in parenthesis to the BS05(AGS05) model (to be discussed in Section 3.3).*

3.2.1 Measurable quantities

Helioseismology observations provide strong constraints for solar models. We have already mentioned that the Sun has a convective envelope and its extension depends on the balance between the radiative and convective transport of energy. Helioseismology can be used to determine the extension of such a convective region and the result is (Christensen-Dalsgaard et al. 1991, Basu & Antia 1997)

$$R_{CZ} = 0.713 \pm 0.001\ \mathrm{R}_\odot. \tag{10}$$

Similarly, the most recent helioseismological determination of the surface helium abundance of the Sun Y_{surf} (Basu & Antia 2004) gives

$$Y_{surf} = 0.2485 \pm 0.0035. \tag{11}$$

These values can be compared to the predictions of the BS05(GS98) model, shown in Table 3. The agreement is very good. We discuss helioseismology results in more detail in later sections.

On the neutrino side, the measured event rate in the chlorine solar neutrino experiment is (Cleveland 1998)

$$\sum \phi(i)\sigma(i)|_{\mathrm{Cl}} = 2.56 \pm 0.16\ (\text{statistical}) \pm 0.16\ (\text{systematic})\ \text{SNU},^5 \tag{12}$$

where the summation is over all eight solar neutrino fluxes. The difference between the predicted standard model value of the chlorine event rate and the measured event rate created the "solar neutrino problem" (Bahcall et al. 1968; Davis et al. 1968). BS05(GS98) predictions, without including neutrino oscillations, are given in Table 3. The weighted average rate measured by the gallium solar neutrino experiments, SAGE, GALLEX and GNO, is (Hampel et al. 1999; Abdurashitov et al. 2003; Altmann et al. 2005)

$$\sum \phi(i)\sigma(i)|_{\mathrm{Cl}} = 68.1 \pm 3.85\ \text{SNU}. \tag{13}$$

[5] 1 SNU = 1 Solar Neutrino Unit = 1 interaction per 10^{36} target atoms per second

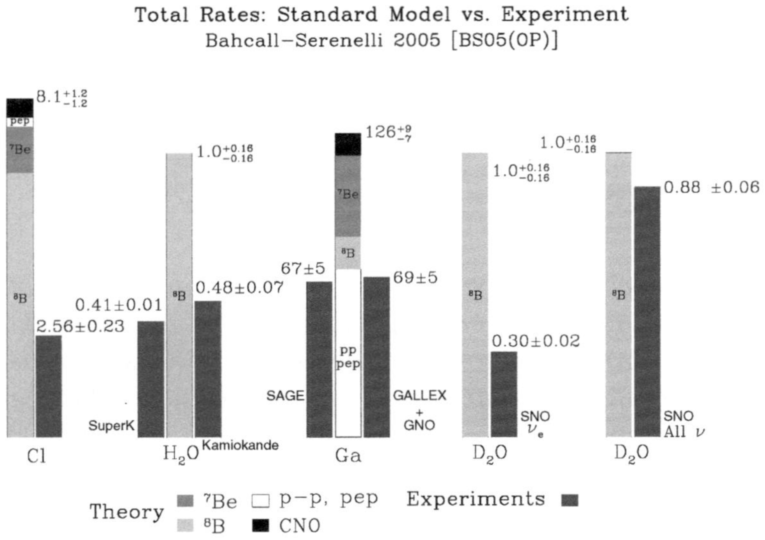

Figure 2. *Solar neutrinos. Comparison between predictions of the SSM BS05 plus the standard model of electroweak interactions with the measured rates in all solar neutrino experiments.*

The flux of electron neutrinos from the ^{8}B neutrino flux measured in the Kamiokande, Super-Kamiokande and SNO experiments is (Aharmim et al. 2005; Ahmed et al. 2004; Fukuda et al. 1996, 2001)

$$\phi(^8\mathrm{B})_e = (1.68 \pm 0.10) \times 10^6\ \mathrm{cm}^{-2}\ \mathrm{s}^{-1}. \tag{14}$$

The rates of electron type neutrinos measured in the chlorine, gallium, Kamiokande, Super-Kamiokande and SNO experiments are, in all cases, much lower than the values predicted by standard solar models, as can be seen by comparing them with the BS05(GS98) predictions listed in Table 3.

The differences between the predicted standard solar model rates and the measured rates mentioned so far can be well explained by the hypothesis of solar neutrino oscillations (Gribov & Pontecorve 1969; Wolfenstein 1978; Mikheyev & Smirnov 1985). The electron type neutrinos that are produced in the Sun have mostly been converted into muon and tau neutrinos by the time they reach the terrestrial detectors.

The SNO experiment has allowed a direct measurement of the total ^{8}B neutrino flux by measuring the neutral current reaction $\nu_x + d \longrightarrow p + n + \nu_x$ (and to a lesser extent by electron scattering reaction $\nu_x + e^- \longrightarrow \nu_x + e^-$). Here ν_x represents either electron, muon or tau neutrinos, d a deuterium atom, p a proton and n a neutron. The total flux measured is (average of Phase I and Phase II measurements, Aharmim et al. 2005)

$$\phi(^8\mathrm{B}) = (4.99 \pm 0.33) \times 10^6\ \mathrm{cm}^{-2}\ \mathrm{s}^{-1}, \tag{15}$$

in excellent agreement with the BS05(GS98) predictions (see Table 3). A detailed comparison between the BS05(GS98) model predictions and all the neutrino experiments is summarized in Fig. 2.

As mentioned before, the "solar neutrino problem" arose around 1968 and it was finally solved by the SNO measurement of the total ^{8}B solar neutrino flux. In the meantime, the solar model results were questioned by many physicists and astrophysicists, especially during the

1980s and 1990s. It is instructive, then, to compare predictions of state-of-the-art solar models, as represented by the BS05(GS98) model, with those given by older models. To this aim, we use the predictions from the solar model presented in Bahcall & Ulrich (1988; hereafter BU88), which represented the first systematic combined investigation of the solar neutrino problem and of helioseismology and which was also the most comprehensive solar model study prior to the inclusion of element diffusion. The 1988 prediction for the rates in the chlorine and gallium experiments were 7.9 and 132 SNU respectively, and the predicted total ^{8}B flux was 5.76×10^6 cm^{-2} s^{-1}. The changes since the BU88 are smaller than the current uncertainties in the theoretical expectations. While it is true that some updates in the solar models have introduced changes of opposite sign that tend to compensate each other, it is remarkable that two decades of improvement in both the modelling and the input physics (radiative opacities, equation of state, nuclear cross sections) of solar models have kept the solar neutrino predictions almost unchanged.

3.2.2 Characteristics of the Standard Solar Model

Important characteristics of the SSM that cannot be directly measured are given in Table 4. Quantities characterizing the center of the Sun are given in the left two columns. The right columns give values at the base of the convective zone and the initial composition of the solar model. Keeping in mind that $X_{\rm ini} = 1 - Y_{\rm ini} - Z_{\rm ini} \approx 0.71$ and comparing this value with $X_{\rm C}$ we see that at its present-day age, 4.57 Gyr, the Sun is about half way along its life as a main sequence (hydrogen core burning) star. The difference between the initial solar metallicity $Z_{\rm ini}$ and the present-day central value $Z_{\rm C}$ is not the result of nuclear burning but rather of element diffusion, that has contributed to increase the metallicity of the solar core by about 8% over its initial value in 4.57 Gyr.

Central values				Other quantities			
Qnt.	Value	Qnt.	Value	Qnt.	Value	Qnt.	Value
$T_{\rm C}$	1.567(1.548)	$X_{\rm C}$	0.346(0.365)	$T_{\rm CZ}$	0.218(0.201)	$M_{\rm CZ}$	0.024(0.020)
$\rho_{\rm C}$	1.529(1.504)	$Y_{\rm C}$	0.634(0.620)	$\rho_{\rm CZ}$	0.0019(0.0016)	$Y_{\rm ini}$	0.273(0.260)
$P_{\rm C}$	2.36(2.34)	$Z_{\rm C}$	0.0202(0.0151)	$P_{\rm CZ}$	$5.58(4.34) \times 10^{-4}$	$Z_{\rm ini}$	0.0188(0.014)

Table 4. *Present-day characteristics of the BS05 model. Left two columns give the central values of temperature ($T_{\rm C}$), density ($\rho_{\rm C}$) and pressure ($P_{\rm C}$) and mass fractions of hydrogen ($X_{\rm C}$), helium ($Y_{\rm C}$) and metals ($Z_{\rm C}$). Right two columns give quantities characterizing the convective envelope, the temperature ($T_{\rm CZ}$), density ($\rho_{\rm CZ}$) and pressure ($P_{\rm CZ}$) at its bottom, i.e. at $R=R_{\rm CZ}$, and the mass contained in the convective envelope ($M_{\rm CZ}$). Also given are the initial helium and metals mass fraction of the solar model, $Y_{\rm ini}$ and $Z_{\rm ini}$ respectively. Temperature, density and pressure are given in units of $10^7 K$, $10^2 g\ cm^{-3}$ and $10^{17} g\ cm^{-1} s^{-2}$ respectively; mass and radius are given in solar units. Values without and within parenthesis refer to the BS05(GS98) and BS05(AGS05) models respectively.*

In the previous section we showed that current predictions for solar neutrino fluxes are similar to those two decades ago. How do current model predictions given in Table 4 compare to those from the BU88 model? In the BU88 model, $T_{\rm C}$, $\rho_{\rm C}$, $P_{\rm C}$, $X_{\rm C}$, $Y_{\rm C}$ and $Z_{\rm C}$ were 1.56, 1.48, 2.29, 0.3411, 0.639, 0.0199 respectively (same units as in Table 4).

By comparing these values to those given in Table 4 we see that, despite important improvements made in the solar model physics, the predicted characteristics of the solar models have changed very little. It should be noted, however, that while the added effects of all the improvements is small, the individual changes (like inclusion of diffusion, improvement in the radiative opacities, change of the solar surface metallicity) are many cases larger than the total effect but tend to compensate because act in opposite directions.

On the other hand, the predicted properties of the convective envelope have changed considerably. For instance, the location of the base of convective envelope, R_{CZ}, in the BS05(GS98) model is 0.713 $R_\odot$ (Table 3), while it was 0.74 $R_\odot$ in the BU88 model. Compare these numbers with the value determined from helioseismology (Eq. 10). The main reason for the change is an increase in the radiative opacity at the base of the convective zone that has two origins: the inclusion of element diffusion (Bahcall & Pinsonneault 1995) that produces an accumulation of metals right below the base of the convective zone (see Fig. 3, particularly panel *d*); improved radiative opacity calculations from OPAL and OP groups that, for a fixed composition, yield higher values for the opacity. Another important change is that of the predicted surface helium abundance $Y_{\rm surf}$, that is 0.243 for the BS05 model and was 0.271 in the BU88 model. Again, compare these values to the one measured by helioseismology (Eq. 11). It is also interesting to note that the initial value of helium, $Y_{\rm ini}$ is 0.273 for the BS05(GS98) model and very similar, 0.271, in BU88. The effect of element diffusion in improving the agreement with helioseismology is evident in the BS05(GS98) model, where the surface helium abundance decreases from the initial value 0.273 down to 0.243 at the present solar age.

3.2.3 Internal Structure

This section is based on results shown in Figure 3, where we show the profile of several relevant quantities as a function of radius for the BS05(GS98) solar model. Panel (*a*) shows the luminosity L, temperature T, density ρ, pressure P and enclosed mass, i.e. $m(R) = \int_0^R m(r)dr$ (units are as in Table 4). L increases very rapidly from the center and at 0.3 $R_\odot$ reaches its surface value, $L_\odot$. The solar model predicts that nuclear reactions are the only relevant energy source in the Sun and the behavior of L shows that almost all nuclear reactions and, consequently, neutrino production, take place within the inner 0.3 $R_\odot$ (we refer to this region as the solar core). Outside this radius the temperature becomes too low for nuclear reactions to be energetically important. In the solar core, T and ρ can be related by the simple analytic relation, accurate to better than 5%, $\rho = 40T_7^3$, where T_7 is the temperature in 10^7K. Panel (*b*) shows, in logarithmic scale, the profiles of the electron (n_e), neutron (n_n) and scatterers of sterile neutrinos ($n_s = n_e - 0.5n_n$) number densities (divided by Avogadro's number). To a very good approximation, this quantities can be expressed in terms of ρ and the hydrogen mass fraction X as $n_e = \rho(1+X)/2$, $n_n = \rho(1-X)/2$ and $n_s = \rho(1+3X)/4$. Analytic approximations to these quantities can be found in Bahcall et al. (2006).

Panels (*c*) and (*d*) show the profiles of the mass fractions of several important nuclear species. In particular, panel (*c*) shows the present-day and initial hydrogen X and helium Y mass fractions. In the solar core, the conversion of hydrogen to helium due to nuclear burning is apparent. The convective envelope, the base of which is indicated by the vertical line denoted R_{CZ}, has a homogeneous composition as a result of the rapid convective mixing. However, present day abundances in the convective envelope differ from the initial ones because of

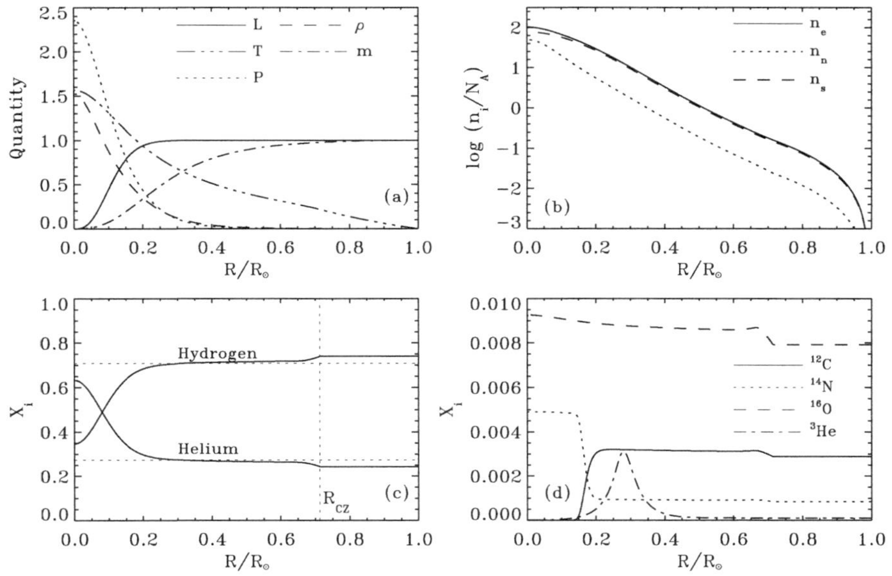

Figure 3. *Structure of the present-day BS05 SSM. Quantities are given as a function of the distance to the solar center, in units of the present-day solar radius. Panel (a) shows the luminosity (L) in solar units, temperature (T), pressure (P), density (ρ), and enclosed mass (m); units are the same as in Table 4. Panel (b) shows the logarithm of the number density of electrons (n_e), neutrons (n_n) and sterile neutrino scatterers (n_s) divided by the Avogadro number in units of cm^{-3}. Panel (c) shows the hydrogen and helium mass fraction profiles. Dotted lines denote the initial ($\tau = 0$) hydrogen and helium mass fractions. The vertical dotted line denotes the location of the inner boundary of the convective envelope. Panel (d) is the same as panel (c) but for ^{12}C, ^{14}N, ^{16}O, and ^{3}He.*

the effect of microscopic diffusion. The change in surface composition due to diffusion is small, about 10% during the solar lifetime, but as mentioned before, the inclusion of element diffusion in solar models has greatly improved the agreement between the models and helioseismology measurements. Panel (*d*) shows, on a different vertical scale, other important species. Let us consider the profile of the ^{3}He mass fraction first, starting from the envelope and moving inwards. In the envelope ^{3}He is depleted by element diffusion similarly (not appreciated in the scale of the plot) as other species like the most abundant ^{4}He and metals. Below 0.45 $R_\odot$ the increase in temperature allows the production of ^{3}He by the reactions $p(p,e^+\nu_e)^2H(p,\gamma)^3He$, but the temperature is not enough to destroy it and its abundance increases inwards. At about 0.3 $R_\odot$ the temperature rises above 7×10^6 K and destruction of ^{3}He by the $^3He(^3He,2p)^4He$ reaction becomes important as the ppI-chain becomes fully operative. Let us now focus on ^{12}C and ^{14}N. Again, the effects of convection in the envelope and element diffusion right below R_{CZ} are evident. The temperature needed to burn ^{12}C is higher than for ^{3}He and consequently ^{12}C is only affected by nuclear burning below 0.2 $R_\odot$. Nuclear reactions in the CN cycle have a very steep dependence on temperature and this is evident in the steep drop and increase in the ^{12}C and ^{14}N abundances, respectively, between 0.2 and 0.15 $R_\odot$. In this narrow region, the CN cycle switches from being completely off to operate in steady state. Carbon and nitrogen isotopes act as catalyzers in the CN cycle but because the $^{14}N(p,\gamma)^{15}O$ is the slowest reaction in cycle, the net effect is to produce ^{14}N basically at the expense of ^{12}C, the most abundant

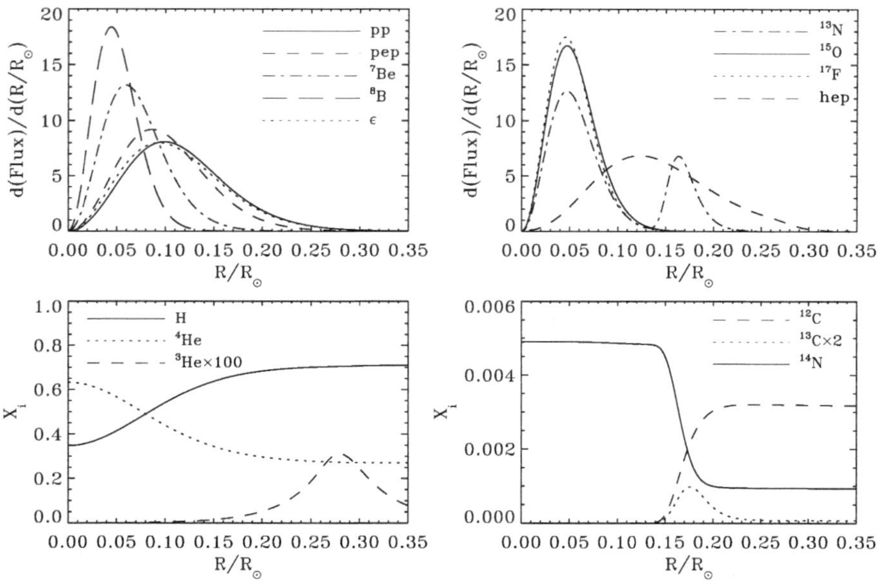

Figure 4. *Panels (a) and (b) show the neutrino production as a function of radius as indicated in the labels for the BS05 SSM. Additionally, in panel (a) ε denotes the production of nuclear energy. All quantities in panels (a) and (b) are normalized such that* $\int q_i dR = 1$. *Panels (c) and (d) show the mass fraction of selected isotopes that are most relevant to the production of neutrinos and nuclear energy generation, also as a function of radius.*

isotope entering the cycle. This is evident in the inner 0.15 $R_\odot$ where ^{12}C is almost completely depleted and the nitrogen abundance has increased to almost 5 times its initial value. Finally, in the same plot we show the profile of ^{16}O, which enters in the NO cycle. Proton captures on this isotope are the responsible for the ^{17}F neutrino flux. However, the predicted ^{17}F neutrino flux (Table 3) is two orders of magnitude smaller than the ^{13}N and ^{15}O fluxes, giving a clear indication that ^{16}O burning must be just marginal in the present-day Sun. In fact, the profile of the ^{16}O mass fraction simply is the result of the action of element diffusion: depletion at the convective envelope and increase towards the center.

3.2.4 Solar Neutrinos

There are eight neutrino fluxes produced in the solar core. It is important to characterize the production of each of these fluxes at different depths in the Sun, in particular because the location of the production region will determine, together with the energy of the neutrinos, how matter oscillations, through the MSW effect, will affect the neutrino propagation. We characterize the production profile for each neutrino flux by defining $d(\phi_i)/d(R/R_\odot)$, a measure of the contribution of a shell of thickness $d(R/R_\odot)$ to the total production of a given neutrino flux. Calling ε_i the number of neutrinos of a given flux that are produced per unit mass in the shell between $R/R_\odot$ and $R/R_\odot + d(R/R_\odot)$, then

$$\frac{d(\phi_i)}{d(R/R_\odot)} \propto \left(\frac{R}{R_\odot}\right)^2 \rho\, \varepsilon_i \tag{16}$$

and the normalization constant can be fixed by imposing the normalization condition $\int_0^{R_\odot} [d(\phi_i)/d(R/R_\odot)]\, d(R/R_\odot) = 1$.

Defined this way, $d(\phi_i)/d(R/R_\odot)$, is simply the probability density function of the production of each neutrino flux. We show $d(\phi_i)/d(R/R_\odot)$ for all the eight neutrino fluxes in panels (*a*) and (*b*) of Figure 4. Additionally, we show for the luminosity an analogous quantity, $d(L/L_\odot)/d(R/R_\odot)$, that traces the production of nuclear energy.

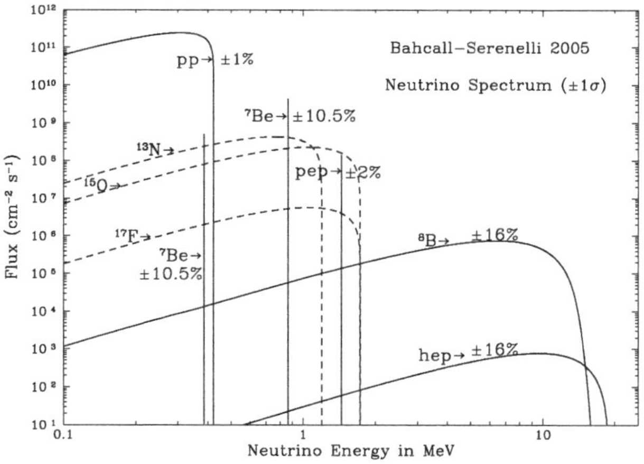

Figure 5. *Solar neutrino energy spectrum for the SSM BS05(GS98). Uncertainties are taken from Bahcall & Serenelli (2005).*

As a general rule, the higher the nuclear charge of the ions involved in the nuclear reaction, the stronger the dependence of the nuclear cross section on temperature. This, in turn, results in a more localized occurrence of the reaction towards regions of higher temperatures, i.e. towards the center. This simple rule explains why the *pp*, *pep* and *hep* fluxes, which depend on T roughly as T^4, have a broad distribution, while the ^{7}Be, ^{8}B, ^{13}N, ^{15}O and ^{17}F fluxes are more concentrated (CNO and ^{8}B fluxes depend with T^{15-20}). It is interesting to note the double-peaked nature of the ^{13}N flux. The inner peak is associated with the region in which the CN cycle operates at steady state, i.e. where the full cycle is operational. The outer peak represents the burning of ^{12}C through the reactions $^{12}C(p,\gamma)^{13}N(\beta^+\nu_e)^{13}C$ at regions where temperature is not high enough to close the CN cycle. This is made clear when looking at panel (*d*) where the abundances of ^{12}C, ^{13}C and ^{14}N are shown. From Table 3 we see that the ^{13}N flux is about 30% higher than the ^{15}O flux. This 30% corresponds to the second peak in the ^{13}N seen in the production profile. In panel (*a*) the profiles corresponding to the *pp* neutrino flux and the total production of nuclear energy $d(L/L_\odot)/d(R/R_\odot)$ almost completely overlap, with the latter being slightly shifted towards the center. The overlap results from the fact that almost all nuclear energy in the Sun is originated in the ppI chain, which is traced by the *pp* neutrino flux production. The small shift towards the center shown by $d(L/L_\odot)/d(R/R_\odot)$, however, is the result of additional energy sources, mostly the ppII chain, which can be associated in this plot to the ^{7}Be neutrino flux.

The spectra of all eight solar neutrino fluxes predicted by the BS05(GS98) model, normalized to their value on Earth, are shown in Figure 5. The plot also includes the theoretical uncertainties as computed in Bahcall & Serenelli (2005).

Finally, we close this section with a brief discussion on how the solar structure and the neutrino energies determine the properties of neutrino propagation in the solar interior. We

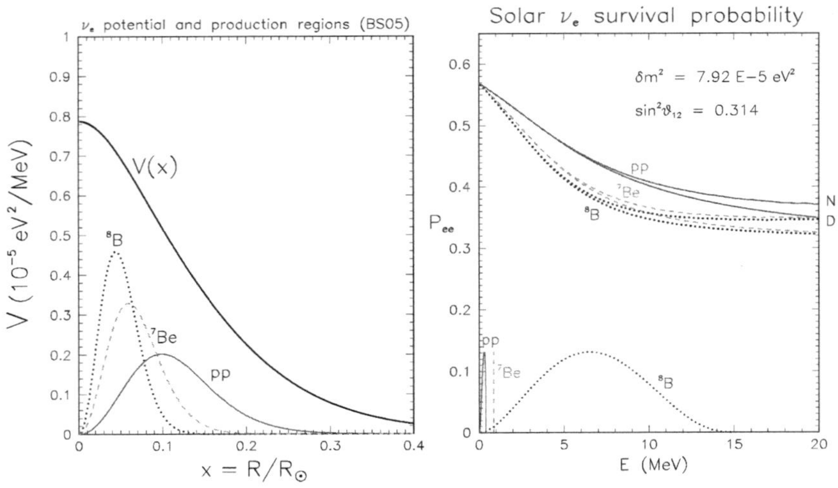

Figure 6. *Left panel: Neutrino potential $V = \sqrt{2}G_F n_e$ as a function of the normalized radius in the Sun. Also shown are the radial production regions for ^8B, ^7Be and pp solar neutrinos (in arbitrary vertical scale). Right panel: energy profile of the solar ν_e survival probability P_{ee} for best-fit LMA values and $\theta_{13} = 0$. $P_{ee}(E)$ shows a smooth transition from vacuum to matter-dominated regime as E increases, with some differences induced by averaging over different production regions for different solar neutrinos and, to a smaller extent, by nighttime (N) Earth effects with respect to daytime (D). Also shown are the corresponding solar neutrino energy (in arbitrary vertical scale). Adapted from Fogli et al. (2006) with permission of the authors. Copyright (2006) Elsevier.*

follow the basic line of argument from Fogli et al. (2006).

A simplifying, but still very good, approximation is to characterize the electron neutrinos produced in the Sun as a mixture of only two mass eigenstates ν_1 and ν_2, i.e. assuming the mixing angle $\theta_{13} = 0$. Then, only two parameters determine the (ν_1, ν_2) oscillations, the mass splitting squared δm^2 and the mixing angle θ_{12}. The solution to the solar neutrino problem has been found in neutrino oscillations to be characterized by the Large Mixing Angle (LMA) solution. Current experiments constrain the neutrino parameters to a region characterized by (Fogli et al. 2006)

$$\delta m^2 = 7.92 \times 10^{-5} eV^2 (1 \pm 0.09) \tag{17}$$

$$\sin^2\theta_{12} = 0.314\left(1^{+0.18}_{-0.15}\right). \tag{18}$$

In the solar interior, matter (MSW) effects are important for the propagation of solar neutrinos (de Holanda et al. 2004). The neutrino potential $V(x) = \sqrt{2}G_F n_e(x)$ is shown in the left panel of Figure 6 (here $x = R/\mathrm{R}_\odot$) together with the neutrino production regions (see Figure 4) for the ^{8}B, ^{7}Be and pp neutrino fluxes. The survival probability for electron neutrinos at the Earth is

$$P_{ee} = \frac{1}{2} + \frac{1}{2}\cos 2\widetilde{\theta}_{12}(x)\cos 2\theta_{12} \tag{19}$$

where

$$\cos 2\widetilde{\theta}_{12} = \frac{\cos 2\theta_{12} - A(x)/\delta m^2}{\sqrt{(\cos 2\theta_{12} - A(x)/\delta m^2)^2 + \sin^2 2\theta_{12}}}. \tag{20}$$

Here, $A(x) = 2EV(x)$ and E is the neutrino energy. There is a smooth transition from the vacuum solution $P_{ee}(0) = 1 - 0.5\sin^2 2\theta_{12}$ to the matter dominated value $P_{ee} \sim sin^2\theta_{12}$ as the energy increases. Matter effects become important when $A(x) \sim \delta m^2$. We can see by examining $V(x)$ in the left panel of Figure 6 that for δm^2 in the LMA range, matter effects are relevant for neutrinos with energies of a few MeV. In the right panel of Figure 6 we show the survival probability as a function of energy for the same neutrino fluxes as before. Curves are different because P_{ee} is averaged over the production region corresponding to each neutrino flux. However, when we take into account the actual solar neutrino spectra (also shown in the plot), it is clear that pp neutrinos are not affected by matter effects and the same is true for the ^{7}Be neutrinos. On the contrary, ^{8}B neutrinos are heavily affected by matter effects. Because $V(x)$ increases towards the solar center, transition from the vacuum to the matter dominated regime is faster (lower energies) for neutrinos produced closer to the center. The plot also shows the day-night difference originated by matter effects in the Earth.

3.2.5 Helioseismology: Sound Speed and Density Profiles

Before the solar neutrino problem was definitely settled by the SNO neutral current experiment, the standard solar model had received strong support from helioseismology measurements. In addition to quantities such as R_{CZ} and Y_{surf} discussed previously, solar oscillations allow to construct a detailed picture of the solar interior or, at least, of some relevant quantities. Solar oscillations are acoustic waves (p-modes) and as such, the sound speed in the solar interior plays a crucial role in their frequencies. Helioseismology can constrain very well the solar sound speed profile. To a very good approximation the equation of state of solar matter is that of an ideal gas, and so the sound speed depends on both temperature T and molecular weight μ as $c \propto \sqrt{T/\mu}$. In Figure 7 we show the sound speed against radius on the left panel for the BS05 models. c increases inwards as temperature does, but in the inner 0.15 $R_\odot$ there is a turn-over in the sound speed profile that then decreases towards the center. This decrease is the effect of an increased μ in regions where nuclear reactions have transformed an appreciable amount of hydrogen into helium. Unfortunately, helioseismology cannot give at the moment a very detailed picture of this region in the Sun because the p-modes do not reach very close to the center. They are limited by a centrifugal potential barrier (think of the term $l(l+1)/r^2$ in the wave equation in spherical coordinates). Future helioseismology experiments are being planned, however, to detect other class of solar oscillations, called g-modes, that do indeed reach the solar core and would be extremely useful in characterizing that solar region. The difficulty lies in that g-modes are very hard to detect as their amplitudes in the solar surface are really very small (velocity amplitudes are expected to be of only a few millimeters per second, compared to several centimeters per second corresponding to amplitudes of p-modes). The relative difference between the model and solar sound speeds as determined from helioseismology inversions (Bahcall et al. 2005) are shown in the right panel of Figure 7 for both the BS05(GS98) and BS05(AGS05) models. Uncertainties shown in the plot are originated in the helioseismology analysis. The agreement between the solar model and helioseismology measurements is excellent. The largest difference occurs right below the convective envelope (CE). We know that SSMs are not a fair description of reality in those regions and it is difficult to identify the origin of the discrepancy. A lot of "action" is going on in a narrow region (usually referred to as the tachocline) below the CE: even if the region is formally stable against convection, i.e. radiatively stratified, convective plumes coming down from the envelope can

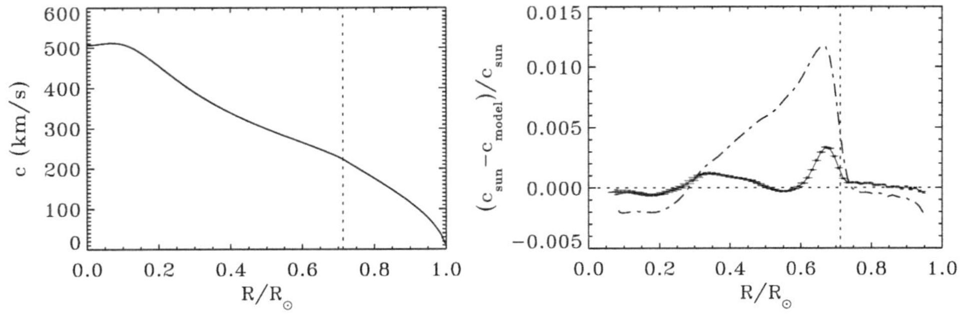

Figure 7. *Left panel: solar sound speed as a function of the normalized radius. Right panel shows the relative difference (δc) between the solar (c_{sun}) and the model sound speed (c_{model}) as inferred from helioseismology inversions for the BS05(GS98) (solid line) and BS05(AGS05) (dash-dotted line) models. Error bars (barely visible at this scale) refer to uncertainties arising from the inversion procedure.*

penetrate certain distance into the radiative region and modify the temperature gradient; solar magnetic fields are thought to be generated and anchored in the tachocline; there are shear effects in the tachocline, where the rotation profile of the Sun changes abruptly; there may be mixing of material between the solar envelope and the interior induced by solar rotation. None of these effects is currently accounted for in standard solar models and may have an influence in the thermodynamic properties of matter, affecting this way the sound speed. However, the largest discrepancy between the BS05(GS98) model and the measurements is only 0.3%. We use the averaged rms difference of the sound speed profile, defined by

$$\langle \delta c \rangle = \sqrt{N^{-1} \sum_{i=1}^{N} \frac{(c_{\odot,i} - c_{\mathrm{model,i}})^2}{c_{\odot,i}^2}}, \tag{21}$$

as a measure of the overall agreement between solar models and helioseismology. Here, $c_\odot$ is the solar sound speed, c_{model} the model sound speed, and the sum is carried out over N shells in the solar model. The density profile can also be determined from helioseismology inversions and an analogous quantity $\langle \delta \rho \rangle$ can be calculated. However, in the case of the density the precision is smaller than for the sound speed. $\langle \delta c \rangle$ and $\langle \delta \rho \rangle$ (see Table 3) are 0.1% and 1.1% respectively for the BS05(GS98) model.

3.2.6 Power-law dependencies

In the previous sections we have presented SSM results. Because the equations of stellar structure are non-linear the effects that variations in the input parameters (e.g. cross sections, element abundances, present-day solar quantities) have on the SSM predictions are sometimes not easy to predict. In the case of the SSM there is an additional difficulty in doing so due to the boundary conditions of matching the present-day solar parameters. A very useful way to understand the relation between input parameters and solar properties, in particular neutrino fluxes, is to compute partial derivatives expressing the dependence of each neutrino flux with respect to a given input parameter (Bahcall 1989).

The procedure is very simple. Let β_j^0 be the best estimate of a given input parameter in the solar model and $\Delta\beta_j$ its fractional uncertainty. By computing a solar model using the value $\beta_j = \beta_j^0(1+\Delta\beta_j)$ as input parameter and another one with $\beta_j = \beta_j^0(1-\Delta\beta_j)$ we can easily calculate the derivative

$$\partial \log \phi_i / \partial \log \beta_j = \alpha_{ij} \tag{22}$$

for each neutrino flux ϕ_i. α_{ij} is a measure of how the flux ϕ_i depends on the input parameter β_j and as such it contains physical information. Numerical tests show that the logarithmic derivatives are a robust measurement of the dependences in the sense that they depend very weakly on the choice of central values for the input parameters in the solar models. In fact, logarithmic derivatives based on the BS05(GS98) solar model are very similar to those obtained with the BU88 solar model (see for instance Table 3 in Bahcall & Serenelli 2005). Additionally, as it will be clear in Section 4, they provide an easy way to estimate total uncertainties for the neutrino fluxes.

By giving a close look to some partial derivatives, we can gain some insight into how input parameters and neutrino fluxes are related. In Table 5 we give the α_{ij} values for some selected input parameters and all the neutrino fluxes. In the following discussion, for simplicity, nuclear reactions are numbered according to Table 1.

Source	S_{11}	S_{33}	S_{34}	$S_{1,14}$	$L_\odot$	C	N	O	Si	Fe
pp	+0.14	+0.03	-0.06	-0.02	+0.73	-0.01	-0.00	-0.01	-0.01	-0.02
pep	-0.17	+0.05	-0.09	-0.02	+0.87	-0.02	-0.01	-0.01	-0.01	-0.06
hep	-0.08	-0.45	-0.08	-0.01	+0.12	-0.01	-0.00	-0.02	-0.04	-0.07
^{7}Be	-0.97	-0.43	+0.86	-0.00	+3.40	-0.00	+0.00	+0.05	+0.10	+0.21
^{8}B	-2.59	-0.40	+0.81	+0.01	+6.76	+0.03	+0.01	+0.12	+0.19	+0.51
^{13}N	-2.53	+0.02	-0.05	+0.85	+5.16	+0.84	+0.18	+0.08	+0.13	+0.34
^{15}O	-2.93	+0.02	-0.05	+1.00	+5.94	+0.83	+0.21	+0.09	+0.15	+0.40
^{17}F	-2.94	+0.02	-0.05	+0.01	+6.25	+0.03	+0.01	+1.10	+0.16	+0.44

Table 5. *Partial derivatives of neutrino fluxes. Each column contains the logarithmic partial derivatives,* α_{ij}, *of the neutrino fluxes with respect to the parameter shown at the top of the column.* S_{11}, S_{33}, S_{34}, $S_{1,14}$ *are the astrophysical factors of the reactions 1, 4, 5 and 14 in Table 1. C, N, O, Si and Fe are the carbon, nitrogen, oxygen, silicon and iron solar abundances.*

Let us consider some important nuclear reaction rates. The slowest reaction in the pp-chains is $p(p, \nu_e e^+)^2H$. This reaction controls the evolutionary timescale of the Sun and about 90% of the total nuclear energy release. An increase in the astrophysical factor S_{11} means more nuclear energy is produced, but since $L_\odot$ is fixed there has to be a reduction in the solar core temperature to keep the total nuclear energy generation rate at a fixed level compatible with $L_\odot$. A reduced core temperature means, of course, lower neutrino fluxes from all other sources, particularly those very sensitive to temperature. S_{33} is a very fast reaction in the ppI chain, the pp and pep neutrino fluxes depend very weakly on it. However, an increased value in S_{33} starves reactions number 5 and 10 from ^{3}He nuclei and thus the neutrino fluxes from the ppII and ppIII chains decrease, together with the hep neutrinos. S_{34} controls the flow of the pp-chains towards the ppII and ppIII chains. As such, an increase in its value directly affects the

^{7}Be and ^{8}B fluxes. Finally, $S_{1,14}$, the cross section of a proton capture by a ^{14}N nuclei, controls the flow of the CN cycle. Logically, there is almost a linear correlation between $S_{1,14}$ and the ^{13}N and ^{15}O neutrino fluxes involved in this cycle.

Nuclear energy accounts for the solar luminosity. Accordingly, variations in $L_\odot$ imply variations in the nuclear energy release and, consequently, variations in the neutrino fluxes. An increase in $L_\odot$ implies an increase in the solar core temperature to enhance nuclear burning. All nuclear reaction rates, and the neutrino production, increase with a temperature increase. Note that the stronger the dependence of the neutrino flux with temperature (the neutrino production distribution from Figure 4 can be used as a proxy for this), the stronger the dependence on $L_\odot$.

We consider now some of the effects of composition changes. In the region where neutrinos are produced, carbon, nitrogen and oxygen do not contribute appreciably to the radiative opacity and, consequently, do not affect the solar core temperature. However, they affect the CNO neutrino fluxes because the rate at which the CN and NO cycles operate depend on the number of CNO nuclei present. This can immediately seen in the table, where C and N abundance only affect the ^{13}N and ^{15}O, while the O strongly affects only the ^{17}F flux. Silicon and iron do not affect directly the neutrino fluxes in the way CNO elements do. However, elements with relatively high abundances and high Z contribute to the radiative opacity at temperatures close to those of the solar core. As we discussed before, larger opacity means steeper temperature gradient and higher core temperature. The fluxes most sensitive to temperature are then preferentially affected by changes in the abundance of these elements.

3.3 SSM and the Low Solar Metallicity Abundance

In sections 3.1 and 8 we have dealt with results from the BS05 Standard Solar Model that adopts the solar composition recommended in GS98. In this section we summarize the predictions of standard solar model BS05(AGS05) that incorporates the Asplund et al. (2005; AGS05) solar abundance determinations. Results regarding the present-day solar model show important differences if the AGS05 composition that is used instead of the older GS98 composition, particularly regarding helioseismology. Results for the BS05(AGS05) model are given in parenthesis in Tables 3 and 4.

Heavy elements are important contributors to the radiative opacity in stellar interiors through bound-bound and bound-free transitions. Even at the conditions in the solar core, heavy metals like iron are not completely ionized. The direct effect of using a lower metallicity value is then to decrease the radiative opacity of the stellar material. But, as we have seen in Section 2.1, the temperature gradient in radiative regions ($\nabla_{\rm rad}$) is proportional to the radiative opacity. How does a lower radiative gradient affect solar model predictions? First, the base of the convective envelope is located where the adiabatic temperature gradient ($\nabla_{\rm ad}$) equals $\nabla_{\rm rad}$. $\nabla_{\rm ad}$ depends only on the equation of state where metals play a minor role under the solar conditions: at first order changing the metallicity does not modify $\nabla_{\rm ad}$. On the other hand, as discussed before, $\nabla_{\rm rad}$ gets lower when the metallicity does. The result is that the point where $\nabla_{\rm ad} = \nabla_{\rm rad}$, i.e. $R_{\rm CZ}$, moves outwards. In the BS05(AGS05) model it is located at 0.728 $R_\odot$ while for BS05(GS98) it is at 0.713 $R_\odot$ (Table 3). The change is many times the uncertainty in $R_{\rm CZ}$ determined by helioseismology. Second, a lower temperature gradient in the radiative interior implies that the temperature in the solar core is lower for models with the AGS05 composition (about 1.2%

at the center in the models presented here). Neutrino fluxes that are particularly sensitive to temperature are consequently lower, as can be seen in the ^{7}Be and ^{8}B fluxes, which are lower by 10% and 20% respectively, in the BS05(AGS05) model. However, neutrino experiments cannot constrain, so far, the solar metallicity. The ^{8}B flux measured by SNO (Eq. 15) is approximately half-way between the BS05(GS98) and BS05(AGS05) predictions. An additional effect of a lower $\nabla_{\rm rad}$ in the solar core is the lower initial and surface present-day values of the solar helium abundance, $Y_{\rm ini}$ and $Y_{\rm surf}$ respectively. For BS05(AGS05), $Y_{\rm surf} = 0.229$ while for BS05(GS98) it is 0.243. As discussed before, $Y_{\rm surf}$ is determined from helioseismology. BS05(AGS05) predictions for both $Y_{\rm surf}$ and $R_{\rm CZ}$ are in disagreement with helioseismology measurements.

Finally, the effect of a lower metallicity in the sound speed profile is clearly seen in the right panel of Figure 7. The averaged rms $\langle \delta c \rangle = 0.5\%$, i.e. 5 times larger than for the BS05(GS98) model, and the maximum difference is 1.2%. For the density profile, results are analogous: $\langle \delta \rho \rangle = 4.4\%$, i.e. 4 times larger than for BS05(GS98).

In summary, standard solar models that incorporate the new AGS05 solar composition show disagreement with helioseismology measurements. New abundance determinations result from improved modelling of the solar atmosphere. Are these abundances actually a better representation of the true solar composition as well? If so, we are led to the question: is the excellent agreement between solar models with older composition and helioseismology a coincidence? On the other hand, the new solar composition determinations need to be confirmed by independent groups of researchers. The relevance of the implications certainly deserve the effort.

4 The Standard Solar Model. Theoretical Uncertainties

Historically, a lot of effort has been put in estimating the uncertainties in the SSM predictions. Initially, the driving force behind this effort was to quantify the magnitude of the "solar neutrino problem". Later, it was the necessity of identifying the most relevant individual uncertainty sources entering the model calculations in order to point out which input parameters had to be determined with better precision by new experiments (e.g. cross section measurements, improved solar abundance determinations).

Quantity	Uncert. (%)	Quantity	Uncert. (%)	Element	Uncert. (%)	Element	Uncert. (%)
p-p	0.4	hep	15.1	C	35 (12)	Si	12 (4.7)
^{3}He+^{3}He	6.0	^{14}N+p	8.4	N	38 (15)	S	9.6 (9.6)
^{3}He+^{4}He	9.4	age	0.44	O	48 (12)	Ar	66 (20)
^{7}Be+e^-	2.0	diffusion	15.0	Ne	74 (15)	Fe	12 (7.2)
^{7}Be+p	3.8	luminosity	0.4	Mg	12 (7.2)		

Table 6. *Sources of uncertainty in the SSM calculations. Cross section uncertainties are identified by the reaction name. For the element abundances two values are given: the so-called conservative uncertainties and the optimistic uncertainties. Refer to Bahcall et al. (2006) for details and relevant references.*

Theoretical uncertainties in the SSM predictions have multiple origins. Bahcall et al. (2006)

identify 21 relevant sources of uncertainty: 7 nuclear cross sections, the age and solar luminosity, microscopic diffusion, 9 individual element abundances and the radiative opacity and equation of state of stellar matter. In Table 6 we list the relevant sources of uncertainty and their magnitude. Relevant references are given in Bahcall et al. (2006).

Power-law dependences developed in Section 3.2.6, provide a flexible way to estimate the total uncertainty while at the same time keeping an eye on the contributions of individual sources. Let us briefly consider power-law dependences again. We can integrate Equation 22 to obtain $\phi_i = \phi_i^0(\beta_j/\beta_j^0)^{\alpha_{ij}}$, where $\beta_j = \beta_j^0(1+\Delta\beta_j)$. The fractional uncertainty in ϕ_i arising from the uncertainty in β_j is easily calculated

$$\frac{\Delta\phi_{ij}}{\phi_i} = \left(1+\frac{\Delta\beta_j}{\beta_j^0}\right)^{\alpha_{ij}} - 1 \approx \alpha_{ij}\frac{\Delta\beta_j}{\beta_j^0}. \tag{23}$$

Assuming that input parameters are independent and have gaussian distributions, we can combine them quadratically and get the total uncertainty

$$\frac{\Delta\phi_i}{\phi_i} = \sqrt{\sum_j\left[\left(1+\frac{\Delta\beta_j}{\beta_j^0}\right)^{\alpha_{ij}} - 1\right]^2}. \tag{24}$$

This expression is a very useful estimator of the total uncertainty of neutrino fluxes. At the same time, it allows a simple way to estimate how an improvement in the uncertainty of a given input parameter will affect the overall uncertainty.

The most robust way to estimate theoretical uncertainties is to perform a Monte Carlo simulation. First, generate a large number of sets of input parameters where the value for each individual parameter is obtained randomly from a distribution function that represents our knowledge of the parameter. Second, calculate one solar model for each parameter set. Third, characterize the resulting distributions of neutrino fluxes, helioseismological quantities, etc. This approach was first applied by Bahcall & Ulrich (1988) and recently extended in Bahcall et al. (2006) where, in addition to neutrino fluxes, it was extensively applied to helioseismological quantities for the first time. A detailed account of the method can be found in that reference. Current neutrino fluxes uncertainties are given in Table 7 for two possible choices of element abundance uncertainties as discussed in Bahcall et al. (2006).

Flux	σ (%)	Flux	σ (%)	Flux	σ_+ (%)	σ_- (%)	Flux	σ_+ (%)	σ_- (%)
pp	0.9 (0.7)	hep	15 (15)	^{8}B	17 (13)	15 (11)	^{15}O	37 (17)	27 (14)
pep	1.5 (1.1)	^{7}Be	11 (9.3)	^{13}N	37 (15)	27 (13)	^{17}F	72 (17)	42 (14)

Table 7. *Total neutrino flux uncertainties as obtained from Monte Carlo simulations from Bahcall et al. (2006). Values without parenthesis refer to the conservative choice of abundance uncertainties. Values in parenthesis refer to the optimistic choice.*

An analysis of the uncertainties can be done by considering the power-law dependences (Table 5), the input parameter uncertainties (Table 6) and expressions (23) and (24). Here, we limit our discussion to a qualitative level. Detailed discussion of uncertainties can be found in Bahcall & Pinsonneault (2004) for the non-composition uncertainties, in Bahcall & Serenelli

(2005) for the composition uncertainties and improved overall uncertainties using power-law dependences, and in Bahcall et al. (2006) for a complete Monte Carlo analysis.

The large uncertainties in the CNO fluxes are dominated by abundance uncertainties. For the hep flux we see by looking at Table 5 that it is not sensitive to the metal abundances. The origin of the large uncertainty must originate in the cross sections. By looking into Table 6 we can see the bulk of the uncertainty comes from the hep cross section ($\partial \log \phi_{\text{hep}} / \log S_{\text{hep}} \approx 1$) and a minor contribution must come from the S_{33} uncertainty. The ^{8}B flux is sensitive to most input parameters. The strongest dependence is with the solar luminosity, but the uncertainty in $L_\odot$ is rather small so it does not play a relevant role in the total ^{8}B flux uncertainty. In Table 5 we identify S_{34} and elements heavier than oxygen as relevant parameters for this flux. The large uncertainties in the element abundances make this source of uncertainty the dominant one for the ^{8}B flux, but uncertainty from S_{34} is also of relevance. For the ^{7}Be flux, S_{34} is the dominant contribution but metal abundances are also relevant (however see next paragraph). The pp and pep neutrino fluxes have small uncertainties to which uncertainties in C, O, Ne, and Fe contribute in a similar way.

A new measurement of the ^{3}He+^{4}He cross section (Bemmerer et al. 2006), not yet included in uncertainty calculations, has lowered its uncertainty to just 3%, compared to the previous 9% uncertainty.

The current situation is that, with the exception of the hep flux, all neutrino fluxes have uncertainties dominated by the solar composition uncertainties. Robust determinations of the solar composition and its uncertainties are needed to refine the SSM predictions. In particular, abundance determinations from multiple groups are desirable in order that systematic uncertainties can be more readily assessed.

5 Conclusions

We have presented the basics of the Standard Solar Model (SSM) calculations and have discussed in some detail its predictions, particularly in relation to solar neutrinos and helioseismology. The link between the SSM and neutrinos has proved fundamental in finding new neutrino physics and determining neutrino parameters. The SSM faces a new problem related to recent solar composition determinations, the "solar abundance problem". If the recent abundance measurements are used, then the SSM gives wrong predictions for helioseismology, whereas if the old solar composition is used, the agreement between the model and the data is very good. Different assumed heavy-element abundances give different predictions for solar neutrino fluxes. Unfortunately, the agreement between solar neutrino fluxes predicted by the SSM and solar neutrino measurements is equally good for any of the two abundance sets (GS98 or AGS05). Future neutrino experiments may help in constraining the solar composition. However, it should be kept in mind that solar neutrino flux uncertainties are dominated by uncertainties in the solar composition.

Acknowledgments

I would like to thank the organizers of the School for the kind invitation, hospitality and the stimulating environment they have created. This work has been supported by a Ralph E. and Doris M. Hansmann Membership.

References

Abdurashitov, J. N., et al. (SAGE Collaboration) (2003), *Nucl. Phys. B (Proc. Suppl.)*, **118**, 39

Adelberger, E. G., et al. (1998), *Rev. Mod. Phys.*, **70**, 1265

Aharmim, B., et al. (SNO Collaboration), *Phys Rev C*, **72**, 055502

Ahmad, Q. R., et al. (SNO collaboration) (2002), *Phys. Rev. Lett.*, **89**, *1302*

Ahmed, S. N., et al. (SNO Collaboration) (2004), *Phys. Rev. Lett.*, **92**, *181301*

Ahmad, Q. R., et al. (2002), *Phys. Rev. Lett.*, **89**, *1302*

Altmann, G., et al. (GNO collaboration) (2005), *Phys. Lett. B*, **616**, 174

Asplund, M., Grevesse, N., & Sauval, A. J. (2005), in ASP Conf. Ser. 336, Cosmic Abundances as Records of Stellar Evolution and Nucleosynthesis, ed. T. G. Barnes III, & F. N. Bash (San Francisco: ASP), 25

Badnell, N. R., et al. (2005), *Monthly Notices of the Royal Astronomical Society,* **360**, *458*

Bahcall, J. N. (1971), *The Astronomical Journal,* **76**, *283*

Bahcall, J. N. (1989), *Neutrino Astrophysics* (Cambridge Eng.: Cambridge University Press)

Bahcall, J. N., Bahcall, N. A., & Shaviv, G. (1968), *Phys. Rev. Lett.,* **20**, *1209*

Bahcall, J. N., & Pinsonneault, M. H. (1995), Rev. Mod. Phys., **67**, 781

Bahcall, J. N., & Pinsonneault, M. H. (2004), *Phys. Rev. Lett.,* **92**, *121301*

Bahcall, J. N., & Serenelli, A. M. (2005), *The Astrophysical Journal, 626, 530*

Bahcall, J. N., Serenelli, A. M., & Basu, S. (2005), *The Astrophysical Journal,* **621**, *L85*

Bahcall, J. N., Serenelli, A. M., & Basu, S. (2006), *The Astroph. Journal Supplement,* **165**, *400*

Bahcall, J. N., & Ulrich, R. K. (1988), *Rev. Mod. Phys., 60*, 297

Basu, S., & Antia, H. M. (1997), *Monthly Notices of the Royal Astronomical Society,* **287**, *189*

Basu, S., & Antia, H. M. (2004), *The Astrophysical Journal,* **606**, *L85*

Bemmerer, D., et al. (LUNA collaboration) (2006), *Phys. Rev. Lett.,* **97**, *122502*

Chapman, S., & Cowling, T. G. (1970), *The Mathematical Theory of Non-uniform Gases* (Cambridge Eng.: Cambridge University Press)

Christensen-Dalsgaard, J., Gough, D. O., & Thompson, M. J. (1991), *The Astrophysical Journal,* **378**, *413*

Clayton, D. D. (1983), *Principles of Stellar Evolution and Nucleosynthesis* (Chicago: University of Chicago Press)

Cleveland, B. T., et al. (1998), *The Astrophysical Journal,* **496**, *505*

Cox, R. P., & Giuli, R. T. (1968), *Principles of Stellar Structure* (New York: Gordon and Breach)

Davis, R. Jr., Harmer, D. S., & Hoffmann, K. C. (1968), *Phys. Rev. Lett.,* **20**, *1205*

de Holanda, P. C., Liao, W., Smirnov A. Yu. (2004), *Nucl. Phys. B*, **702**, 307

Dziembowski, W. A., Pamyatnykh, A. A., & Sienkiewicz, R. (1990), *Monthly Notices of the Royal Astronomical Society,* **244**, *542*

Fogli, G. L., et al. (2006), *Prog. Part. and Nucl. Phys.*, **57**, 742

Fukuda, S., et al. (Super-Kamiokande Collaboration) (2001), *Phys. Rev. Lett.,* **86**, *5651*

Fukuda, Y., et al. (Kamiokande Collaboration) (1996), *Phys. Rev. Lett.,* **77**, *1683*

Grevesse, N., & Sauval, A. J. (1998), *Space Sci. Rev.*, **85**, 161

Gribov, V. N., & Pontecorvo, B. M. (1969), *Phys. Lett. B*, **28**, 493

Hampel, W., et al. (GALLEX collaboration) (1999), *Phys. Lett. B*, **447**, 127
Kippenhahn, R., & Weigert, A. (1990), *Stellar Structure and Evolution* (Berlin: Springer-Verlag)
Mikheyev, S. P., & Smirnov, A. Y. (1985), *Soviet J. Nucl. Phys.*, **42**, 913
Rogers, F. J., & Iglesias, C. A. (1996), *The Astrophysical Journal,* **464***, 943*
Rogers, F. J., Swenson, F. J., & Iglesias, C. A. (1996), *The Astrophysical Journal,* **456***, 902*
Seaton, M. J. (2005), *Monthly Notices of the Royal Astronomical Society,* **362***, L1*
Thoul, A., Bahcall, J. N., & Loeb, A. (1994), *The Astrophysical Journal,* **421***, 828*
Wolfenstein, L. (1978), *Phys. Rev. D*, **17**, 2369

Solar Neutrino Experiments - Results and Prospects

Based on Lectures given by David Wark

Imperial College London and STFC, Rutherford Appleton Laboratory

1 Introduction

The idea of the neutrino arose due to problems encountered during the early days of beta decay. The beta decay spectra were continuous instead of discrete and the the spins did not add up. For example in the decay $^{14}C \longrightarrow ^{14} N + e^-$, the initial particle is in a spin 0 state, while the two decay states are in spin 1 and spin 1/2 respectively (Scott 1935). Bohr suggested that maybe energy and momentum were not conserved in beta decay. In a famous letter to Lise Meitner presented at a worshop in Tübingen in 1930, Pauli made a radical proposal:

"I have hit upon a desperate remedy to save the "exchange theorem" of statistics and the law of conservation of energy. Namely, the possibility that there could exist in the nuclei electrically neutral particles, that I wish to call neutrons, which have spin 1/2 and obey the exclusion principle and which further differ from light quanta in that they do not travel with the velocity of light. The mass of the neutrons should be of the same order of magnitude as the electron mass and in any event not larger than 0.01 proton masses."

The detection of neutrinos (the name later given by Fermi to Pauli's "neutrons") was an extreme challenge for the experiments of the mid twentieth century. Pauli, in fact, apologized for hypothesizing a particle that could not be detected. In a Chalk River report in 1946, Bruno Pontecorvo pointed out the advantages of a radiochemical experiment based on $\nu_e +^{37} Cl \longrightarrow ^{37} Ar + e^-$ (and even mentioned solar neutrino detection using this method), long before the Reines-Cowan experiment that finally discovered the neutrino.

The first application of this method of Pontecorvo, however, was by Ray Davis using reactor anti-neutrinos (Davis 1955). Davis deployed large tanks containing carbon tetrachloride near the Brookhaven nuclear reactor to search for evidence that the antineutrino and the neutrino are the same particle. If $\nu = \overline{\nu}$ you would expect to see ^{37}Ar produced by the reaction: $\overline{\nu_e} +^{37} Cl \longrightarrow ^{37} Ar + e^-$. By 1957, enough sensitivity had been reached to show that the rate was too small, from which it was concluded that the neutrino and antineutrino were different particles. We now know that this is not correct because parity is violated in weak interactions and the neutrino could be a Majorana particle.

The emission of neutrinos from the sun are a direct probe of the fusion reactions that occur in the centre of the sun. Photons take 10^4 years to escape the sun, the energy produced by the fusion reactions takes 10^7 years to get out, while the neutrinos come out at the speed of light.

The net fusion reaction that powers the sun is: $4p \longrightarrow^4 He + 2e^+ + 2\nu_e$. This releases 25.7 MeV of energy per reaction, or 4.12×10^{-12} J, per helium nucleus produced (half that per neutrino). The solar constant is 1370 Watts m^{-2} at the orbit of the Earth, thus the neutrino flux should be: $1370/(2.06 \times 10^{-12})$ m^{-2} s^{-1} $= 6.65 \times 10^{10}$ cm^{-2} s^{-1}. This calculation is accurate to better than 10%. However, the neutrino flux model is much more complicated than this simple calculation, see (Serenelli 2008) in these proceedings and Fig. 1. The dominant reaction is the low energy pp reaction, with a continuous neutrino spectrum up to 0.420 MeV. The ^7Be reaction produces two neutrino line spectra of 0.383 MeV and 0.861 MeV, while the pep reaction also produces a line spectrum at 1.442 MeV. The high energy ^8B and hep reactions produce continuous neutrino spectra with end points 14.06 MeV and 18.77 MeV respectively.

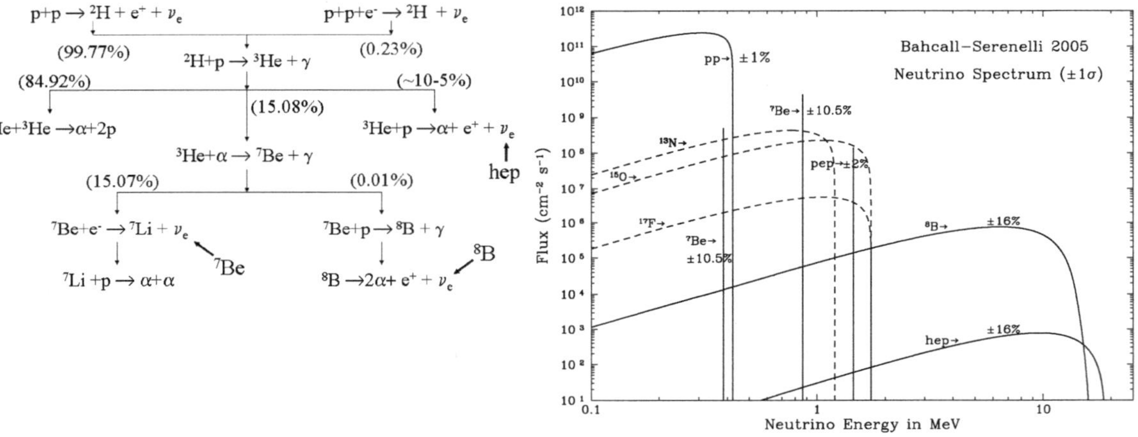

Figure 1. *Left: Standard solar model reactions in the sun. Right: Standard Solar Model neutrino spectra as a function of energy.*

The remainder of this article will describe the experiments that were designed for the detection of neutrinos from the sun, the interpretation of the results derived from those experiments and a future outlook in the field of solar neutrinos. Section 2 describes the history of solar neutrino experiments and how the solar neutrino problem arose. Section 3 describes how neutrino oscillations were considered to be the solution to the solar neutrino problem. Sections 4 and 5 describe the two high statistics solar neutrino experiments (carried out at SuperKamiokaNDE and at the Sudbury Neutrino Observatory) that finally confirmed that neutrino oscillations are responsible for the solar neutrino deficit. Finally, section 6 will review future solar neutrino experiments and the results that we might expect to obtain from them.

2 Solar Neutrino Experiments

2.1 The Davis Chlorine Experiment

The search for solar neutrinos began through the pioneering experiments carried out by Ray Davis and his collaborators at Brookhaven National Laboratory. Davis used the same technique of the inverse beta decay reaction of neutrinos on chlorine in a large tank of carbon tetrachloride: $\nu_e + ^{37}Cl \longrightarrow ^{37}Ar + e^-$. The large tank (6.1 m in diameter and 14.6 m long) containing 100,000 gallons of carbon tetrachloride (615 tonnes) was installed at a depth of 1478 m below the surface, to protect it from cosmic rays, in the Homestake Gold Mine in Lead, South Dakota. This experiment was known as the Brookhaven Solar Neutrino Experiment (Fig. 2).

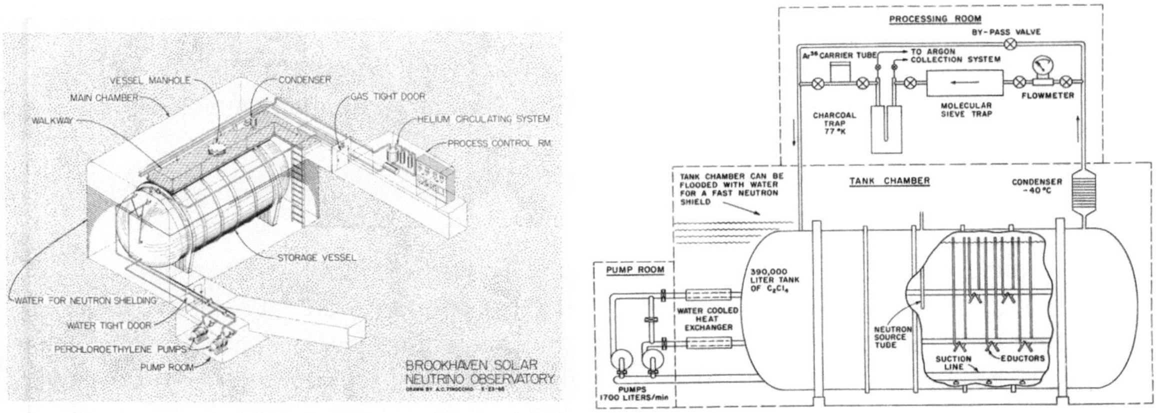

Figure 2. *Left: Diagram of the Brookhaven Solar Neutrino Experiment in Lead, South Dakota. Right: Schematic diagram of the tanks for the Brookhaven Solar Neutrino Experiment. From (Cleveland, 1998). Copyright (1998) The American Astronomical Society.*

The small number of argon atoms produced by solar neutrinos are removed by purging the tank with helium and collecting the argon atoms with a charcoal trap. The radioactive argon atoms are counted by allowing them to decay inside a series of low noise proportional counters that record the K orbital electron capture followed by Auger electrons of 2.823 keV. Results from the Homestake Chlorine Solar Neutrino Experiment between 1970 and 1994 are shown in Fig. 3 (Cleveland 1998). The average rate is $0.478 \pm 0.030(stat) \pm 0.029(syst)$ ^{37}Ar atoms produced per day, which is $2.56 \pm 0.16(stat) \pm 0.16(syst)$ Solar Neutrino Units or SNU (1 SNU = one interaction per 10^{36} target atoms per second). The standard solar model of Bahcall and Serenelli (Serenelli 2008) predicts a rate between 6.6 and 8.1 SNU, depending on the model of elemental abundances found in the sun. Clearly, the experiment by Davis confirmed that the sun produces neutrinos, but the number of neutrinos detected were only only about one-third of the number of neutrinos predicted by theory. This was known as the "solar neutrino puzzle", and gave birth to a number of different experiments, all of which were trying to confirm the solar neutrino deficit discovered by Davis and his collaborators.

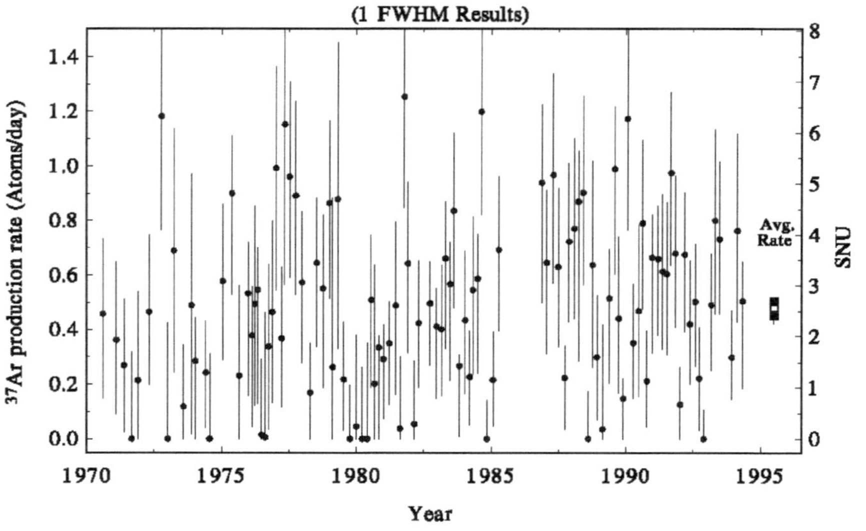

Figure 3. *Results from the Homestake Chlorine Solar Neutrino Experiment between 1970 and 1994. From (Cleveland, 1998). Copyright (1998) The American Astronomical Society.*

2.2 KamiokaNDE

The Kamioka Nucleon Decay Experiment (KamiokaNDE) was a large water Cherenkov tank of 15.6 m in diameter and 16.1 m height, containing 3000 tons of pure water at a depth of 1000 m underground in the Kamioka mine in Japan (Fig. 4). The original nucleon decay experiment was upgraded to be able to detect solar neutrinos. The experiment ran between 1990 and 1995. It detected the elastic scattering of neutrinos from electrons in the water: $\nu_e + e^- \longrightarrow \nu_e + e^-$. The recoil electrons were detected from the Cherenkov light they emit in a fiducial volume inside the tank containing 680 tons of water, detected by 948 photomultiplier tubes. The detection threshold reached a 50% efficiency at 5 MeV in the later stages of the experiment. The angular resolution of the electrons is mainly determined by multiple Coulomb scattering in the water and is about 26° at 10 MeV. Fig. 4 shows the results for a 1036 day run between 1990 and 1995, where the number of events as a function of the cosine of the zenith angle with respect to the position of the sun, clearly shows an excess of events above background in the direction of the sun (Fukuda et al. 1996), thereby proving that electron neutrinos indeed arrive from the sun. The number of events observed in the direction of the sun (390 ± 35 events) correspond to a flux of $(2.80 \pm 0.19(stat) \pm 0.33(syst)) \times 10^6$ cm^{-2} s^{-1}, which was found to be $0.496 \pm 0.044(stat) \pm 0.045(syst)$ of the standard solar model at the time (Bahcall & Pinsonneault 1996). The latter result showed that, indeed, there was a deficit of solar neutrinos, thus confirming the solar neutrino puzzle.

2.3 Gallium Experiments

The next stage in the process was to find a reaction that would be sensitive to the dominant low energy solar neutrino pp component. The reaction, proposed by Kuzmin (Kuzmin 1966) to detect pp neutrinos was $\nu_e +^{71} Ga \longrightarrow^{71} Ge + e^-$, with a reaction threshold of 0.233 MeV, below

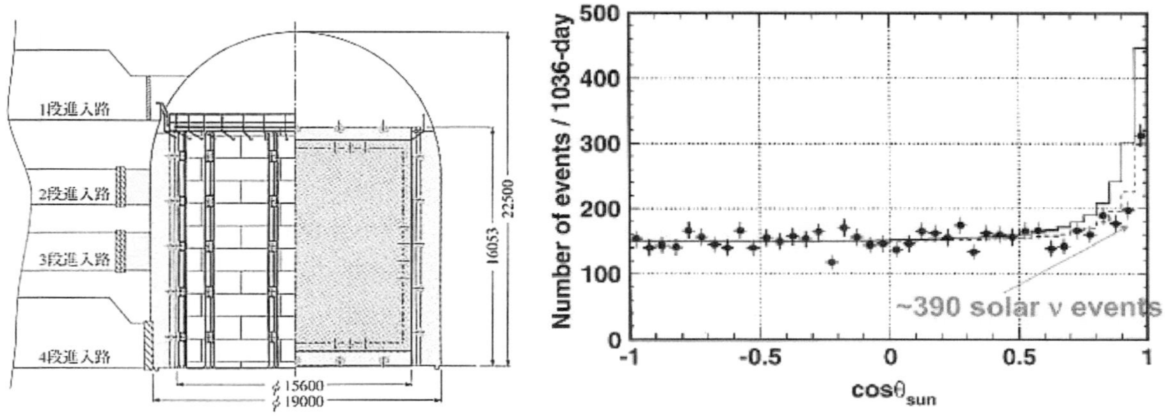

Figure 4. *Left:Diagram of the KamiokaNDE water Cherenkov detector. Right: Number of low energy electron events in the KamiokaNDE detector during a period of 1036 days as a function of the zenith angle with respect to the position of the sun. From (Fukuda et al. 1996). Copyright (1996) The American Physical Society.*

the 0.42 MeV end point. The Standard Solar Model of Bahcall and Serenelli predicted between 119 and 126 SNU (depending on the solar elemental abundances) for this reaction capture rate, with the pp neutrinos contributing 59%, the ^7Be neutrinos 29% and the ^8B neutrinos 12% of this rate, respectively.

Two radio-chemical gallium experiments, similar to the chlorine experiment, started taking data in the 1990s: the Russian-American Gallium Experiment (SAGE) established in the Baksan tunnel under Mount Andyrchi in the Caucusus Mountains in Russia at a depth of 4700 m water equivalent (mwe) and the GALLium EXperiment (GALLEX), with its successor the Gallium Neutrino Observatory (GNO), which was established at the Laboratori Nazionale di Gran Sasso in the Gran Sasso tunnel in Italy. The main difference between the experiments is that the target is metallic gallium (50 tons) in SAGE and is a 100 ton aqueous gallium chloride solution (with 30 tons of natural gallium) in GALLEX/GNO. Otherwise, the two experiments are very similar.

We will describe the SAGE procedure, as representative of both experiments. The SAGE target is divided in 7 reactors. A Ga-Ge alloy with a known Ge content of approximately 350 μg is distributed equally amongst the reactors and stirred thoroughly to disperse the Ge throughout the Ga mass. After a typical exposure interval of 27 days, the Ge carrier and ^{71}Ge atoms produced by solar neutrinos and background sources are chemically extracted from the Ga. A small amount of germane gas (GeH_4) is synthesised and used as the proportional counter fill gas, with an admixture of 80% to 90% of Xe. The total efficiency of extraction is the ratio of mass of Ge in the germane to the mass of initial Ge carrier and is typically in the range of 80% to 90%. The ^{71}Ge decays via electron capture to ^{71}Ga with a half-life of 11.43 days, yielding low energy K shell (10.4 keV) and L shell (1.2 KeV) Auger electrons and X rays.

Furthermore, both the gallium experiments were calibrated with $\sim$ 1 MCi ^{51}Cr sources. The ^{51}Cr isotope decays via electron capture to ^{51}V with neutrino energies of 751 keV (90.12%) and 426 keV (9.88%). This provided a detailed calibration with sources of known activity. For the GALLEX experiment, two exposures with sources of 1.17 ± 0.04 MCi and 1.87 ± 0.07 MCi yielded the expected ^{71}Ge production rates of 11.7 ± 0.2 and 12.7 ± 0.2 atoms per day. The ratio of measured to expected yield was 0.93 ± 0.08, thereby verifying all the radio-chemical extraction and counting procedures carried out by the experiment (Hampel et al. 1998). A similar calibration with a 0.517 ± 0.007 MCi source was performed in SAGE, yielding a ratio of 0.95 ± 0.12 in the measured to expected number of ^{71}Ge atoms (Abdurashitov et al. 1999). Additionally, SAGE have carried out another calibration using a 409 ± 2 kCi source of ^{37}Ar (Barsanov et al. 2007) yielding a ratio of measured to predicted event rate of 0.79 ± 0.10.

The results from both the SAGE and GALLEX/GNO experiments are shown in Fig. 5. SAGE has taken 145 runs between January 1990 and December 2005. The average measurement of solar neutrinos on a gallium target by the SAGE experiment yields $66.5^{+3.5}_{-3.4}(stat)$ $^{+3.5}_{-3.2}(syst)$ SNU (Abdurashitov et al. 2002). GALLEX had 65 runs between May 1990 and January 1997, whle GNO had 58 runs between May 1998 and September 2003, with a combined yield from both GALLEX and GNO of $69.3 \pm 4.1(stat) \pm 3.6(syst)$ SNU (Altmann et al. 2005). The combined result from SAGE and GALLEX/GNO is 67.7 ± 3.6, which is 52% of the Standard Solar Model (Gorbachev 2006), hence confirming the low energy pp origin of solar neutrinos but also confirming the deficit of neutrinos from the sun.

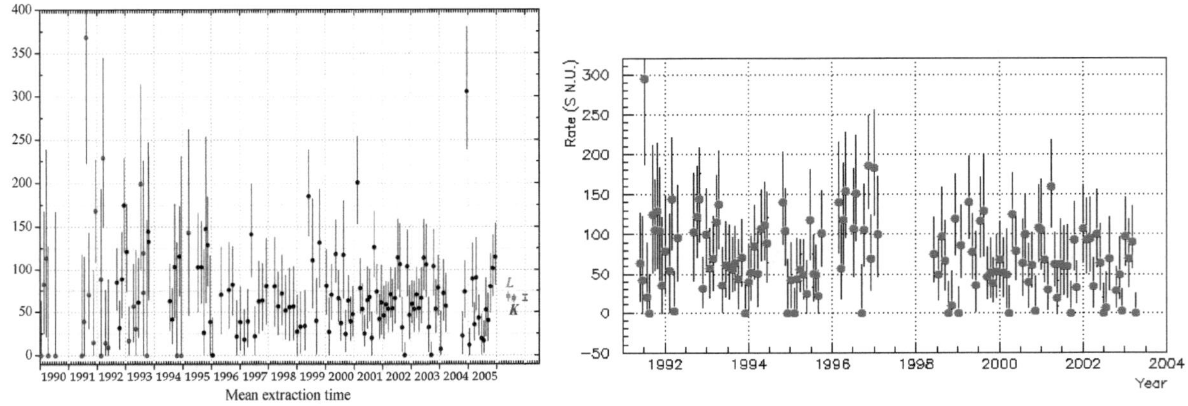

Figure 5. *Results from the SAGE experiment (left) (Copyright 2002 Springer) and results from the GALLEX/GNO experiments (right) (Copyright 2005 Elsevier) in Solar Neutrino Units (SNU) as a function of the run time.*

3 Neutrino Oscillations

Before the Standard Model was even formalized, Bruno Pontecorvo in 1957 wondered if there were any other particles that could undergo oscillations, analogous to K^0 - $\overline{K^0}$ oscillations, and hit upon the idea of neutrino - antineutrino oscillations (Pontecorvo 1958). In 1962, Maki, Nakagawa, and Sakata proposed that the "weak neutrinos", as they were known at the time, were superpositions of "true neutrinos" with definite masses, and that this could lead to transitions between the different weak neutrino states (Maki, Nakagawa & Sakata 1962). In 1968, Pontecorvo then considered the effects of all different types of oscillations in light of what was then known, and pointed out before any results from the Davis experiment were known that the rate in that experiment could be expected to be reduced by a factor of two (Pontecorvo 1968).

By 1972, Pontecorvo is informed by John Bahcall that Davis does indeed see a reduced rate in his chlorine solar neutrino experiment, and Pontecorvo responds with a letter:

"Dear Prof. Bahcall,

Thank you very much for your letter and the abstract of the new Davis investigation the numerical results of which I did not know. It starts to be really interesting! It would be nice if all this will end with something unexpected from the point of view of particle physics. Unfortunately, it will not be easy to demonstrate this, even if nature works that way."

For two neutrino flavours in vacuum, oscillations lead to the appearance of a new neutrino flavour with the corresponding disappearance of the original neutrino flavour:

$$P\left(\nu_e \longrightarrow \nu_\mu\right) = \sin^2 2\theta \sin^2\left(1.27\frac{\Delta m^2 L}{E}\right), \tag{1}$$

with $\Delta m^2 = m_2^2 - m_1^2$ in eV^2, L in metres and E in MeV (Kayser 2008). These oscillations can be significantly modified by the MSW effect (Wolfenstein 1978), (Mikheyev & Smirnov 1989) when the neutrinos pass through matter. The Schrödinger equation for two neutrino flavours is:

$$i\frac{d}{dt}\begin{pmatrix}\nu_e\\ \nu_x\end{pmatrix} = H\begin{pmatrix}\nu_e\\ \nu_x\end{pmatrix}. \tag{2}$$

In vacuum:

$$H = \begin{pmatrix} -\frac{\Delta m^2 L}{4E}\cos 2\theta & \frac{\Delta m^2 L}{4E}\sin 2\theta \\ \frac{\Delta m^2 L}{4E}\sin 2\theta & \frac{\Delta m^2 L}{4E}\cos 2\theta \end{pmatrix}, \tag{3}$$

but in the presence of matter, the Hamiltonian is modified to:

$$H = \begin{pmatrix} -\frac{\Delta m^2 L}{4E}\cos 2\theta + \sqrt{2}G_F N_e & \frac{\Delta m^2 L}{4E}\sin 2\theta \\ \frac{\Delta m^2 L}{4E}\sin 2\theta & \frac{\Delta m^2 L}{4E}\cos 2\theta \end{pmatrix}, \tag{4}$$

due to the fact that all neutrino flavours can undergo neutral current interactions, while only ν_e can undergo charged current interactions with the electrons in matter (G_F is the Fermi coupling constant, while N_e is the number density of electrons in matter). One recovers the Hamiltonian in Eq. 2, by substituting θ for the mixing angle in matter θ_M, where:

$$\sin 2\theta_M = \frac{\sin 2\theta}{(\omega - \cos 2\theta)^2 + \sin^2 2\theta} \tag{5}$$

with $\omega = -2\sqrt{2}G_F N_e/\Delta m^2$.

Based on the observed neutrino survival probability from solar neutrino experiments, and assuming the MSW effect as neutrinos escape from the sun, we can deduce contours of survival probability as a function of mixing angle and Δm^2 (for example, Fig. 6 shows the survival probability for ^7Be solar neutrinos). Three different solutions satisfied the results from the solar neutrino experiments: the Small Mixing Angle (SMA) solution, with $\tan\theta \sim 10^{-4}$, the Large Mixing Angle (LMA) with $\tan\theta \sim 1$ and $\Delta m^2 \sim 10^{-5}$ eV^2 and the Low Δm^2 (LOW) solution at $\Delta m^2 \sim 10^{-8}$ eV^2. New experiments would be needed to determine which of these solutions were the correct ones.

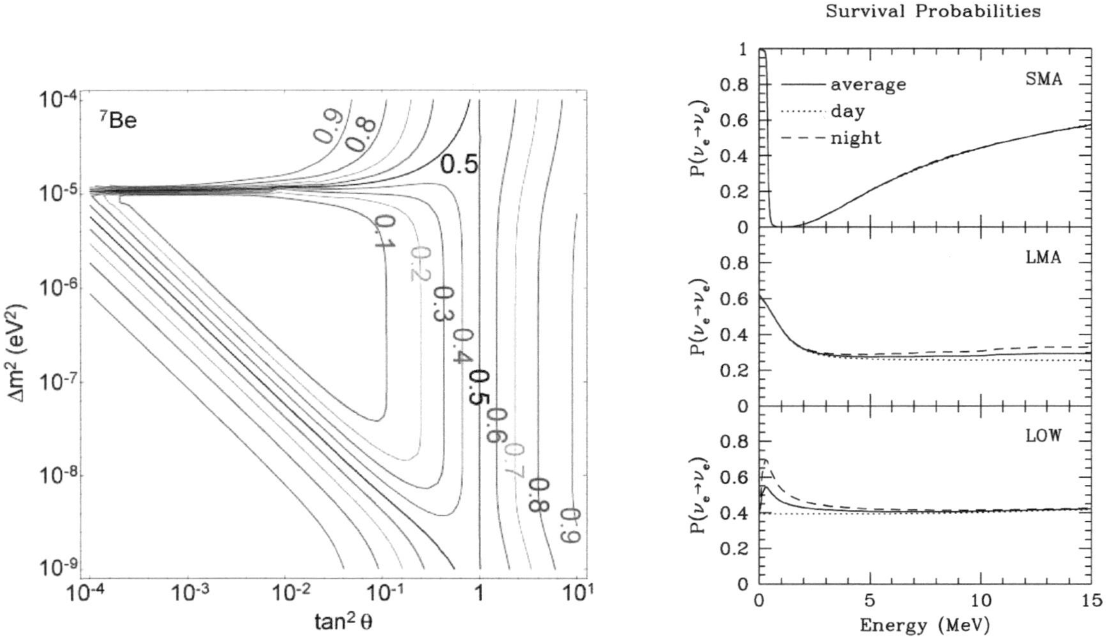

Figure 6. *Left: Contours of survival probability as a function of mixing angle and Δm^2 for ^7Be neutrinos. Right: Survival probabilities for ^7Be neutrinos for the Small Mixing Angle (SMA), the Large Mixing Angle (LMA) and the Low Δm^2 (LOW) solution.*

4 Super-Kamiokande

The Super-Kamiokande detector is located 1000 meters underground (2700 meters of water equivalent) in the Kamioka mine in Gifu Prefecture in Japan. This detector has produced the highest statistics measurement of solar neutrinos to date. It has been able to determine, with high precision, measurements of the solar neutrino flux, energy spectrum, and possible time variations of the flux in a large water Cherenkov detector from the $\nu_e + e^- \longrightarrow \nu_e + e^-$ elastic scattering reaction.

The detector consists of about 50000 tons of ultra-pure water in a stainless steel cylindrical tank of diameter 39.3 m and height 41.4 m (Fig. 7), with 11146 20-inch photomultiplier tubes in the inner detector and 1885 8-inch photomultiplier tubes in the outer detector to serve as a veto (Fukuda et al. 2003). In the inner detector, the active photodetector coverage is 40.4%. The fiducial size used for the solar neutrino measurement is 22.5 kilotons and the energy range for solar neutrino events was 5.0-20.0 keV.

The angular distribution of the solar neutrino event candidates in the Super-Kamiokande detector during a period of 1496 days is shown in Fig. 7 (Hosaka et al. 2006). The dotted area is the contribution from background events, while the shaded area indicates the elastic scattering peak, with a best fit measurement of $22404 \pm 226(stat)^{+784}_{-717}(syst)$ candidate events. This corresponds to a ^8B flux of $2.35 \pm 0.02(stat) \pm 0.08(syst) \times 10^6\,\text{cm}^{-2}\text{s}^{-1}$, which is $0.465 \pm 0.005(stat) \pm 0.0016(syst)$ of the Standard Solar Model rate, thus confirming once more the solar neutrino deficit.

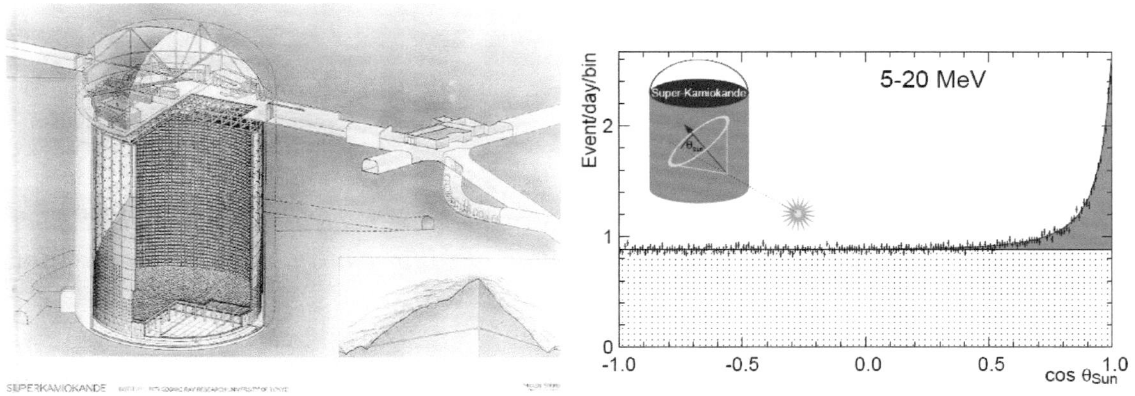

Figure 7. *Left: Super-Kamiokande detector. Right: Number of solar neutrino events in the Super-Kamiokande detector during a period of 1496 days as a function of the zenith angle with respect to the position of the sun. Copyright (2006) American Physical Society.*

The ratio of the solar neutrino energy spectrum with respect to the Standard Solar Model is shown in Fig. 8. The lack of a distortion of the spectrum favours the Large Mixing Angle (LMA) or the Low Δm^2 solution. Fig. 8 also shows the seasonal variation of the solar neutrino flux. The 1.7% orbital eccentricity of the Earth, which causes a 7% flux variation is included in the flux prediction as a solid line. The observed flux variation is consistent with the predicted annual modulation and constitutes the world's first observation of the eccentricity of the orbit of the Earth made with neutrinos.

As solar neutrinos cross the Earth at night, oscillated neutrinos may regenerate into ν_es through the MSW effect. The day time flux and night time flux of solar neutrinos in Super-Kamiokande were calculated using events which occurred when the solar zenith angle cosine was less than and greater than zero, respectively and the data was binned in bins where the neutrinos cross different lengths of the inside of the Earth (Fig. 9). The observed fluxes are $2.32 \pm 0.03(stat) \pm 0.08(syst) \times 10^6\ \text{cm}^{-2}\text{s}^{-1}$ during the day and $2.37 \pm 0.03(stat) \pm 0.08(syst) \times 10^6\ \text{cm}^{-2}\text{s}^{-1}$ at night, which also favour the Large Mixing Angle solution.

The best fit region of the two neutrino oscillation parameter space combining the Super-Kamiokande, Chlorine and Gallium solar neutrino data and data from the CHOOZ reactor experiment, gave values in the Large Mixing Angle (LMA) region at $tan^2\theta = 0.52$ and $\Delta m^2 = 6.3 \times 10^{-5}\ \text{eV}^2$. However, the two other regions at Small Mixing Angle (SMA) and the Low Δm^2 region (LOW) could not be ruled out. Similarly, there was no distinctive "smoking gun" signal that determined that neutrino oscillations were responsible for the solar neutrino deficit. This was provided by the Sudbury Neutrino Observatory, which will be described in the next section.

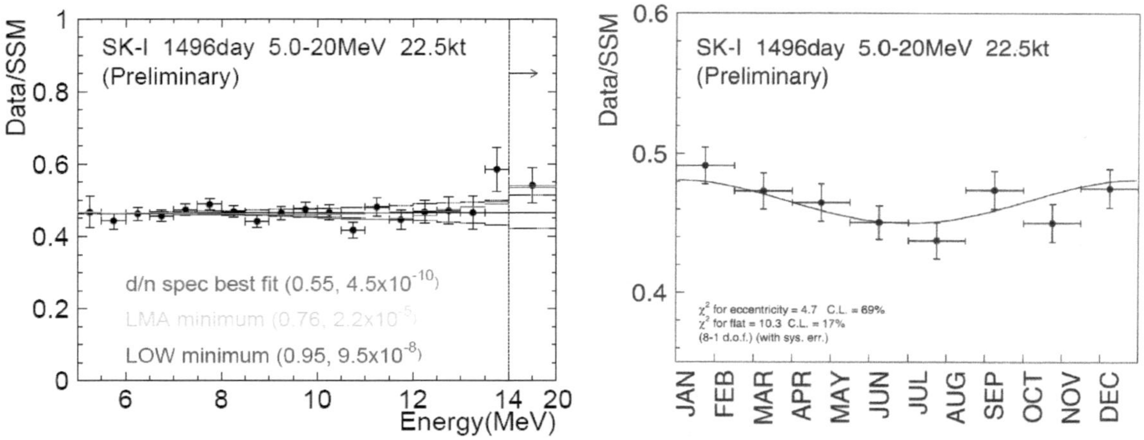

Figure 8. *Left: Spectral distortion of Super-Kamiokande solar neutrino events with respect to the Standard solar Model. Right: Seasonal variation of solar neutrino rate due to distance between the Sun and the Earth.*

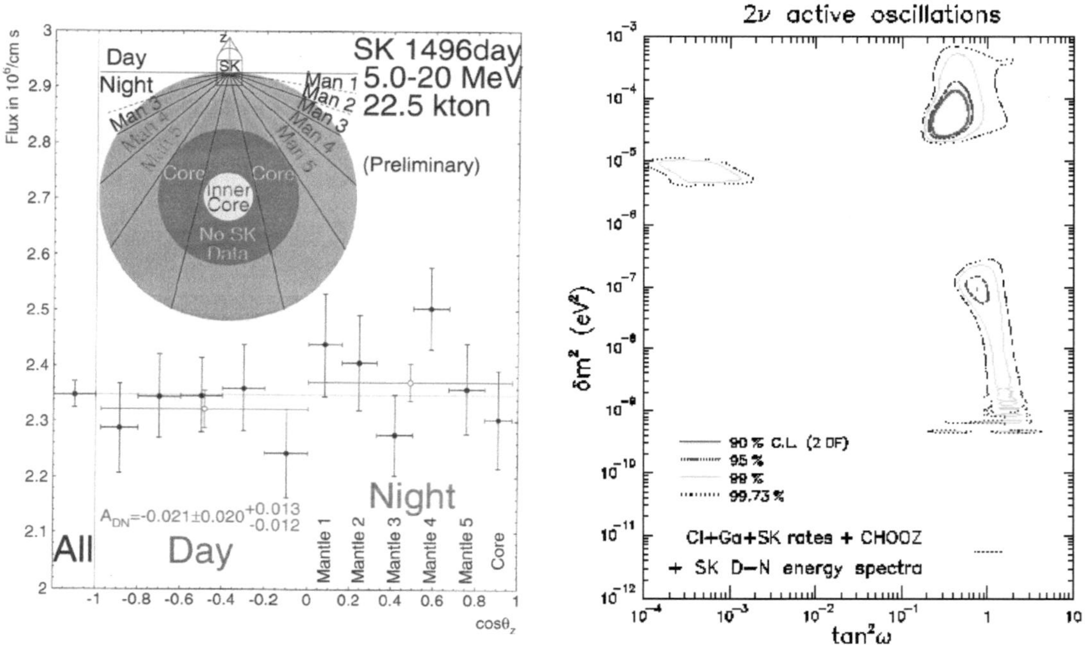

Figure 9. *Left: Day-night effect from Super-Kamiokande solar neutrino events. Right: Oscillation parameters derived from a combination of Super-Kamiokande, Chlorine and Gallium solar neutrino data and data from the CHOOZ reactor experiment.*

5 Sudbury Neutrino Observatory (SNO)

5.1 Description of SNO

The Sudbury Neutrino Observatory (SNO) is a second generation water Cherenkov detector designed to determine whether the currently observed solar neutrino deficit is a result of neutrino oscillations. The detector uses heavy water (D_2O) as a detection medium, which allows it to compare the charged current (CC) and neutral current (NC) interaction rates. The experiment is located at a depth of 6010 m of water equivalent in the INCO, Ltd. Creighton Mine in Sudbury, Ontario, Canada.

The target is inside a 12 m diameter spherical low radioactivity acrylic vessel containing 1000 tonnes of heavy water, which permits detection of neutrinos through the reactions:

$$\nu_x + e^- \longrightarrow \nu_x + e^- \tag{6}$$

$$\nu_e + d \longrightarrow e^- + p + p \tag{7}$$

$$\nu_x + d \longrightarrow \nu_x + n + p \tag{8}$$

where ν_x refers to any active flavor of neutrino, d is a deuteron nucleus and p is a proton. The heavy water vessel is inside a 22 m diameter, 34 m high cavity and is surrounded by a 6500 tonne shield of pure water(Fig. 10). Each of these interactions is detected when one or more electrons produce Cherenkov light that impinges on an array of photomultiplier tubes. The 9438 inward facing 8 inch photomultiplier tubes, with a 27 cm diameter light concentrator to enhance the collection of Cherenkov photons, provide a photocathode coverage of 54%. Another 91 photomultiplier tubes without concentrators are mounted facing outward to detect light from muons and other sources in the region exterior to the photomultiplier tube support structure (Boger et al. 2000).

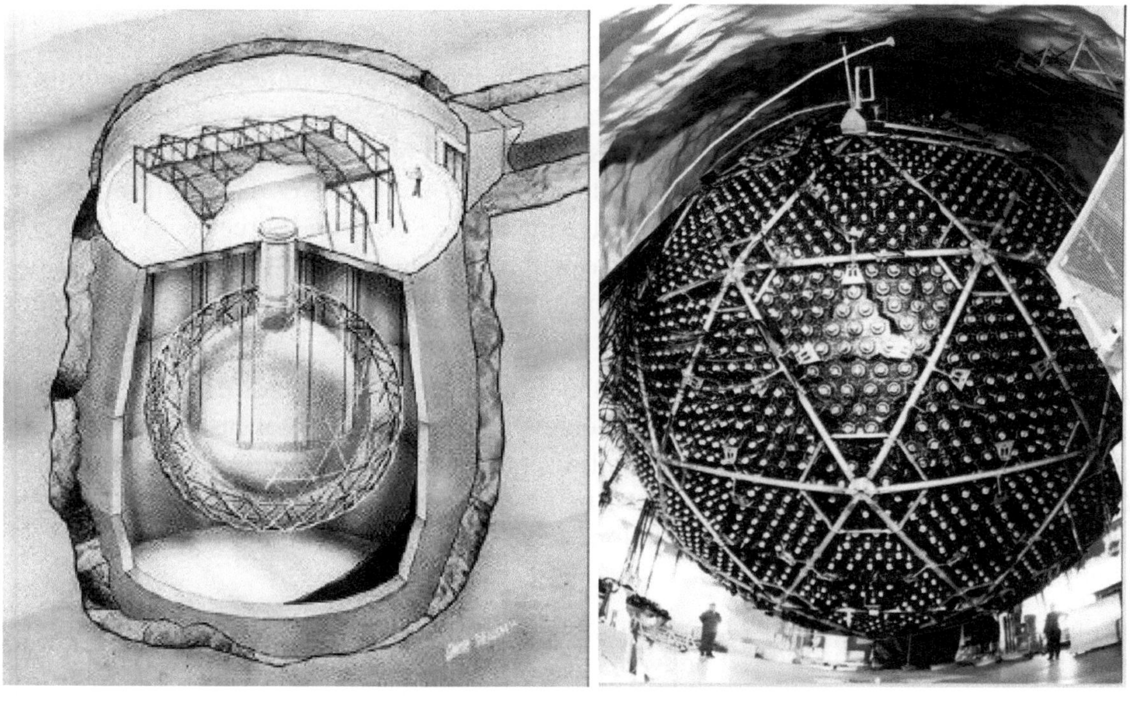

Figure 10. *Left: Sudbury Neutrino Observatory (SNO). Right: Acrylic vessel with photomultiplier tubes for SNO. Photos courtesy of SNO.*

The elastic scattering (ES) of electrons by neutrinos (Eq. 6) is the same reaction as seen by the Super-Kamiokande detector, is highly directional, and establishes the sun as the source of the detected neutrinos. The cross-section for ν_e ES interactions with electrons is 6.5 times larger than for ν_μ and ν_τ, so is predominantly sensitive to ν_es from the sun. The charged current (CC) reaction of ν_e on deuterons (Eq. 7) produces an electron with an energy highly correlated with that of the neutrino. This reaction is sensitive to the energy spectrum of ν_e and hence to deviations from the parent spectrum. It also has an angular distribution that follows $\left(1-\frac{1}{3}\cos\theta\right)$, where θ is the angle between the incoming neutrino and the recoil electron. The neutral current (NC) disintegration of the deuteron by neutrinos (Eq. 8) is independent of neutrino flavor and has a threshold of 2.2 MeV. To be detected, the resulting neutron must be absorbed by a neutron on deuterium reaction, giving a 6.25 MeV photon: $n+d \longrightarrow t+\gamma(6.25\ MeV)$. This constitutes phase I of the experiment. In a second phase of the experiment $NaCl$ is added to the D_2O (salt phase) for absorption of the neutrons on ^{35}Cl giving photons totalling 8.6 MeV. The photons subsequently Compton scatter, imparting enough energy to electrons to create Cherenkov light. In a third phase of the experiment, special purpose neutron detectors consisting of ^3He counters have been installed to provide an alterante way of measuring the neutral current reactions. Measurement of the rate of the NC reaction determines the total flux of ^8B neutrinos, even if their flavor has been transformed to another active flavor. The ability to measure the CC and NC reactions separately is unique to SNO and makes the interpretation of the results of the experiment independent of theoretical astrophysics calculations.

5.2 Calibration and data reduction of SNO

The calibration of the photomultiplier tubes is carried out using electronic pulsers and pulsed laser light sources between 337 and 620 nm wavelength. The absolute energy scale and uncertainties are established with a triggered ^{16}N source, which produces gamma rays of energy predominantly ~ 6.13 MeV deployed over a grid within the D_2O and H_2O. The resulting Monte Carlo predictions of detector response are tested using a ^{252}Cf neutron source (with 6.25 MeV gamma rays from neutron capture) a ^8Li source with a Q value of 16.0 MeV (providing the main beta decay branch with a central end point energy of 12.96 MeV) and a proton-tritium (pT) $^3H(p,\gamma)^4He$ source (that provides 19.8 MeV gamma rays).

Fig. 11 shows the data reduction from the 435,721,068 raw data events down to the 3055 events of solar neutrino signal. The first of these cuts is the elimination of instrumental background, including electrical pickup or electrical discharges in photomultiplier tubes that produce light. These backgrounds are eliminated using cuts based on the photomultiplier tube position, time and charge data, time correlations from one event to another and veto phototmultiplier tubes.

The data reduction is verified by comparing results from two independent background rejection analyses (Ahmad et al. 2001). For events passing the first stage, the calibrated times and positions of the hit PMTs are used to reconstruct the vertex position and the direction of the particle. The reconstruction accuracy and resolution are measured using Compton electrons from the ^{16}N and the proton-tritium (pT) $^3H(p,\gamma)^4He$ sources, while the energy and source variation of reconstruction are checked with the ^8Li source (Fig. 11). Angular resolution is measured using Compton electrons produced more than 150 cm from the ^{16}N source. At these energies, the vertex resolution is 16 cm and the angular resolution is 26.7°. A comparison between the source calibration and Monte Carlo is shown in Fig. 12.

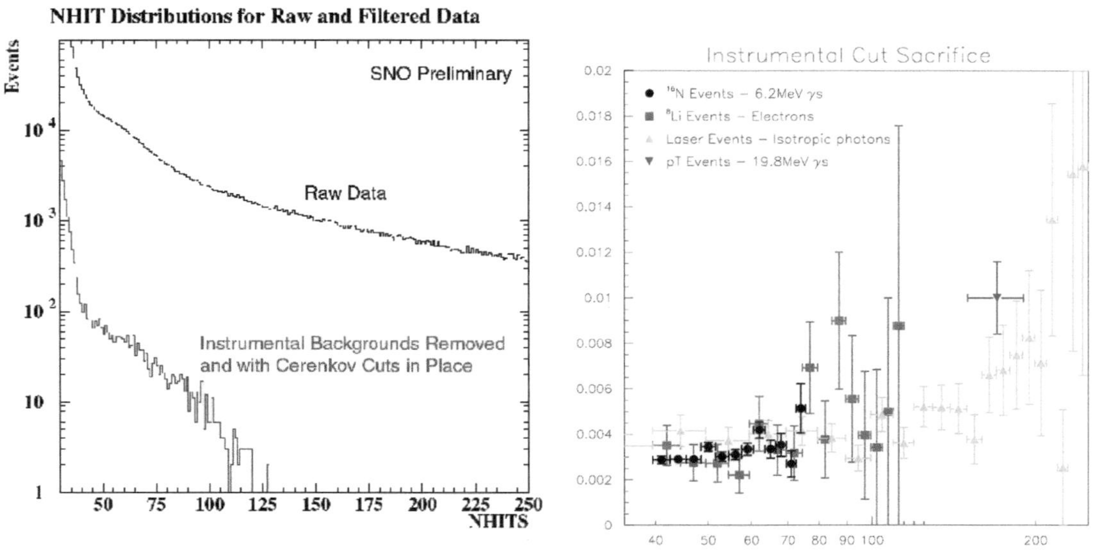

Figure 11. *Left: Data reduction from raw data to signal events through selection cuts. Right: Verification of instrumental cut efficiencies through a variety of calibration sources, a laser light source and* ^{16}N, ^{8}Li *and* $^{3}H(p,\gamma)^{4}He$ *(pT) radioactive sources.*

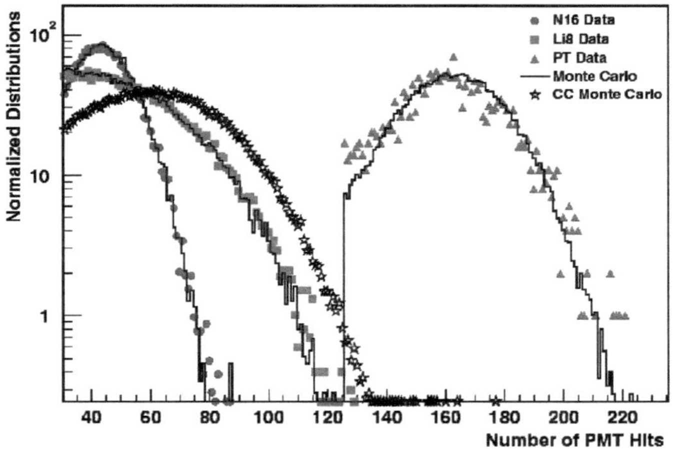

Figure 12. *Left: Calibration from radioactive* ^{16}N, ^{8}Li *and* $^{3}H(p,\gamma)^{4}He$ *(pT) radioactive sources compared to Monte Carlo and solar neutrino charged current (CC) events.*

5.3 Background in SNO

The main background to the Sudbury solar neutrino signals is from βs and γs from the radioactive decay chains of Uranium and Thorium (Fig. 13). These chains interfere with the signals at low energies, and γs over 2.2 MeV cause $d+\gamma \longrightarrow n+p$ that can mimic the signal. The design in SNO called for very stringent concentrations of U and Th in all the materials in the experiment.

- In the D_2O: concentration of U and Th $< 10^{-15}$ g g^{-1}
- In the H_2O: concentration of U and Th $< 10^{-14}$ g g^{-1}

- In the acrylic vessel: concentration of U and Th $< 10^{-12}$ g g^{-1}

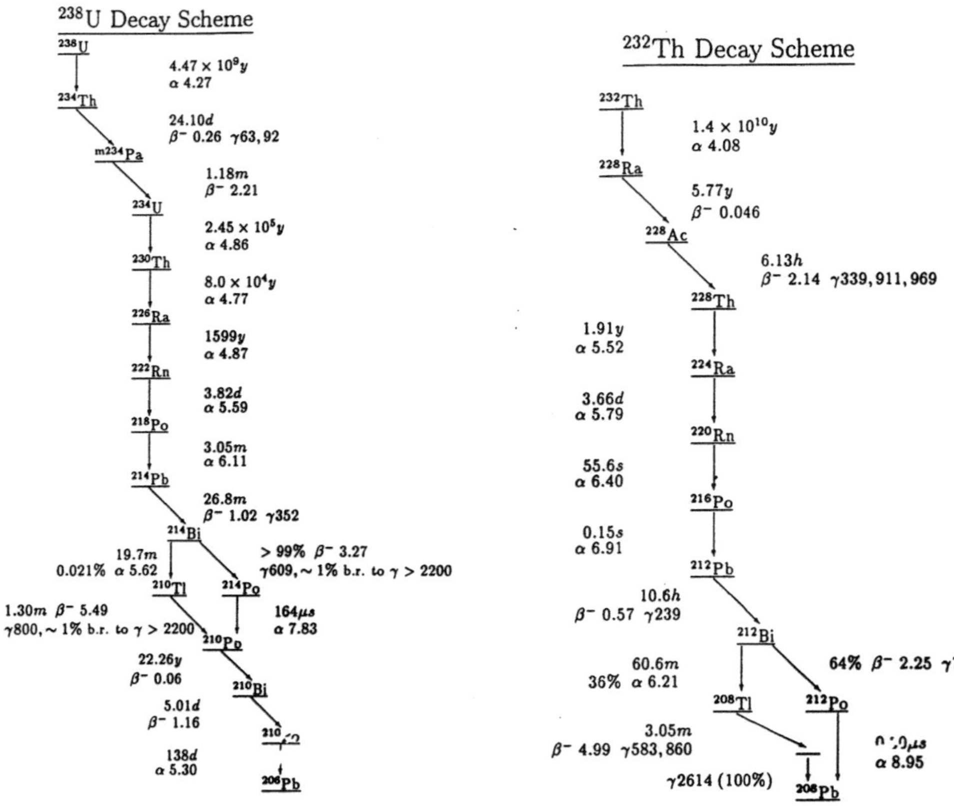

Figure 13. *Left: Uranium chain of radioactive decays contributing to the SNO background. Right: Thorium chain of radioactive decays contributing to the SNO background.*

Fig. 14 shows the internal and external background remaining after instrumental cuts above a threshold of 4.5 MeV. The internal background from radioactivity in the D_2O and H_2O are measured by regular low level radioactive assays of U and Th decay chain products, and compared to Cherenkov light signals of this radioactivity to monitor backgrounds. The high threshold imposed ultimately reduces contributions of low energy ($<$ 4 MeV) gamma rays from the remnant radioactivity. The external background is expressed as a function of the volume weighted radial variable $(R/R_{AV})^3$, where $R_{AV} = 6.00$ m is the radius of the acrylic vessel. For $(R/R_{AV})^3 > 1.0$ there are background events from external gamma rays detected in the H_2O, from the photomultiplier tubes (PMT) and from the acrylic vessel (AV).

5.4 Solar neutrino results from SNO

The data are resolved into contributions from Charged Current (CC), Elastic Scattering (ES) and Neutral Current (NC) events above threshold using probability density functions in the kinetic energy (T_{eff}), $\cos\theta_\odot$ (where $\theta_\odot$ is the angle difference between the recoil particle and the direction to the sun) and $(R/R_{AV})^3$, assuming the shape of the standard ^8B spectrum (Ahmad

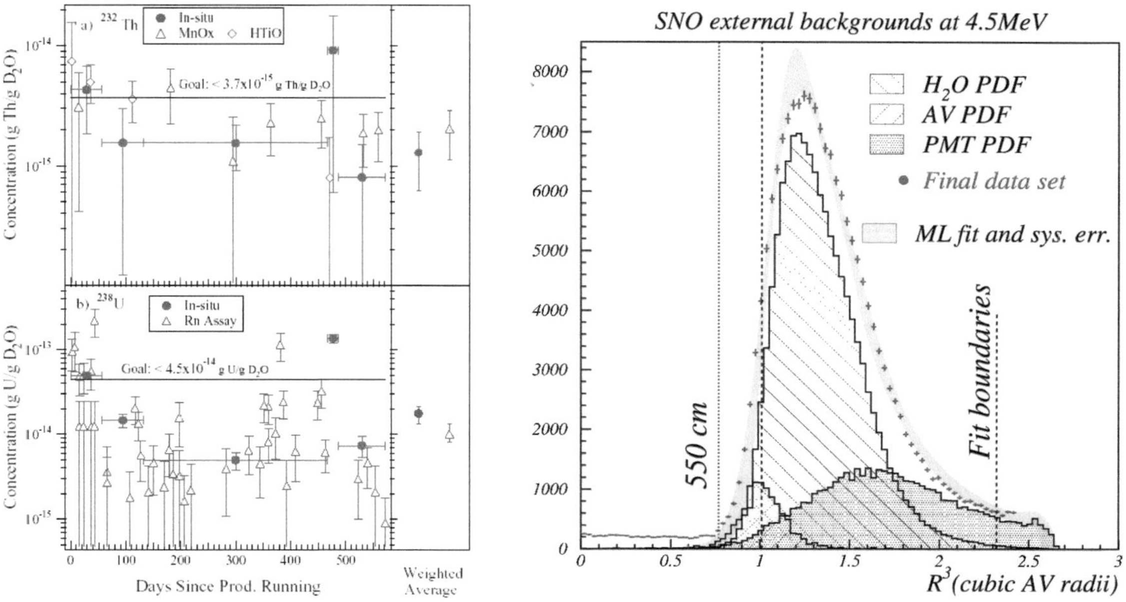

Figure 14. *Left: Internal SNO background. From (Ahmad et al. 2002). Copyright (2002) American Physical Society. Right: External SNO background.*

et al. 2002). The maximum likelihood method used in the signal extraction yielded 1967.7 ± 61.9 CC events, 263.6 ± 26.4 ES events and 576.5 ± 49.5 NC events for the fiducial volume in the first extraction of the solar neutrino flux. Fig. 15 shows the zenith angular distribution and the energy spectrum for the ^8B solar neutrino events obtained by the Sudbury Neutrino Observatory. The measured ^8B neutrino fluxes are then extracted to be:

$$\Phi^{CC}_{SNO}(\nu_e) = (1.76 \pm 0.06(stat) \pm 0.09(syst)) \times 10^6 \text{ cm}^{-2}\text{s}^{-1},$$
$$\Phi^{ES}_{SNO}(\nu_x) = (2.39 \pm 0.24(stat) \pm 0.12(syst)) \times 10^6 \text{ cm}^{-2}\text{s}^{-1},$$
$$\Phi^{NC}_{SNO}(\nu_x) = (5.09 \pm 0.44(stat) \pm 0.46(syst)) \times 10^6 \text{ cm}^{-2}\text{s}^{-1}.$$

The flux of CC reactions is smaller than the flux of ES interactions, due to the contribution to the ES reactions from ν_μ and ν_τ neutrinos from oscillations. The ES flux is consistent with the flux measured by Super-Kamiokande, but with larger errors. The NC flux is in agreement with the Standard Solar Model flux of ^8B neutrinos: $\Phi_{SSM}(\nu_e) = \left(5.05^{+1.01}_{-0.81}\right) \times 10^6 \text{ cm}^{-2}\text{s}^{-1}$, since all neutrino flavours contribute to this flux. Any ν_e neutrinos that would have oscillated to either ν_μ or ν_τ would also be visible in this channel, and hence the neutral current measurement provides a model independent way of determining the ^8B neutrino spectrum. This result provided conclusive proof that neutrino oscillations were responsible for the deficits observed by all previous experiments.

5.5 NaCl data from SNO

The second phase of the Sudbury Neutrino Observatory began in June of 2001 with the addition of ~ 2000 kg of NaCl to the 1000 tonnes of D_2O, for a total of 391.4 days of live data taking and ended in October 2003 when the NaCl was removed (Aharmim et al. 2005). The addition of the salt enhanced the sensitivity of SNO to measure the ^8B neutral current neutrino spectrum: $\nu_x +$

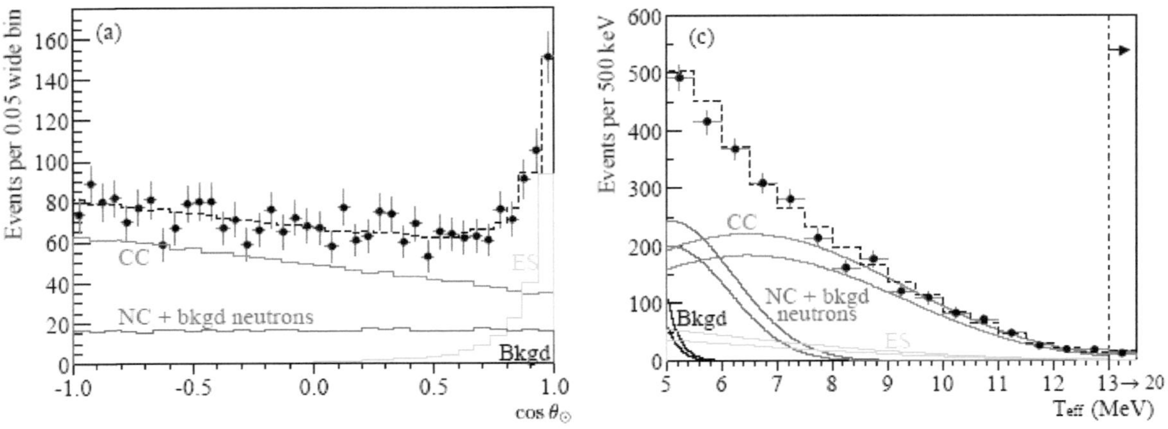

Figure 15. *Left: Zenith distribution with respect to the direction of the sun of SNO solar neutrino events. Right: Solar neutrino spectrum measured by SNO. From (Ahmad et al. 2002). Copyright (2002) American Physical Society.*

$d \longrightarrow \nu_x + n + p$, since it increased the neutron capture efficiency by a factor of three, through the reaction: $n + ^{35}Cl \longrightarrow ^{36}Cl + \gamma(8.6\,MeV)$. The thermal neutron capture cross section of ^{35}Cl is 44 b, significantly higher than that of the deuteron at 0.5 mb. Multiple gamma rays ($\sim$2.5 per capture) are produced, totalling an energy of 8.6 MeV, higher than the gamma ray energy of 6.25 MeV from neutron capture on deuteron (Fig. 16). The combination of the increased cross section and the higher energy released resulted in a larger neutron detection efficiency for a given energy threshold. Additionally, since the Compton scattering electrons from these gamma rays are more isotropic compared to the Cherenkov light emitted by the electrons from the CC and ES events, it allows good statistical separation of the event types. The variable $\beta_{14} = \beta_1 + 4\beta_4$, was used to separate neutron events (more isotropic) from the electron events (more directional) in SNO (Fig. 16). The β_l variable is defined as:

$$\beta_l = \frac{2}{N(N-1)} \sum_{i=l}^{N-1} \sum_{j=i+1}^{N} P_l(\cos\theta_{ij}), \tag{9}$$

where P_l is the Legendre polynomial of order l and θ_{ij} is the angle formed between any pair $(i,\, j)$ of photomultiplier tubes (PMT) hit and the event vertex, and N the total number of PMTs triggered.

A blind analysis technique was used to extract the ^8B neutrino spectra. An unknown fraction of muon followers were included in the data set for analysis, the NC cross-section was tampered with in the Monte Carlo and the data was prescaled by an unkown amount ($80 \pm 10\%$). This procedure was introduced to remove any bias in the event selection and fits. The final results for the extended maximum likelihood analysis yielded 2176 ± 78 CC, 279 ± 26 ES, 2010 ± 85 NC, and 128 ± 42 external-source neutron events. This translates into the following fluxes:

$$\Phi_{CC} = (1.68 \pm 0.06(stat) \pm 0.09(syst)) \times 10^6\ \mathrm{cm^{-2}s^{-1}},$$
$$\Phi_{ES} = (2.35 \pm 0.22(stat) \pm 0.15(syst)) \times 10^6\ \mathrm{cm^{-2}s^{-1}},$$
$$\Phi_{NC} = \left(4.94 \pm 0.21(stat)^{+0.38}_{-0.34}(syst)\right) \times 10^6\ \mathrm{cm^{-2}s^{-1}}.$$

This confirmed the results from Phase I of SNO with higher statistics and lower systematic errors. The ratio of fluxes compared to the standard solar model is shown in Fig. 17 and

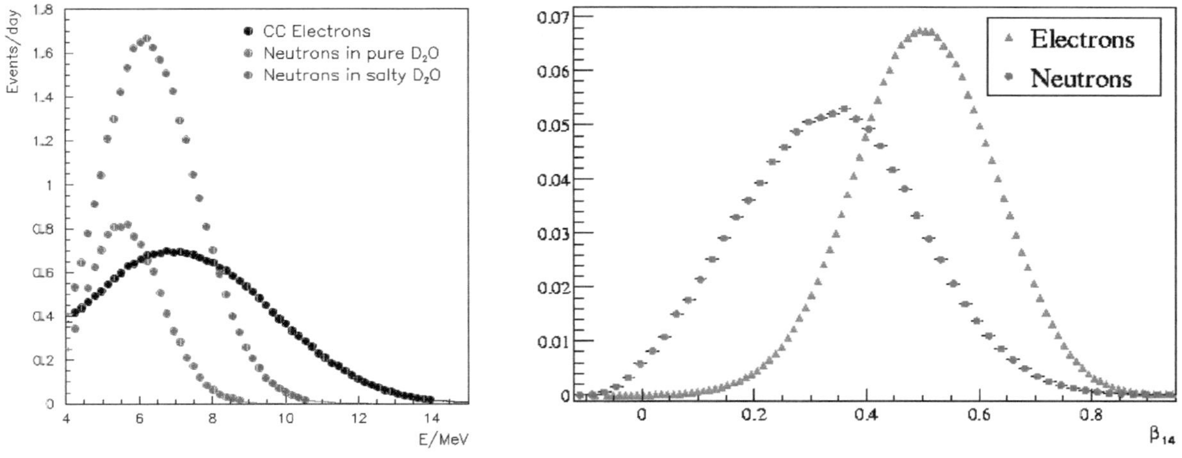

Figure 16. *Left: Comparison of spectra from CC electrons, neutron capture on D_2O and neutron capture on NaCl salt in SNO. Right: Comparison of the isotropy variable β_{14} for electron and neutron events in SNO.*

shows that the NC result is in perfect agreement with the expected ^8B flux, while there remains a deficit in the CC and ES events. The ratio of CC to NC events is $\frac{\Phi_{CC}}{\Phi_{NC}} = 0.340 \pm 0.023(stat)^{+0.029}_{-0.031}(syst)$. Also the spectrum of ^8B neutrinos from the SNO salt phase can be seen in Fig. 17, including all systematic error bands. It is clear from this plot that the predicted shape of the ^8B spectrum by the standard solar model is in perfect agreement with the data assuming the MSW Large Mixing Angle (LMA) solution to the solar neutrino problem and ruling out the Small Mixing Angle (SMA) solution that would have predicted a distortion in this shape.

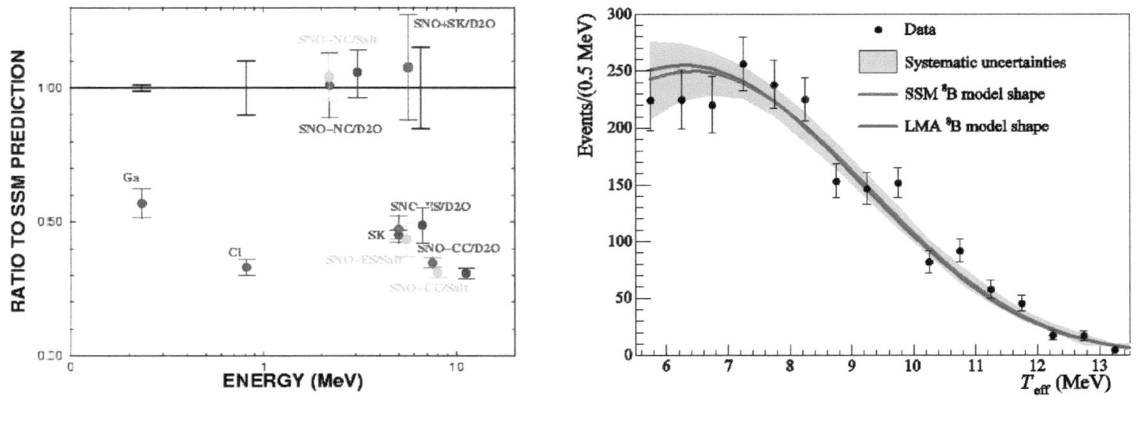

Figure 17. *Left: Comparison of measured solar neutrino fluxes with the standard solar model for different solar neutrino experiments (Chlorine, Gallium, Super-Kamiokande, SNO Phase I and SNO salt phase) as a function of neutrino threshold. Right: Spectrum of ^8B neutrinos from the SNO salt phase. From (Aharmim et al. 2005). Copyright (2005) American Physical Society.*

These results can be interpreted as resonant neutrino oscillations in the solar medium (MSW effect). A global neutrino oscillation analysis that included data from all solar neutrino experiments gives the best-fit oscillation parameters as shown in Fig. 18, with 1σ uncertainties on the 2-dimensional parameter region given by $\Delta m^2 = 6.5^{+4.4}_{-2.3} \times 10^{-5}$ eV^2 and $\tan^2\theta = 0.45^{+0.09}_{-0.08}$. There is a solar suppression factor of more than a factor of 2 which implies that the vacuum oscillation length is $L_V \sim 100$ km, much smaller than the length L from the sun to the earth. A search for a day-night effect was also carried out to determine whether the MSW effect is also observable from neutrinos traversing through matter on Earth. A total of 2134 events were recorded during the day (day rate = 12.09 ± 0.26 day^{-1}) while a total of 2588 events were recorded at night (night rate = 12.04 ± 0.24 day^{-1}), thus implying that there is no evidence for a day-night asymmetry.

5.6 Confirmation of Solar Neutrino Results: KamLAND

The results from SNO were spectacularly confirmed by the KamLAND neutrino reactor experiment, as described by (Blucher 2008) in these proceedings. KamLAND is an experiment that observed anti-neutrino oscillations from a large number of nuclear reactors in Japan to a 1 kton detector isoparaffine based liquid scintillator contained in a 13-m-diameter transparent balloon suspended in non-scintillating oil. The balloon is surrounded by 1879 photomultiplier tubes mounted on the inner surface of an 18 m diameter spherical stainless steel vessel. Electron anti-neutrinos are detected via inverse β decay, $\nu_e + p \longrightarrow e^+ + n$, with a 1.8 MeV ν_e energy threshold. The prompt scintillation light from the e^+ gives an estimate of the incident energy, while the ~ 180 μs delayed 2.2MeV gamma ray from neutron capture on hydrogen is used to eliminate background. The 766 ton-year exposure from KamLAND (Araki et al. 2005) is consistent with neutrino oscillations with a vacuum oscillation length of $L_\nu \sim 180$ km, consistent with the solar data and reduces the error on the value of Δm^2. The best fit parameters from the global solar plus KamLAND analysis gives the neutrino oscillation parameters: $\Delta m^2 = 8.0^{+0.4}_{-0.3} \times 10^{-5}$ eV^2 and $\theta = 33.9^{+2.4}_{-2.2}$ degrees. The total systematic uncertainty for both Δm^2 and θ_{12} in KamLAND will be reduced by expanding the fiducial radius of the detector and through calibration sources.

5.7 Future prospects for Sudbury Neutrino Observatory

SNO has been in its third data taking phase from November 2004 to November 2006. The salt was removed and an array of dedicated Neutral Current Detectors (NCD) were installed. The NCDs are ^3He counters to detect the neutrons from the neutral current reactions directly. This final phase of the analysis will be carried out by blinding the data, to minimise any bias in the results. The goal of the third phase is to provide a systematically different measurement of the solar neutrino flux. This measurement will provide a much reduced uncertainty when combined with the previous data.

There are plans to use the SNO cavity and the detector for a pep solar neutrino experiment by replacing the heavy water with liquid scintillator. This experiment, called SNO+, aims to perform a precision measurement of the pep neutrino rate in an energy region sensitive to distortions from the MSW effect (1-4 MeV), to verify the MSW effect or to find any new mechanisms. This experiment would be part of the new international laboratory, SNOLAB,

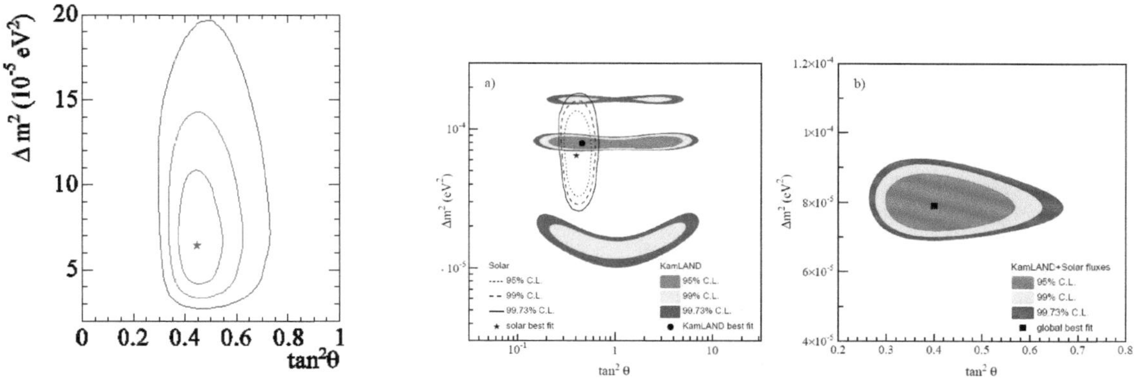

Figure 18. *Left: Global neutrino oscillation analysis using only solar neutrino data. From (Aharmim et al. 2005). Copyright (2005) American Physical Society. Middle: Neutrino oscillation analysis using solar and KamLAND data separately. Right: Combined neutrino oscillation analysis using solar and KamLAND data. From (Araki et al. 2005). Copyright (2005) American Physical Society.*

expected to be completed in 2007. Other physics goals of SNO+ are the measurement of the geo-neutrino flux and carrying out a high sensitivity double beta decay experiment, by loading the scintilator with Neodymium.

6 Prospects for future solar neutrino experiments

6.1 Borexino

Borexino is an experiment that aims to perform a real-time measurement of the monochromatic ^7Be 863 keV neutrino flux from the sun at the Gran Sasso Underground Laboratory in Italy (with about 3500 meters of water equivalent overburden). The Borexino design is based on the concept of a graded shield of progressively lower intrinsic radioactivity as one approaches the sensitive volume of the detector. The detector contains 300 tons of well shielded ultrapure scintillator viewed by 2200 photomultipliers (Fig. 19). The detector core is inside a transparent spherical nylon inner vessel, 8.5 m in diameter, filled with 300 tons of liquid scintillator (pseudocumene, PC, doped with 1.5 g/l of PPO fluorescent dye) and surrounded by 1000 tons of a high purity buffer liquid (PC). The photomultipliers are supported by a stainless steel sphere which also separates the inner part of the detector from the external shielding, provided by 2400 tons of pure water. A nylon outer vessel is interposed between the inner nylon sphere and the photomultipliers, to reduce radon diffusion. The outer water shield is instrumented with 200 outward pointing photomultipliers serving as a veto for penetrating muons. The ultimate background will be dominated by the intrinsic contamination of the scintillator, which is measured to be below 5×10^{-15} g g^{-1} of U and Th equivalent.

The detection reaction is the electroweak elastic scattering of neutrinos on electrons, in which the recoil electron has a maximum of 664 keV kinetic energy. The detection threshold for Borexino is 250 keV, so that a rate of 46 events/day is predicted by the Standard Solar Models

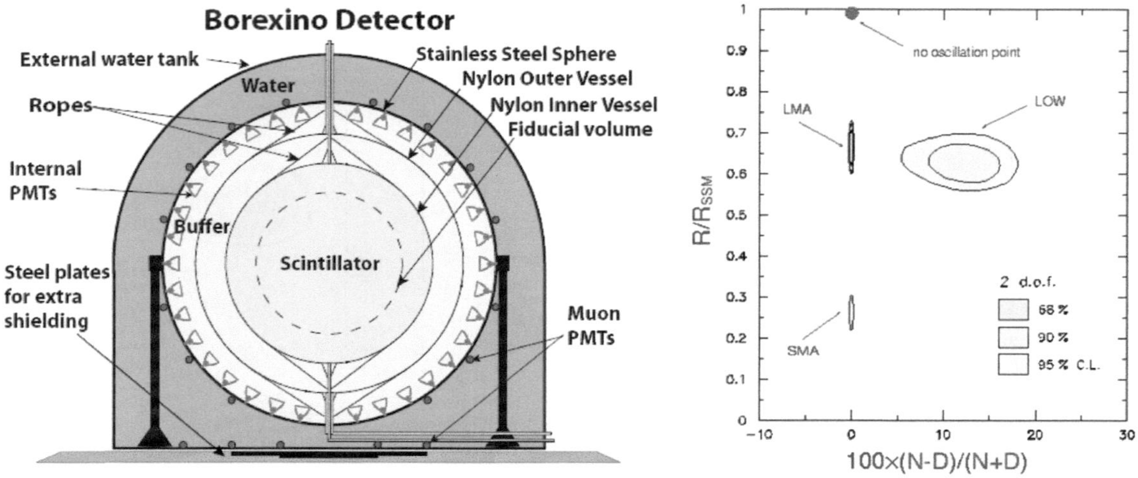

Figure 19. *Left: Schematic of Borexino detector. Right: Expected sensitivity of the Borexino experiment. From (Arpesella et al. 2008). Copyright (2008) Elsevier.*

in a fiducial volume of 100 tons in Borexino. The recoil electron profile for a monoenergetic neutrino is similar to that of Compton scattering of a single gamma ray, so the recoil electron profile is basically a rectangular shape with a sharp cut-off edge at 665 keV.

The first results from the Borexino experiment were announced after this Summer School (Arpesella et al. 2008). This is the first real-time spectral measurement of sub-MeV solar neutrinos. The result for 0.862 MeV ^7Be neutrinos is $47 \pm 7(stat) \pm 12(syst)$ counts/(day $\cdot$ 100 ton), consistent with predictions of the standard solar model and neutrino oscillations with LMA-MSW parameters. Two different analyses were found to be consistent: one which did not separate α and β candidates and another analysis which carried out α/β separation by pulse shape discrimination (Fig. 20). The rate averaged over the earth orbit based on oscillations assuming these parameters is expected to be 49 ± 4 counts/(day $\cdot$ 100 ton), while the rate expected without oscillations is 75 ± 4 counts/(day $\cdot$ 100 ton).

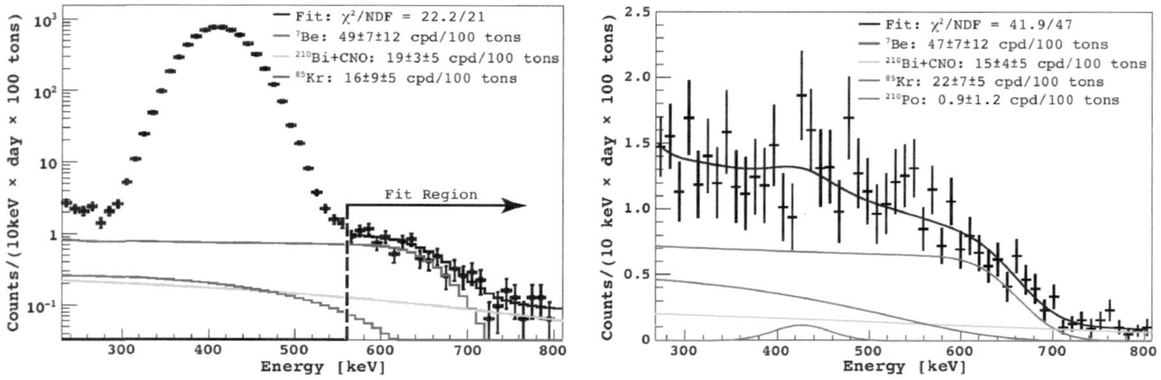

Figure 20. *^7Be neutrino results from Borexino detector without pulse-shape discrimination (left) and with α/β separation through pulse-shape discrimination (right). From (Arpesella et al. 2008). Copyright (2008) Elsevier.*

The future programme to be carried out by Borexino includes more detailed statistical error of the solar ^7Be neutrino line intensity and its time (seasonal, day-night) variation, extraction of the ^8B and neutrino spectra from the CNO cycle, studies of the neutrino magnetic moment using radiactive sources (either ^{90}Sr or ^{51}Cr), antineutrino searches of geophysical neutrinos from the Earth and possible double beta decay searches.

6.2 Low Energy Solar Neutrinos

There are a number of other experiments that are being proposed to carry out real-time low energy neutrino searches. For example, these can be used to constrain models of the core of the sun and non-standard solar physics. Producing high-statistics, real time measurements of the lower energy solar neutrinos would allow the verification of the transition from vacuum oscillations to MSW oscillations (the transition region is around 1 MeV for the current best fit values) and more accurate determination of the oscillation parameters, in particular, θ_{12} (Fig. 21). Performing high statistics solar neutrino spectroscopy allows strong tests of the solar models. Observation of CNO neutrinos (Fig. 21), even though it is currently constrained to be less than 7.3% from solar neutrino data (Bahcall, Gonzalez-Garcia & Peña Garay 2003), would produce the first direct experimental test of the main nucleosynthesis reaction in the Universe.

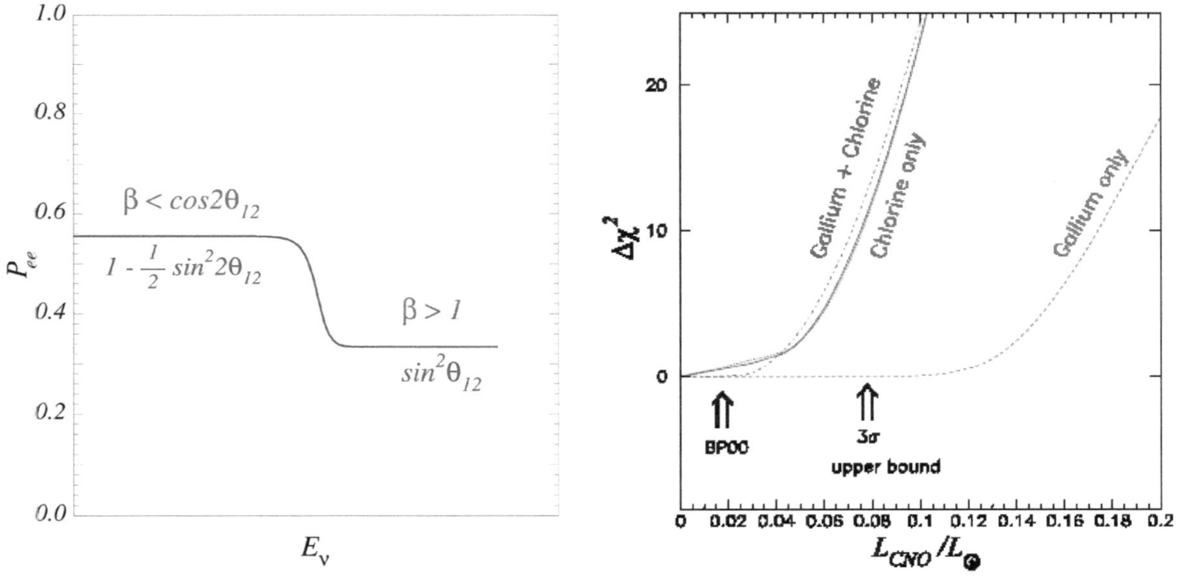

Figure 21. *Left: Transition from vacuum neutrino oscillations (low energy) below resonance to MSW effect at higher neutrino energies (~ 1 MeV in the sun). Right: Limits on the CNO flux based on solar neutrino results.*

Many experiments are being proposed. In addition to Borexino and SNO+ already covered, there are LENS (Grieb 2007), HERON (Huang et al. 2007) and CLEAN (McKinsey et al. 2005). LENS is being proposed as a $\sim 100 - 150$ tonne indium loaded liquid scintillator detector, in which the low energy (pp) neutrinos are captured by inverse beta decay: $\nu_e + ^{115}In \longrightarrow ^{115}Sn^* + e^-$, with a threshold of 114 KeV, followed by the decay: $^{115}Sn^* \longrightarrow ^{115}Sn + \gamma(116\ keV) + \gamma(497.3\ keV)$ (Raghavan 1976), with a decay time of 4.76 μs. HERON relies on the idea of detecting low energy solar neutrinos with 21.6 tonnes of superfluid helium (Lanou, Maris & Seidel 1987) at cryogenic temperatures (30-50 mK). The concept of CLEAN is that of a stainless steel tank holding approximately 135 metric tons of liquid neon, with several thousand photomultipliers to record the scintillation light in the neon from elastic scattering of low energy solar (pp) neutrinos. If any of these proposals get built, real-time low energy solar neutrino experiments will enter into a precision era that will return to the original aim of measuring the parameters of the dominant nuclear reactions of the sun and will test astrophysical models of the sun.

7 Conclusions

Solar neutrinos provided the first evidence for what was subsequently confirmed as physics beyond the Standard Model. The Davis chlorine experiment first demonstrated a deficit of solar electron neutrinos, subsequently confirmed by Kamiokande that this result was not instrumental. The gallium experiments SAGE and GALLEX demonstrated that this also applied to the low energy solar neutrinos, eliminating any credible astrophysical alternative. Dozens of empirically adequate particle physics models were proposed to explain this "solar neutrino problem". Super-Kamiokande then confirmed that neutrinos have mass through neutrino oscillations from atmospheric neutrinos. However, a large number of empirically adequate models still existed. The Sudbury Neutrino Observatory then confirmed that electron neutrinos from the sun change flavour, thus eliminating many models, but many others remained. The reactor neutrino experiment KamLAND eliminated all but the most contrived of these models other than neutrino oscillations. There has been further confirmation of the reduction of the ^7Be flux in Borexino, consistent with neutrino oscillations. Since then, the long baseline neutrino experiments, K2K and MINOS, have confirmed more predictions of neutrino oscillations. Currently, neutrino oscillations is the simplest hypothesis to explain all the above phenomena. It does not mean that the simplest possible form is the correct explanation, but it does provide compelling evidence that neutrinos indeed do mix and that they have mass.

Acknowledgements

This article is dedicated to the late memory of John Norris Bahcall (1934-2005) and Raymond Davis Jr. (1914-2006).

This work was partly supported by the Science and Technology Facilities Council (STFC), UK, formerly, the Particle Physics and Astronomy Research Council (PPARC) and the Council for the Combined Laboratories of the Research Councils (CCLRC).

References

Abdurashitov, J. N. et al. (1999), *Phys. Rev. C* **59**, 2246.

Abdurashitov, J. N. et al. (2002), *J. Exp. Theor. Phys.* **95**, 181–193.

Aharmim, B. et al. (2005), *Phys. Rev.* **C72**, 055502.

Ahmad, Q. R. et al. (2001), *Phys. Rev. Lett.* **87**, 071301.

Ahmad, Q. R. et al. (2002), *Phys. Rev. Lett.* **89**, 011301.

Altmann, M. et al. (2005), *Phys. Lett. B* **616**, 174–190.

Araki, T. et al. (2005), *Phys. Rev. Lett.* **94**, 081801.

Arpesella, C. et al. (2008), *Phys. Lett. B* **658**, 101–108.

Bahcall, J., Gonzalez-Garcia, M. & Peña Garay, C. (2003), *Phys. Rev. Lett.* **90**, 131301.

Bahcall, J. & Pinsonneault, M. (1996), *Phys. Rev. Lett.* **77**, 1683–1686.

Barsanov, V. I. et al. (2007), *Phys. Atom. Nucl.* **70**, 300–310.

Blucher, E. (2008), *These proceedings* .

Boger, J. et al. (2000), *Nucl. Instrum. Meth.* **A449**, 172–207.

Cleveland, B. T. et al. (1998), *The Astrophysical Journal*, **496**, 505.

Davis, R. (1955), *Phys. Rev.* **97**, 766.

Fukuda, S. et al. (2003), *Nucl. Instr. and Meth. A* **501**, 418.

Fukuda, Y. et al. (1996), *Phys. Rev. Lett.* **77**, 1683–1686.

Gorbachev, V. (2006), *XXXIII International Conference on High Energy Physics, Moscow* .

Grieb, C. (2007), *Nucl. Phys. B (Proc. Suppl.)* **168**, 122.

Hampel, W. et al. (1998), *Phys. Lett. B* **420**, 114–126.

Hosaka, J. et al. (2006), *Phys. Rev.* **D73**, 112001.

Huang, Y. H. et al. (2007).

Kayser, B. (2008), *These proceedings* .

Kuzmin, V. A. (1966), *Sov. Phys. JETP* **22**, 1051–1056.

Lanou, R. E., Maris, H. J. & Seidel, G. M. (1987), *Phys. Rev. Lett.* **58**, 2498.

Maki, Z., Nakagawa, M. & Sakata, S. (1962), *Prog. Theor. Phys.* **28**, 870.

McKinsey, D. et al. (2005), *Astroparticle Phys.* **22**, 355–368.

Mikheyev, S. P. & Smirnov, A. Y. (1989), *Prog. Part. Nucl. Phys.* **23**, 41–136.

Pontecorvo, B. (1958), *Sov. Phys. JETP* **7**, 172–173.

Pontecorvo, B. (1968), *Sov. Phys. JETP* **26**, 984–988.

Raghavan, R. (1976), *Phys. Rev. Lett.* **37**, 259.

Scott, F. (1935), *Phys. Rev.* **48**, 391.

Serenelli, A. (2008), *These proceedings*.

Wolfenstein, L. (1978), *Phys. Rev.* **D17**, 2369.

Neutrinos and Stars

Georg G. Raffelt

Max-Planck-Institut für Physik (Werner-Heisenberg-Institut)
Föhringer Ring 6, 80805 München, Germany

1 Introduction

Neutrinos and astrophysics, including cosmology, have three main areas of interface. The first is the standard role of neutrinos as radiation in the early universe, in ordinary stars and in core-collapse supernovae. Sometimes they play a dominant dynamical role (early universe, supernovae) or simply carry away energy (ordinary stars). In addition, they play a crucial role for nucleosynthesis in some of these environments because they participate in beta interactions of the type $\nu_e + n \leftrightarrow p + e^-$ and $\bar{\nu}_e + p \leftrightarrow n + e^+$. Their broad role as radiation is possible because of their small masses. Actually, in a typical stellar plasma or in the early universe, the dispersive photon mass far exceeds that of neutrinos, allowing for the plasmon decay process $\gamma \to \nu\bar{\nu}$, i.e., in many astrophysical situations, neutrinos are effectively the lowest-mass particles except for gravitons. Of course, neutrinos do have small vacuum masses and therefore contribute a small fraction of the cosmic dark matter. Moreover, their small masses may be responsible for creating the cosmic matter-antimatter asymmetry by virtue of the leptogenesis mechanism.

The second broad role of neutrinos is that of astrophysical messengers. Neutrinos can reach us from sites that are opaque to photons, in particular from the interior of the Sun and of core-collapse supernovae, allowing us to study these objects in the "light of neutrinos." It may even become possible to study the Earth's crust and interior by observing the geophysical $\bar{\nu}_e$ flux from natural radioactive elements. Observing the diffuse cosmic neutrino background from all past supernovae could provide an independent measure of the cosmic star-formation rate. High-energy neutrinos that are produced in the context of cosmic-ray acceleration and propagation may become observable in future large-scale neutrino telescopes. The key point here is that electric charge neutrality prevents neutrino deflection in galactic and intergalactic magnetic fields so that they point back to their sources, in contrast to protons and nuclei, hopefully allowing one to identify the mysterious cosmic-ray accelerators. Annihilations of the putative massive weakly interacting dark-matter particles would also produce high-energy neutrinos that could help us to identify the physical nature of dark matter.

Finally, one may use the "heavenly laboratories" to learn about the properties of neutrinos themselves. Solar and atmospheric neutrino observations have revealed the first evidence for flavor oscillations while cosmic precision observables provide limits on the overall neutrino mass scale. In addition, astrophysical and cosmological arguments constrain a broad range of non-standard neutrino properties such as electromagnetic form factors, secret neutrino-neutrino interactions, or sterile neutrinos.

Some of these topics are covered by other lecturers at this school. I restrict my lectures to the role of neutrinos in ordinary stars (Sect. 2) and in supernovae (Sect. 3) with a focus on useful lessons for particle physics. For neutrino cosmology I refer to the reviews by Dolgov (2002), Lesgourgues and Pastor (2006) and Hannestad (2006) and for leptogenesis to Buchmüller, Di Bari and Plümacher (2005). Nuclear astrophysics in the early universe and in stars is treated by Rolfs and Rodney (1988), Arnett (1996), and Käppeler, Thielemann and Wiescher (1998). High-energy neutrinos and neutrino astronomy are reviewed by Learned and Mannheim (2000), Stanev (2004) and Halzen (2006). Neutrino astrophysics and cosmology in a broad sense is covered by Fukugita and Yanagida (2003). I have previously reviewed the material of the following lectures in much more detail elsewhere (Raffelt 1990a, 1996, 1999, 2005).

2 Neutrinos in Ordinary Stars

The idea that neutrinos could play an important role in stars was first launched by Gamow and Schoenberg (1940). Today, neutrinos are recognized as a standard ingredient of the theory of stellar structure and evolution (Clayton 1983, Kippenhahn and Weigert 1990). In normal stars, neutrinos are produced in nuclear reactions and by thermal plasma processes. Hydrogen-burning stars such as our Sun produce energy by the reaction $4p+2e^- \to {}^4\mathrm{He}+2\nu_e$ that proceeds through the pp chains and the CNO cycle (Bahcall 1989). The neutrino energies are a few MeV, reflecting the relevant nuclear energy release. It is these neutrinos from nuclear reactions that have been measured from the Sun. Advanced burning phases essentially combine α particles (^{4}He nuclei) to higher isotopes without significant $n \leftrightarrow p$ conversions and thus without much nuclear neutrino production.

The second source of neutrinos are those produced by reactions among thermal plasma participants, in particular the Compton-like process $\gamma + e^- \to e^- + \nu + \bar{\nu}$, bremsstrahlung by electron scattering on nuclei $e^- + Ze \to Ze + e^- + \bar{\nu} + \nu$ as well as free-bound and bound-free processes, plasmon decay $\gamma \to \nu + \bar{\nu}$, and in very hot plasmas pair annihilation $e^- + e^+ \to \nu + \bar{\nu}$. The importance of these non-nuclear reactions was recognized in the early 1960's, shortly after the $V-A$ theory of weak interactions had been formulated, implying a direct interaction between neutrinos and electrons. These thermal processes do not have a threshold so that the neutrino energies are of order the stellar temperature T. For the Sun, the central temperature is about 1.3 keV so that the plasma neutrinos have keV energies, too small to be detected. The energy carried away is negligible.

After hydrogen is exhausted in the central region of a normal star, helium burning is ignited. Subsequently further burning stages proceed at higher and higher temperatures. The thermal neutrino emission rates depend steeply on T. Carbon burns at $T \sim 50$ keV where neutrino emission of the stellar core is comparable to surface photon emission. In the final silicon-burning phase of a massive star at $T \sim 270$ keV, the neutrino energy losses exceed the photon luminosity by a factor of about 10^6. Therefore, the fast evolutionary time scales of these advanced evolution phases is dictated by thermal neutrino losses. If this were not the case, these stars would shine in photons for much longer periods, i.e., many more of them should be around than are actually observed. The scarcity of bright giant stars led Stothers (1970) to estimate the strength of the direct neutrino-electron coupling several years before it was actually measured in the laboratory.

This general "energy-loss argument" remains useful to constrain non-standard neutrino interactions, notably possible electromagnetic form factors. It was recognized by Bernstein, Ruderman and Feinberg (1963) that a direct neutrino-photon interaction caused either by a small electric charge ("milli charge") or by a magnetic dipole or transition moment would accelerate the rate of the plasmon decay process $\gamma \to \nu\bar{\nu}$. This process is kinematically allowed because in a plasma photons acquire an effective mass that is identical to the plasma frequency. If the electrons are non-relativistic, it is

$$\omega_{\rm pl} = \left(\frac{4\pi\alpha n_e}{m_e}\right)^{1/2} = 28.7\ {\rm eV}\left(\frac{Y_e\rho}{1\ {\rm g\ cm^{-3}}}\right)^{1/2}, \tag{1}$$

where $\alpha \approx 1/137$ is the fine-structure constant, m_e the electron mass, n_e the electron number density, ρ the mass density, and Y_e the number of electrons per baryon, a quantity that depends on the chemical composition. The energy-loss rate per unit volume of a non-relativistic stellar plasma for a single neutrino species is approximately (Raffelt 1996)

$$Q_{\gamma\to\nu\bar{\nu}} \approx \frac{8\zeta_3}{3\pi}T^3 \times \begin{cases} \alpha_\nu & \left(\dfrac{\omega_{\rm pl}^2}{4\pi}\right) & \text{Millicharge,} \\ \dfrac{\mu_\nu^2}{2} & \left(\dfrac{\omega_{\rm pl}^2}{4\pi}\right)^2 & \text{Dipole moment,} \\ \dfrac{C_V^2 G_{\rm F}^2}{\alpha} & \left(\dfrac{\omega_{\rm pl}^2}{4\pi}\right)^3 & \text{Standard model.} \end{cases} \tag{2}$$

where $\zeta_3 \approx 1.202$, $\alpha_\nu = e_\nu^2/4\pi$ is the putative neutrino fine structure constant, μ_ν the magnetic dipole moment, $G_{\rm F}$ the Fermi constant, and C_V the effective neutral-current neutrino-electron vector-coupling constant; the axial-vector contribution is negligible. The standard-model photon decay process is enabled by the presence of the electrons of the medium that give the photons an effective mass and mediate an effective neutrino-photon interaction. For ν_e we have $C_V^2 = \frac{1}{2} + 2\sin^2\Theta_{\rm W} \approx 0.96$ with $\sin^2\Theta_{\rm W} = 0.2315$, including both Z^0 and W exchange. For ν_μ and ν_τ we have only Z^0 exchange so that $C_V^2 = -\frac{1}{2} + 2\sin^2\Theta_{\rm W} \ll 1$. The peculiar value of the weak mixing angle implies that only the electron flavor contributes significantly because ν_μ and ν_τ have a very small neutral-current vector coupling to electrons.

The plasmon decay process dominates in the degenerate cores of low-mass stars before helium ignition. The evolution theory for low-mass stars can be tested by using a globular cluster as an ensemble of coeval stars that differ primarily by their initial mass. In a color-magnitude diagram such as Figure 1 the stars occupy characteristic loci that reveal different evolutionary phases. Notably the red giant branch (RGB) corresponds to stars that burn hydrogen in a thin shell surrounding a degenerate helium core. Helium ignition would be delayed by excessive neutrino cooling, implying that the tip of the RGB would extend to brighter stars. The observed brightness of the brightest red giants in several globular clusters reveals that the true neutrino loss rate does not exceed the standard-model rate by a significant amount (Raffelt 1990b). Doubling the neutrino losses roughly increases the brightness of the RGB tip by 0.25 mag (Raffelt and Weiss 1992), i.e., by 0.25 in the units on the vertical axis of Figure 1.

With the requirement that a new energy-loss rate should not significantly exceed the standard plasmon decay rate in low-mass red giants before helium ignition, together with Eq. (2) and $\omega_{\rm pl} \approx 9$ keV, implies (Raffelt 1996)

$$e_\nu \lesssim 2\times 10^{-14}e \quad \text{and} \quad \mu_\nu \lesssim 3\times 10^{-12}\mu_{\rm B}\,, \tag{3}$$

where $\mu_{\rm B} = e/2m_e$ is the Bohr magneton. This is the most restrictive limit on neutrino dipole moments. More restrictive limits on neutrino charges exist, but the limit on e_ν applies to any putative milli-charged particle and as such remains very useful.

Globular-cluster stars can be used in similar ways to constrain the coupling strengths of other low-mass weakly interacting particles that can be emitted by plasma processes, notably of axions that are produced by the Primakoff process $\gamma + Ze \to Ze + a$ because they have a two-photon vertex that allows for photon-axion transitions in the presence of external electric and magnetic fields. A recent update of astrophysical axion limits was given by Raffelt (2006).

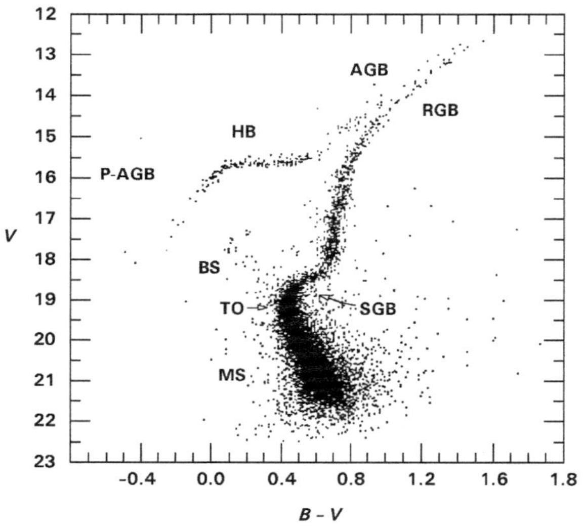

Figure 1. *Color-magnitude diagram for the globular cluster M3, based on 10,637 stars (Buonanno et al. 1986). Vertically is the brightness in the visual (V) band, horizontally the difference between B (blue) and V brightness, i.e. a measure of the color and thus surface temperature, where blue (hot) stars lie toward the left. The classification for the evolutionary phases is as follows (Renzini and Fusi Pecci 1988). MS (main sequence): core hydrogen burning. BS (blue stragglers). TO (main-sequence turnoff): central hydrogen is exhausted. SGB (subgiant branch): hydrogen burning in a thick shell. RGB (red-giant branch): hydrogen burning in a thin shell with a growing core until helium ignites. HB (horizontal branch): helium burning in the core and hydrogen burning in a shell. AGB (asymptotic giant branch): helium and hydrogen shell burning. P-AGB (post-asymptotic giant branch): final evolution from the AGB to the white-dwarf stage.*

3 Supernova Neutrinos

A core-collapse supernova (SN) is the only system besides the early universe that is dynamically dominated by neutrinos. Moreover, stellar-collapse neutrinos have been observed once from SN 1987A and may be observed again from a future galactic SN. A core-collapse SN marks the evolutionary end of a massive star ($M \gtrsim 8 M_\odot$) that has reached the usual onion structure with several burning shells, an expanded envelope, and a degenerate iron core. The core mass grows by nuclear burning at its edge until it reaches the Chandrasekhar limit. The collapse cannot ignite nuclear fusion because iron is the most tightly bound nucleus. Therefore, the collapse continues until the equation of state stiffens at about nuclear density (3×10^{14} g cm^{-3}). At this "bounce" a shock wave forms, moving outward and expelling the stellar mantle and envelope, i.e., the explosion is a reversed implosion. Within the expanding nebula, a compact object remains in the form of a neutron star or perhaps sometimes a black hole. The kinetic energy of the explosion carries about 1% of the liberated gravitational binding energy of about 3×10^{53} erg, the remaining 99% going into neutrinos. The main phases of this sequence of events are illustrated in Figure 2. The physics of core-collapse SNe is reviewed in Burrows (2000), Woosley and Janka (2005) and Fryer (2004).

Almost twenty years ago, the neutrino burst of SN 1987A in the Large Magellanic Cloud

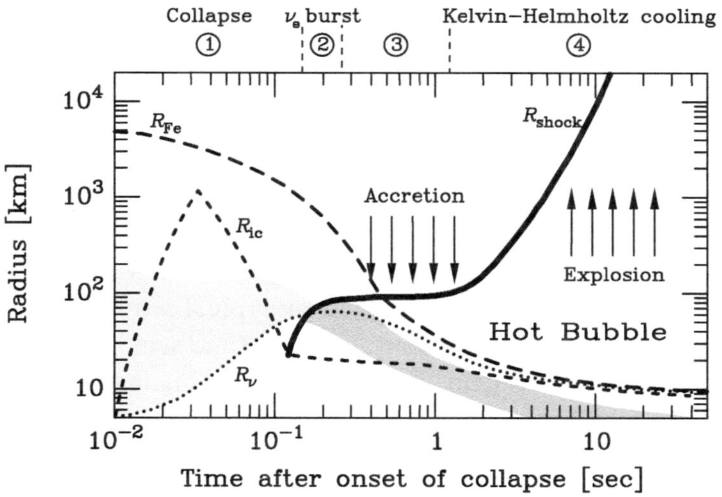

Figure 2. *Schematic picture of the core collapse of a massive star ($M \gtrsim 8\ M_\odot$) and the beginning of a SN explosion. The four main phases are: 1. Collapse. 2. Prompt-shock propagation and break out, release of prompt ν_e burst. 3. Matter accretion and mantle cooling. 4. Kelvin-Helmholtz cooling of proto-neutron star. The curves mark the evolution of several characteristic radii: The stellar iron core (R_{Fe}). The neutrino sphere (R_ν) with diffusive transport inside and free streaming outside. The "inner core" (R_{ic}) for $t \lesssim 0.1$ s is the region of subsonic collapse, later it is the settled, compact inner region of the nascent neutron star. The SN shock wave (R_{shock}) is formed at core bounce, stagnates for as much as several 100 ms, and then propagates outward. The neutrino source region is shaded.*

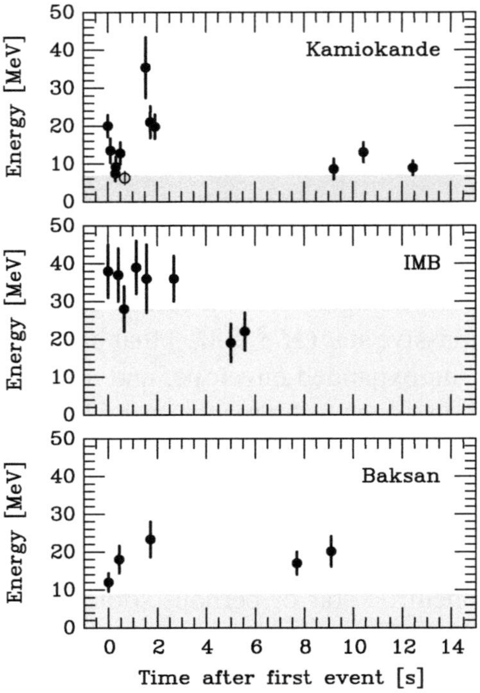

Figure 3. *SN 1987A neutrino observations at Kamiokande (Hirata et al. 1988), IMB (Bratton et al. 1988) and Baksan (Alexeyev et al. 1987). The energies refer to the secondary positrons from the reaction $\bar{\nu}_e p \to n e^+$. In the shaded area the trigger efficiency is less than 30%. The clocks have unknown relative offsets; in each case the first event was shifted to $t = 0$. In Kamiokande, the event marked as an open circle is attributed to background.*

(LMC), a small satellite galaxy of our Milky Way at a distance of about 50 kpc, was observed in several detectors (Figure 3). A lively first-hand account of this exciting and sometimes confusing discovery is given in Koshiba (1992). The burst was measured by the inverse β reaction on protons, $\bar{\nu}_e + p \to n + e^+$, where the positron is registered by its Cherenkov or scintillation light. The energies and overall time scale of emission roughly agree with expectations, although the inferred average energies are lower than implied by typical numerical simulations, a discrepancy that has never been fully resolved. Certainly, the time scale of emission over several seconds confirms the usual picture of neutrino trapping in the nuclear-density SN core where the energy slowly leaks out by diffusive neutrino transport.

The neutrino burst is emitted a few hours before the optical explosion. This feature may serve as an early warning for astronomers that the next galactic SN is imminent if a neutrino burst is seen in one of the operating detectors (see SNEWS web page at http://snews.bnl.gov). In the case of SN 1987A, the first photons indeed arrived a few hours after the neutrino burst. A generous uncertainty of 3 h between the relative arrival times implies that the transit time of 1.6×10^5 years was the same for photons and neutrinos within 2×10^{-9}.

Both signals traversed a gravitational potential $\Phi_{\rm grav}$ so that one expects a Shapiro time delay for each of them,

$$\Delta t_{\rm Shapiro} = -2 \int \Phi_{\rm grav}(\mathbf{r}_t) dt\,, \tag{4}$$

where the integral is to be performed over the trajectory $\mathbf{r}_t$ of the signal. With different models of the galactic gravitational potential, Krauss and Tremaine (1988) estimated $\Delta t_{\rm Shapiro} = 1$–5 months, implying that it is equal between photons and neutrinos within $(0.7–4) \times 10^{-3}$. Gravitational lensing experiments prove that photons respond to gravity as expected so that the SN 1987A comparison with neutrinos reveals that neutrinos respond to gravity also as expected. This is the only direct evidence for the gravitational interaction of neutrinos.

The intrinsic dispersion of the $\bar{\nu}_e$ flux was used to derive time-of-flight limits on the neutrino mass that are only of historical interest. However, one also obtains a limit on a putative neutrino electric charge because the deflection in the galactic magnetic field causes an energy-dependent arrival-time delay similar to that caused by a neutrino mass. For typical values of the galactic magnetic field and its coherence length one finds $e_\nu/e \lesssim 3 \times 10^{-17}$ (Bahcall 1989).

The most useful information for particle physics is obtained from the overall burst duration. The emission of more weakly interacting particles than neutrinos that can freely escape from the SN core would remove energy that is lost for the late-time neutrino signal. Cases in point are the emission of sterile neutrinos by active-sterile mixing, of Majoron emission in neutrino-neutrino interactions, or Kaluza-Klein graviton or axion emission by nucleon-nucleon bremsstrahlung. Frequently this energy-loss argument provides the most restrictive limits on the underlying physical model (Raffelt 1990a, 1996, 1999, 2006).

The statistical significance of these results suffers from the small number of neutrino events. If a galactic SN at a typical distance of 10 kpc were observed today with the Super-Kamiokande detector, one would measure around 10^4 events. Likewise, IceCube that is being built at the South Pole would provide a high-statistics SN neutrino light curve. The ongoing experimental programme for long-baseline neutrino oscillation experiments almost guarantees the operation of large neutrino detectors for many years so that the eventual observation of a high-statistics SN neutrino signal is plausible, even though galactic SNe are rare, perhaps a few per century (Diehl 2006). Such an observation would allow one to follow the detailed evolution of stellar collapse and the cooling of a proto neutron star and thus would provide a detailed test of the current core-collapse paradigm. Of course, it is possible that the neutrino signal will show new and unexpected features. A sudden termination, for example, would signify black-hole formation.

A lot of effort has been put into understanding if a SN neutrino observation could reveal new information about neutrino mixing parameters or conversely, if shock-wave propagation effects would manifest themselves in a characteristic time variation of the neutrino signal by time-dependent matter effects on flavor oscillations (Dighe 2005, Raffelt 2005, Fogli et al. 2006a). Several issues complicate this question. First, oscillation effects depend upon the initial neutrino fluxes and spectra being significantly different, something that is not to be taken for granted, at least not during the cooling phase (Keil, Raffelt and Janka 2003). Second, oscillation signatures can be washed out by stochastic density fluctuations in the SN medium (Fogli et al. 2006b, Friedland and Gruzinov 2006). Most recently it was recognized that the matter effect of neutrinos causes nonlinear oscillation effects over the first hundred kilometers above the neutrino sphere that completely change the previous picture of neutrino propagation in this region (Duan et al. 2006, Hannestad et al. 2006). These issues need to be better understood before meaningful forecasts of observable oscillation effects can be made. Of course, it would be extremely useful for SN neutrino physics if the mixing parameters, notably the 13-mixing angle and the sign of Δm^2_{13}, would be measured in laboratory experiments.

As a final topic, the next SN neutrinos to be measured could well be the cosmic diffuse

SN neutrino background (DSNB) from all past SNe. While SNe in a typical galaxy are rare on a human life span, the average energy output in core-collapse neutrinos is comparable to the galaxy's photon luminosity. Therefore, the cosmic energy density in diffuse SN neutrinos is comparable to the energy density in star light. One finds a typical diffuse $\bar{\nu}_e$ flux of about $10\ \mathrm{cm}^{-2}\ \mathrm{s}^{-1}$, with considerable uncertainties both with regard to the number flux and the energy spectrum and thus the detection rate (Lunardini 2006).

The upper limit on the DSNB from Super-Kamiokande is close to the predicted level (Malek et al. 2003). Improving the sensitivity requires background rejection by tagging the neutron in $\bar{\nu}_e + p \to n + e^+$ in order to distinguish this detection reaction from the decay of sub-Cherenkov muons produced by atmospheric neutrinos. Neutron tagging can be achieved by loading Super-Kamiokande with gadolinium (Gd), an efficient neutron absorber (Beacom and Vagins 2004). The feasibility of this approach is being studied by the Super-Kamiokande collaboration with promising results (private communication by M. Vagins). As an alternative one can use a large-scale scintillator detector such as the proposed LENA experiment (Marrodán Undagoitia et al. 2006). The eventual measurement of the DSNB appears to be a realistic prospect for the not too distant future.

4 Summary and Conclusions

Ordinary stars as well as core-collapse SNe produce neutrinos with energies ranging from keV to tens of MeV. The measurement of solar neutrinos and the neutrino burst from SN 1987A have both produced a wealth of useful information for particle physics. Besides the measurement of flavor oscillations in the solar neutrino flux, the SN 1987A signal duration has provided many restrictive limits on the properties of neutrinos or hypothetical particles. Likewise, the energy loss by thermal neutrinos produced in ordinary stars can cause observable modifications of stellar evolution, allowing one to derive useful constraints, for example on neutrino electromagnetic form factors.

While galactic SNe are rare, the long-term perspective of operating large neutrino observatories for other purposes is excellent, notably Super-Kamiokande and IceCube, but also the smaller liquid scintillator detector LVD or in future a possible megatonne-class water-Cherenkov detector or a large scintillator detector such as the proposed LENA. Therefore, observing a high-statistics neutrino light curve from a future galactic SN is a realistic possibility that would provide an immense scientific harvest.

We have warned, however, that the theoretical situation concerning SN neutrino oscillations is currently too confusing to come up with realistic forecasts for measuring neutrino mixing parameters. Conversely, determining the missing elements of the neutrino mixing matrix in laboratory experiments, besides the obvious particle-physics importance, is called for by SN physics as necessary input information. Core-collapse SNe are the one case where flavor oscillations can make a real difference in a macroscopic physical system. Therefore, the benefit for SN physics from the ongoing long-baseline oscillation programme is two-pronged. The detectors are excellent SN neutrino observatories and the eventual determination of the mixing parameters provides microscopic input information for SN physics.

Acknowledgments

This work was partly supported by the Deutsche Forschungsgemeinschaft under Grants No. SFB-375 and TR-27 and by the European Union under the ILIAS project, contract No. RII3-CT-2004-506222.

References

Alexeyev E N et al., 1988, "Detection of the neutrino signal from SN 1987A in the LMC using the INR Baksan underground scintillation telescope," *Phys. Lett. B* **205**, 209.

Arnett D, 1996, *Supernovae and nucleosynthesis* (Princeton University Press).

Bahcall J N, 1989, *Neutrino astrophysics* (Cambridge University Press).

Beacom J F and Vagins M R, 2004, "Antineutrino spectroscopy with large water Cherenkov detectors," *Phys. Rev. Lett.* **93**, 171101 [hep-ph/0309300].

Bernstein J, Ruderman M and Feinberg G, 1963, "Electromagnetic properties of the neutrino," *Phys. Rev.* **132**, 1227.

Bratton C B et al., 1988, "Angular distribution of events from SN 1987A," *Phys. Rev. D* **37**, 3361.

Buchmüller W, Di Bari P and Plümacher M, 2005, "Leptogenesis for pedestrians," *Annals Phys.* **315**, 305 [hep-ph/0401240].

Buonanno R, Buzzoni A, Corsi C E, Fusi Pecci F and Sandage A R, 1986, "High precision photometry of 10000 stars in M3," *Mem. Soc. Astron. Ital.* **57**, 391.

Burrows A, 2000, "Supernova explosions in the universe," Nature **403**, 727.

Clayton D D, 1983, *Principles of stellar evolution and nucleosynthesis* (University of Chicago Press).

Diehl R et al., 2006, "Radioactive ^{26}Al and massive stars in the galaxy," Nature **439**, 45 [astro-ph/0601015].

Dighe A, 2005, "Supernova neutrinos: Production, propagation and oscillations," *Nucl. Phys. B (Proc. Suppl.)* **143**, 449 (2005) [hep-ph/0409268].

Dolgov A D, 2002, "Neutrinos in cosmology," *Phys. Rept.* **370**, 333 [hep-ph/0202122].

Duan H, Fuller G M, Carlson J and Qian Y Z, 2006, "Simulation of coherent non-linear neutrino flavor transformation in the supernova environment. I: Correlated neutrino trajectories," Phys. Rev. D **74**, 105014 [astro-ph/0606616].

Fogli G, Lisi E, Mirizzi A and Montanino D, 2006a, "Supernova neutrino physics in future water-Cherenkov detectors," *Proceedings of Science Server (PoS)* **HEP2005**, 184.

Fogli G, Lisi E, Mirizzi A and Montanino D, 2006b, "Damping of supernova neutrino transitions in stochastic shock-wave density profiles," *JCAP* **0606**, 012 [hep-ph/0603033].

Friedland A and Gruzinov A, 2006, "Neutrino signatures of supernova turbulence," astro-ph/0607244.

Fryer C (ed.), 2004, *Stellar collapse* (Kluwer Academic Publishers, Dordrecht).

Fukugita M and Yanagida T, 1993, *Physics of neutrinos and applications to astrophysics* (Springer-Verlag, Berlin).

Gamow G and Schoenberg M, 1940, "The possible role of neutrinos in stellar evolution," Phys. Rev. **58**, 1117.

Hannestad S, 2006, "Primordial neutrinos," to be published in *Annual Review of Nuclear and Particle Science* [hep-ph/0602058].

Hannestad S, Raffelt G G, Sigl G and Wong Y Y Y, 2006, "Self-induced conversion in dense neutrino gases: Pendulum in flavor space," *Phys. Rev. D* **74**, 105010 [astro-ph/0608695].

Halzen F, 2006, "Lectures on high-energy neutrino astronomy," *AIP Conf. Proc.* **809**, 130.

Hirata K S et al., 1988, "Observation in the Kamiokande-II detector of the neutrino burst from supernova SN 1987A," *Phys. Rev. D* **38**, 448.

Käppeler F, Thielemann F K and Wiescher M, 1998, "Current quests in nuclear astrophysics and experimental approaches," *Annu. Rev. Nucl. Part. Sci.* **48**, 175.

Keil M T, Raffelt G G and Janka H-T, 2003, "Monte Carlo study of supernova neutrino spectra formation," *Astrophys. J.* **590**, 971 [astro-ph/0208035].

Kippenhahn R and Weigert A, 1990, *Stellar structure and evolution* (Springer-Verlag, Berlin).

Koshiba M, 1992, "Observational neutrino astrophysics," *Phys. Rept.* **220**, 229.

Krauss L M and Tremaine S, 1988, "Test of the weak equivalence principle for neutrinos and photons," *Phys. Rev. Lett.* **60**, 176).

Learned J G and Mannheim K, 2000, "High-energy neutrino astrophysics," *Annu. Rev. Nucl. Part. Sci.* **50**, 679.

Lesgourgues J and Pastor S, 2006, "Massive neutrinos and cosmology," *Phys. Rept.* **429**, 307 [astro-ph/0603494].

Lunardini C, 2006, "The diffuse supernova neutrino flux," astro-ph/0610534.

Malek M et al. (Super-Kamiokande Collaboration), 2003, "Search for supernova relic neutrinos at Super-Kamiokande," *Phys. Rev. Lett.* **90**, 061101 [hep-ex/0209028].

Marrodán Undagoitia T, von Feilitzsch F, Göger-Neff M, Hochmuth K A, Oberauer L, Potzel W and Wurm M, 2006, "Low energy neutrino astronomy with the large liquid-scintillation detector LENA," *J. Phys. Conf. Ser.* **39**, 287.

Renzini A and Fusi Pecci F, 1988, "Tests of evolutionary sequences using color-magnitude diagrams of globular clusters," *Annu. Rev. Astron. Astrophys.* **26**, 199.

Raffelt G G, 1990a, "Astrophysical methods to constrain axions and other novel particle phenomena," *Phys. Rept.* **198**, 1.

Raffelt G G, 1990b, "New bound on neutrino dipole moments from globular cluster stars," *Phys. Rev. Lett.* **64**, 2856 (1990).

Raffelt G G, 1996, *Stars as laboratories for fundamental physics* (University of Chicago Press).

Raffelt G G, 1999, "Particle physics from stars," *Annu. Rev. Nucl. Part. Sci.* **49**, 163 [hep-ph/9903472].

Raffelt G, 2005, "Supernova neutrino oscillations," *Phys. Scripta* **T121**, 102 [hep-ph/0501049].

Raffelt G, 2006, "Astrophysical axion bounds," hep-ph/0611350.

Raffelt G and Weiss A, 1992, "Nonstandard neutrino interactions and the evolution of red giants," *Astron. Astrophys.* **264**, 536.

Rolfs C E and Rodney W S, 1988, *Cauldrons in the cosmos—nuclear astrophysics* (University of Chicago Press).

Stanev T, 2004, *High energy cosmic rays* (Springer-Verlag, Berlin).

Stothers R B, 1970, "Astrophysical determination of the coupling constant for the electron-neutrino weak interaction," *Phys. Rev. Lett.* **24**, 538.

Woosley S and Janka H-T, 2005, "The physics of core-collapse supernovae," Nature Physics **1**, 147 [astro-ph/0601261].

Section III: Experimental Neutrino Physics

Accelerator-Based Neutrino Oscillation Experiments

Deborah A. Harris

Fermi National Accelerator Laboratory, Batavia, Illinois USA

1 Introduction

Neutrino oscillations were first discovered by experiments looking at neutrinos coming from extra-terrestrial sources, namely the sun (Fukuda et al. 1998*b*) and the atmosphere (Fukuda et al. 1998*a*), but we will be depending on earth-based sources to take many of the next steps in this field. This article will describe what has been learned so far from accelerator-based neutrino oscillation experiments, and then describe very generally what the next accelerator-based steps are. In section 2 the article will discuss how one uses an accelerator to make a neutrino beam, in particular, one made from decays in flight of charged pions. There are several different neutrino detection methods currently in use, or under development. In section 3 these will be presented, with a description of the general concept, an example of such a detector, and then a brief discussion of the outstanding issues associated with this detection technique. Finally, section 4 will describe how the measurements of oscillation probabilities are made. This includes a description of the near detector technique and how it can be used to make the most precise measurements of neutrino oscillations.

1.1 What we know so far from accelerators

The first accelerator-based neutrino oscillation experiments were designed to look for oscillations in the neutrino sector that were assumed to be similar to quark mixing: that is, very small mixing angles. CHORUS (Eskut et al. 2007) and NOMAD (Astier et al. 2001) are two recent such experiments that found no oscillations between muon neutrinos and tau neutrinos at high mass squared splittings (in other words, the two mass eigenstates would have squared mass differences, $m_1^2 - m_2^2 = \Delta m^2$ above 1 eV^2), and set limits on the oscillation probabilities at the sub per cent level. Accelerator-based neutrino oscillation experiments have also provided confirmation of the "atmospheric neutrino anomaly". They have unambiguously measured the disappearance of muon neutrinos which would correspond to a squared mass splitting of $3 \times 10^{-3} eV^2$. The two experiments that have provided these measurements so far are K2K (Aliu et al. 2005) and MINOS (Michael et al. 2006), and at the time of this writing the mass splitting is just starting to be known well enough that results are being plotted on linear scales rather than on logarithmic scales.

The accelerator-based LSND experiment has reported evidence of muon to electron transitions at a large mass splitting ($0.1 eV^2$), and this would assuredly have to be confirmed or refuted by another accelerator-based neutrino experiment (Athanassopoulos et al. 1996). Given the solar and the atmospheric neutrino oscillation signatures, if the LSND result were to be confirmed this would change drastically our picture of neutrinos and call into question even the number of neutrinos there are in the first place. For the remainder of this chapter we will assume there are three generations of neutrinos but the reader should keep in mind that if LSND were to be due to oscillations then the measurements that the next generation of neutrino experiments would be making would need to be completely re-evaluated in this totally new framework. At the time of this writing, the LSND signature has been excluded as being due to a simple oscillation scenario by the MiniBooNE experiment (Aguilar-Arevalo et al. 2007).

1.2 What is next to learn?

In the three-generation neutrino model, the mass and flavor eigenstates are related one to the other by a three-by-three mixing matrix, defined by three mixing angles $\theta_{12}, \theta_{23}, \theta_{13}$ and a CP-violating phase is δ. If $s_{ij} = \sin\theta_{ij}, c_{ij} = \cos\theta_{ij}$, then the matrix can be parameterized in the following way (Kobayashi & Maskawa 1973):

$$U = \begin{pmatrix} 1 & 0 & 0 \\ 0 & c_{23} & s_{23} \\ 0 & -s_{23} & c_{23} \end{pmatrix} \begin{pmatrix} c_{13} & 0 & s_{13}e^{i\delta} \\ 0 & 1 & 0 \\ -s_{13}e^{-i\delta} & 0 & c_{13} \end{pmatrix} \begin{pmatrix} c_{12} & s_{12} & 0 \\ -s_{12} & c_{12} & 0 \\ 0 & 0 & 1 \end{pmatrix}$$

There are several (sometimes competing) goals of the accelerator-based neutrino oscillation program. Clearly the biggest issue in the field of oscillation physics that must be addressed is that of the number of neutrinos. We want to know whether or not sterile neutrinos, which would be needed to explain the LSND signature, exist. If in fact there are three generations of neutrinos mixing, then we already know from the solar and atmospheric neutrino experiments that two of the three mixing angles are large. There is a third mixing angle whose magnitude has not been measured, although the CHOOZ experiment has determined it to be less than about 8 degrees (Apollonio et al. 2003). If this third mixing angle is found to be non-zero then the possibility of CP-violation in the lepton sector arises. Finally, in the three-generation model, there are two possibilities for the mass eigenstates: either they are "normal" or "inverted", where "normal" means like the charged fermion sector where the mass splitting between the two heaviest fermions is much larger than the mass splitting of the two lightest.

When one starts with one flavor eigenstate and searches for other flavor eigenstates, clearly there are two oscillation frequencies, corresponding to the two independent mass differences between the three eigenstates. The relevant quantities are actually the differences between the square of the masses, and the standard notation is given by $\Delta m^2_{ij} = m^2_i - m^2_j$. For a derivation of the oscillation probabilities see reference (Fisher, Kayser & McFarland 1999). Simply put, for a transition between two different flavors, there is a term that would be proportional to $\sin^2 \Delta m_{23}^2 L/E_\nu$, a term that is proportional to $\sin^2 \Delta m_{12}^2 L/E_\nu$ and two terms that are proportional to $\sin \Delta m_{12}^2 L/E_\nu \times \sin \Delta m_{23}^2 L/E_\nu$. These last two terms are the "interference terms" which include a CP-conserving and a CP-violating piece. For the case of $\nu_\mu \rightarrow \nu_e$ and $\bar{\nu}_\mu \rightarrow \bar{\nu}_e$, the four terms can be expressed as follows, using the above definition of the mixing matrix: Starting with the above notation, these four terms can be expressed as:

$$\begin{aligned}
P(\nu_\mu \rightarrow \nu_e) &= P_1 + P_2 + P_3 + P_4 \\
P_1 &= \sin^2 \theta_{23} \sin^2 2\theta_{13} \left(\frac{\Delta_{13}}{B_\pm}\right)^2 \sin^2 \frac{B_\pm L}{2} \\
P_2 &= \cos^2 \theta_{23} \sin^2 2\theta_{12} \left(\frac{\Delta_{12}}{A}\right)^2 \sin^2 \frac{AL}{2} \\
P_3 &= J \cos\delta \left(\frac{\Delta_{12}}{A}\right) \left(\frac{\Delta_{13}}{B_\pm}\right) \cos \frac{\Delta_{13} L}{2} \sin \frac{AL}{2} \sin \frac{B_\pm L}{2} \\
P_4 &= \mp J \sin\delta \left(\frac{\Delta_{12}}{A}\right) \left(\frac{\Delta_{13}}{B_\pm}\right) \sin \frac{\Delta_{13} L}{2} \sin \frac{AL}{2} \sin \frac{B_\pm L}{2}
\end{aligned}$$

where

$$\begin{aligned}
\Delta_{ij} &= \frac{\Delta m_{ij}^2}{2E_\nu} \\
A &= \sqrt{2} G_F n_e \\
B_\pm &= |A \pm \Delta_{13}| \\
J &= \cos\theta_{13} \sin 2\theta_{12} \sin 2\theta_{13} \sin 2\theta_{23}
\end{aligned}$$

and the $\pm$ signifies neutrinos or antineutrinos. G_F is the Fermi coupling constant, n_e is the electron density in the earth, L is the distance the neutrino has travelled between production and detection, and E_ν is the neutrino energy.

These four terms are shown for a single neutrino energy, baseline, and value of θ_{13} in Figure 1, but for both possible mass hierarchies and all possible values of the CP violating phase δ. If there is one experiment at one energy and baseline, it is clear that even in the presence of a very precise neutrino and antineutrino oscillation probability measurement, one cannot completely determine the CP violating phase or the mass hierarchy.

Although there are these four different terms in the oscillation probability, all of the experiments so far have been only been sensitive to one of the oscillation frequencies. In particular, the atmospheric neutrino experiments measure the θ_{23} mixing angle and the magnitude of the mass splitting Δm_{23}^2. The solar neutrino experiments measure the θ_{12} mixing angle and the mass splitting Δm_{12}^2. The last unmeasured angle is θ_{13}, and in the above parameterization it is particularly clear that if θ_{13} is non-zero then there arises the possibility of a CP-violating phase δ. But to see this difference one must be sensitive to the interference term, which means that the first two terms must not be too large. It turns out that for the transitions between muon and electron neutrinos, the mixing angle θ_{13} being small means that the "first term" in the expansion will be small, giving experiments a chance to see the interference term, and therefore CP-violation.

Two simplifications that are often described are to consider the difference divided by the sum for neutrino and antineutrino oscillations: in the absence of matter effects and terms to

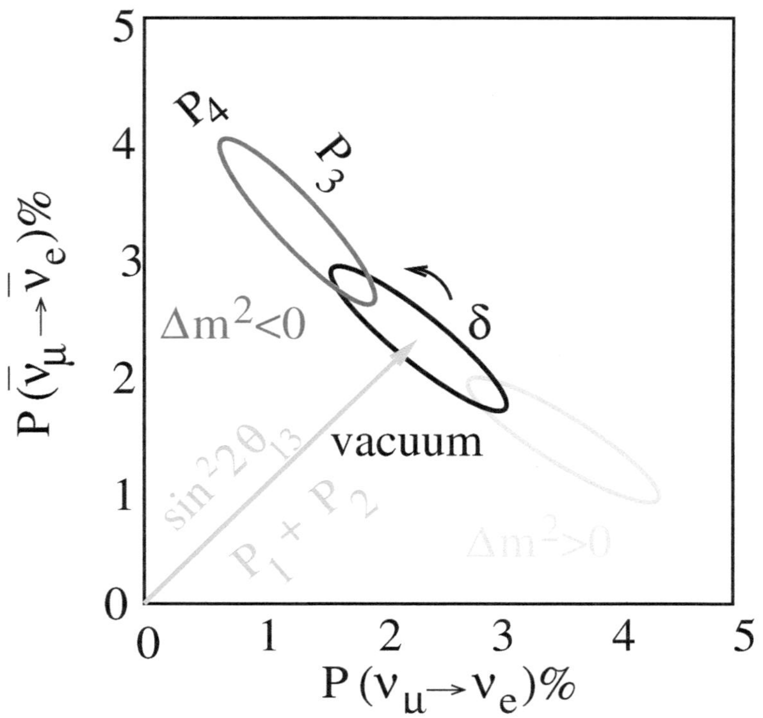

Figure 1. *Plot of the neutrino versus anti-neutrino oscillation probability for one baseline and energy. The ellipse centered on the diagonal represents the range for all possible values of δ and a given $sin^2 2\theta_{13}$ and but assuming no matter effects, and then the two other ellipses correspond to the actual ranges of possibilities for two different signs of the mass hierarchy.*

second order in the solar mass squared difference:

$$\frac{P(\nu_\mu \to \nu_e) - P(\bar{\nu}_\mu \to \bar{\nu}_e)}{P(\nu_\mu \to \nu_e) + P(\bar{\nu}_\mu \to \bar{\nu}_e)} \propto \frac{\Delta m^2_{sol} L}{E} \frac{\sin\delta}{\sin\theta_{13}} \tag{1}$$

And at the other extreme, the asymmetry in the presence of matter effects at the δm^2_{23} oscillation maximum but ignoring for the moment CP violation:

$$\frac{P(\nu_\mu \to \nu_e) - P(\bar{\nu}_\mu \to \bar{\nu}_e)}{P(\nu_\mu \to \nu_e) + P(\bar{\nu}_\mu \to \bar{\nu}_e)} = \frac{2E_\nu}{E_R} \tag{2}$$

for neutrino energies below about 5 GeV, where E_R, the Earth's resonant neutrino energy, is equal to $\frac{\Delta m^2_{atm}}{2\sqrt{2}G_F n_e}$ and is $\approx 11\,\text{GeV}$.

The remaining sections of this article will describe some of the details of how one measures a neutrino oscillation probability using an accelerator-based neutrino source. Section 2 will describe how to make a beam of neutrinos using pion decays. Section 3 will describe the various detector technologies that are available for measuring neutrino interactions. Finally, section 4 will describe the strategies associated with making the most accurate measurements: the strategy of using near detectors, and the challenges associated with both muon neutrino disappearance measurements and electron appearance measurements.

2 Making Neutrino Beams at Accelerators

Simply put, the way to make neutrino beams at accelerators is to first make charged particles that will decay to neutrinos, magnetically focus those particles so that they are all travelling in the same direction, and then finally, give the particles a space in which to decay in flight to neutrinos. For conventional neutrino beams, the parent particles are predominantly pions, and because of the short lifetime of the pion the focused beam is simply sent through a decay volume followed by an absorber to catch the remaining particles. This results in a neutrino beam that is predominantly muon neutrino or muon antineutrino, with a small electron neutrino component that comes from kaon and muon decays. See (Kopp 2007) for a more detailed description of the techniques and challenges associated with accelerator neutrino beams, what follows here is an abbreviated summary.

A different way to make neutrino beams is from focusing and then letting decay either muons (Geer 1998) or radioactive ions (Zucchelli 2002). In both cases these particles are re-circulated several times in a race-track shaped storage ring. The straight sections in the storage rings can point to one or more neutrino detectors, where the beam at the far detector is created mostly while the parent particles are decaying in those straight sections. Muon-based storage rings would provide beams that are an almost even mixture of muon neutrinos and electron anti-neutrinos (or their CP conjugates), whereas beams from radioactive ion decays, so-called "beta-beams" would be purely electron neutrino or electron anti-neutrino, depending on the ion in the storage ring. The beamline elements of these newer kinds of beams are completely different from those of conventional neutrino beams, and they will not be discussed in this article. The motivations for developing these other kinds of beamlines can be found in other proceedings at this school (Kuno 2007). In summary, the motivation is due to the fact that the fluxes can be as high or higher than conventional beams while the backgrounds can be significantly lower.

2.1 Proton Beam

The first step in producing a neutrino beam is to have an extremely intense beam of protons. Certainly the more protons in the beam, the more pions that are produced, and the number of pions produced scales as the number of protons times the proton energy, or the proton power. The requirement that these experiments look for small oscillation probabilities over very large distances has translated into enormous jumps in the required proton power to be directed towards a target in a neutrino beamline. Between the late 1990's and the early 2000's the jump in proton power has been almost a factor of 100: Table 1 shows the previous, current and next generation of proton sources and the proton power at each of these sources.

2.2 Targets

The next step in producing a neutrino beam is to put a target into the path of these protons. The longer the target, the larger fraction of protons will interact to produce pions, yet the more the produced pions will re-interact with the target either through multiple scattering or through showering. So targets for neutrino beamlines tend to be long thin objects which allow pions to exit out of the sides of the target into the focusing region. Another consideration

Proton Source	Experiment	Proton Energy (GeV)	protons per year	Proton Power (MW)	Neutrino Energy (GeV)
KEK	K2K	12	$1 \times 10^{20}/4$	0.0052	1.4
FNAL Booster	MiniBooNE	8	5×10^{20}	0.05	1
FNAL Main Injector	MINOS	120	2.5×10^{20}	0.25	3-17
FNAL Main Injector	NOvA	120	6×10^{20}	1	3-17
CERN	OPERA	400	0.45×10^{20}	0.12	25
J-PARC	T2K	40-50	11×10^{20}	0.75	0.77

Table 1. *Comparison of proton sources: their energies, proton powers, and associated oscillation experiments.*

Experiment	Target Material	Shape	Transverse Size (mm)	Length (cm)
MiniBooNE	Be	cylinder	10	70
K2K	Al	cylinder	30	66
MINOS	graphite	ruler	6.4×20	90
NOvA	graphite	ruler	> 6.4	90
CNGS	carbon	ruler	4 mm wide	200
T2K	graphite	cylinder	12-15	90

Table 2. *Comparison of different targets in use or planned to be used in oscillation experiments, including their composition, size, and shape.*

which has become important in the era of these high power proton sources is the cooling of the targets themselves. If one wants to avoid having pions reinteract or multiple scatter one wants to minimize material in the surrounding area, which puts tight constraints on the cooling system. Some targets are air cooled, while others are cooled by flowing water through nearby pipes, and finally the T2K target is designed to be cooled by flowing 300 degree Kelvin helium through the target itself (Nakadaira 2005, Fitton, M. 2006).

The general rule of thumb which seems to be applied in most target designs is that the transverse target size is roughly three times one standard deviation in the proton spot size, in order to catch a large enough fraction of the proton beam but to not introduce too much additional material for the pions to traverse. Table 2 gives a comparison of several different targets in use (or being designed) for past, current, and future oscillation experiments.

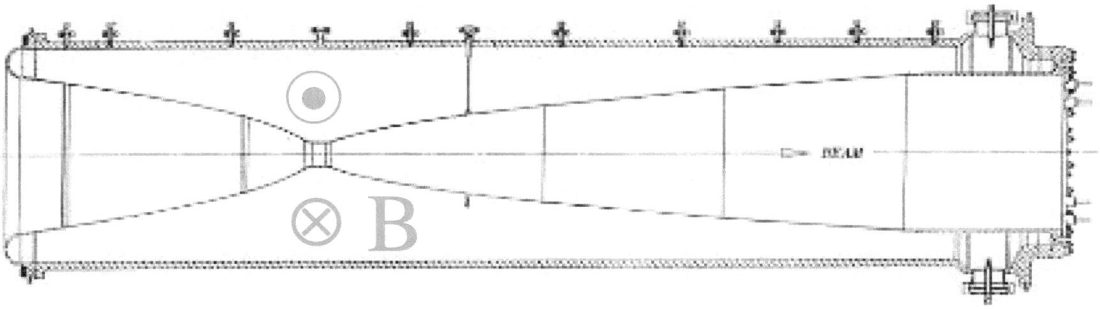

Figure 2. *Schematic diagram of a horn to focus pions coming from the left.*

2.3 Focusing Systems

The challenge of designing a focusing system for a neutrino beam is not unlike the challenge of designing a target: one wants a strong magnetic field in order to get the best focusing of the secondary particles, yet one wants to minimize the material that these secondary particles must go through. The typical transverse momentum of the pions is approximately the strong interaction energy scale, Λ_{QCD}. Although previous experiments have used alternating sets of quadrupole magnets to focus secondary particles, and solenoids have also been proposed to focus particles with a minimum of material in the path of the pions, these have the disadvantage of focusing both negative and positive pions simultaneously. This sounds like an advantage in terms of overall rate, but it would mean that in order to measure both neutrino and anti-neutrino oscillation probabilities precisely one would need to measure the charge of the final state electrons in a detector, which would be prohibitively expensive.

Consider the case of particles flying out from a single target: those particles that reach the focusing element at a higher radius are the ones with the largest transverse momentum, so they need the largest transverse momentum kick. Those particles leaving on the proton axis would need almost no transverse momentum kick. Clearly the best focusing system would be one whose integrated path length times magnetic field were proportional to the radius at which the particle entered the focusing system. The magnetic field from a line current source falls like the inverse radius, so in order to focus all the particles of a given transverse momentum, one would want the time spent in the magnetic field (or distance traversed) to be proportional to the square of the radius at which the particle entered the focusing system. This is the principle of the parabolic horn: an inner conductor is shaped in a parabola so that the length of field region at a given radius is proportional to the square of that radius, and then an outer conductor provides the return for the current. The magnetic field is completely contained between the inner and outer conductors, and there is no field at radii smaller than the inner conductor. Figure 2 shows a schematic diagram of a parabolic horn as viewed from the side (the pions enter the horn from the left in this case): note that the magnetic field is between the "horn-shaped" inner conductor and the cylindrical outer conductor. The transverse momentum kick p_t is therefore

$$\delta p_t \approx \frac{e\mu_o I}{2\pi c r} \times \frac{r^2 l}{r_{outer}^2} \approx p_{tune}\theta \tag{3}$$

where μ_0 is the permittivity of free space, I is the current in the horn, r is the average radius at which the particle traverses the horn, l is the length of the horn, and r_{outer} is the radius of the outer conductor of the horn.

It is interesting to consider how high the current must be in a typical horn: consider two horns that are each 3 m long and 16 cm in diameter, what kind of current would be needed to give a 200 MeV/c momentum kick to the produced secondary particles?

$$\delta p_t(\mathrm{MeV/c}) = 0.3B(\mathrm{T})l(\mathrm{m}) \quad (4)$$

$$B(\mathrm{T}) = \frac{\mu_0 I}{2\pi r} \quad (5)$$

$$B(\mathrm{T}) = 2\times 10^{-7}\frac{I(\mathrm{A})}{r(\mathrm{m})}(\frac{r}{r_{outer}})^2 \quad (6)$$

And assuming that the pions would go through on average half the radius of the outer conductor, one finds

$$I(\mathrm{A}) = \frac{\delta p_t(\mathrm{GeV/c})}{0.3}\frac{2r_{max}}{l}\frac{1}{2\times 10^{-7}} \quad (7)$$

where B is the magnetic field in Tesla, l is again the length of the horn, I is the current and r is the average radius at which the particle crosses the horn.

So the currents in a horn are of the order of hundreds of thousands of Amps!

2.4 Decay Kinematics

For pions at rest, there is no preferred direction since they are spin zero particles, and they will decay isotropically. In this case the neutrino produced in the decay $\pi \rightarrow \mu\nu_\mu$ only has a single energy which is determined by the difference in the pion and muon masses. When one boosts to the lab frame, however, the neutrino energy depends on the angle between the neutrino direction in the center of mass frame and the pion boost direction which translated to the lab frame is denoted by the angle θ. It can be shown that the neutrino energy, E_ν, is related to the neutrino-pion angle (θ) and energy (E_π) in the lab frame in the following way:

$$E_\nu = E_\pi \frac{1-\frac{m_\mu^2}{m_\pi^2}}{1+\gamma^2\theta^2} \quad (8)$$

where γ is the relativistic boost of the pion, or E_π/m_π and m_μ and m_π are the muon and pion mass, respectively.

The flux of neutrinos (Φ_ν) can be derived by boosting the isotropic decays of pion at rest into the lab frame. We obtain:

$$\Phi_\nu = BR\frac{1}{4\pi L^2}\left(\frac{2\gamma}{1+\gamma^2\theta^2}\right) \quad (9)$$

where BR is the branching fraction for that 2-body decay, and L is the distance between the detector and where the pion decayed. The interesting thing to note here is how the neutrino energy increases with increasing pion energy, and the flux increases as the square of the pion energy. Given that the neutrino cross section increases linearly with neutrino energy, one would naively think that the number of muon neutrino events at a far detector simply increases as the cube of the pion energy, and as such one should try to design as high a pion beam energy as possible. However, the higher the neutrino energy, the longer the distance one must put the far detector to probe a given mass splitting. Also, for a given incoming proton energy, the number of pions produced for a given proton energy is a steeply falling function of pion momentum. So in designing beamlines one must fold in many factors, not simply the decay kinematics.

2.5 Decay Region Strategies

As was mentioned above, there are not only pions produced when protons strike a target. Kaons are also produced, and although most of them will also undergo a two-body decay to muon neutrinos, there is a significant fraction of both charged (and neutral) kaons that will decay to electron neutrinos. This is one source of beam impurity, the other source is due to the fact that some small fraction of muons produced in pion decays will themselves decay before they reach the end of the decay region.

Suppose one wanted to design a decay region for an experiment looking for muon to electron neutrino oscillations: one would want to minimize the fraction of electron neutrinos per muon neutrino. The longer the decay region, the larger a fraction of pions that would decay, but the larger the fraction of muons that would also decay. By considering the average decay length of a beam of pions of energy E_π and calculating the fraction of muons that would decay in the remaining length of the decay pipe, one can show that the ratio of electron to muon neutrino fluxes is dependent on the decay pipe length (L), the average pion energy, and the pion and muon lifetimes:

$$\frac{\Phi(\nu_e)}{\Phi(\nu_\mu)} = \frac{Lm_\mu c}{E_\pi \tau_\mu}\left(\frac{1}{e^{y_\pi}-1}+1+\frac{1}{y_\pi}\right) \tag{10}$$

where y_π is the number of pion lifetimes in the decay pipe:

$$y_\pi = \frac{Lm_\pi c^2}{E_\pi c \tau_\pi} \tag{11}$$

In this equation τ_i is the lifetime of particle i, m_i is the mass of particle i, and E_i is the energy of particle i in the lab frame. One can similarly define y_μ as the number of muon lifetimes in the decay pipe

$$y_\mu = \frac{Lm_\mu c^2}{E_\mu c \tau_\mu} \tag{12}$$

Table 3 gives the lengths of several decay pipes of past, current, and future oscillation experiments, as well as the average pion energy and the calculated theoretical ratio of electron to muon neutrino fluxes. Keep in mind that these calculations are for the integrals of the fluxes: clearly the fluxes from 2-body decays producing muon neutrinos will be more collimated than those from the three body decays. Similarly, the neutrino energies from the two body decays carry on average a larger fraction of the pion energy than those from the three-body muon decays.

2.6 Off Axis Strategy

For the future searches for ν_μ to ν_e appearance, one wants to minimize the fraction of electron neutrinos in the beamline. One strategy might be to make the decay region very short so that there are no tertiary muon decays, another strategy might be to also make the incoming proton energy very low so that there are a minimum of kaons produced. However, both of those strategies will also reduce the muon neutrino rate at least as much as the electron neutrino rate. One technique that reduces the ratio of electron neutrinos to muon neutrinos at a given energy is to put the far detector slightly off of the pion beamline axis. For a two-body pion decay, the

Experiment	Length	E_π (GeV)	y_π	y_μ	$\frac{\Phi(\nu_e)}{\Phi(\nu_\mu)}$ (approx)
MiniBooNE	50m	2.5	0.36	0.3%	0.15%
K2K	200m	3.5	1.0	0.9%	0.5%
MINOS	675m	9	1.3	1.2%	0.8%
CNGS	1000m	50	0.36	0.3%	0.15%
T2K	130m	9	0.47	0.2%	0.10%

Table 3. *Decay pipe characteristics of several decay volumes of currently operating or planned neutrino beamlines. For the MINOS experiment, the beamline decay pipe parameters are defined for that experiment, while for the NOvA experiment, which uses the same beamline but a different peak of focused pion energy, y_π and y_μ will be slightly different.*

neutrino energy is completely determined by the pion boost and the angle between the pion momentum and the neutrino direction. Figure 3 shows the transverse versus longitudinal neutrino momentum for pions at different energies, represented by different ellipses. Note that for on-axis decays the neutrino longitudinal momentum, and hence the neutrino energy, is linear with the pion energy. This corresponds to the case in Equation 8 where θ_μ or the angle between the pion and the neutrino (detector) direction, is equal to zero. However, as the far detector angle moves slightly off axis, there is a broad range of pion energies that will contribute neutrinos in a much narrower range of neutrino energies. Because the electron neutrinos come primarily from three-body decays instead of two-body decays, this peaking in energy is not nearly as pronounced.

So although putting a far detector off the axis to which the pions are focused will reduce the peak muon neutrino energy as well as the muon neutrino rate the ratio of electron to muon neutrino fluxes drops substantially. This technique will be put to use by both the T2K and NOvA experiments to optimize their searches for ν_e appearance.

3 Detecting Neutrinos from Accelerators

In conventional neutrino beams, one has a predominantly muon neutrino beam with a small admixture of electron neutrino contamination, which will be less than a per cent for future experiments. So to measure $\nu_\mu \rightarrow \nu_e$ transitions one must first and foremost be able to distinguish between electrons and muons, but one must also be able to distinguish between electrons and neutral pions, which are made in a large fraction of neutrino interactions above several hundred MeV. Neutrino beams focused with horns are predominantly neutrinos or anti-neutrinos, depending on the polarity of the horn current, so a detector in a conventional beam does not need to identify the charge of the final state lepton, merely the flavor.

Neutrino beams made from decays of beams of muons (so-called "neutrino factories") present another challenge entirely. Since the beam will have roughly equal numbers of electron neutrinos and muon antineutrinos, (or vice versa), the detector will need to again distinguish between electrons and muons, but will also need to measure accurately the charge of that final

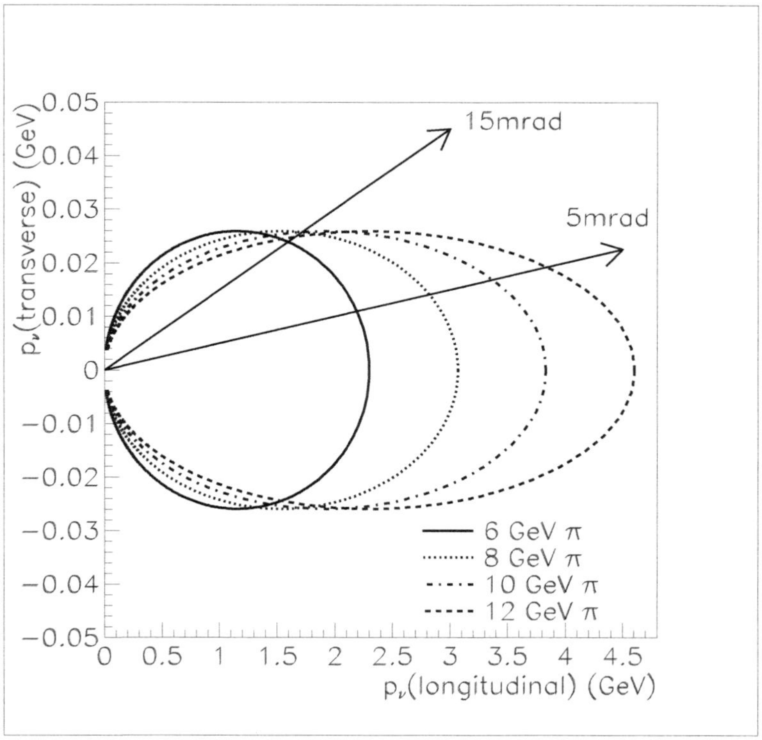

Figure 3. *Plot of the transverse versus longitudinal neutrino momenta. Different ellipses correspond to different parent pion energies. An off-axis detector located at a given angle with respect to the beamline axis will see neutrinos of energy corresponding to the intersect of that angle and the set of ellipses corresponding to the focused pion peak energies.*

state muon: if the muon charge is the same as the charge of the muon circulating in the storage ring, it means that the muon was produced from a muon neutrino from that ring. However, if the charge of the final state muon is the opposite of the charge of the muons in the ring, then it must have come from an electron neutrino that has oscillated into a muon neutrino. When this idea was first developed it was appreciated that the probability of mis-identifying the charge of a muon could be reduced to well below the intrinsic electron neutrino contamination of a conventional beam. This means that one could have access to much lower backgrounds, which would in turn allow access to much lower mixing angles than with a conventional beam (Geer 1998).

Neutrino beams made from the decays of radioactive ions are purely electron neutrino or antineutrino, depending on the ion in the ring (Zucchelli 2002). For this reason they are the closest one could get to "preparing a beam in a flavor eigenstate". The measurement in a far detector of a muon neutrino would unambiguously signal oscillations. The challenge of these "beta-beams" is that to get above the threshold for muon neutrino charged current interactions the neutrino beam energy must be above a few hundred MeV (Burguet-Castell, Casper, Gomez-Cadenas, Hernandez & Sanchez 2004). Because the difference in total mass between the initial state and the final state of these decays is so small this means an enormous energy per nucleon would be required for a given neutrino energy. Similar to the neutrino factories, the detector challenge here would simply be to distinguish muons from electrons, and the energies would likely be low enough that not many pions would also be produced in the final state neutrino interactions.

Total Detector Mass	Events Per Detector Background ν_e	NC	Total	Signal Events	Figure of Merit (Significance)
5 kton	2.5×5	$400\times.0025\times 5$	17.5	25	3.8
15 kton	1.5×15	$400\times.005\times 15$	52.5	45	4.6

Table 4. *Signal and background event calculations for the sample experiments described in the text. The neutral current rates are* 400 *events per kiloton of detector, and the nu*$_\mu$ *charged current rates are assumed to be 1000 events per kiloton of detector. The oscillation probability is assumed to be 1% in this example. The significance is defined as the signal divided by the square root of the number of signal and background events.*

3.1 Backgrounds

To understand why it is not only the signal rate but also the background rates which are important, consider the following two scenarios: a 5 kton detector has a 50% acceptance for ν_e charged current events, and only a 0.25% probability of mis-identifying a neutral current event as a background event. Another detector is only 30% efficienct for ν_e charged current events, and has a higher background mis-identification probability of 0.5%. But because the second detector is less fine-grained, it is significantly cheaper and for the same money one could build 15 kton of the second detector. Assume in both cases that there is also a 0.5% intrinsic ν_e component in the ν_μ beam. Which detector would see a larger signal, if the $\nu_\mu \to \nu_e$ oscillation probability were 1.0%? Assume 1000 ν_μ charged current events per kiloton, and 400 neutral current events per kiloton. If the ν_e contamination is 0.5%, then this implies 5 ν_e charged current events per kiloton as an intrinsic background (before acceptance efficiency is applied).

So although on the surface the more coarse-grained detector seems worse, by being able to build more of it one can design a more sensitive experiment, see Table 4.

Now to do the same exercise for a neutrino factory, imagine you have a factory which produces 500 (200) $\bar{\nu}_\mu$ charged (neutral) current events per kiloton, and 1000 (400) ν_e charged (neutral) current events per kiloton, and again assume a 1% oscillation probability, and a background rate of 0.1% for all kinds of interactions, and a 50% ν_μ signal efficiency and 15kton of detector (because it is less challenging than a ν_e appearance detector): The background in this case is $0.0001\times 2100\times 15 = 3$, while the signal events would number $1000 * 0.01 * 0.5\times 15 = 150$, and the signal over the square root of the background would be 12 instead of only 3 or 4 in the above case. Another way of thinking of this is that the neutrino factory would see a 12 sigma result rather than a 3 or 4 sigma result.

3.2 Particles passing through Material

Before discussing the various detectors used for neutrino oscillation experiments it is worth reviewing briefly what happens when particles pass through material. Because neutrinos interact so rarely one is usually faced with using the active material in the detector as a neutrino target itself. If one is trying to search for electron neutrinos, then one would care about the detector

Material	Radiation Length (cm)	Interaction Length (cm)	dE/dx (MeV/cm)	Density (g/cm^3)
Liquid Argon	14	83.5	2.1	1.4
Water	37	83.6	2.0	1
Steel	1.76	17	11.4	7.87
Scintillator (CH)	42	80	1.9	1
Lead	0.56	17	12.7	11.4

Table 5. *Summary of material properties used in neutrino detectors.*

segmentation in terms of radiation lengths. If one is looking for muon neutrinos, then the issue is more the segmentation in terms of the length of a typical muon track, which is determined by dE/dx in that material. Finally, if one is looking for a τ neutrino interaction, then the important distance to keep in mind is the decay length of a τ, which is the usually small boost of the tau multiplied by 87 microns. Table 5 gives a summary of some of the materials that are used for neutrino detectors (Yao, W.M. et al. 2006).

3.2.1 Charged Particle Energy Loss

The Bethe-Bloch equation describes how much energy per unit distance is lost as a particle traverses a medium. Energy loss is a function of the particle's charge (z), velocity normalized to the speed of light (β) and Lorentz boost (γ) of the particle, the electron mass m_e, as well as the atomic number (Z) and atomic mass (A, in units of grams per mol) of the medium in question. Because dE/dx depends on γ and β it can be used at some level for particle identification, especially if a particle ranges out in that material. The Bethe-Bloch equation is expressed as follows:

$$\frac{dE}{dx} \propto z^2 \frac{Z}{A} \frac{1}{\beta} \frac{1}{\beta^2} \left[\frac{1}{2} \frac{\ln 2m_e c^2 \beta^2 \gamma^2 T_{max}}{I^2} - \beta^2 - \frac{\delta}{2} \right] \tag{13}$$

where x is expressed in units of g/cm^2, T_{max} is the maximum kinetic energy imparted to a free electron in a single collision, and δ is a density effect correction (Yao, W.M. et al. 2006). This equation is shown for different particle species in Figure 4. Note that at high particle momenta the energy loss per unit distance is comparable for all the particle types, but that as the particles get to very low momenta (for example just before they range out) the energy loss levels are very different.

3.3 Liquid Argon TPC

A liquid argon TPC is made by filling a vessel with ultra-pure liquid argon and 2 wire planes at right angles to each other. When an electric field is applied a charged particle that passes through the detector will ionize the Argon, and the free electrons will drift along the direction of the electric field. This will produce a signal on the planes of wires, and by measuring the

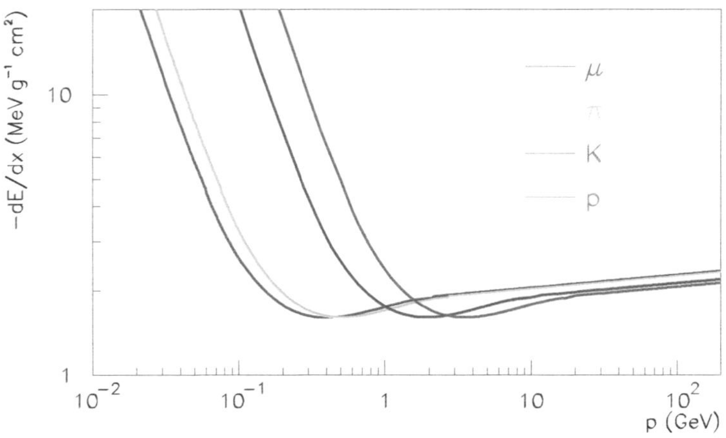

Figure 4. *Plot showing the energy loss per unit distance as a function of momentum for different particle species.*

signal on the two planes of wires as a function of time one can reconstruct a 3-dimensional (3-d) picture of the charged particles that produced the electrons. The higher the energy deposition the more electrons are produced, so you get not only a 3-d picture of the event, but you can use the Bethe-Bloch equation to get back to the type of particle that ionized the electrons.

Figure 5 shows this idea in practice: this is from the cosmic ray run in Pavia of a 600 Ton Liquid Argon TPC, where there is clearly a charged particle decay chain (Rubbia, A. 2005).

One promising way to discriminate between showers that originate from photons (from neutral pions) and those that originate from electrons (from electron neutrino charged current events) is to look at the first few radiation lengths of the interaction. At the very start of the interaction the dE/dx is nearly constant with momentum (see Figure 4), but a photon would convert to an electron and a positron, so the dE/dx would be twice as high for a converted photon as for an electron. By requiring all signal electrons to not start showering in the first 2 radiation lengths of material one loses only a small amount of signal and rejects all but a few per mil of the incoming π^0's, as shown in Figure 6.

At the time of this writing there are a few outstanding issues for the Liquid Argon TPC: the only downside of having so much information about each event is that the event reconstruction can be extremely challenging. The more experience the field gets with these detectors in known neutrino beams the more the reconstruction techniques can be improved. Right now there is substantial work on simulations of this kind of detector, but it is extremely important to show that the simulations agree with the actual detector performance, by using a known (neutrino) source. One very interesting possibility is that of putting one of these detectors in a magnetic field: in that way it could be used for both a conventional beam and a neutrino factory experiment and have extremely high signal acceptance.

Another issue has to do with the cost of this detector: clearly the fewer wire planes that are needed, the less expensive this detector would be. The ICARUS detector ran with wire planes separated by 1.5 m and has made Argon pure enough to use a 3 m spacing, but if the spacing could be 5 m or longer then there would be significant cost savings. Finally, regarding the cost an important issue for most of these detector technologies is one of granularity: the finer

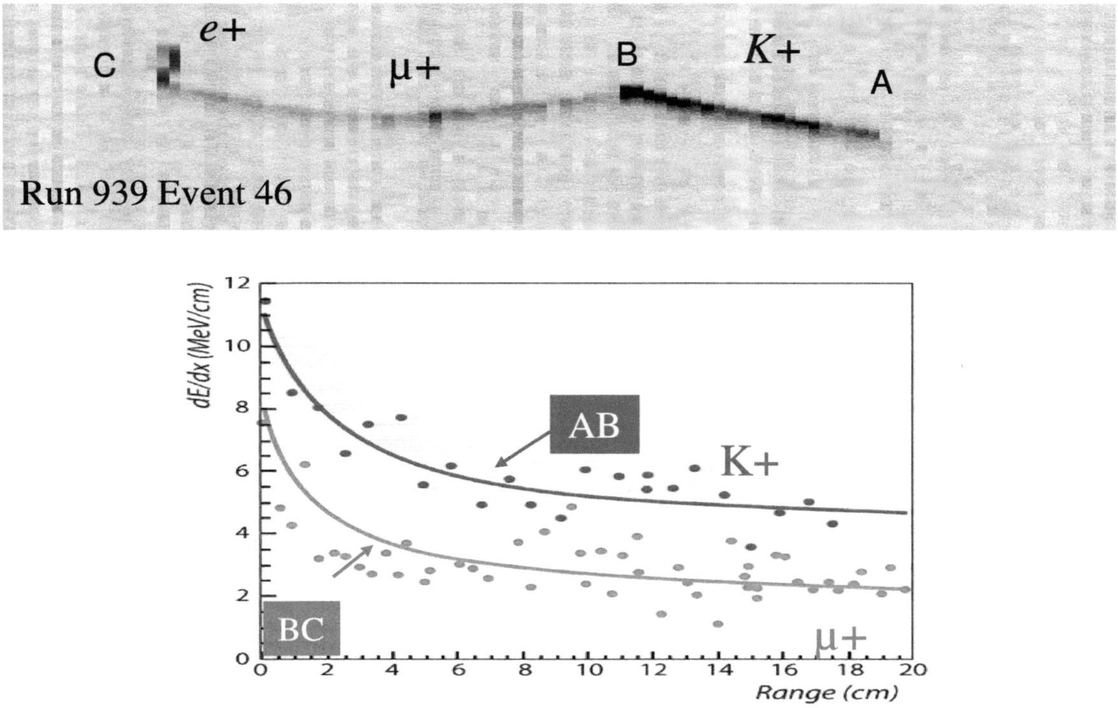

Figure 5. *An event display from a cosmic ray test run of a Liquid Argon TPC (top) and a plot of the energy deposition as a function of the distance from the end of the track (bottom). Figures courtesy of the ICARUS collaboration.*

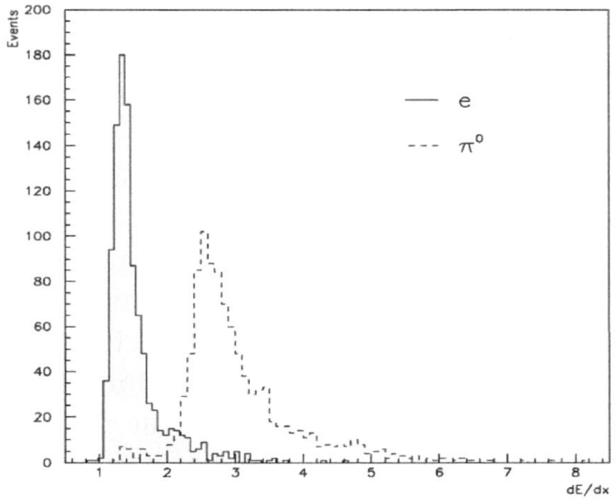

Figure 6. *The distribution of energy deposited in the first two radiation lengths of a liquid argon TPC for electrons and for photons coming from the decays of neutral pions.*

the wire spacing the more precise the 3-d picture is. However, for accelerator-based neutrino energies it is not clear that 3 mm wire spacing is critical, and the detector would be substantially less expensive if a wider pitch could be used.

3.4 Cerenkov Detectors

As particles move faster than the speed of light in a given material, a "shock wave" of light is produced. This shock wave propagates out at the Cerenkov angle θ_c, which is defined as $\cos\theta_c = 1/n(\lambda)$ where $n(\lambda)$ is the index of refraction for the wavelength λ of the Cerenkov light in that material. For water, this angle is about 43°. The momentum threshold for a particle to emit Cerenkov radiation is equal to the particle's mass divided by $\sqrt{n(\lambda)^2 - 1}$. Given what the index of refraction is for "typical" materials, this means that the thresholds are when particles have momenta that are roughly 35% above their rest mass.

This light is collected by phototubes placed on the edge of the active volume, and since a cone of light is produced, one only has to measure photons in enough places that one can reconstruct a circle as the cone passes through the surface of the detector.

Although this technique has very low threshold for electrons, the threshold for muons is much higher, and the thresholds for protons and pions are higher still. Nevertheless it remains an extremely powerful detector technique, because you can collect this light by putting detectors on the surface of this vessel, so the cost per unit mass is potentially lower than other techniques where one has to instrument the entire volume of a detector.

In order to determine which signals come from a given interaction, one needs the timing information associated with each phototube hit. The time of flight information is used to determine the origin of the neutrino interaction, and then all the tubes that are determined to be within a specified time window are used to find rings. A Hough Transformation (Shiozawa 1999) is used to associate given phototube hits with a set of rings. Once the rings are determined then the particle identification can begin, as well as energy reconstruction. In many cases, the decay electron coming from a stopped muon can also be found, although it will make a signal later in time than the primary muon signal.

Given the different ways in which particles interact in materials, different particles will produce different patterns of Cerenkov radiation. Therefore this technique can provide flavor information about the incoming neutrino. For example, muons can travel a long distance in material and lose energy slowly, so that means that they will produce many concentric rings of Cerenkov light before they go below threshold. Also, compared to electrons they have much less multiple scattering, so muons (and pions) will produce much sharper outer rings than electrons will produce. Electrons only go a short distance in material before producing showers and losing energy, so their rings may be brighter but have much fuzzier edges. Finally, photons are visible when they convert to an electron positron pair, and for the energies of interest those two particles are collinear, so each photon will look like a single electron-like ring. Neutral pions will decay to two photons, so depending on how large the angle is between those two photons, a neutral pion will produce two separate electron-like rings. Figure 7 shows events from the Super-Kamiokande detector (Nakahata et al. 1999) with roughly 0.5 GeV of visible energy, but one (on the top) that is produced by an electron and one (on the bottom) by a muon, respectively.

The most significant outstanding issue associated with water Cerenkov detectors is how far in size this detector technology can be extended. The largest neutrino detector to date that has been constructed, Super-Kamiokande, uses water Cerenkov technology. Making a still larger single-volume Cerenkov detector appears to be the most economical way to increase the detector mass, since one only needs to instrument the surface of the vessel rather than the volume.

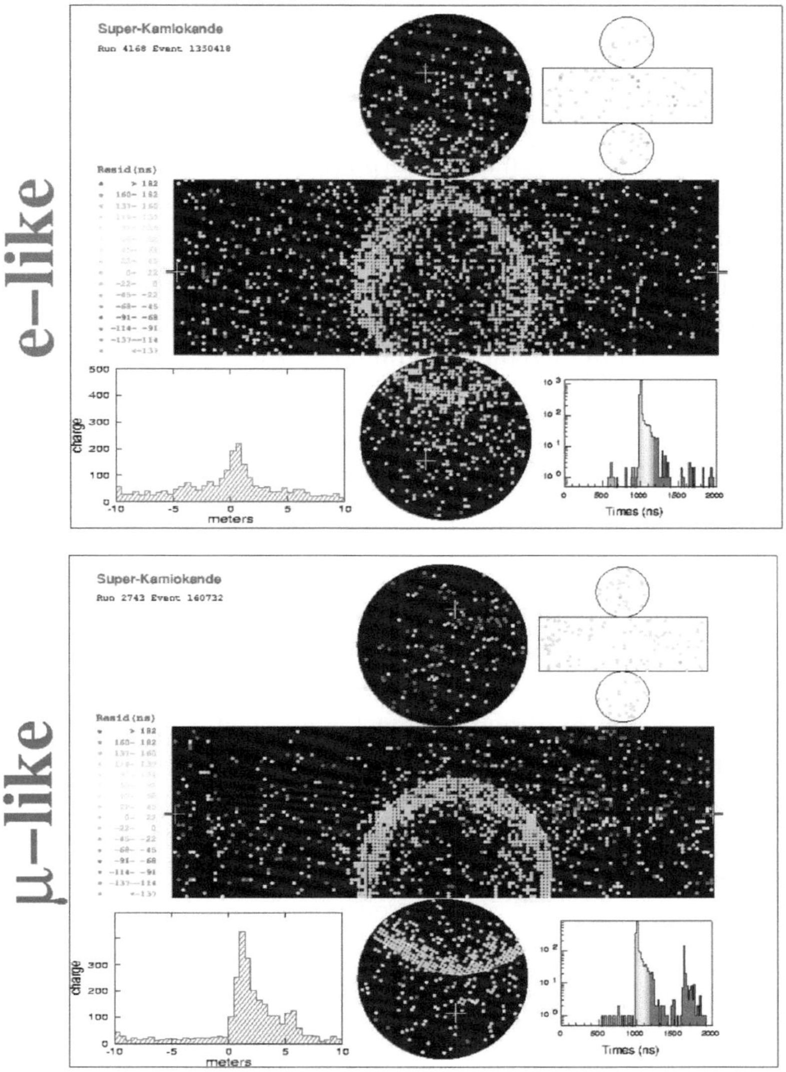

Figure 7. *Two event displays from the Super-Kamiokande Water Cerenkov Detector: one for an electron neutrino interaction (top) and one from a muon neutrino interaction (bottom). Figure courtesy of the Super-Kamiokande collaboration.*

Eventually, however, the absorption length of water at the Cerenkov light frequencies (60 m) will limit the ultimate size of a single vessel. Also, there are concerns that the water pressure on too large a vessel would be damaging to the phototubes, so in fact more physically robust methods of collecting light over large areas are being pursued.

Another outstanding issue associated with water Cerenkov detectors is the appropriate energy regime. Although water Cerenkov detectors have proven extremely effective at the 0.5-1.5 GeV neutrino energy regime, they have yet to be fully exploited at higher energies, where there are several rings in the detector that must be identified and analyzed. As the neutrino energy gets higher, there are more and more particles produced in the interaction: the more particles that are above Cerenkov threshold the more rings must be found, yet a larger fraction of the neutrino energy will also be lost to particles that get produced below Cerenkov threshold. Currently the Super-Kamiokande detector has 10% of the vessel's surface covered with photo-tubes; if a higher fraction were covered then more rings could be distinguished, allowing the detector to function better at higher energies. One remaining issue is that for muons that do not range out of the water Cerenkov detector, an energy measurement is not possible without an additional detector downstream. Given the absorption length of 60 m mentioned above, this translates into 12 GeV muon energy loss across the full detector, or an average of 6 GeV energy loss for particles originating in the center of the detector.

Finally, the sensitivity of phototubes to a magnetic field means that it would be extremely difficult to use this kind of detector technology in a neutrino factory experiment, where charge identification is necessary to distinguish the incoming neutrino flavor.

3.5 Sampling Detectors Overview

Another powerful technique for neutrino detector is to use sampling detectors. The benefit is that by having a significant fraction of material in the detector that is not active, there are fewer readout channels, and therefore the readout costs are lower, but also this also means that denser material can be used which would provide more detector mass in a given enclosure. By adding steel to the list of possible detectors then one can also envision large detectors that are easily magnetized, which is a requirement for a detector used at a neutrino factory. The disadvantages to sampling are of course that there is a loss of information, and particle identification is more challenging because of that loss. For a sampling detector, the energy resolution is usually dominated by the number of samples that are made in a given shower: for electromagnetic showers, the resolution scales as the square root of the samples per radiation length, while for hadronic showers it's the square root of the samples per interaction length.

It is worth noting a few differences between high atomic number and low atomic number, Z, sampling calorimeters to understand why in some cases the detector mass is made of lead or steel, while in other cases the detector mass is made of hydrocarbons. In high Z materials the electromagnetic and hadronic showers are far more compact. For example if one is trying to see small distances scales such as a kink in a $\tau \rightarrow \mu$ decay then it is natural to require a small transverse segmentation. It is also clear that the higher Z materials are far more dense than the low Z materials, so the size of the building or cavern excavation would be considerably smaller for high Z detectors.

The main advantage of the lower Z target materials is that if one considers the number of target nuclei per radiation length then low Z materials are superior. The fact that the hadronic and electromagnetic showers are far more spread out also means that the transverse segmentation does not need to be as small, which again points towards fewer readout channels. On the other side, if the showers are large in transverse extent, this will mean that one cannot accept events that originate too close to the edge of the detector. In these cases the "fiducial volume" might be significantly lower than the actual detector volume.

3.6 Emulsion-Lead

In order to cleanly identify individual ν_τ charged current events originating from a ν_μ beam produced at CNGS, the OPERA experiment is building a sampling calorimeter made of emul-

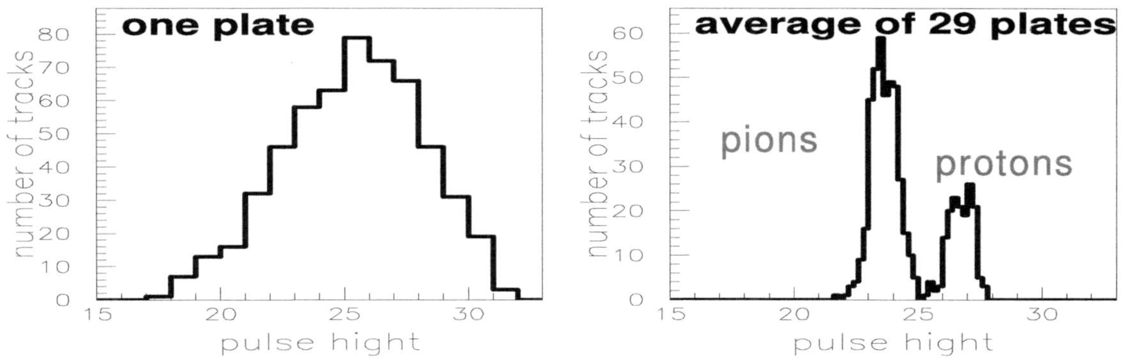

Figure 8. *Pion and proton signals in a single plate (left) and the average signal level when the particles cross 29 plates (right).*

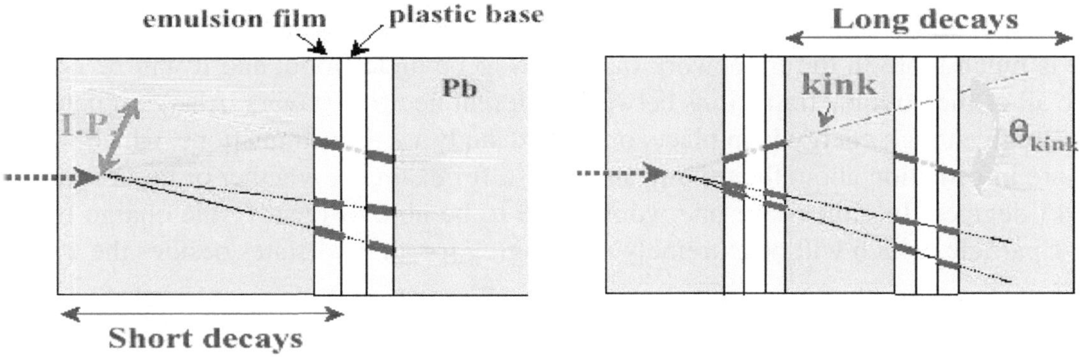

Figure 9. *Two examples of signatures that a fine grained lead emulsion detector can use to look for ν_τ charged current events through τ decays. Figure courtesy of the OPERA collaboration.*

sion interleaved with thin sheets of lead. Although the lead sheets are 1 mm thick and the emulsion layers are 44 μm thick, the emulsion acts as a tracking medium as well as a shower reconstruction medium. To build up enough detector mass these lead/emulsion sandwiches are packaged as 8.3 kg bricks which are then stacked to form walls of target material. The emulsion must be developed once a neutrino interaction is identified as having taken place in that brick. This is achieved by placing electronic tracking devices downstream of the target walls, which can point back to the bricks. Although the signal in one plate cannot distinguish between pions and protons, by looking at the energy loss from different tracks in a series of tracks, one can distinguish between pions and protons, as shown in Figure 8 (De Serio, M. 2005).

The signals in an emulsion detector can come from two sources. The first source, called short decays, are events where the τ is produced in the lead, travels a short distance in the lead and then decays, causing a track with a very small impact parameter compared to other tracks in the event. The second source (long decays) are events where the τ is produced, it crosses through a pair of emulsion sheets, and then decays in the next lead sheet. In this case the kink is more clearly defined. Because of the fine granularity, τ decays to both electrons and muons can be used as signal events. Figure 9 shows these two examples of decays.

The backgrounds in this detector come primarily from charm production: if the incoming neutrinos are energetic enough to undergo ν_τ charged current interactions, they are also energetic enough to produce D mesons which will also travel a short distance before decaying. These events can be eliminated, however, is by looking at the reconstructed invariant mass of the primary tracks: since these are high energy neutrino interactions in a very segmented detector, there will always be more than one track in a candidate event. The signal events will have on average a much larger invariant mass than the background events, and by cutting at an invariant mass of 3 GeV^2 99.8% of the background can be removed, at a cost of only 20% of the signal events.

Although the backgrounds are low, this detection technique is still a challenging one: the total signal detection efficiency is approximately 9.1%, split roughly evenly between the three possible decay channels: $\tau \rightarrow e$, $\tau \rightarrow \mu$ and $\tau \rightarrow$ hadrons. Nevertheless, this detector may well see more ν_τ events after 730 km of travel than the DONUT experiment saw in its short baseline experiment designed to detect ν_τ interactions in the first place!

The outstanding issues associated with this detection technique depend somewhat on what the current generation of experiments sees: if the LSND signature is really due to oscillations, then there is much more in the framework that needs to be understood, and it will be crucial to measure all of the possible transitions between different neutrino flavors. Also, if a neutrino factory of a high enough energy is in place, one could study $\nu_e \rightarrow \nu_\tau$ transitions, which would provide more information about the mixing angles θ_{23}, for example, whether or not it is above or below 45 degrees. In either case one would need to be able to identify the charge of the final state τ particle, which will be extremely challenging for all final states besides the $\tau \rightarrow \mu$ decay.

3.7 Segmented Scintillator

In order to cleanly identify ν_e charged current events originating from a ν_μ beam produced at Fermilab, the NOvA experiment is planning to build an almost entirely active detector made of plastic scintillator. Since 85% of the detector is active one might describe this not as a sampling calorimeter but more like a coarse-grained bubble chamber: the segmentation is 6 cm thick in the longitudinal direction, and 3.87 cm thick in the transverse direction. Examples of signal and background event signatures are shown in Figure 10.

The detector itself is constructed of extruded PVC tubes which are then filled with liquid scintillator, and then filled with a loop of wave-length shifting fiber which is read out at one end by avalanche photo-diodes (APD's) which are like 2-stage phototubes which are much smaller and less expensive than regular phototubes, although a little more noisy and as such must be operated at reduced temperatures.

Because this detector is built of such a low mass material and because it is a long baseline experiment looking for a small oscillation probability, the physical size of the detector is matched only by that of the Super-Kamiokande detector in Japan. The NOvA detector will weigh roughly 25 ktons, and will measure approximately 15.7 m tall by 15.7 m wide by 132 m long. Events in this detector will measure several meters long and may be one meter or more wide. Similar to the previous detectors discussed, particle identification is possible by looking at the energy loss as each particle traverses many cells of the detector. An electron track will look "fuzzy" while a muon or pion track will be much sharper. Neutral pions are often iden-

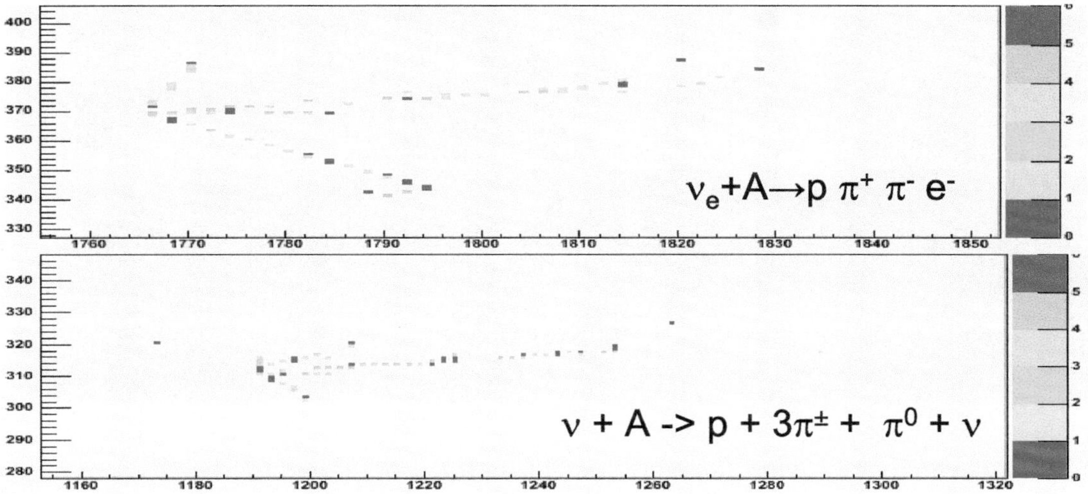

Figure 10. *Segmented Scintillator event displays for signal and background events in a $\nu_\mu \rightarrow \nu_e$ search. Event displays courtesy of the NOvA collaboration.*

tified by the fact that the two fuzzy showers are found which point back to the same location. There is also often a several cell gap between where the neutrino interacted and either of the two photons coming from the neutral pions. In the case where there are other tracks pointing back to the primary vertex, the segmentation is a powerful method to reduce the neutral current backgrounds that would otherwise swamp the oscillation signal.

Due to the fine granularity of the NOvA detector, the energy resolution for electron neutrino charged current events is about $10\%/\sqrt{E(\text{GeV})}$. For an experiment that is using a narrow-band beam, having good energy resolution is an important component of reducing the backgrounds, especially from electron neutrinos from kaon or muon decay.

The outstanding issues for this detector technology are simply associated with cost: the field of particle physics has a long history of using scintillator detectors, and the new technology associated with this detector is avalanche photo diodes. This technology choice was made to reduce the cost of the detector. The tests so far show that these detectors can operate at a low enough noise level. The biggest uncertainty is not how well this detector can reject a given background process or if the detector simulations can be made to match the detector performance, but is more related to what the cross sections for the background processes are in the first place.

3.8 Steel-Scintillator

There are many steel-scintillator detectors that have been used in neutrino experiments in the past. The most recent version is currently in use by the MINOS experiment. This detector has alternating planes of 2.54 cm steel and 1 cm scintillator plates, where as the scintillators are segmented transverse to the beam at a 4 m pitch. The MINOS far detector is an 8 m tall octagon, with a coil running through the center of the detector to provide a magnetic field in the steel which is on average 1.3 Tesla. The total mass of the detector is 5.4 ktons, and because of the shape of the detector and the density of steel, the fiducial volume of this detector is a significant

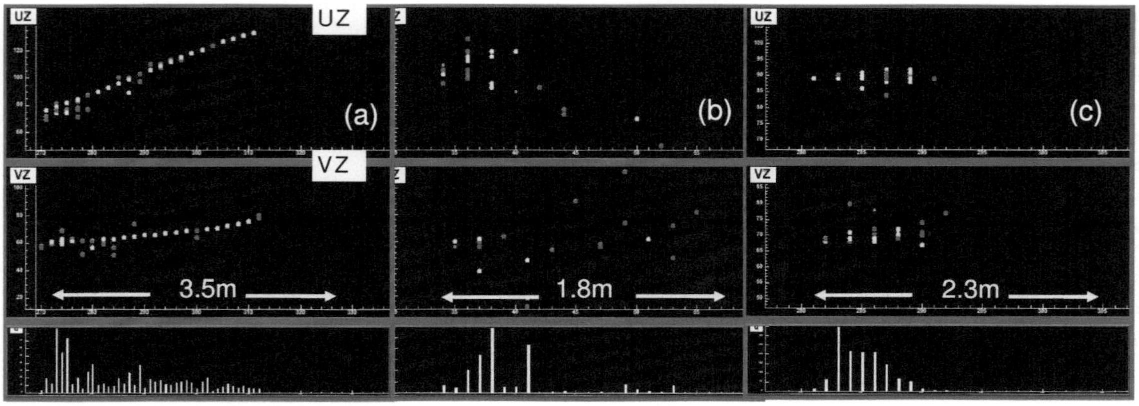

Figure 11. *Segmented steel-scintillator event displays for a simulated (a) ν_μ charged current event, (b) neutral current event, and (c) ν_e charged current event. Event displays courtesy of the MINOS collaboration.*

fraction of the total volume.

Particle identification in a steel-scintillator detector is done by looking at the longitudinal and transverse profiles of the energy deposition, but because of the typically coarse segmentation the discrimination between electrons and neutral pions is far more challenging. Discrimination between muons and anything else is very straightforward, especially for muon momenta above a few GeV. At low muon momenta the muon/charged pion discrimination becomes more difficult, since at low energies pions have higher probabilities of simply ranging out before inducing hadronic showers. Figure 11 shows longitudinal and transverse profiles for three different kinds of events: ν_μ charged current events (left), neutral current events (center), and ν_e charged current events (right). Notice that with this granularity it is difficult to distinguish between these last two kinds of events.

The energy resolution of the MINOS detector is $55\%/\sqrt{E(\mathrm{GeV})}$ for hadronic showers. Muon momenta are measured with a resolution of 6% for muons that range out inside the detector, or 13% through measurement of the curvature in the muon track. The outstanding issues for steel scintillator detectors are mostly associated with their use in neutrino factories. So far this is the one established technology that can easily incorporate a magnetic field. The low background levels described in section 3.1 are only available with this technology. Since this detector is modular in nature, the cost of scaling up the size of the detector is well understood. So the question of how to make this detector cheaper is mostly associated with understanding the minimum amount of segmentation and therefore the readout requirements. Clearly the higher the neutrino factory energy, the more coarse-grained the detector can be.

3.9 Detector Summary

The detector technologies that have been described in the previous sections are summarized in table 6. The important points to compare are what the largest mass is that has (at the time of writing) been constructed, whether or not there is event-by-event flavor discrimination, whether or not the charge of the final state lepton can be distinguished, and the energy range that each of these detector technologies is best suited. Note that there is no single detector technology

Detector Technology	Largest Mass to date (kton)	Event by Event Identification ν_e	ν_μ	ν_τ	charge identification	Ideal ν energy Range
Liquid Argon TPC	0.6	yes	yes		not yet	huge
Water Cerenkov	50	yes	yes			$< 2GeV$
Emulsion/Pb/Fe	0.27	yes	yes	yes		$> 0.5GeV$
Scintillator	1	yes	yes			huge
Steel/Scintillator	5.4		yes		yes	$> 0.5GeV$

Table 6. *Summary of detector technologies currently in use or planned for future accelerator-based oscillation experiments. Although lepton charge identification has been proposed as possible in a liquid argon TPC, this has yet to be demonstrated.*

which can do every possible measurement.

Although there are several technologies that have been developed for observing neutrino interactions, the demands on the next generation of detectors are large: the biggest question is how to increase the product of the detector mass and signal efficiency without similar increases in the detector cost, while keeping the background rejection high. Another point to remember is that by increasing the detector mass for accelerator-based experiments, we may gain sensitivity to other non-accelerator based physics. It is important to try to keep the functionality of these sensitive detectors high to get the most physics output from the significant investment that will be required.

4 Measuring Neutrino Oscillation Probabilities

Now that the beamlines and detectors have been described, it is worth spending some time to understand in detail how oscillation probabilities are actually measured, given a combination of these two components. Although we are currently in the "first discovery and confirmation" stage of the field of accelerator-based oscillation measurements, we will have to progress to the "precision measurements" stage in order to compare small neutrino and antineutrino probabilities and even smaller differences in probabilities. This section will describe the strategies that are being developed to arrive at this precision measurement stage.

4.1 Overview

Recall that the number of events at a far detector (N_{far}) is related to the neutrino flux (Φ_{ν_μ}, cross section (σ_{ν_x}, detector mass (M_{far}), and oscillation probability ($P(\nu_\mu \to \nu_x)$)in the following way:

$$N_{far} = \Phi_{\nu_\mu}\sigma_{\nu_x}P(\nu_\mu \to \nu_x)\varepsilon_x M_{far} + B_{far} \tag{14}$$

where ε_x is the detector efficiency for the signal to be detected and to pass all analysis cuts, and B_{far} is the number of background events predicted. These backgrounds can come from not only the muon neutrinos that are arriving at the far detector, but also from the intrinsic electron neutrinos in the beam or the tau neutrinos that are a result of the muon neutrinos having travelled a far distance. The background events can be expressed in the following way:

$$B_{far} = \Sigma_{i=\mu,e\tau}\Phi_{\nu_i}\sigma_{\nu_i}\varepsilon_{ix}M_{far} \tag{15}$$

where ε_{ix} is the efficiency for identifying neutrino type i as the signal neutrino flavor x. Clearly the efficiency for detecting an electron neutrino of a given energy is the same regardless of if it's an from muon or kaon decay at the source or if it's from $\nu_\mu \to \nu_e$ oscillations.

By solving for the oscillation probability

$$P(\nu_\mu \to \nu_x) = \frac{N_{far} - B_{far}}{\Phi_{\nu_\mu}\sigma_{\nu_x}\varepsilon_x M_{far}} \tag{16}$$

and then taking the derivative to look at uncertainties on that probability one arrives at the following expression:

$$\left(\frac{\delta P}{P}\right)^2 = \frac{(N_{far} + (\delta B_{far})^2)}{(\Phi_{\nu_\mu}\sigma_{\nu_x}\varepsilon_x M_{far})^2} + (N_{far} - B_{far})\left(\left[\frac{\delta\phi_{\nu_\mu}}{\phi_{\nu_\mu}}\right]^2 + \left[\frac{\delta\sigma_{\nu_x}}{\sigma_{\nu_x}}\right]^2 + \left[\frac{\delta\varepsilon_{\nu_x}}{\varepsilon_{\nu_x}}\right]^2\right) \tag{17}$$

From this expression it is clear that there are two limits in the uncertainty on the oscillation probability: one limit where the number of signal events is comparable to the number of background events. In that case the uncertainty is dominated by the uncertainty on the background prediction and by the statistical uncertainty of the final event sample. But the other limit is in the situation where the signal events far outnumber the background events. In this case the uncertainty on the neutrino flux, the signal cross sections, and the signal efficiencies become much more important and may become larger than the statistical uncertainty.

4.2 Near Detector Justification

In order to reduce the systematic uncertainties, most oscillation experiments also place an additional detector in the same beamline, but very close to the neutrino source before the oscillation has had a chance to occur. For an appearance experiment, the near detector will have no signal events but can put constraints on the background events if the same cuts are used to search for the signal in the near and far detector. For a disappearance experiment, the near detector can provide a constraint on the prediction of the signal events in the far detector in the absence of oscillations.

Consider an appearance experiment, where the near detector is used to look for background events: the number of events in the near detector, N_{near} is given by the following expression (assuming the same efficiency in the near and far detectors):

$$N_{near} = \Sigma_{i=\mu,e}\Phi_{\nu_i}\sigma_{\nu_i}\varepsilon_{ix}M_{near} \tag{18}$$

Then, you can predict the number of events at the far detector by taking the ratio between near and far detector background expressions:

$$B_{far} = N_{near}\frac{\Sigma_{i=\mu,e,\tau}\Phi_{\nu_i}\sigma_{\nu_i}\varepsilon_{ix}M_{far}}{\Sigma_{i=\mu,e}\Phi_{\nu_i}\sigma_{\nu_i}\varepsilon_{ix}M_{near}} \tag{19}$$

and take advantage of the fact that detector simulations are better at predicting ratios of efficiencies and cross sections than the absolute levels. Although the claim is often made that cross sections and acceptances cancel between the near and far detector, this is only an approximation which is valid in the case for a detector with perfect energy resolution and only one process (i.e. no backgrounds).

Unfortunately, the cross sections and detector efficiencies do not cancel completely for a few reasons: first of all, because the near detector has different neutrino energy spectra than the far detector, and one must integrate the numerator and denominator both over all energies and processes. Second of all, the near detector has no ν_τ component, and because of this, the possible backgrounds from τ decay cannot be measured at all at a near detector, and furthermore, the lack of ν_μ backgrounds because of those that have oscillated to ν_τ's are also not measured. An example of how cross sections can fail to cancel using a near detector of identical capabilities is described in reference (Harris et al. 2004). Systematic uncertainties on CP violation in neutrino experiments are discussed in (Huber, Mezzetto & Schwetz 2007).

Other important differences between the near and far detector locations are the event rates and detector size. Because of cost constraints the near detector is usually much smaller than the far detector, and because of this the fiducial volume cuts will have very different effects on the near and far detectors. The close location of the near detector means that the instantaneous neutrino event rates differ from the far detector rate by a factor of 10^4 to 10^5. For neutrino beams which occur in spills lasting a few microseconds, the near detector may require much faster and therefore different electronics and data acquisition systems than the far detector. Also, for far detectors that are located deep underground, the near detectors will by definition have different cosmic ray rates than the far detectors, and hence those backgrounds will also be different.

Nevertheless, measurements at a near detector facility are a must for precision oscillation measurements, and they do put important constraints on far detector predictions. In this next section we will describe briefly the near detector designs for previous, current, and future neutrino oscillation experiments.

4.3 Near Detector Strategies

There are two different strategies for designing near detectors for oscillation experiments: one strategy is to build a near detector that is as identical as possible to the far detector, given the constraints listed above. Since the far detector is by definition only as segmented as one can afford, this means that with this strategy the near detector is also only as segmented as the far detector. So once similar kinds of cuts are made, one cannot distinguish how much of the "near detector background events" come from the various sources of backgrounds. For ν_e oscillation experiments, the three sources (neutral currents, ν_μ charged currents, and intrinsic ν_e's in the beam) all have very different far detector to near detector extrapolations, so the far detector prediction will suffer from uncertainties in the various cross sections that make up the near detector signal rates.

The second strategy is to have a near detector that is much more fine-grained and gives much more information than the far detector, and then rely on Monte Carlo simulations to predict the far detector signal efficiencies. In this way the near detector measures the product of the flux and cross sections well. The uncertainty arises from the predictions on the efficiencies of the far detector.

4.3.1 K2K

The K2K experiment used the 50 kton Super-Kamiokande Water Cerenkov detector for the far detector (Nakahata et al. 1999), and used several near detectors including a 1 kton Water Cerenkov detector at a nearby location (Suzuki et al. 2000, Ishii et al. 2002, Maesaka 2003). Because the near water Cerenkov detector was much smaller the events were on average much closer to the phototube wall than at the far detector, and understanding the efficiency as a function of distance from that wall was extremely important. To understand the cross sections and the population of events that would be below Cerenkov threshold, the K2K experiment also had a scintillating fiber tracker with a 6 ton fiducial mass water target, as well as a 9.38 ton fiducial mass fully active scintillator tracker detector. Finally, because only very low energy muons would range out in any of the upstream near detectors, there was also a muon range detector of 330 tons of fiducial mass. This last detector was large enough in transverse dimensions to see the high energy muons and measure the direction of the neutrino beam that produced those high energy muons after interacting in the steel.

4.3.2 T2K

The T2K experiment, which will also use the Super-Kamiokande Water Cerenkov detector as a far detector, has a slightly different program of near detectors planned(Itow et al. 2001). There are two planned locations for near detectors: one location is at 280 m from the proton target, and the other is at 2 km from the proton target. The 280 m detector location will have two neutrino detectors: one that is located on the beamline axis whose function is strictly to measure the beam direction using iron and scintillator stacks in a grid of locations. The other 280 m detector will be a fine-grained detector that has two detector technologies: one a water target instrumented with electromagnetic calorimetry to understand sources of electrons and photons in an off axis neutrino beam, and the second region will be a TPC combined with fine grained detectors, also with a water target, to measure the secondary particle energy spectra from neutrino-water interactions.

At the 2km location there are plans to use both a water Cerenkov detector and a liquid argon TPC, followed by a muon range detector. This more downstream site has a muon and electron neutrino flux that is much closer to the far detector flux at least before oscillations, and should be extremely useful for precise muon neutrino disappearance measurements.

4.3.3 MINOS

The MINOS experiment (Adamson et al. 2006), which uses a 5.4 kton steel scintillator far detector that is 8 m in transverse dimension to the beam axis, has a 1 kton near detector that is also steel scintillator with the same longitudinal segmentation for the target area of the detector, although it measures roughly 3.8 m $\times$ 4.8 m in transverse size. The downstream region of the MINOS near detector is instrumented with scintillator only once for every five planes of steel and is used as a muon spectrometer. Because of the very high event rate the MINOS detector uses much faster electronics to get continuous sampling in the spill, which it must do in order to untangle the many events that occur in the MINOS near detector for every 10 microsecond spill.

4.3.4 NOvA

The NOvA experiment, which will use a 25 kton segmented scintillator detector for its far detector, uses an almost identical near detector but one that is significantly smaller in the transverse direction (Ayres et al. 2004). The far detector measures roughly 16 m $\times$ 16 m, while the near detector measures 4.1 m $\times$ 2.1 m Because the near detector is constrained in size based on the fact that it will be in an underground enclosure, the muon spectrometer region of this detector will be made of steel and scintillator planes rather than only scintillator, but the total detector size will be long enough to contain 2 GeV muons that enter the upstream region of the detector. Because the decay pipe for the NOvA experiment is so long and the near detector is so near to the end of the decay pipe, there is a large difference between the near and far detector spectrum, so the NOvA experiment is considering making measurements at several off axis angles, not only the off axis angle that is identical to the far detector. This would help reduce the uncertainties on the far detector prediction since they will have a few slightly different near detector spectra to compare.

4.4 Muon Neutrino Disappearance Measurements

According to the atmospheric neutrino measurements, the mixing angle between muon neutrinos and tau neutrinos is near or at maximal. This means that the disappearance phenomenon is very large which permits statistically a very precise measurement of the mass difference between the two neutrino species, given a well-understood muon neutrino beam. In order to measure the mass difference one must measure the oscillation probability versus neutrino energy for a given distance or alternatively for the same energy at several distances. The challenge is to measure the incoming neutrino energy as accurately as required. How well do we really know what the incoming neutrino energy is? We can measure the final state particle energies, but this might not be the whole story.

Quasi-elastic events have a great feature where you can measure the lepton angle θ_ℓ and momentum (p_ℓ) and derive the neutrino energy (E_ν) if you assume a quasi-elastic interaction off a proton. In this case the neutrino energy is given by the following expression:

$$E_\nu = \frac{m_N E_\ell - m_\ell^2/2}{m_N - E_\ell + p_\ell \cos\theta_\ell} \tag{20}$$

where m_N is the nucleon mass. But what happens when there is an interaction with a higher hadronic mass? If you miss the other final state particles, you will reconstruct a lower energy than what the original neutrino energy was. So you need to know your detector's ability to see those other particles. You also need to know how often those other particles are produced. This is much harder and goes back to knowing cross sections and differential distributions of the secondary lepton. So the dominant uncertainty in the energy measurement for a water Cerenkov detector is the ratio of quasi-elastic to non-quasi-elastic cross sections. This is the justification for building several near detectors, at least one of which has pion and proton detection well below the Cerenkov threshold.

Fully active detectors have another problem: they have lower detection thresholds so that is an advantage, but on the other hand they still have to deal with the nuclear environment. The center of a nucleus is a very dense place, and there is a substantial probability that a pion produced there will be absorbed or at least lose a lot of energy as it leaves the nucleus. If no particles lost energy in this way, you don't have to worry so much about how many final state particles, assuming your detector is linear: if your detector produces ten 1 GeV pion or three 3 GeV pions the detector energy measured might be the same. But if the probability of absorbing the 300 MeV pions is 5 times the probability of absorbing the 1 GeV pion, then you do care how many times a neutrino makes the 300 MeV pions compared to the 1 GeV pions.

Another important concern in a muon neutrino disappearance measurement comes from knowing the neutral current background accurately. Since the mixing angle for disappearance is near maximal, at the oscillation minimum most of the events may well be neutral current events which have been mis-identified as charged current events. The uncertainty on that contamination is very difficult to constrain at the near detector because the charged current fraction is so much higher at the near detector, and only the tails of the neutral current events will contaminate the charged current event selection. Measuring those tails when they are even more hidden by the near detector charged current sample is difficult.

One technique that the MINOS experiment uses to better constrain their far detector predictions is to look at the near detector data in several different beam configurations. The NuMI beamline was designed specifically to accommodate different target locations relative to the positions of the two horns, and the movement from one target location to another was possible by remote operations and only took a few minutes. Because of this the MINOS experiment took several weeks of data in conditions that were not optimal for seeing an oscillation signal at the far detector, but could be used as tests of the experiments understanding of the near detector events. The data taken in some of these configurations are shown in Figure 12. By moving the neutrino target back and forth with respect to the horn location, they select different momentum for the "tune" momentum of the beam which results in different near (and far) detector neutrino energy spectra. These events will all have different neutral current and charged current fractions, which can better constrain the model than only one target position data. Similarly, they also took data at several different horn currents: this means that although the event rate coming from pions that went through the inside of the inner conductor does not change, the focused peak can be dramatically reduced. This also lowers the number of signal events in the low energy charged current sample and so the ratio of neutral to charged currents can be closer to that expected in the case of oscillations at the far detector.

4.5 Electron Neutrino Appearance in a Muon Neutrino Beam

The most important information that accelerator-based experiments can tell us about neutrino oscillations will come from comparisons of neutrino and antineutrino oscillations between muon and electron neutrinos. Measurements of CP violation and measurements of the neutrino mass hierarchy simply cannot be done by any other kind of experiment. Therefore it is important that we understand just how to get to precision measurements of these quantities, and what the challenges are. Unfortunately there are several different backgrounds that arise in

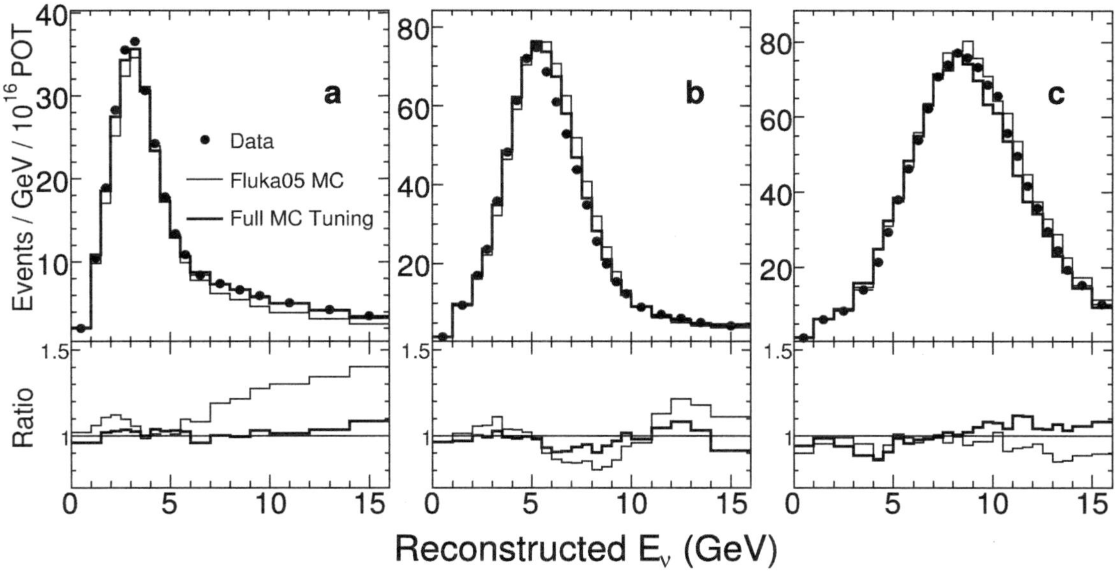

Figure 12. *MINOS near detector events for different beamline configurations, as described in the text. From (Michael et al. 2006). Copyright (2006) American Physical Society.*

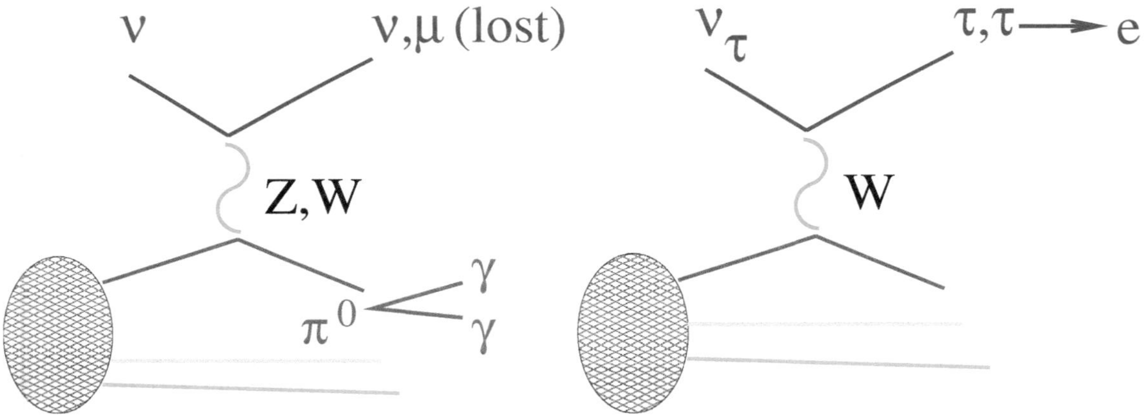

Figure 13. *Feynman diagrams for backgrounds to electron neutrino searches in muon neutrino beams.*

a conventional neutrino beam when one is searching for electron neutrino appearance.

The first background, as mentioned above, comes from electron neutrinos produced at the accelerator itself. These are produced in muon and kaon decays in the beamline. Event by event these neutrinos will look in every way like the signal neutrinos, the only difference is that their total energy distribution will be very different from that of the signal neutrinos from muon neutrino oscillations. The second background, shown at the left in Figure 13 comes from neutral current events that contain neutral pions, where one or more of the photons from the pion decay is mis-reconstructed as a single electron in the far detector. Again the measured neutrino (and "electron candidate") energy distributions will be different from the signal neutrinos, since the neutral current visible energy spectrum is steeply falling.

Experiment	Background					Signal	Figure
	$\nu_\mu CC$	NC	ν_e	ν_τ CC	Total		of Merit
K2K (Yamamoto et al. 2006)	0	1.3	0.4	0	1.7	1	0.6
MINOS (Smith 2006)	5.6	39	8.7	4.7	58	29.1	3.1
OPERA (Komatsu et al. 2003)	1	5.2	18	4.5	28.7	10	1.6
T2K (Itow et al. 2001)	1.8	9.3	11.1	0	22.2	103	9
NOvA (Ayres et al. 2004)	0.5	7	11	0	18.5	148	11.5

Table 7. *Table describing the signal and background rates for a typical exposure for long baseline ν_e appearance search. The "figure of merit" is defined as the signal divided by the square root of the total number of signal and background events expected.*

The third background, also shown at the left in Figure 13 comes from ν_μ charged current events where the outgoing muon is so low in energy that a neutral pion again from the hadronic shower is mis-reconstructed as a single electron and is tagged as the leading lepton in the event. This background is particularly dangerous not because it is large, but because due to $\nu_\mu \to \nu_\tau$ oscillations, the background energy distribution from this source will be extremely different between the near and far detectors.

The last background, shown at the right in Figure 13 is relevant only for higher energy neutrino beams, comes from ν_τ charged current events where the τ decays into an electron. The neutrino and lepton energy distributions will be different between signal and background events, but the challenge here is that this background cannot be measured at all at a near detector since by definition the τ neutrinos have not yet begun to appear.

The initial relative levels of these four backgrounds differ depending on what detector technology is employed, and depending on how optimized the beamline has been for the ratio between muon and electron neutrino production. Table 7 lists the number of events for a "standard exposure" for many of the experiments that have been described in this article, for an optimized analysis. Although the intrinsic ν_e fractions of the different beamlines are all different (see Table 3) it is clear that the experiments optimized for ν_e appearance have a background dominated by the intrinsic ν_e component in the beamline. The signal rates shown assume that $\sin^2\theta_{13} = 0.1$ and that $\delta = 0$, and the NOvA signal assumes that the neutrino mass hierarchy is the normal one. All these signals assume slightly different δm^2 values but within $2.5 - 3 \times 10^{-3} eV^2$.

5 Summary

Although the field of neutrino oscillations was started by measurements of naturally occurring sources of neutrinos, namely the sun and the atmosphere, the ultimate steps that need to be taken depend on accelerator-based sources of neutrinos. The search for CP violation in the lepton sector and a determination of the neutrino mass hierarchy may ultimately require extremely powerful neutrino beams and massive detectors that are beyond what is currently planned. How challenging these measurements will be depends on the size of the last unmeasured mixing angle in the leptonic mixing matrix.

Because of the inherent contamination in conventional neutrino beams, new beamline strategies such as beta beams or neutrino factories are being investigated. Similarly, because of the difficulty of distinguishing between neutral pions made in neutral current interactions and electrons in electron neutrino charged current interactions, new detector strategies such as liquid argon are being investigated. Both examples offer potentially large breakthroughs in the level of backgrounds that would be seen in these future experiments.

Finally, it will take more than massive detectors and powerful beamlines to make these measurements. The goals of discovering CP violation and mass hierarchy will be achieved by comparing the differences of oscillation probabilities that are already limited to being on the order of a few percent. To make these comparisons we will need a detailed understanding of how the neutrino beam is produced, and how the neutrinos themselves interact in these massive detectors. The field of oscillations was born from two very surprising results, and we should not presume to know the origin of the next surprise.

References

Adamson, P. et al. (2006), 'The MINOS calibration detector', *Nucl. Instrum. Meth.* **A556**, 119–133.

Aguilar-Arevalo, A. A. et al. (2007), 'A search for electron neutrino appearance at the $\Delta m^2 \sim 1$ eV^2 scale', *Phys. Rev. Lett.* **98**, 231801.

Aliu, E. et al. (2005), 'Evidence for muon neutrino oscillation in an accelerator-based experiment', *Phys. Rev. Lett.* **94**, 081802.

Apollonio, M. et al. (2003), 'Search for neutrino oscillations on a long base-line at the CHOOZ nuclear power station', *Eur. Phys. J.* **C27**, 331–374.

Astier, P. et al. (2001), 'Final NOMAD results on $\nu_\mu \rightarrow \nu_\tau$ and $\nu_e \rightarrow \nu_\tau$ oscillations including a new search for ν_τ appearance using hadronic tau decays', *Nucl. Phys.* **B611**, 3–39.

Athanassopoulos, C. et al. (1996), 'Evidence for $\bar{\nu_\mu} \rightarrow \bar{\nu_e}$ oscillation from the LSND experiment at the Los Alamos Meson Physics Facility', *Phys. Rev. Lett.* **77**, 3082–3085.

Ayres, D. S. et al. (2004), 'NOvA proposal to build a 30-kiloton off-axis detector to study neutrino oscillations in the Fermilab NuMI beamline', *hep-ex/0503053* .

Burguet-Castell, J., Casper, D., Gomez-Cadenas, J. J., Hernandez, P. & Sanchez, F. (2004), 'Neutrino oscillation physics with a higher gamma beta- beam', *Nucl. Phys.* **B695**, 217–240.

De Serio, M. (2005), The OPERA experiment, Weak Interations and Neutrinos, CERN.

Eskut, E. et al. (2007), 'Final results on $\nu_\mu \rightarrow \nu_\tau$ oscillation from the CHORUS experiment', *arXiv:0710.3361 [hep-ex]* .

Fisher, P., Kayser, B. & McFarland, K. S. (1999), 'Neutrino mass and oscillation', *Ann. Rev. Nucl. Part. Sci.* **49**, 481–528.

Fitton, M. (2006), Design and computational fluid dynamic analysis of the T2K target, Neutrino Beams and Instrumentation, CERN.

Fukuda, Y. et al. (1998*a*), 'Evidence for oscillation of atmospheric neutrinos', *Phys. Rev. Lett.* **81**, 1562–1567.

Fukuda, Y. et al. (1998*b*), 'Measurements of the solar neutrino flux from Super-Kamiokande's first 300 days', *Phys. Rev. Lett.* **81**, 1158–1162.

Geer, S. (1998), 'Neutrino beams from muon storage rings: Characteristics and physics potential', *Phys. Rev.* **D57**, 6989–6997.

Harris, D. A. et al. (2004), 'Neutrino scattering uncertainties and their role in long baseline oscillation experiments', *hep-ex/0410005* .

Huber, P., Mezzetto, M. & Schwetz, T. (2007), 'On the impact of systematical uncertainties for the CP violation measurement in superbeam experiments', *arXiv:0711.2950 [hep-ph]* .

Ishii, T. et al. (2002), 'Near muon range detector for the K2K experiment: Construction and performance', *Nucl. Instrum. Meth.* **A482**, 244–253.

Itow, Y. et al. (2001), 'The JHF-Kamioka neutrino project', *hep-ex/0106019* .

Kobayashi, M. & Maskawa, T. (1973), 'CP Violation in the renormalizable theory of weak interaction', *Prog. Theor. Phys.* **49**, 652–657.

Komatsu, M. et al. (2003), 'Sensitivity to theta(13) of the CERN to Gran Sasso neutrino beam', *J. Phys.* **G29**, 443.

Kopp, S. E. (2007), 'Accelerator neutrino beams', *Phys. Rept.* **439**, 101–159.

Kuno, Y. (2007), This volume.

Maesaka, H. (2003), 'The K2K SciBar detector'. Prepared for KEK - RCNP International School and Miniworkshop for Scintillating Crystals and their Applications in Particle and Nuclear Physics, Tsukuba, Japan, 17-18 Nov 2003.

Michael, D. G. et al. (2006), 'Observation of muon neutrino disappearance with the MINOS detectors and the NuMI neutrino beam', *Phys. Rev. Lett.* **97**, 191801.

Nakadaira, T. (2005), 'J-PARC neutrino beam line and target development', *Nucl. Phys. Proc. Suppl.* **149**, 303–305.

Nakahata, M. et al. (1999), 'Calibration of Super-Kamiokande using an electron linac', *Nucl. Instrum. Meth.* **A421**, 113–129.

Rubbia, A. (2005), Concepts and R&D for very large liquid argon time project chambers, International Neutrino Factory and Superbeam Scoping Study Meeting, CERN.

Shiozawa, M. (1999), 'Reconstruction algorithms in the Super-Kamiokande large water Cherenkov detector', *Nucl. Instrum. Meth.* **A433**, 240–246.

Smith, C. (2006), MINOS results from the first year of NuMI operations, Fermilab Joint Theoretical and Experimental Seminar, FERMILAB.

Suzuki, A. et al. (2000), 'Design, construction, and operation of SciFi tracking detector for K2K experiment', *Nucl. Instrum. Meth.* **A453**, 165–176.

Yamamoto, S. et al. (2006), 'An improved search for $\nu_\mu \rightarrow \nu_e$ oscillation in a long-baseline accelerator experiment', *Phys. Rev. Lett.* **96**, 181801.

Yao, W.M. et al. (2006), 'Particle Data Group', *Journal of Physics G* **33**, 1.

Zucchelli, P. (2002), 'A novel concept for a $\bar{\nu}_e/\nu_e$ neutrino factory: The beta beam', *Phys. Lett.* **B532**, 166–172.

Neutrino Oscillation Studies with Atmospheric Neutrinos

Takaaki Kajita

Institute for Cosmic Ray Research, Univ. of Tokyo

1 Introduction

Neutrinos are the only known electrically neutral fermions, and exist in three separate flavors. Neutrinos are much lighter than all other known fermions, and in fact all attempts to measure their mass have only yielded upper limits. However if neutrinos have non-vanishing masses, they are able to change their flavor during propagation. This phenomenon is known as neutrino (flavor) oscillations In general, the neutrino states with well defined flavor: (ν_e, ν_μ, ν_τ) and the states with well defined mass:(ν_1, ν_2, ν_3) do not coincide. The two triplets are related by a Unitary matrix.

For simplicity let us discuss two flavor neutrino oscillations. If neutrinos have finite masses, each flavor eigenstate (for example, ν_μ) can be expressed by a combination of mass eigenstates (ν_2 and ν_3). The relation between the mass eigenstates (ν_2, ν_3) and the flavor eigenstates (ν_μ, ν_τ) can be expressed by;

$$\begin{pmatrix} \nu_\mu \\ \nu_\tau \end{pmatrix} = \begin{pmatrix} \cos\theta & \sin\theta \\ -\sin\theta & cos\theta \end{pmatrix} \begin{pmatrix} \nu_2 \\ \nu_3 \end{pmatrix}, \tag{1}$$

where θ is the mixing angle. The probability for a neutrino produced in a flavor state ν_μ to be observed in a flavor state ν_μ after traveling a distance L through the vacuum is:

$$P(\nu_\mu \to \nu_\mu) = 1 - \sin^2 2\theta \sin^2\left(\frac{1.27\Delta m^2(\mathrm{eV}^2)L(\mathrm{km})}{E_\nu(\mathrm{GeV})}\right), \tag{2}$$

where E_ν is the neutrino energy, θ is the mixing angle between the flavor eigenstates and the mass eigenstates, and Δm^2 is the mass-squared difference of the neutrino mass eigenstates.

Since there are 3 neutrino flavors, the above description has to be generalized to three-flavor oscillations. In the three-flavor oscillation framework, neutrino oscillations are parameterized by three mixing angles (θ_{12}, θ_{23}, and θ_{13}), three mass squared differences (Δm^2_{12}, Δm^2_{23}, and

Δm^2_{13}; among the three Δm^2's, only two are independent) and one CP phase (δ). (For more details of neutrino oscillations and references, see (Kayser 2006).) If a neutrino mass hierarchy is assumed, the three Δm^2's are approximated by two Δm^2's, and neutrino oscillation lengths are significantly different for the two Δm^2's. One $\Delta m^2 (\Delta m^2_{12})$ is related to solar neutrino experiments and the KamLAND reactor experiment. The other $\Delta m^2 (\Delta m^2_{23}$ or $\Delta m^2_{13})$ is related to atmospheric, reactor and long baseline neutrino oscillation experiments. It is known that it is approximately correct to assume two-flavor oscillations for analyses of the present atmospheric neutrino data. Therefore, in this article, we mostly discuss two flavor neutrino oscillations as described in Eqs. 1 and 2.

Atmospheric neutrinos arise from the decay of secondaries (π, K and μ) produced by primary cosmic-ray interactions in the atmosphere. These neutrinos can be detected by underground neutrino detectors. Interactions of low energy neutrinos, around 1 GeV, have all of the final state particles "fully contained (FC)" in the detector. Higher energy charged current (CC) ν_μ interactions may result in the muon exiting the detector; these are referred to as "partially contained (PC)". In order to reject background from cosmic ray particles, as well as to cleanly reconstruct the details of the event, the vertex position of the interaction is typically defined to be within some fiducial volume.

There is a third category of atmospheric neutrino events, where the CC ν_μ interaction occurs outside the detector, and the muon enters and either passes through the detector or stops in the detector. These are referred to as "upward-going muons" because one generally requires they originate from below the horizon to ensure that a sufficient amount of rock absorbs ordinary cosmic ray muons. The typical energies for fully contained, partially contained, and upward-going stopping and through-going muon event samples are 1, 10, 10, and 100 GeV, respectively.

Atmospheric neutrino experiments started in the 1960's. One experiment was carried out in the Kolar Gold Field in India (Achar et al. 1965, Krishnaswamy et al. 1971). Other experiments were carried out at the East Rand Proprietary Mine in South Africa (Reines et al. 1965, Crouch et al. 1978). In these experiments, neutrino events occurring in the rock surrounding a neutrino detector were measured. Since the experiments were carried out in extremely deep underground (about 8000 meters water equivalent (m.w.e.)), charged particles traversing the detectors almost horizontally were essentially of atmospheric neutrino origin. Also, since it was required that the particle should penetrate through the rock and the detector, most of these neutrinos were expected to be CC ν_μ events.

In the early 1980's, the first massive underground detectors (of the order 1 kton) were constructed, primarily to search for proton decay with a lifetime of less than 10^{32} years. These detectors were the first large volume detectors that can contain particles with $\sim$1 GeV and study details of these particles. The most serious background for proton decay searches is atmospheric neutrino events, at a rate of approximately 10^2 events/yr/kt. Therefore, these experiments studied details of the observed atmospheric neutrino events.

The current interest in the atmospheric neutrinos was initiated by the study of the ν_μ/ν_e flux ratio in the late 1980's (Hirata et al. 1988). The Kamiokande water Cherenkov experiment measured the number of e-like and μ-like events, which were mostly CC ν_e and ν_μ interactions, respectively. They found that the number of μ-like events had a significant deficit compared with the Monte Carlo prediction, while the number of e-like events were in good agreement with the prediction. The flavor ratio of the atmospheric neutrino flux, $(\nu_\mu + \bar{\nu}_\mu)/(\nu_e + \bar{\nu}_e)$, has been calculated to an accuracy of better than 5% in the relevant energy range. Therefore, if $(\mu/e)_{Data}/(\mu/e)_{Prediction}$ is taken, this double-ratio must be sensitive to the change in the neutrino flavor. Because of the small $(\mu/e)_{Data}/(\mu/e)_{Prediction}$ ratio, it was concluded: "We are unable to explain the data as the result of the systematic detector effects or uncertainties in the atmospheric neutrino fluxes. Some as-yet-accounted-for physics such as neutrino oscillations might explain the data." This result triggered the interest in atmospheric neutrinos. A consistent result was reported in the early 1990's from the IMB water Cherenkov experiment (Casper et al. 1991). On the other hand, fine grained iron calorimeter experiments, NUSEX (Aglietta et al. 1989) and Frejus (Berger et al. 1990), did not observe any significant ν_μ deficit within their statistics. Hence, the situation was unclear in the early 1990's.

Other important data toward the understanding of the atmospheric neutrino anomaly were reported in the mid 1990's (Fukuda et al. 1994). Zenith angle distributions for multi-GeV fully-contained events and partially contained events were studied in Kamiokande. For detectors near the surface of the Earth, the neutrino flight distance, and thus the neutrino oscillation probability, is a function of the zenith angle of the neutrino direction. Vertically downward-going neutrinos travel about 15 km while vertically upward-going neutrinos travel about 13,000 km before interacting in the detector. The data showed that the deficit of μ-like events depended on the neutrino zenith angle. The *up*/*down* ratio (where *up* and *down* represent the number of events with $\cos(zenith\text{-}angle) < -0.2$ and > 0.2, respectively) was $0.58^{+0.13}_{-0.11}$. However, due to the relatively poor event statistics, the statistical significance of the up-down asymmetry was 2.9 standard deviations, and therefore the data were not conclusive. In 1998, the Super-Kamiokande experiment, with substantially larger data statistics than those in the previous experiments, concluded that the atmospheric neutrino data gave evidence for neutrinos oscillations (Fukuda et al. 1998). Various studies of neutrino oscillations have been made using atmospheric neutrino data. The atmospheric neutrino experiments are still contributing substantially to our understanding of neutrino masses and mixing angles.

In the following sections, details of the studies of atmospheric neutrino interactions are described. Section 2 describes the calculation of the atmospheric neutrino fluxes. Section 3 outlines the simulation of neutrino interactions relevant to the energy range of atmospheric neutrinos. Sections 4 and 5 describe the details of the atmospheric neutrino data and the neutrino oscillation analyses. Then, in Section 6, future prospects of the atmospheric neutrino experiments are briefly described. Section 7 summarizes this article. In most part of this article, when mention is made of neutrinos, we imply both neutrinos and anti-neutrinos.

2 Atmospheric Neutrino Flux

To carry out detailed studies of neutrino oscillations using atmospheric neutrinos, it is important to know the expected flux in the absence of neutrino oscillations. We outline the methods and results of the three most detailed atmospheric neutrino flux calculations (Honda et al. 2004, Barr et al. 2004, Battistoni et al. 2003). More detailed description of the flux calculation can be found in Ref. (Gaisser & Honda 2002). In these calculations, primary cosmic ray flux data are used, solar modulation and geomagnetic field effects are taken into account, and the interaction of cosmic ray particles with the air nuclei, the propagation of the secondary particles in the air, and the decay of them are simulated. In recent calculations, the 3 dimensional nature

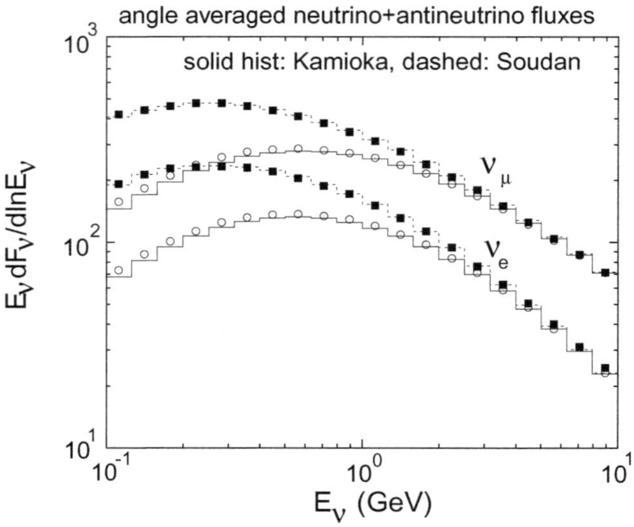

Figure 1. *(a) The atmospheric neutrino energy spectrum calculated for the Kamioka (Japan) and Soudan-2 (North America) sites (Barr et al. 2004). Copyright (2004) American Physical Society. The ($\nu_\mu + \overline{\nu}_\mu$) and ($\nu_e + \overline{\nu}_e$) fluxes are plotted for the three dimensional (points) and one dimensional (histograms) calculations. The solid histograms are for the Kamioka site and the dashed histograms are for the Soudan-2 site. The difference in the absolute fluxes for the 2 sites is due to the cutoff rigidity to the primary cosmic rays.*

of the particles produced by cosmic ray interactions are simulated properly (3 dimensional calculation), while the older calculations assumed that the secondary particles have the same momentum direction as that of the primary cosmic rays (1 dimensional calculation).

The calculated energy spectra of atmospheric neutrinos, by the one and the three dimensional methods, at Kamioka (Japan) and the Soudan mine (USA) are shown in Fig. 1. The accuracy in recent primary cosmic ray flux measurements (Alcaraz et al. 2000, Sanuki et al. 2000) below 100 GeV is about 5% and the data from independent experiments agree within the quoted uncertainties. However the primary cosmic ray data are much less accurate above 1 TeV. Therefore, for neutrino energies below about 10 GeV where large fraction of the neutrinos are produced by primaries with the energies less than 100 GeV, the calculated absolute flux should have small uncertainty of about 10%. Indeed calculated fluxes by (Honda et al. 2004, Barr et al. 2004, Battistoni et al. 2003) agree within 10%. However, at much higher energies than 10 GeV, the uncertainties in the absolute neutrino flux must be substantially larger.

Figure 2 shows the calculated $\nu_\mu + \overline{\nu}_\mu$ over $\nu_e + \overline{\nu}_e$ flux ratio as a function of the neutrino energy, integrated over solid angle. This ratio is essentially independent of the primary cosmic ray spectrum. The calculated flux ratio has a value of about 2 for energies less than a few GeV and increases with increasing neutrino energy. This is because a π-decay produces a ν_μ and a μ; the μ when it decays, produces another ν_μ and a ν_e. Furthermore, the 3 neutrinos produced in the chain decay of a π have approximately the same average energy. In the higher energy regions, the probability of a muon not to decay before reaching the ground increases with increasing muon energy. In the energy region of less than about 10 GeV, a large fraction of the neutrinos are produced by the decay chain of pions and the uncertainty in the π/K production ratio does not contribute significantly to the uncertainty in the $\nu_\mu + \overline{\nu}_\mu$ over $\nu_e + \overline{\nu}_e$

flux ratio. Therefore, the expected uncertainty of this ratio is about 3%. In the higher energy region (>10 GeV), the contribution of K decay in the neutrino production is more important. There, the ratio depends more on the K production cross sections and the uncertainty of the ratio is larger.

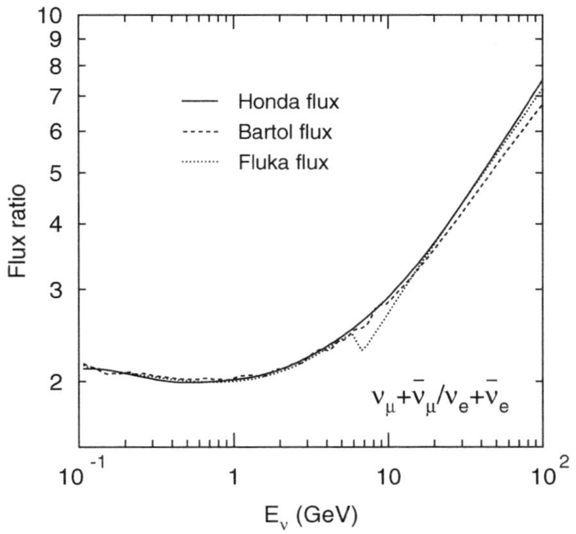

Figure 2. *The flux ratio of* $\nu_\mu + \overline{\nu}_\mu$ *to* $\nu_e + \overline{\nu}_e$ *versus neutrino energy. Solid, dashed and dotted lines show the prediction by (Honda et al. 2004), (Barr et al. 2004), and (Battistoni et al. 2003), respectively. From (Ashie et al. 2005). Copyright (2005) American Physical Society.*

Figure 3 shows the zenith angle dependence of the atmospheric neutrino fluxes for several neutrino energies for the Kamioka site. At low energies, the fluxes of downward-going neutrinos are lower than those of upward-going neutrinos. This is due to the cutoff of primary cosmic rays by the geomagnetic field. The cutoff rigidity for primary particles varies from less than 1 GeV to more than 10 GeV depending on the location in the earth. For neutrino energies higher than a few GeV, the calculated fluxes are essentially up-down symmetric, because the primary particles, which produce these neutrinos, are more energetic than the rigidity cutoff. The enhancement of the flux near the horizon for low energy neutrinos is a feature characteristic of low energy atmospheric neutrinos, first reported in a full three dimensional calculation of (Battistoni et al. 2000) and confirmed by other calculations. The mechanism for the enhancement is discussed in (Lipari 2000). The 3-dimensional effect is only important below about 1 GeV as seen in Fig. 3. However, the horizontal enhancement cannot be seen experimentally due to the relatively poor angular correlation between neutrinos and leptons below 1 GeV.

In summary of the atmospheric neutrino flux, we remark that, while the absolute flux value has a relatively large, energy-dependent uncertainty of about 10 to 30 %, the $(\nu_\mu + \overline{\nu}_\mu)$ over $(\nu_e + \overline{\nu}_e)$ flux ratio is predicted to an accuracy of about 3%. The zenith angle dependence of the flux is well understood, and especially, above a few GeV neutrino energies, the flux is predicted to be up-down symmetric.

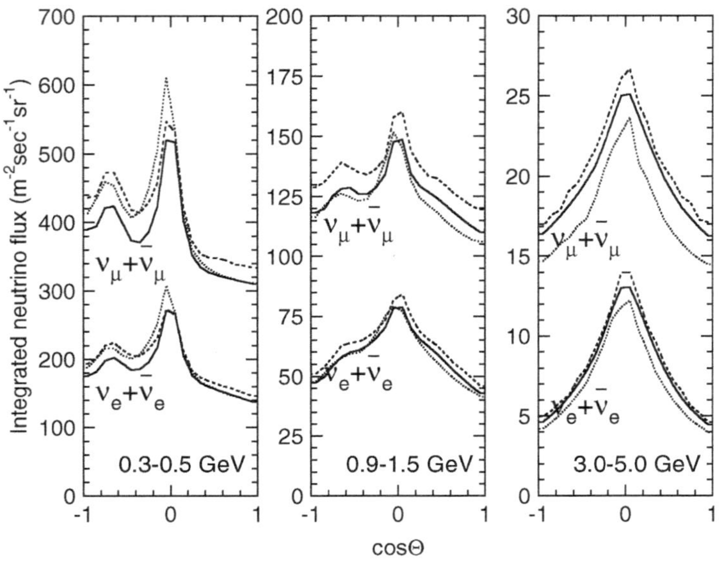

Figure 3. *The flux of atmospheric neutrinos versus zenith angle for 3 neutrino energy ranges. Solid, dashed and dotted lines show the prediction by (Honda et al. 2004), (Barr et al. 2004), and (Battistoni et al. 2003), respectively. From (Ashie et al. 2005). Copyright (2005) American Physical Society.*

3 Neutrino Interactions

The important energy range for atmospheric neutrino interactions is rather wide; between 0.1 GeV and 10 TeV. The Monte Carlo technique is used to simulate the neutrino interactions.

Usually, the following charged and neutral current (NC) neutrino interactions are considered in the simulation of atmospheric neutrino events.

- (quasi-)elastic scattering, $\nu N \longrightarrow l N'$,
- single meson production, $\nu N \longrightarrow l N' m$,
- coherent π production, $\nu X \longrightarrow l \pi X$,
- deep inelastic scattering, $\nu N \longrightarrow l N'$ *hadrons*.

Where N and N' are the nucleons (proton or neutron), X is the nucleus, l is the lepton, and m is the meson. The most dominantly produced mesons are pions. If the neutrino interaction occurs in a nucleus, generated particles like pions and kaons interact with the nucleus before escaping from the nucleus.

In the lowest energy range of less than 1 GeV, the quasi-elastic scattering is dominant. For scattering off nucleons in a nucleus, the Fermi motion of the nucleons and Pauli's Exclusion Principle must be taken into account. Usually, these effects are treated based on the relativistic Fermi gas model. In the GeV energy range, the single pion production processes are also important. In the multi-GeV or higher energy ranges, the deep inelastic scattering is dominant. For more details of the neutrino interactions, see (McFarland 2006).

4 Atmospheric neutrino experiments

To date, two significantly different techniques, water Cherenkov and fine grained tracking detectors, have been used to observe atmospheric neutrino events.

In water Cherenkov detectors, an atmospheric neutrino event is detected by observing Cherenkov radiation from relativistic charged particles produced by the neutrino interaction off the nucleus. A two dimensional array of photomultiplier tubes on the inside surface of the detector detects the photons. The hit time and the pulse height from each photomultiplier tube are recorded. The timing information, with a typical resolution of a few nsec for a single photo-electron pulse, is useful for reconstructing the vertex position. The total number of photo-electrons gives information on the energy of the particles above Cherenkov threshold. There have been 3 large water Cherenkov detectors. IMB and Kamiokande were operational in the 1980's and the 1990's. Super-Kamiokande started taking data in 1996, and is still in operation. It is the largest atmospheric neutrino detector ever built.

The second category of atmospheric neutrino detectors consists of comparatively fine resolution tracking detectors. These detectors have generally been smaller than the water Cherenkov detectors. However, tracking detectors have an advantage in sensitivity, because they can detect low momentum charged particles that would be below Cherenkov threshold in water. In particular, the Soudan 2 detector is able to reconstruct the short and heavily ionizing trajectory of recoil protons from atmospheric neutrino events such as $\nu n \rightarrow \mu p$. MINOS is primary a detector for long baseline neutrino oscillation experiment. However, it can also detect atmospheric neutrinos. It is the first magnetized tracking detector for atmospheric neutrinos. MINOS is able to get information on the track direction, the charge and the momentum.

There is a second type of fine grained tracking detector that is mostly sensitive to muon neutrinos in the form of upward-going muons. These detectors identify the direction of the muon by resolving the time-of-flight as it traverses two or more layers of liquid scintillator. The MACRO detector is composed three horizontal planes with the lower section filled with crushed rock absorber and a hollow upper section. In addition to through-going muons, MACRO has analyzed partially contained and stopping muon events, where the crushed rock in the lower section acts as neutrino target or muon stopper, respectively.

5 Atmospheric neutrino data and oscillations

5.1 μ/e ratio

Although the uncertainty in the absolute flux of atmospheric neutrinos is large, the (ν_μ/ν_e) flux ratio is predicted much more accurately. Therefore, the measurement of this ratio is a sensitive test for neutrino oscillations.

The atmospheric (ν_μ/ν_e) flux ratio has been measured by identifying electrons and muons produced by CC ν_e and CC ν_μ interactions, respectively. Electrons produce electromagnetic showers while propagating in matter. On the other hand, muons slowly lose their energy by the ionization loss in matter. The different propagation of these particles in matter is used to separate electrons and muons.

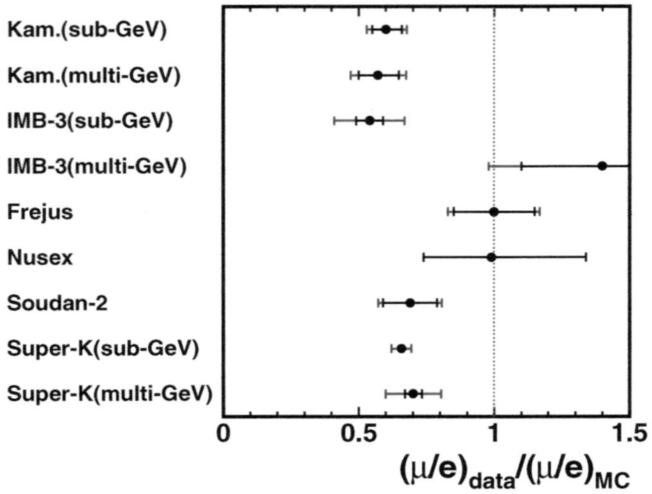

Figure 4. *Summary of the measurements of the double ratio μ-like/e-like(data) to μ-like/e-like(MC) by various atmospheric neutrino experiments. The inner and outer error bars show the statistical and statistical+systematic errors, respectively.*

In tracking calorimeters, the event pattern is recorded in a 3-dimensional view, and the separation of electrons and muons is carried out by event pattern recognition. A CC ν_μ event is recognized by a long straight track, which is a candidate muon. A CC ν_e event, on the other hand, has an electromagnetic shower linked to the vertex point.

In water Cherenkov detectors, CC ν_e and ν_μ events are distinguished by the difference in the characteristic shape of the Cherenkov ring. A clear outer edge of the Cherenkov light is a characteristic of a μ-like event, while the edge is relatively unclear for an e-like event due to an electromagnetic shower and the multiple scattering of low-energy electrons.

Based on the particle identification technique, the flavor ratio in the atmospheric neutrino flux was measured. The $(\nu_\mu + \overline{\nu}_\mu)/(\nu_e + \overline{\nu}_e)$ flux ratio has been calculated to an accuracy of about 3%. Experimentally, the μ-like/e-like ratio for the data is compared with the same ratio for the MC. Therefore, $(\mu\text{-like}/e\text{-like})_{Data}/(\mu\text{-like}/e\text{-like})_{MC}$ should be consistent with unity within the errors for the case of no neutrino oscillations. Figure 4 summarizes the $(\mu/e)_{Data}/(\mu/e)_{MC}$ measurements from various experiments. Measurements from Kamiokande (both sub- and multi-GeV data samples) (Hirata et al. 1992, Fukuda et al. 1994), IMB-3 (sub-GeV) (Becker-Szendy et al. 1992), Soudan-2 (Allison et al. 1999) and Super-Kamiokande (Ashie et al. 2005) (both sub- and multi-GeV data samples) showed $(\mu/e)_{Data}/(\mu/e)_{MC}$ ratios which were smaller than unity. The sub-GeV(multi-GeV) sample in Kamiokande and Super-Kamiokande was defined to include events with E_{vis} <1.33 (>1.33) GeV, where E_{vis} is the visible energy. E_{vis} =1.33 GeV corresponds to an electron(muon) momentum of 1.33 (about 1.4) GeV/c. There were a few measurements (Aglietta et al. 1989, Berger et al. 1990, Clark et al. 1997) suggesting a ratio consistent with unity. However, these measurements had relatively large statistical errors.

Since many of the uncertainties of the prediction and the experiment cancel by taking this ratio, the systematic error of this measurement was relatively small. Neither a statistical fluctuation nor systematic uncertainties was able to explain the small (μ/e) ratio of the data. The small μ/e ratio was interpreted as due to $\nu_\mu \leftrightarrow \nu_\tau$ or $\nu_\mu \leftrightarrow \nu_e$ oscillations with a large mixing angle. However, the other possibilities of non-standard physics were not excluded with this double-ratio only. For example, authors in (Mann, Kafka & Leeson 1992) discussed that the data could be explained by proton decay into $e^+\nu\nu$, since the observed small μ/e ratio could be interpreted to be due to an excess of e-like signal by these positrons.

5.2 Zenith angle distribution

The atmospheric neutrino flux is predicted to be up-down symmetric above a few GeV neutrino energies. For downward-going neutrinos the flight length is about 15 km, while that for vertically upward-going neutrinos is 13,000 km. Therefore, if the neutrino oscillation length is approximately in the range of 100 to 1,000 km, it should be possible to observe up-down asymmetry of the flux. Early data from Kamiokande (Fukuda et al. 1994) showed that the the event rate for upward-going multi-GeV μ-like events had substantially lower than that for the downward-going ones.

The zenith angle distributions for e-like and μ-like events observed in Super-Kamiokande (141 kton·yr) are shown in Fig.5. The μ-like data have exhibited a strong deficit of upward-going events, while no significant deficit has been observed in the e-like data. Many systematic errors cancel for the up-down ratio. The observed up-down ratio for multi-GeV events suggests a near maximal neutrino mixing, because the up-down ratio can be approximately expressed as $(up/down) = 1 - 0.5 \cdot \sin^2 2\theta_{23}$, where 0.5 is due to the average of $\sin^2(1.27\Delta m^2 L/E)$ over many oscillations. For sub-GeV μ-like events, the observed ratio is larger than that in the multi-GeV sample due to the poorer angular correlation between neutrinos and leptons.

The observed number of events for upward through-going muons was slightly higher than the no-oscillation prediction. The predicted neutrino flux has more than 20% uncertainty in the energy range relevant to upward through going muons ($\sim$100 GeV). On the other hand, the observed zenith angle distribution is explained well by neutrino oscillations, if the large systematic uncertainty in the absolute flux is taken into account. [Note added: After the school, a new calculation of the atmospheric neutrino flux has been reported (Sanuki et al. 2006, Honda et al. 2006). In this calculation, the absolute neutrino fluxes are calibrated with recent high precision atmospheric muon data up to about 1 TeV. As a result, the newly calculated fluxes of the atmospheric ν_μ (anti-ν_μ) with the energies higher than 100 GeV are higher by 15% (5%) or more compared with the previous calculation (Honda et al. 2004). Therefore, it is likely that the problem of low predicted event rate for upward through-going muons is fixed.]

Also shown in Fig. 5 are the zenith angle distributions observed in Soudan-2 (Sanchez et al. 2003, Allison et al. 2005) and MINOS (Adamson et al. 2006). Although the statistics of these data are limited, the observed deficit of upward-going μ-like events is consistent with the Super-Kamiokande data.

Shown at the bottom of Fig. 5 are the zenith angle distributions for three data samples observed in MACRO (Ambrosio et al. 2004). The predicted and observed numbers of events near horizon are small due to the lower efficiency for near horizontal events. The zenith angle distribution for upward-going PC events shows a relatively large deficit of the data events, while that for upward-going stopping μ plus downward-going PC events shows only a small deficit of the data events. This can be understood by neutrino oscillations, since about a half of the upward-going neutrinos should be oscillated away, while the downward-going neutrinos

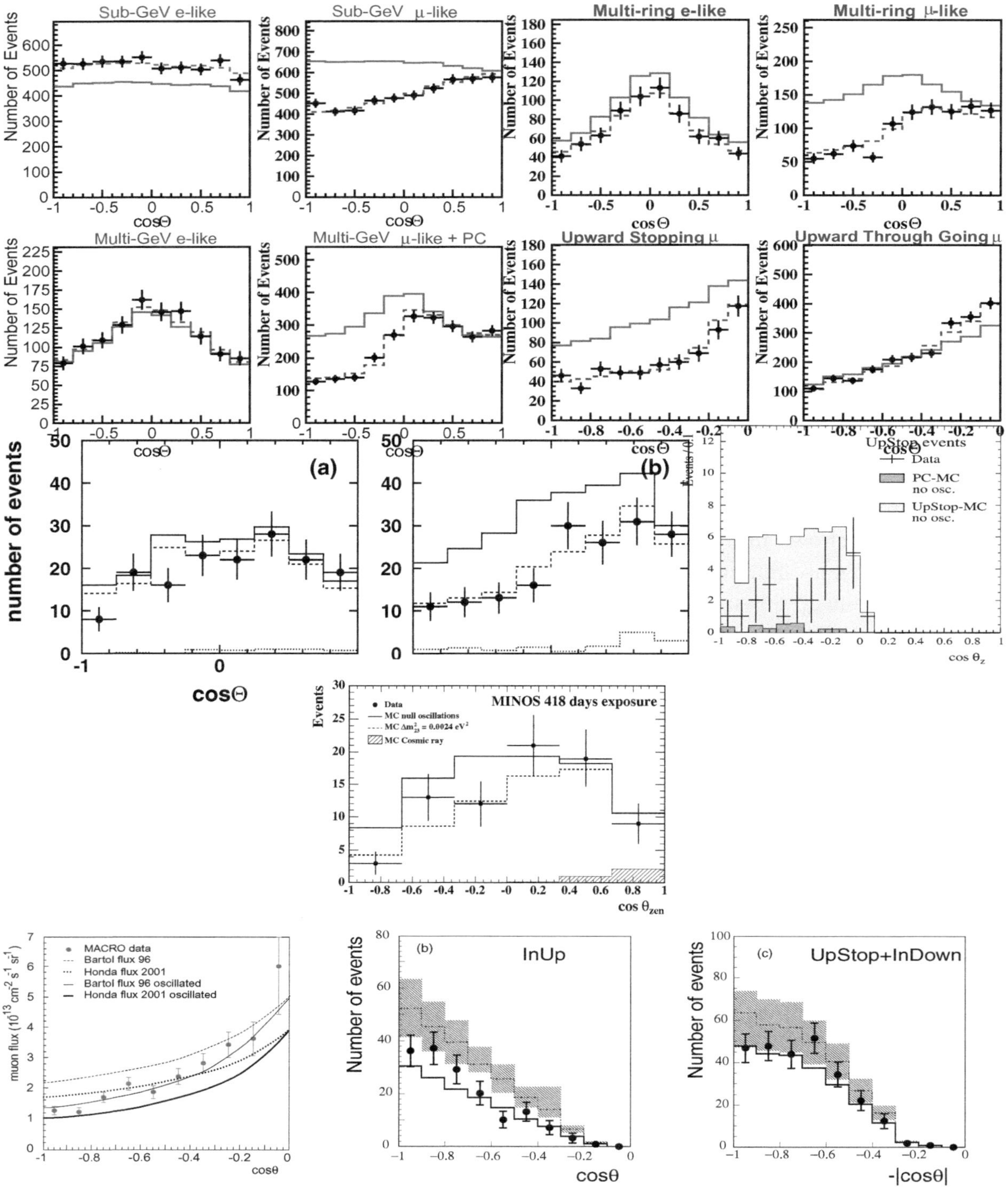

Figure 5. *Zenith angle distributions for atmospheric neutrino events observed in Super-Kamiokande (Copyright 2005 American Physical Society) Soudan-2 (left: e-like, middle: μ-like, right: upward-stopping) (Copyright 2003, 2005 American Physical Society), MINOS (Copyright 2006 American Physical Society) and MACRO (left: through-going muon flux, Copyright 2001 Elsevier; middle: upward-going PC events, right: upward-going stopping muons + downward-going PC events, Copyright 2004 Springer), from top to bottom.* $\cos\Theta = 1(-1)$ *means down-going (up-going). The histograms show the prediction with and without* $\nu_\mu \to \nu_\tau$ *oscillations.*

have very little oscillation effects, and since the number of downward-going PC events and upward-going stopping muon events are expected to be approximately equal for no oscillation. Thus the MACRO data are also consistent with neutrino oscillations.

5.3 Two flavor neutrino oscillation analysis

While e-like events show no evidence for inconsistency between data and non-oscillated Monte Carlo, μ-like events show the zenith angle and energy dependent deficit of events. Hence it can be concluded that the oscillation could be between ν_μ and ν_τ. Indeed, detailed oscillation analyses from various atmospheric neutrino experiments have concluded that the data are consistent with two-flavor $\nu_\mu \leftrightarrow \nu_\tau$ oscillations. From here on, the discussion will be mostly based on the Super-Kamiokande data, since the statistics are dominated by this experiment. In the analysis of the atmospheric neutrino data observed in Super-Kamiokande, a χ^2 method that takes into account various systematic uncertainties is used. It is defined to be:

$$L = \prod_{i=1}^{760} \frac{\exp(-N_i^{\mathrm{exp}})(N_i^{\mathrm{exp}})^{N_i^{\mathrm{obs}}}}{N_i^{\mathrm{obs}}!} \times \prod_{j=1}^{70} \exp\left(\frac{\varepsilon^j}{\sigma_j}\right), \tag{3}$$

$$N_i^{\mathrm{exp}} = N_i^0 \cdot P(\nu_\mu \to \nu_\mu) \cdot (1 + \sum_{j=1}^{71} f_j^i \cdot \varepsilon^j), \tag{4}$$

$$\chi^2 \equiv -2\ln\left(\frac{L(N^{exp}, N^{obs})}{L(N^{obs}, N^{obs})}\right). \tag{5}$$

Events are divided to 760 bins (380 from both Super-K-I and -II) based on the event type, momentum and zenith angle. N_i^{obs} is the number of observed events in the *i-th* bin and N_i^{exp} is the expected number of events based on the Monte Carlo simulation. During the fit, the values of N_i^{exp} are recalculated to account for neutrino oscillations, and systematic variations in the predicted rates due to uncertainties in the neutrino flux model, neutrino cross-section model, and detector response. The likelihood (L) definition takes into account the contributions from 70 variables which parameterized the systematic uncertainties in the expected neutrino rates. σ_j represents the systematic error for the *j-th* term. The absolute normalization is treated as a free parameter, and is not included in the 70 terms in the L definition. During the fit, these parameters are varied to minimize χ^2 for each choice of oscillation parameters; $\sin^2 2\theta$ and Δm^2.

The estimated oscillation parameters ($\sin^2 2\theta$, Δm^2) for two flavor $\nu_\mu \leftrightarrow \nu_\tau$ oscillation from various experiments (Ashie et al. 2005, Allison et al. 2005, Adamson et al. 2006, Ambrosio et al. 2004, Hatakeyama et al. 1998) are shown in Fig. 6(left). All results are essentially consistent. If the most accurate result from Super-Kamiokande is referred, the oscillation parameters are determined as; $\sin^2 2\theta > 0.93$ and $1.9 \times 10^{-3} < \Delta m^2 < 3.1 \times 10^{-3}\mathrm{eV}^2$ at 90% C.L. as shown in Fig. 6(right). Finally, we remark that consistent results have been obtained by a recent long baseline neutrino oscillation experiment based on accelerator neutrino beams (Harris 2006).

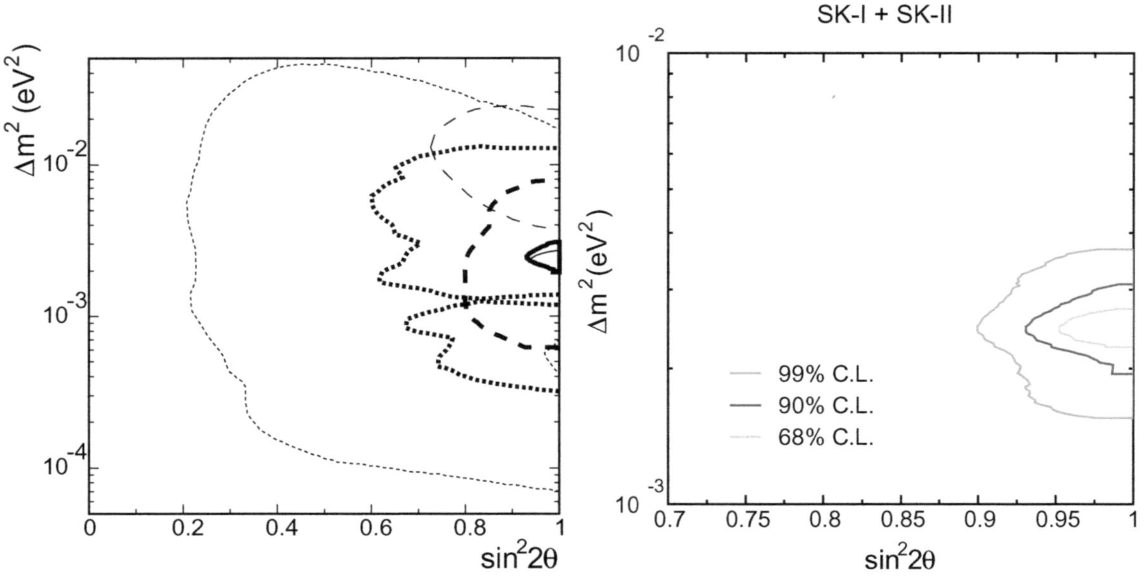

Figure 6. *Left: 90% C.L. allowed parameter regions of oscillation parameters from the atmospheric neutrino data observed in Super-Kamiokande(thick line), Kamiokande(thin dashed line), Soudan-2(thick dotted line), MINOS(thin dotted line) and MACRO(thick dashed line). Two flavor $\nu_\mu \to \nu_\tau$ oscillation is assumed. Also shown is the allowed region from the L/E analysis in Super-Kamiokande (thin line). Right: 68, 90 and 99% C.L. allowed oscillation parameter regions for 2-flavor $\nu_\mu \leftrightarrow \nu_\tau$ oscillations obtained by the zenith-angle analysis from Super-K-I+II.*

5.4 Constraints on $\nu_\mu \leftrightarrow \nu_{sterile}$ oscillations

It is important to ask whether $\nu_\mu \leftrightarrow \nu_\tau$ oscillation is the only possible explanation for the atmospheric neutrino data. Since the observed effect was the energy and zenith-angle dependent deficit of CC ν_μ events, there were several proposals for alternative explanations. One proposal was neutrino oscillations between ν_μ and $\nu_{sterile}$, where $\nu_{sterile}$ is a neutrino-like particle that does not interact with matter by either CC or NC weak interactions. Since the deficit was seen for CC ν_μ events, $\nu_\mu \leftrightarrow \nu_{sterile}$ oscillation could explain the atmospheric neutrino data. Because the possible existence of $\nu_{sterile}$ has significant impact on particle physics and cosmology, it was studied seriously whether $\nu_\mu \to \nu_{sterile}$ oscillation is really favored by the atmospheric neutrino data.

There are several ways to discriminate between the two possibilities. One possibility is to use a matter effect for upward going neutrino events (Liu, Mikheyev & Smirnov 1998, Lipari & Lusignoli 1998). In the case of $\nu_\mu \to \nu_{sterile}$ oscillations, the matter effect could change the oscillation probability (Eq.2) significantly, while there is no change in the $\nu_\mu \to \nu_\tau$ oscillation probability. For $\Delta m^2 \sim (2 \sim 3) \times 10^{-3}\text{eV}^2$, the oscillation probability is expected to be significantly different only for high energy ($> 10 \sim 20$ GeV) atmospheric neutrinos traveling through the Earth. In addition, the zenith angle distribution for a NC enriched sample is useful to discriminate between the two possibilities, because the NC events should be affected only for $\nu_\mu \leftrightarrow \nu_{sterile}$ oscillations.

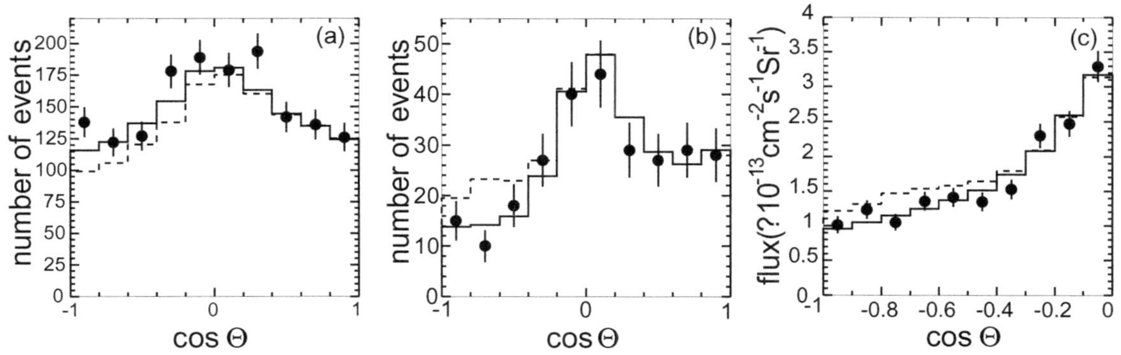

Figure 7. *Zenith-angle distributions for: (a)multi-ring events with the most energetic ring being e-like and E_{vis} >400 MeV, (b)PC events with E_{vis} >5 GeV, and (c)upward through-going muon events (Fukuda et al. 2000). Copyright 2000 American Physical Society. The detector exposure was 70.5 kton·yr for FC and PC events, and 1138 days for upward through-going muons. In these figures, solid (dashed) histograms show predictions for $\nu_\mu \leftrightarrow \nu_\tau$ ($\nu_\mu \leftrightarrow \nu_{sterile}$) oscillations with $\Delta m^2 = 3\times10^{-3}eV^2$ and $\sin^2 2\theta = 1.0$. The predictions are normalized so that the number of observed and predicted events are equal at $0.4 < \cos\Theta < 1.0$ for (a) and (b) and at $-0.4 < \cos\Theta < 0.0$ for (c).*

Super-Kamiokande analyzed NC-enriched multi-ring, high-energy PC with E_{vis} >5 GeV, and upward through-going muon events (Fukuda et al. 2000). Multi-ring events with the most energetic ring being *e*-like and E_{vis} >400 MeV were used. The estimated fraction of NC events in this sample (in the absence of neutrino oscillations) was 29%. Figure 7 shows the zenith angle distributions for these samples. It is clear that all the data samples disfavor $\nu_\mu \to \nu_{sterile}$ oscillations. Pure $\nu_\mu \to \nu_{sterile}$ oscillation has been excluded at more than 99% C.L. (Fukuda et al. 2000). A consistent result was obtained by MACRO using upward through-going muons (Ambrosio et al. 2001).

It has been proposed to analyze the data assuming neutrino oscillations between ν_μ and ν_x, where ν_x is a mixed flavor state ($\cos\xi\,\nu_\tau + \sin\xi\,\nu_{sterile}$)(Fogli, Lisi & Marrone 2001). Using all the atmospheric neutrino data from Super-Kamiokande, the analysis on this scheme was carried out. No evidence for a finite sterile neutrino component as a partner of the ν_μ oscillations has been observed. The upper limit on the $\nu_{sterile}$ admixture is $\sin^2\xi < 0.25$ at 90%C.L.

5.5 L/E analysis

The deficit of ν_μ events observed by the atmospheric neutrino experiments is explained very well by $\nu_\mu \leftrightarrow \nu_\tau$ oscillations. However, models that explain the zenith angle and energy dependent deficit of ν_μ events have been proposed. They are neutrino decay (Barger et al. 1999) and neutrino decoherence (Lisi, Marrone & Montanino 2000) models. These models may not be particularly appealing theoretically. However, it is still important to experimentally determine which is the right explanation for the atmospheric ν_μ deficit. One of the key feature of neutrino oscillations is the sinusoidal ν_μ disappearance probability as a function of L/E. On the other hand, the ν_μ survival probability for the neutrino decay and decoherence models are not sinusoidal. Therefore, in order to further confirm neutrino oscillations, it is important to measure

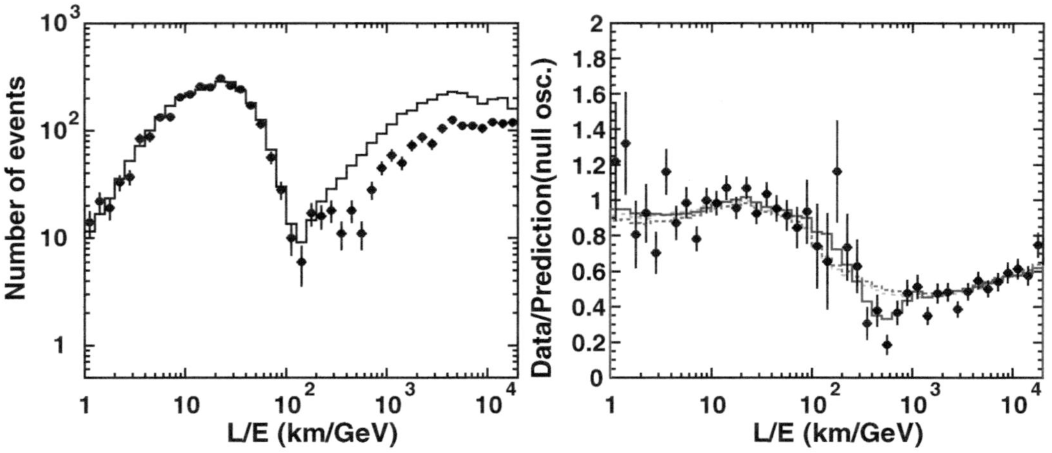

Figure 8. *Left: Number of events as a function of L/E for the Super-Kamiokande-I+II data (points) and the atmospheric neutrino MC events without oscillations (solid histogram). The MC is normalized by the detector live-time. Right: Ratio of the data to the MC events without neutrino oscillation (points) as a function of the reconstructed L/E together with the best-fit expectation for 2-flavor $\nu_\mu \leftrightarrow \nu_\tau$ oscillations (solid line). Also shown are the best-fit expectation for neutrino decay (dashed line) and neutrino decoherence (dotted line). From (Ashie et al. 2004). Copyright 2004 American Physical Society.*

the sinusoidal neutrino oscillation feature.

Super-Kamiokande carried out an L/E analysis (Ashie et al. 2004) and updated with the Super-K-I+II data. FC and PC events, which satisfy the (L/E) resolution better than 70%, were used. Since the L/E resolution depends on the zenith angle, the energy and the type of events, the events were selected using these information. Essentially, low-energy events are not used, because of the bad angular correlation between the neutrino and outgoing particles. Also, horizontal-going events are not used, because L changes significantly with a small change in the zenith angle near horizon. Figure 8 (left) shows the number of events as a function of the reconstructed L/E, together with the null-oscillation MC prediction. Two clusters of events are visible below and above 150 km/GeV. They mostly correspond to downward-going and upward-going events, respectively.

Figure 8 (right) shows the ratio of the data over non-oscillated MC as a function of L/E together with the best-fit expectation for 2-flavor $\nu_\mu \leftrightarrow \nu_\tau$ oscillations with systematic errors. A dip, which should correspond to the first maximum oscillation, is observed around L/E = 500 km/GeV. Due to the L/E resolution of the detector, the second and higher maximum oscillation points should not be observable in the Super-Kamiokande experiment.

In the neutrino decay and decoherence models, it is assumed that a neutrino flavor eigenstate is a mixture of mass eigenstates. However, the mechanisms for generating ν_μ disappearance are different. In particular, these models do not predict sinusoidal ν_μ disappearance probability. Also shown in Fig. 8 are the L/E distributions for the best-fit expectation for the neutrino decay and decoherence models. Since these models cannot predict the dip observed in the data, the χ^2 values for these models were worse. The pure neutrino decay and decoherence models were disfavored at 4.8 and 5.3 standard deviation levels, respectively (preliminary).

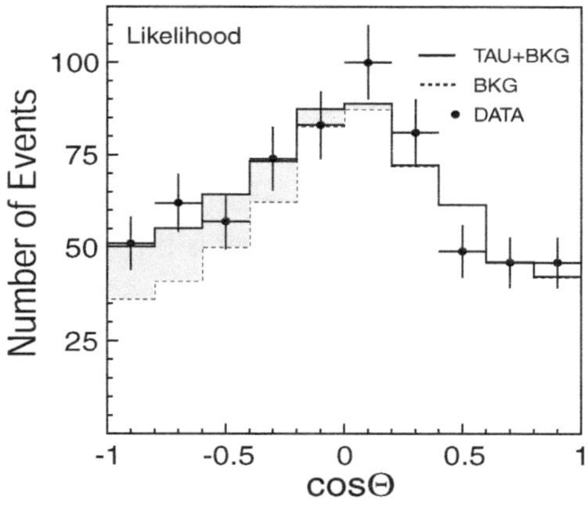

Figure 9. *Zenith angle distribution for the candidate tau neutrino interactions observed in Super-Kamiokande (92 kton·yr). The gray and while regions show the fitted tau neutrino contribution and the non-tau atmospheric neutrino interactions, respectively.*

The observed L/E distribution gives evidence that the neutrino flavor transition probability obeys the sinusoidal function as predicted by neutrino oscillations.

5.6 Detecting CC ν_τ events

If $\nu_\mu \to \nu_\tau$ is the dominant oscillation channel, about one CC ν_τ event per kiloton year exposure is expected to occur in an atmospheric neutrino detector. The low event rate is due to the soft energy spectrum of the atmospheric neutrino flux and the threshold effect of the τ production process which requires a ν_τ energy of at least 3.5 GeV to produce a τ lepton. These τ typically decay to hadrons (branching ratio is 64%) within 1 mm from the vertex point. These events should be upward-going but otherwise similar to energetic NC events, hence it is difficult to isolate ν_τ events in the on-going atmospheric neutrino experiments.

Super-Kamiokande searched for CC ν_τ events in the multi-GeV FC events. First, it was required that the most energetic ring is e-like, in order to reduce the CC ν_μ background. Then, the candidate ν_τ events were selected by a maximum likelihood or a neural network method. Several kinematical information were used as the inputs to these analyses. Even with these analyses, the signal to noise ratio was about 10%. However, the zenith-angle distribution can be used to estimate the number of ν_τ events statistically, because both the ν_τ signal and background events have accurately predicted zenith angle distributions. See Fig. 9. The best fit number of ν_τ interactions that occurred in the fiducial volume of the detector during the 92 kton·yr exposure was $138\pm48(\text{stat.})^{+15}_{-32}(\text{syst.})$. The expected number was 78 ± 26. The observed number of ν_τ interactions was consistent with the $\nu_\mu \to \nu_\tau$ expectation. Zero CC ν_τ assumption was excluded at the 2.4 standard deviation level.

6 Prospects for future atmospheric neutrino experiments: beyond two flavor oscillations

So far, all the analyses suggest that the atmospheric neutrino data are fully consistent with pure $\nu_\mu \leftrightarrow \nu_\tau$ oscillations. However, since there are 3 neutrino flavors, 3 flavor neutrino oscillation effects might be visible at some level of statistics in the atmospheric neutrino data.

At the time of writing this article, there have been several considerations on future detectors that can improve our knowledge on the neutrino masses and mixings through studies of atmospheric neutrinos. One type of the detectors is huge water Cherenkov detectors. These detectors are assumed to have the masses of about 0.5 to 1 Mton. The other type of detectors are magnetized tracking detectors. The masses for these detectors are assumed to be several tens of kton. The other possibility is a large liquid Argon detector with the mass similar to the other types of detectors and possibly a magnetic field inside the detector. We discuss the physics potential for future atmospheric neutrino experiments.

6.1 Three flavor oscillation effect: θ_{13}

As an extension of the 2 flavor oscillation analysis, we discuss 3 flavor neutrino oscillations. Δm^2_{12} is much smaller than $\Delta m^2_{23}(\Delta m^2_{13})$ as measured by the solar neutrino (Wark 2006) and KamLAND experiments (Blucher 2006), and therefore we assume the effect of the oscillations related to Δm^2_{12} is negligible in atmospheric neutrino experiments. The oscillation length relevant to Δm^2_{12} is shorter than the diameter of the earth for E_ν below 1 GeV. However, due to the accidental feature that the atmospheric ν_μ/ν_e ratio is approximately 2 in the absence of $\nu_\mu \rightarrow \nu_\tau$ oscillations in the relevant energy range and due to the (near) maximal θ_{23} mixing angle, the effect of the oscillations related to Δm^2_{12} is predicted to be small (see Subsection 6.3). Under this approximation, there are only three oscillation parameters; θ_{13}, θ_{23} and $\Delta m^2 (\equiv \Delta m^2_{13} = \Delta m^2_{23})$. In this case, for example, the $\nu_\mu \leftrightarrow \nu_e$ oscillation probability in the vacuum can be written as:

$$P(\nu_\mu \rightarrow \nu_e) = \sin^2\theta_{23}\sin^2 2\theta_{13}\sin^2\left(\frac{1.27\Delta m^2_{23}L}{E_\nu}\right). \tag{6}$$

The oscillation probability in matter is different from that in vacuum for oscillations involving ν_e. This is due to the difference in the forward scattering amplitude for ν_e and ν_μ (or ν_τ). Assuming $\Delta m^3_{23} > 0$, the instantaneous mixing parameter ($\sin^2 2\theta^m_{13}$) for neutrinos in matter with the electron number density N_e is:

$$\sin^2 2\theta^m_{13} = \frac{\sin^2 2\theta_{13}}{(A/\Delta m^2_{23} - \cos 2\theta_{13})^2 + \sin^2 2\theta_{13}}, \tag{7}$$

where $A = 2\sqrt{2}G_F N_e E_\nu$ (G_F is the Fermi coupling constant). For anti-neutrinos, A must be replaced by $-A$. If a condition $A = \Delta m^2 \cos 2\theta_{13}$ is satisfied, $\sin^2 2\theta^m_{13}$ is maximum (=1) even though the vacuum mixing angle θ_{13} is small. For Δm^2_{23} =2 to 3 $\times 10^{-3}\text{eV}^2$ with a small θ_{13}, the resonance could occur for neutrinos passing through the earth with their energies between 5 and 10 GeV. Therefore, the effect of a non-zero θ_{13} could be observed as an excess of multi-GeV electron neutrinos in the upward-going direction. Figure 10 shows the $\nu_e \leftrightarrow \nu_\mu$ oscillation probability as a function of the neutrino energy and zenith angle. A clear resonance effect is seen for upward-going neutrinos near 7 GeV. For neutrinos passing through the core of the earth ($\cos\Theta < -0.83$), resonances will occur in slightly lower neutrino energies.

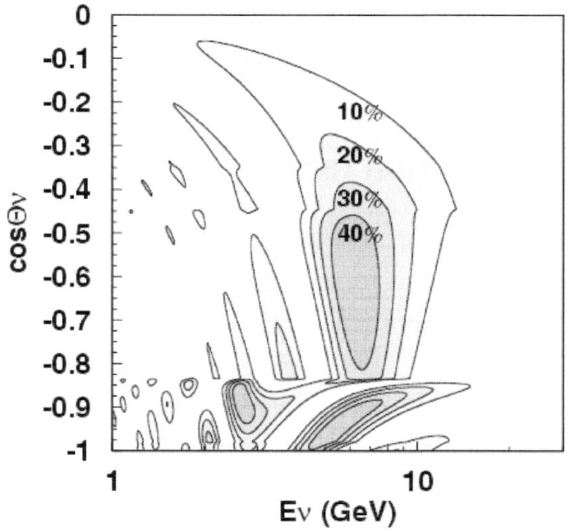

Figure 10. $\nu_e \leftrightarrow \nu_\mu$ *oscillation probability for neutrinos passing through the earth as a function of the neutrino energy and zenith angle for* $\Delta m^2_{23} = 2.5 \times 10^{-3} eV^2$, $\sin^2\theta_{23} = 0.50$ *and* $\sin^2\theta_{13} = 0.04$. *From (Hosaka et al. 2006). Copyright (2006) American Physical Society.*

The present data from Super-Kamiokande and Soudan-2 (Fig.5) show no evidence for excesses of e-like events in the upward-going direction. The result on the analysis of $\sin^2\theta_{13}$ has been published based on 92 kton·yr data (Hosaka et al. 2006). No evidence for non-zero $\sin^2\theta_{13}$ has been observed. The constraints on θ_{13} from reactor experiments (Blucher 2006) are still more stringent than that from the present atmospheric neutrino data.

The sensitivity to θ_{13} in a large water Cherenkov detector was discussed in detail in Ref. (Kajita et al. 2004). For 450 kton·yr exposure, it will be possible to observe more than 3 standard deviation effect if $\sin^2 2\theta_{13} > 0.05$ and if $\theta_{23} = 45°$. The chance of observing finite θ_{13} rapidly improves for larger $\sin^2\theta_{23}$. Also the sensitivity does not depend strongly on Δm^2_{23}.

6.2 Sign of Δm^2

The resonance effect occurs only for neutrinos for positive Δm^2, and therefore only appears for the e^- and μ^- spectrum. This, in turn, suggests that the sign of Δm^2_{23} could be measured by atmospheric neutrino experiments. It is generally believed that a massive magnetized detector, which can measure the charge of a muon, is necessary to measure the sign of Δm^2_{23}. With a 150 kton·yr exposure of a magnetized detector, it will be possible to determine the sign of Δm^2 if $\sin^2 2\theta_{13} > 0.04$ (Indumathi & Murthy 2005).

Super-Kamiokande and other water Cherenkov detectors are unable to distinguish ν_e and $\overline{\nu}_e$ interactions event-by-event bases. However, the cross section and the y $(= (E_\nu - E_{lepton})/E_\nu)$ dependence of the cross section are different between ν and $\overline{\nu}$, and therefore it may be possible to distinguish the positive and negative Δm^2_{23}. For positive Δm^2_{23}, the resonance effect occurs only for neutrinos. Since the neutrino interactions produce more high-y events (i.e., more multi-hadron events) than the anti-neutrino interactions, a larger effect of the finite θ_{13} can be seen in multi-ring e-like events for positive Δm^2_{23} than for negative Δm^2_{23}. Detailed Monte Carlo studies showed that it is possible to measure the sign of Δm^2_{23} in water Cherenkov detectors with very high exposure (more than 1 Mton·yr), if the $\sin^2 2\theta_{13}$ is near the present limit ($\sin^2 2\theta_{13} > (0.05-0.10)$) and and if $\sin^2\theta_{23} \geq 0.5$ (Kajita et al. 2004).

6.3 Effects of the solar oscillation terms

So far, we have neglected the oscillation terms that are related to solar neutrinos (θ_{12} and Δm^2_{12}). It has been pointed out that these terms could play unique roles in the atmospheric neutrino oscillations, such as the possible measurement of $\sin^2\theta_{23}$, (i.e., the discrimination of the octant of θ_{23}) (Peres & Smirnov 1999, Peres & Smirnov 2004). For simplicity, we assume that $\theta_{13} = 0$. The change in the atmospheric ν_e flux due to oscillations driven by the solar oscillation (1–2 oscillation) terms is written as;

$$\frac{F^{osc}_{\nu e}}{F^0_{\nu e}} - 1 = P_2(r\cos^2\theta_{23} - 1), \tag{8}$$

where $F^{osc}_{\nu e}$ and $F^0_{\nu e}$ are the atmospheric ν_e fluxes with and without oscillations, $r(\equiv F^0_{\nu\mu}/F^0_{\nu e})$ is the ratio of the un-oscillated atmospheric ν_μ and ν_e fluxes, and P_2 is the two neutrino transition probability ($\nu_e \to \nu_x$) in matter driven by the solar oscillation terms. P_2 is large for neutrinos passing through the earth with the energies below 1 GeV. Thus the sub-GeV atmospheric neutrinos play an important role in observing the solar term effect. Since the ν_μ and ν_e flux ratio (r) is approximately 2 in the sub-GeV neutrino energy region, the $F^{osc}_{\nu e}/F^0_{\nu e}$ value in Eq.8 is very close to 1 in the case of the maximal 2–3 mixing (i.e., $\theta_{23} = 45°$). However, according to Eq.8, if θ_{23} is in the first octant ($\theta_{23} < 45°$), an excess of the sub-GeV e-like events is expected. If θ_{23} is in the second octant ($\theta_{23} > 45°$), a deficit of the sub-GeV e-like events is expected. This will be a unique possibility to measure the octant of θ_{23}. The preliminary result from Super-Kamiokande did not show any significant change in $\sin^2\theta_{23}$ with and without the solar terms.

7 Summary

Atmospheric neutrinos have played essential roles in the discovery of neutrino oscillations, and are still contributing significantly in the study of neutrino oscillations. The atmospheric neutrino data from various experiments are well explained by $\nu_\mu \to \nu_\tau$ oscillations. Many proposed non-standard explanations have been excluded by detailed studies of the high statistics atmospheric neutrino data. Especially, the L/E analysis from Super-Kamiokande gave the direct evidence that the neutrino survival probability obeys the sinusoidal function as predicted by neutrino oscillations. The 90%C.L. allowed region of neutrino oscillation parameters from the Super-Kamiokande is: $1.9\times10^{-3} < \Delta m^2_{23} < 3.1\times10^{-3}\text{eV}^2$ and $\sin^2 2\theta_{23} > 0.93$.

Atmospheric neutrino experiments are also sensitive to the other oscillation parameters. No evidence for these effects has been observed yet. However, it is expected that higher statistics atmospheric neutrino data will give information on these parameters. The study of neutrino oscillations will continue to be an important and exciting field and atmospheric neutrino experiments are likely to continue to contribute to this field substantially.

Acknowledgments

The author thank the organizers of the school for their hospitality. The author gratefully acknowledges the members of the Super-Kamiokande collaboration for useful discussions and information. This work was partly supported by the Japanese Ministry of Education, Culture, Sports, Science and Technology and by the Japan Society for the Promotion of Science.

References

Achar, C. V. et al. (1965), *Phys. Lett.* **18**, 196.

Adamson, P. et al. (2006), *Phys. Rev.* **D73**, 072002.

Aglietta, M. et al. (1989), *Europhys. Lett.* **8**, 611–614.

Alcaraz, J. et al. (2000), *Phys. Lett.* **B490**, 27–35.

Allison, W. W. M. et al. (1999), *Phys. Lett.* **B449**, 137–144.

Allison, W. W. M. et al. (2005), *Phys. Rev.* **D72**, 052005.

Ambrosio, M. et al. (2001), *Phys. Lett.* **B517**, 59–66.

Ambrosio, M. et al. (2004), *Eur. Phys. J.* **C36**, 323–339.

Ashie, Y. et al. (2004), *Phys. Rev. Lett.* **93**, 101801.

Ashie, Y. et al. (2005), *Phys. Rev.* **D71**, 112005.

Barger, V. D. et al. (1999), *Phys. Lett.* **B462**, 109–114.

Barr, G. D. et al. (2004), *Phys. Rev.* **D70**, 023006.

Battistoni, G. et al. (2000), *Astropart. Phys.* **12**, 315–333.

Battistoni, G. et al. (2003), *hep-ph/0305208* .

Becker-Szendy, R. et al. (1992), *Phys. Rev.* **D46**, 3720–3724.

Berger, C. et al. (1990), *Phys. Lett.* **B245**, 305–310.

Blucher, E. (2006), *in these Lecture Notes* .

Casper, D. et al. (1991), *Phys. Rev. Lett.* **66**, 2561–2564.

Clark, R. et al. (1997), *Phys. Rev. Lett.* **79**, 345–348.

Crouch, M. F. et al. (1978), *Phys. Rev.* **D18**, 2239–2252.

Fogli, G. L., Lisi, E. & Marrone, A. (2001), *Phys. Rev.* **D63**, 053008.

Fukuda, S. et al. (2000), *Phys. Rev. Lett.* **85**, 3999–4003.

Fukuda, Y. et al. (1994), *Phys. Lett.* **B335**, 237–245.

Fukuda, Y. et al. (1998), *Phys. Rev. Lett.* **81**, 1562–1567.

Gaisser, T. K. & Honda, M. (2002), *Ann. Rev. Nucl. Part. Sci.* **52**, 153–199.

Harris, D. (2006), *in these Lecture Notes* .

Hatakeyama, S. et al. (1998), *Phys. Rev. Lett.* **81**, 2016–2019.

Hirata, K. S. et al. (1988), *Phys. Lett.* **B205**, 416.

Hirata, K. S. et al. (1992), *Phys. Lett.* **B280**, 146–152.

Honda, M. et al. (2004), *Phys. Rev.* **D70**, 043008.

Honda, M. et al. (2006), *astro-ph/0611418* .

Hosaka, J. et al. (2006), *Phys. Rev.* **D74**, 032002.

Indumathi, D. & Murthy, M. V. N. (2005), *Phys. Rev.* **D71**, 013001.

Kajita, T. et al. (2004). Proc.of the 5th Workshop on Neutrino Oscillations and their Origin (NOONE2004), Tokyo, Japan, 11-15 Feb 2004.

Kayser, B. (2006), *in these Lecture Notes* .

Krishnaswamy, M. R. et al. (1971), *Proc. Roy. Soc. Lond. A* **323**, 489.

Lipari, P. (2000), *Astropart. Phys.* **14**, 153–170.

Lipari, P. & Lusignoli, M. (1998), *Phys. Rev.* **D58**, 073005.

Lisi, E., Marrone, A. & Montanino, D. (2000), *Phys. Rev. Lett.* **85**, 1166–1169.

Liu, Q. Y., Mikheyev, S. P. & Smirnov, A. Y. (1998), *Phys. Lett.* **B440**, 319–326.

Mann, W. A., Kafka, T. & Leeson, W. (1992), *Phys. Lett.* **B291**, 200–205.

McFarland, K. (2006), *in these Lecture Notes* .

Peres, O. L. G. & Smirnov, A. Y. (1999), *Phys. Lett.* **B456**, 204–213.

Peres, O. L. G. & Smirnov, A. Y. (2004), *Nucl. Phys.* **B680**, 479–509.

Reines, F. et al. (1965), *Phys. Rev. Lett.* **15**, 429–433.

Sanchez, M. C. et al. (2003), *Phys. Rev.* **D68**, 113004.

Sanuki, T. et al. (2000), *Astrophys. J.* **545**, 1135.

Sanuki, T. et al. (2006), *astro-ph/0611201* .

Wark, D. (2006), *in these Lecture Notes* .

Neutrino Experiments with Reactors

Edwin Blucher

The University of Chicago, USA

1 Introduction

Nuclear reactors have played a critical role in exploring the properties of neutrinos, from the first direct observation of the neutrino in 1956 to current neutrino oscillation experiments. In these lectures, first we will review some properties of reactors as antineutrino sources and the techniques used to detect antineutrinos in experiments. Next, we will discuss the Reines-Cowan experiments , which illustrate many issues relevant for all reactor neutrino experiments. The major part of the lectures is devoted to the use of reactor neutrinos as a tool to investigate neutrino oscillations. Additional details on many topics presented here can be found in the excellent review article by Bemporad, Gratta, and Vogel (Bemporad, Gratta & Vogel 2002).

2 Reactors as Antineutrino Sources

Nuclear reactors are high intensity, isotropic sources of $\overline{\nu}_e$. The antineutrinos are the result of the β^- decay of neutron rich fragments of fission of uranium and plutonium. For example, consider the fission of ^{235}U:

$$^{235}_{92}U + n \rightarrow X_1 + X_2 + 2n. \tag{1}$$

The stable nuclei with the most probable atomic numbers from the fission of ^{235}U are $^{94}_{40}Zr$ and $^{140}_{58}Ce$. Together, these nuclei have 98 protons and 136 neutrons, while the fission fragments (X_1 and X_2) have 92 protons and 142 neutrons. Therefore, on average, 6 neutrons must decay to 6 protons to reach stable matter. Since each fission releases about 200 MeV of energy, the average of 6 $\overline{\nu}_e$ per fission implies that a typical 3 GW$_{thermal}$ power reeactor produces about 6×10^{20} $\overline{\nu}_e/$ sec.

A more careful calculation of the antineutrino yield from a reactor must consider the other nuclei present in the reactor core. More than 99.9% of the antineutrinos result from four nuclei: ^{235}U, ^{238}U, ^{239}Pu, and ^{241}Pu. Direct measurements of the electron spectrum from thin layers of ^{235}U, ^{239}Pu, and ^{241}Pu have been done using a beam of thermal neutrons (Schreckenbach,

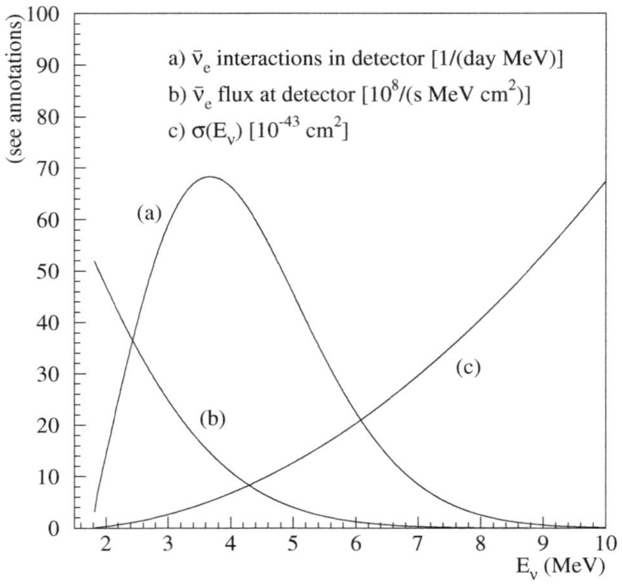

Figure 1. *Energy spectrum of $\overline{\nu}_e$ from a reactor, the inverse beta decay cross section, and the observed spectrum of detected IBD events (Bemporad et al. 2002). Copyright (2002) American Physical Society.*

Colvin, Gelletly & Von Feilitzsch 1985, Hahn et al. 1989). These measurements are then converted to neutrino spectra. For ^{238}U, which only undergoes fast neutron fission, no measurements are available and a calculation is used. The resulting uncertainty in the total antineutrino flux is 2 to 3%. Short baseline reactor neutrino experiments (Declais et al. 1995, Zacek et al. 1986), have been used to check these predictions.

A few features of nuclear power plants are relevant to the experiments we will discuss later. Each reactor core is an extended, cylindrical neutrino source, about 3 m in diameter and 4 m high. Commercial reactors typically shut down for 1 month every 12-18 months to replace 1/3 of the fuel assemblies and reposition the remaining assemblies. Plutonium breeding over each 12-18 month fuel cycle ($\sim$ 250 kg per fuel cycle) changes the antineutrino energy spectrum and reduces the antineutrino rate by 5-10%.

3 Antineutrino Detection

Almost all reactor experiments detect anti-electron neutrinos by inverse beta decay (IBD): $\overline{\nu}_e + p \rightarrow e^+ + n$. Figure 1 shows the energy spectrum of $\overline{\nu}_e$ from a reactor, the IBD cross section as a function of energy, and the observed energy spectrum for IBD events. The observed IBD spectrum peaks around 3.6 MeV. The IBD reaction can occur only for $\overline{\nu}_e$ with energy greater than $\sim$ 1.804 MeV ($E_{threshold} \sim m_n + m_{e+} - m_p = 1.804$ MeV), so only about 1.5 of the $\sim$ 6 $\overline{\nu}_e$/fission can be detected. A rough rule of thumb is that one IBD event will be detected per ton of liquid scintillator (the typical target material) per GW thermal of reactor power at a distance of 1 km.

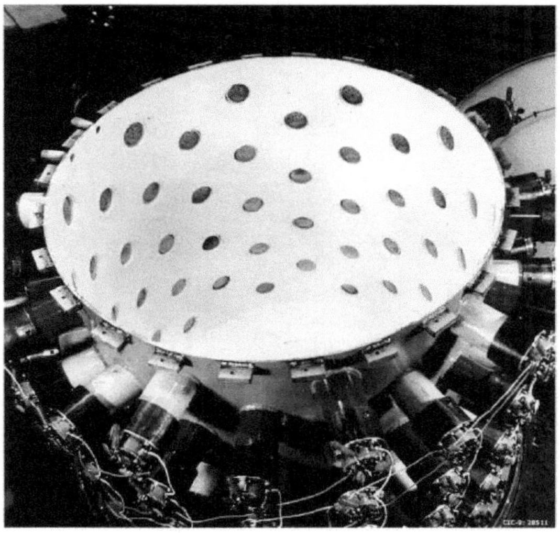

Figure 2. *Photograph of the detector used for the 1953 Hanford Experiment (Arms 2001). Copyright 2001 Springer.*

Experiments detect the coincidence between a prompt signal from the positron and a delayed signal from neutron capture on hydrogen (or cadmium, gadolinium, etc.). The mean time for capture on hydrogen is about 200 μs and results in a 2.2 MeV γ: $n+p \rightarrow d+\gamma(2.2MeV)$. As will be discussed, neutron capture on cadmium (Cd) or gadolinium (Gd) results in a higher energy signal and shorter capture time. The antineutrino energy can be inferred from the positron energy deposit. Applying energy conservation to the IBD reaction:

$$E_{\overline{\nu}} \simeq E_{e^+} + E_n + (M_n - M_p) + M_{e^+}. \tag{2}$$

The neutron energy is very small (10-40 keV) and $M_n - M_p + M_{e^+} = 1.8$ MeV. Therefore, including the 1022 keV from the e^+ annihilation, $E_{prompt} \simeq E_{\overline{\nu}} - 0.8$ MeV.

4 The Reines-Cowan Experiments

The first detection of the neutrino was reported by Reines and Cowan about 50 years ago using antineutrinos from a nuclear reactor. (An interesting summary of the Reines-Cowan experiments is given by Arms (Arms, R.G. 2001).) A first experiment was performed in 1953 at the Hanford Engineering Works in Washington using 300 tons of liquid scintillator loaded with cadmium viewed by ninety 2-inch photomultiplier tubes (PMTs). Aside from its relatively small size, the detector used for this experiment (shown in Figure 2) was very similar to those being built for current experiments at reactors. The antineutrino signal from IBD was the delayed coincidence between a positron and neutron capture on cadmium (releasing about 8 MeV in gammas). The experiment had a very high background ($S/N \sim 1/20$) and was inconclusive: after background subtraction, they found 0.41 ± 0.20 IBD events per minute.

In 1956, Reines and Cowan performed a better experiment at Savannah River. The detector for this experiment is shown in Figure 3. Tanks I, II, and III were filled with liquid scintillator (LS) and instrumented with 5" PMTs. The target tanks, placed between the three LS tanks,

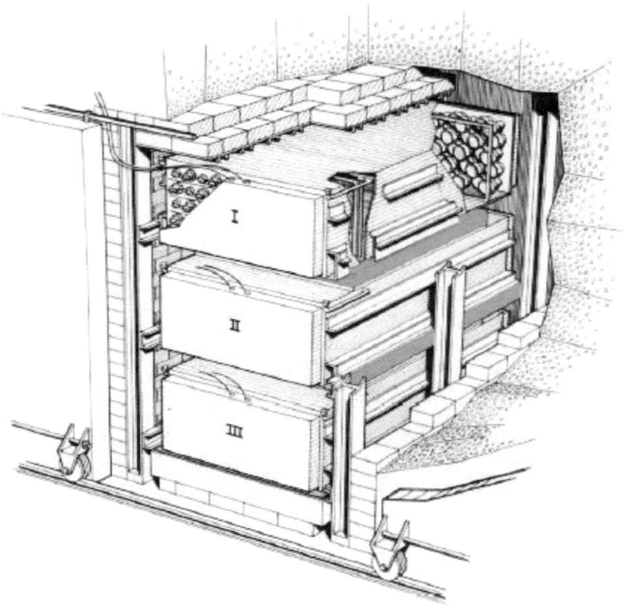

Figure 3. *Diagram of the 1956 Savannah River experiment (Reines et al. 1960). Copyright (1960) American Physical Society.*

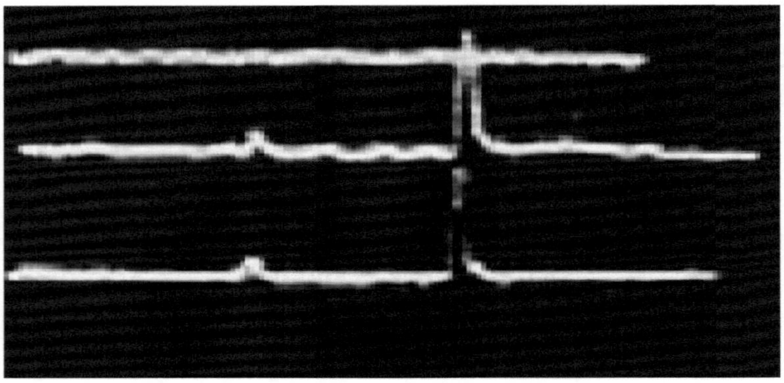

Figure 4. *Example of inverse beta decay signal from Savannah River experiment recorded on an oscilloscope (Reines et al. 1960). Copyright (1960) American Physical Society. The three traces show signals from tanks I, II, and III (see Figure 3).*

were filled with water plus cadmium choloride. Invervse β decay would produce two signals in neighboring tanks (I and II or II and III):

- a prompt signal from e^+ annihilation producing two 0.511 MeV γs (one in each tank);
- a delayed signal from n capture on cadmium producing $\sim$ 8 MeV in γs.

Data were recorded photographically from oscilliscope traces. Figure 4 shows the oscilliscope trace for a candidate IBD event from the target tank between LS tanks II and III.

By April of 1956, a clear reactor-dependent signal had been observed (Reines & Cowan 1956). In contrast to the earlier experiment, the signal to reactor-independent background was

3 to 1. In addition to reporting the observation of a neutrino signal, the group also reported a cross section that was within 5% of the 6.3×10^{-44} cm^2 expected (although the predicted cross section had a 25% uncertainty). In hindsight, the experiment was not well enough understood at the time to make the cross section measurement meaningful. In 1959, following the discovery of parity violation, the theoretical cross section was increased by a factor of 2 to $(10 \pm 1.7) \times 10^{-44}$ cm^2. In 1960, Reines and Cowan reported a reanalysis of the 1956 experiment (Reines et al. 1960) and quoted a cross section of $\sigma = (12^{+7}_{-4}) \times 10^{-44}$ cm^2 that agreed with the revised theoretical prediction.

5 Neutrino Oscillation Experiments

As has been described in detail in the lectures of Kayser (Kayser 2007), oscillations between different flavors of neutrinos have been established with two distinct mass differences. The observations are elegantly described by a picture in which the neutrino flavor eigenstates (ν_e, ν_μ, and ν_τ) are mixtures of mass eigenstates (ν_1, ν_2, and ν_3), analogous to mixing in the quark sector. This mixing is described by a 3×3 unitary matrix:

$$\begin{pmatrix} U_{e1} & U_{e2} & U_{e3} \\ U_{\mu 1} & U_{\mu 2} & U_{\mu 3} \\ U_{\tau 1} & U_{\tau 2} & U_{\tau 3} \end{pmatrix} = \begin{pmatrix} \cos\theta_{12} & \sin\theta_{12} & 0 \\ -\sin\theta_{12} & \cos\theta_{12} & 0 \\ 0 & 0 & 1 \end{pmatrix} \begin{pmatrix} \cos\theta_{13} & 0 & e^{-i\delta_{CP}}\sin\theta_{13} \\ 0 & 1 & 0 \\ -e^{-i\delta_{CP}}\sin\theta_{13} & 0 & \cos\theta_{13} \end{pmatrix} \begin{pmatrix} 1 & 0 & 0 \\ 0 & \cos\theta_{23} & \sin\theta_{23} \\ 0 & -\sin\theta_{23} & \cos\theta_{23} \end{pmatrix} \quad (3)$$

As shown, this matrix often is written in terms of 3 Euler rotations (θ_{12}, θ_{13}, and θ_{23}) and a CP violating phase (δ_{CP}).

Antineutrinos from reactors can be used to study neutrino oscillations with the "solar" $\Delta m^2_{12} \sim 8 \times 10^{-5}$ eV^2 and the "atmospheric" $\Delta m^2_{13} \sim 2.5 \times 10^{-3}$ eV^2 ($\Delta m^2_{ij} \equiv m^2_i - m^2_j$). The mean energy of detected antineutrinos from a reactor is 3.6 MeV, so only disappearance experiments are possible (i.e., there is not enough energy to produce a muon or tau):

$$P(\overline{\nu}_e \to \overline{\nu}_e) \simeq 1 - \sin^2 2\theta_{13} \sin^2 \frac{\Delta m^2_{13} L}{4E} - \cos^4\theta_{13} \sin^2 2\theta_{12} \sin^2 \frac{\Delta m^2_{12} L}{4E}. \quad (4)$$

Experiments search for a deviation from the $1/r^2$ fall-off of the antineutrino rate. Using the mean antineutrino energy of 3.6 MeV, we can calculate the baseline (L) for the oscillation maxima corresponding to the solar and atmospheric Δm^2 values:

$$\Delta m^2_{12} \sim 8 \times 10^{-5} \quad eV^2: \quad L \sim 60 \text{ km} \quad (5)$$
$$\Delta m^2_{13} \sim 2.5 \times 10^{-3} \quad eV^2: \quad L \sim 1.8 \text{ km}. \quad (6)$$

In addition to a simple counting analysis, in which one compares the detected neutrino flux to that expected without oscillations, the observed neutrino energy spectrum, inferred from the

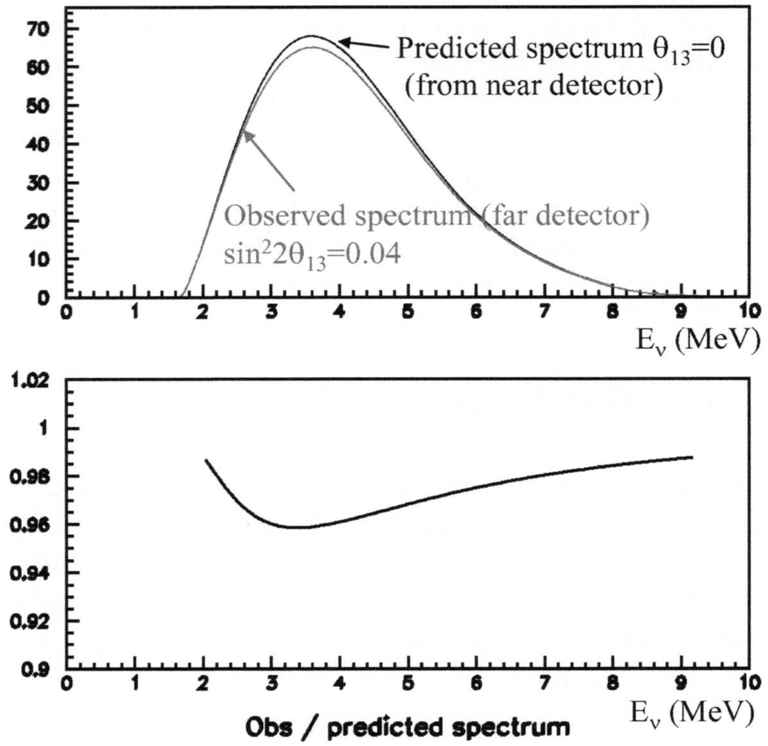

Figure 5. *(top)Energy spectrum of* $\overline{\nu}_e$ *without oscillations and with* $\sin^2 2\theta_{13} = 0.04$ *at a distance of 1.5 km from a reactor. (bottom) ratio of spectrum with oscillations to spectrum without oscillations.*

positron energy spectrum, also contains information about neutrino oscillations. Figure 5 illustrates this effect, which results from the energy dependence of the oscillation probability. For most experiments, the counting and energy spectrum analyses are complementary because they are sensitive to different systematic uncertainties. The counting analysis is directly sensitive to the acceptance of the detector(s), but is relatively insensitive to energy calibration. The energy spectrum analysis, on the other hand, is sensitive to the energy scale and linearity, but is relatively insensitive to overall detector acceptance.

Several issues are common to all reactor-based neutrino oscillation experiments.

- *Knowledge of the antineutrino flux and spectrum.* In a single detector experiment, these quantities must be calculated from the reactor's thermal power and detailed information about the reactor's fuel composition. In a two-detector experiment, a detector close to the reactor can be used to measure the unoscillated neutrino flux and spectrum directly.

- *Detector acceptance.* For a given flux of neutrinos, one must determine how many inverse beta decay events will be detected by the experiment. The acceptance calculation depends on the mass and chemical composition of the target, in addition to the effects of all event selection requirements.

- *Backgrounds.* Understanding and subtracting background events is a key requirement for reactor experiments. The backgrounds that mimic the inverse beta decay signal may be separated into two categories: uncorrelated and correlated. Uncorrelated background events result from random coincidences between two different events, that together mimic the IBD signal. For example, a coincidence between a radioactive decay and a neutron produced by a cosmic ray muon could produce a fake signal event. Fortunately, the level of these events can be measured easily by changing the required time coincidence window (*e.g.* by exchanging the required sequence for the positron-like and neutron-like signals).

Correlated backgrounds, in which both the positron and neutron-like signals result from the same parent particle, are more problematic. Three important correlated backgrounds are:

1. *Fast neutron.* Fast neutrons produced by cosmic ray muons can fake the IBD signal by elastically scattering and subsequently capturing on Gd. The recoil protons from the elastic scatter can mimic the prompt positron signal from inverse beta decay, followed by the delayed n+Gd capture signal.
2. *Muon capture.* Muon capture, particularly on nuclei in dead material along the muon track, can produce fast neutrons that mimic the IBD signal as described above.
3. ^{9}*Li.* Spallation from cosmic ray muons can create ^{9}Li inside the detector. About 50% of ^{9}Li β decays produce a neutron that can capture on Gd, which along with the prompt signal from the β mimics the IBD signal. The 178 ms half life makes these events difficult to veto without producing unacceptable dead time.

5.1 Solar Δm^2 Experiments

The $\mathcal{O}(100)$ km baseline needed to be sensitive to the solar Δm^2 requires a very large neutrino source, a very large detector, and a very deep site (to make the signal-to-background acceptable). The layout of nuclear power plants on the coast of Japan (see Figure 6)and the existing excavation from the Kamiokande experiment offered an ideal site to perform such an experiment, called KamLAND. Figure 6(right) shows that only a limited number of baselines contribute to the flux of reactor antineutrinos at Kamioka (the distribution is sharply peaked at $L \sim 180$ km), making it an ideal site for an oscillation experiment.

A diagram of the KamLAND detector, located under 2700 mwe shielding, is shown in Figure 7. The neutrino detector/ target consists of 1 kton of liquid scintillator inside a 13 m diameter ballon made of tranparent nylon. The inside of an 18 m diameter stainless steel vessel is tiled with 1879 PMTs (554 20" PMTs and 1325 17" PMTs). The region between the steel vessel and the nylon balloon is filled with mineral oil to shield the liquid scintillator from external radiation (e.g., from phototube glass). The steel containment vessel is surrounded by a water-Cerenkov detector to absorb γ rays and neutrons from the surrounding rock, and to tag cosmic-ray muons.

The antinuetrino signature is the coincidence between a prompt e^+ signal and a delayed 2.2 MeV signal from the neutron capture on hydrogen. Figure 8 shows a plot of delayed energy versus prompt energy; the accepted energy range of 1.8 MeV $< E_{delay} <$ 2.6 MeV for the delayed signal is indicated.

The first results from KamLAND were reported in 2003 (Eguchi et al. 2003). Based on data

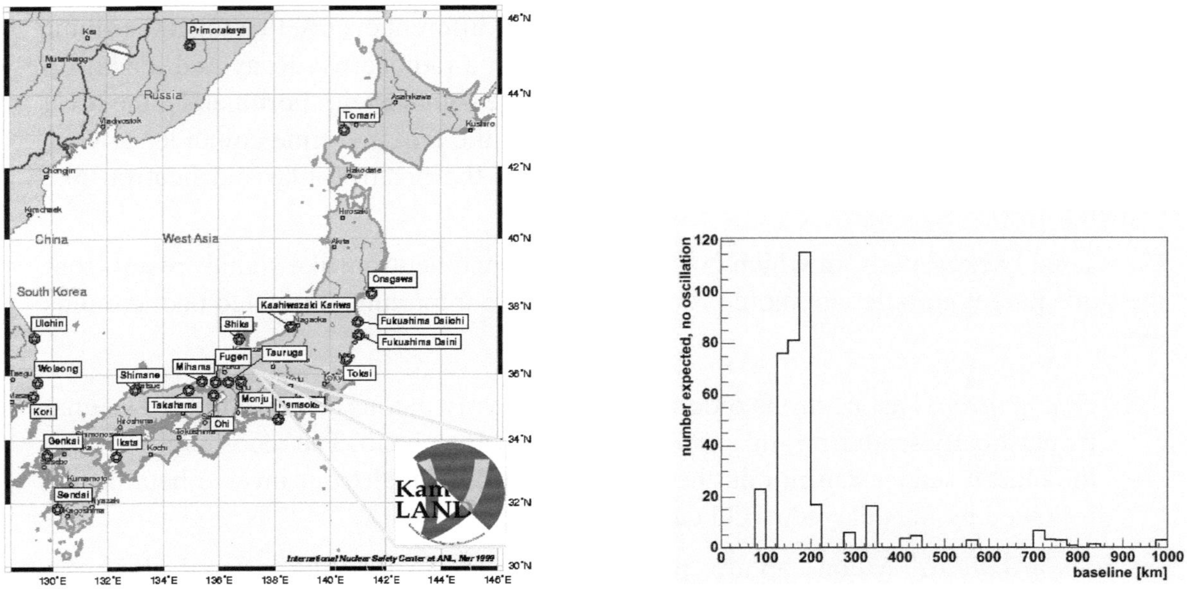

Figure 6. *(left) Location of the KamLAND experiment and large nuclear power plants in Japan, Korea, and Far East Russia. (right) Expected baseline distribution for detected IBD events with no oscillations.*

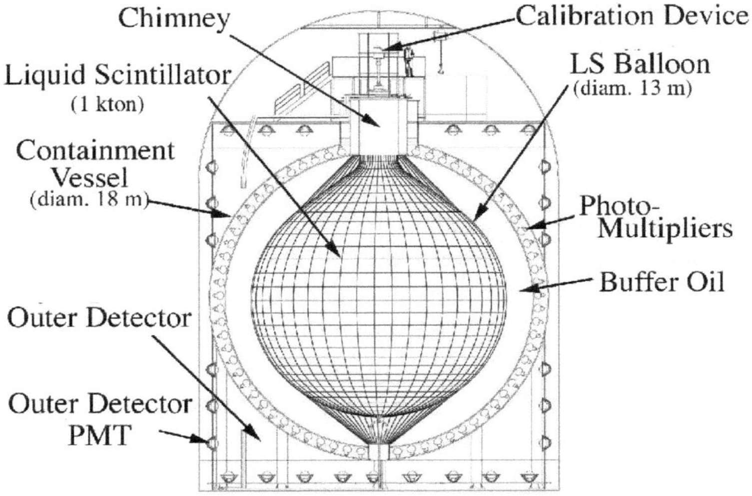

Figure 7. *Schematic diagram of the KamLAND detector (Eguchi et al. 2003). Copyright (2003) American Physical Society.*

collected from March to October of 2002, the KamLAND experiment observed 54 inverse beta decay events with an estimated background of 1 ± 1 events. The expected number of events without neutrino oscillations was 86.8 ± 5.6 events. Combining these numbers gives:

$$\frac{N_{obs} - N_{bckg}}{N_{expected}} = 0.611 \pm 0.085(stat) \pm 0.041(syst), \tag{7}$$

inconsistent with a $1/R^2$ flux dependence at 99.95% confidence level. Figure 9 (Eguchi et al. 2003) shows this result with the large mixing angle prediction for neutrino oscillations. The

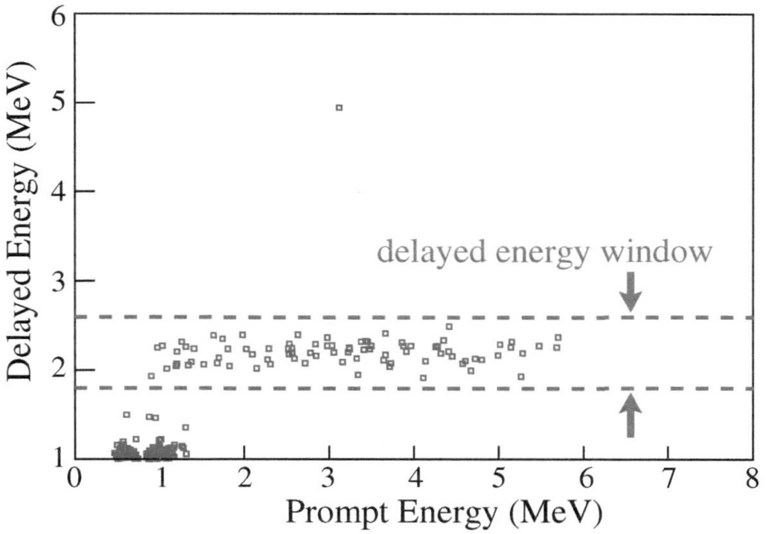

Figure 8. *Delayed energy versus prompt energy for the KamLAND experiment. The band shows the accepted range of delayed energies from neutron capture on hydrogen. The single point with a delayed energy of 5 MeV is from neutron capture on ^{12}C (Eguchi et al. 2003). Copyright (2003) American Physical Society.*

prompt energy spectrum for this data sample was consistent with oscillations at the 93% c.l., but was also consistent with the no oscillation shape at 53% confidence level.

In 2005, after including data taken through January 2004, the group reported clear evidence for a spectral distortion from neutrino oscillations (Araki et al. 2005). Figure 10 shows the ratio of the observed $\overline{\nu}_e$ spectrum to the expectation with no oscillation versus L/E. The data and predictions are plotted with L = 180 km. The oscillatory behavior of the data is clear. Including both spectral and rate information, the group reported $\Delta m^2 = (7.9^{+0.6}_{-0.5}) \times 10^{-5}$ eV^2 and $tan^2\theta = 0.40^{+0.10}_{-0.07}$.

5.2 Atmospheric Δm^2 Experiments

Several new reactor neutrino experiments are focused on the study of the atmospheric Δm^2 region, which is sensitive to the value of θ_{13}, the last unmeasured mixing angle in the neutrino mixing matrix. The value of this angle is central to several important questions in neutrino physics, including:

- What is the mass hierarchy (i.e., what is the sign of $m_1^2 - m_3^2$)?
- Do neutrino oscillations violate CP symmetry?
- Why are the quark and neutrino mixing matrices so different?

The value of θ_{13} sets the scale for experiments needed to resolve the mass hierarchy and to search for CP violation; if θ_{13} is zero, there is no possibility of CP violation in neutrino oscillations. In addition, the size of θ_{13} with respect to the other mixing angles may give insights into the orgin of these angles and the source of neutrino mass.

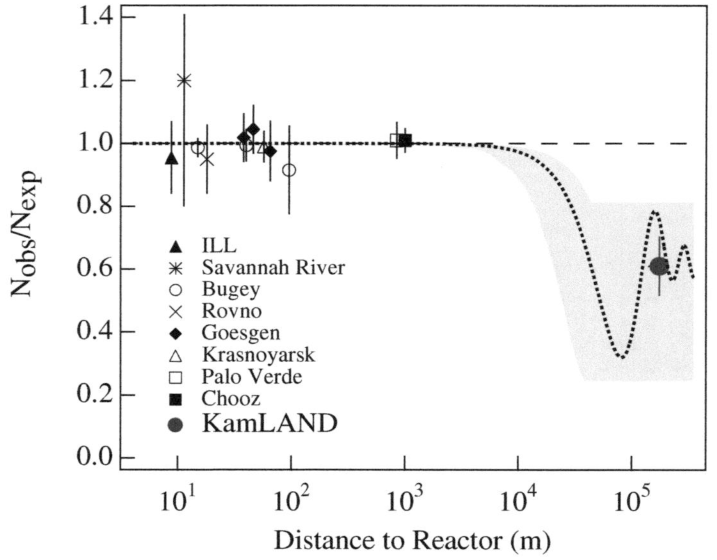

Figure 9. *The ratio of measured to expected $\overline{\nu}_e$ flux from reactor experiments. The solid circle is the KamLAND result plotted at a flux-weighted average distance of ~ 180 km. The shaded region indicates the range of flux predictions corresponding to the 95% C.L. LMA region from a global analysis of the solar neutrino data. The dotted curve is representative of a best-fit LMA prediction and the dashed curve is expected for no oscillations (Eguchi et al. 2003). Copyright (2003) American Physical Society.*

Accelerator-based and reactor-based experiments now under construction will search for this mixing angle with an order-of-magnitude better precision than previous experiments. The accelerator and reactor experiments are complementary. Accelerator experiments look for the appearance of electron neutrinos in a muon neutrino beam:

$$P(\nu_\mu \rightarrow \nu_e) = \sin^2\theta_{23}\sin^2 2\theta_{13}\sin^2\frac{\Delta m^2_{13}L}{4E} + \text{not small terms}(\delta_{CP}, sign(\Delta m^2_{13})). \quad (8)$$

The observed oscillation probability is sensitive to θ_{13}, but is also sensitive to the CP phase and the mass hierarchy. The NOνA and T2K experiments, discussed in the lectures of Harris (Harris 2007), are pursuing this approach.

As discussed earlier, reactor experiments search for the disappearance of anti-electron neutrinos. At distances appropriate to the atmospheric Δm^2, Equation 4 becomes

$$P(\overline{\nu}_e \rightarrow \overline{\nu}_e) \simeq 1 - \sin^2 2\theta_{13}\sin^2\frac{\Delta m^2_{13}L}{4E} + \text{very small terms}. \quad (9)$$

This oscillation would appear as a short distance scale oscillation superimposed on the curve in Figure 9. The oscillation probability in a reactor experiment allows a clean measurement of $\sin^2 2\theta_{13}$. It does not depend on matter effects or CP violation, and contains almost no correlation with other parameters.

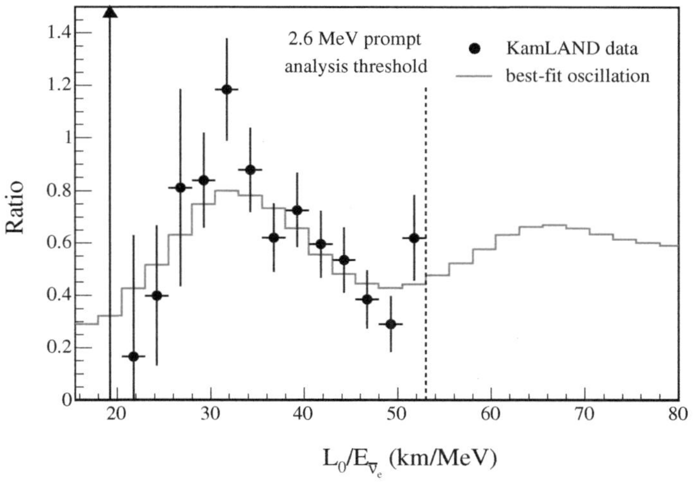

Figure 10. *Ratio of the observed $\overline{\nu_e}$ spectrum to the expectation with no oscillation versus L/E. The data and predictions are ploted with $L_0 = 180km$. The curve shows the expectation for the best-fit oscillation parameters (Araki et al. 2005). Copyright (2005) American Physical Society.*

5.2.1 The Chooz Experiment

The best current limit on $\sin^2 2\theta_{13}$ comes from the Chooz experiment (Apollonio et al. 2003), shown schematically in Figure 11. The Chooz experiment uses a single detector with a 5 ton target of Gd-loaded liquid scintillator (LS) at a distance of 1.05 km from two 4.2 GW_{th} reactors. As in other reactor experiments, the neutrinos are detected through inverse β decay: $\overline{\nu_e} + p \rightarrow e^+ + n$. The experiment detects the prompt signal from the positron and the delayed signal from neutron capture on Gadolinium (Gd), which gives about 8 MeV in 4 or 5 gammas:

$$n +^m Gd \rightarrow^{m+1} Gd^* \rightarrow^{m+1} Gd + \gamma s(8 \text{ MeV}). \tag{10}$$

The Gd serves the same purpose as the cadmium in the Reines-Cowan experiments: it shortens the neutron capture time (to 30μsec from 200μsec for capture on hydrogen), reducing backgrounds, and results in a higher energy signal from the neutron capture ($\sim$ 8 MeV instead of 2.2 MeV from capture on H). Table 1 shows the cross sections and natural isotopic abundances for the two relevant isotopes of Gd. The cross section for neutron capture on Gd is so much higher than that for hydrogen, that even though the Gd-loaded scintillator is only 0.1% Gd by weight, 85% of the neutrons capture on Gd instead of hydrogen. All of the currently proposed experiments add a small amount of Gd to the LS to improve neutron detection.

The Chooz experiment collected data between April 1997 and July 1998. They began data collection before the reactor began to operate, and therefore had the unique opportunity to measure backgrounds directly. Future experiments will not have this possibility. The experiment collected about 2.2 IBD events per day per ton with 0.2 - 0.4 background events per day per ton, depending on trigger thresholds. The total data sample included about 3600 IBD events. Figure 12 shows a plot of neutron-like and positron-like energy for reactor on and reactor off data. Subtracting the properly scaled reactor-off data yields the positron energy distribution shown

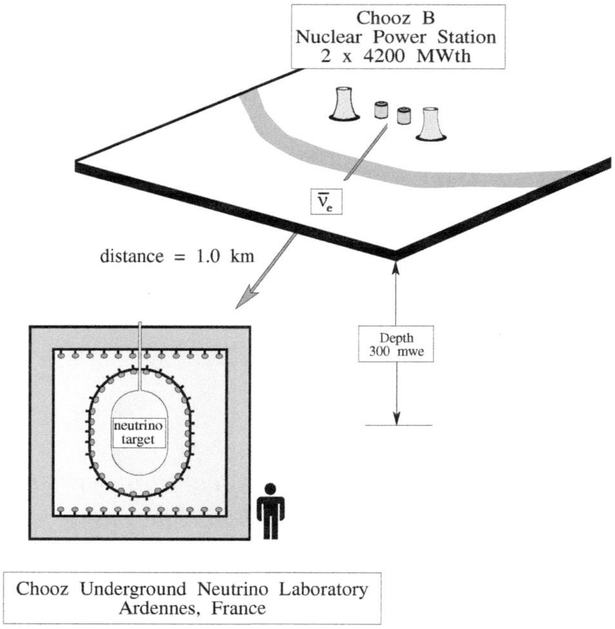

Figure 11. *Overview of Chooz neutrino oscillation experiment (Apollonio et al. 2003). Copyright (2003) Elsevier.*

Element	σ (barns)	Isotopic abundance (%)
^{155}Gd	61,400	14.8
^{157}Gd	255,000	15.7
Gd (natural)	49,100	–
H	0.328	–

Table 1. *Neutron capture cross sections for gadolinium and hydrogen.*

in Figure 13. The ratio of the observed to predicted flux is $1.01 \pm 0.028(\text{stat}) \pm 0.027(\text{syst})$. Table 2 summarizes the two main sources of systematic uncertainty. The Chooz result, together with the result from the Palo Verde experiment (Boehm et al. 2001), is shown in Figure 14; the results also are plotted in Figure 9. For $\Delta m^2 = 2.5 \times 10^{-5}$ eV^2, the Chooz result corresponds to $\sin^2 2\theta_{13} < 0.15$ at 90% confidence level.

5.2.2 How can one improve on the Chooz experiment?

To improve on the sensitivity of the Chooz experiment, both statisitical and systematic errors must be reduced. The possible strategies include the following:

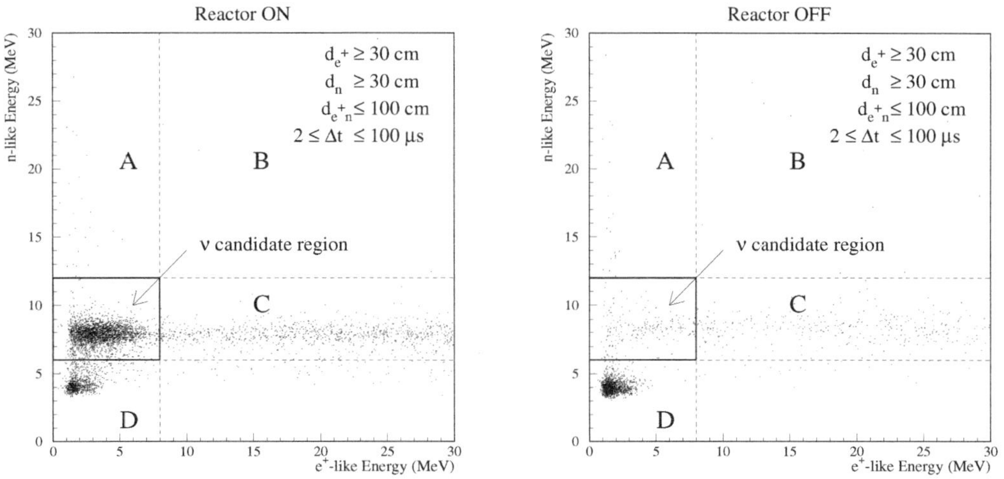

Figure 12. *Chooz distribution for neutron-like versus positron-like energy for IBD candidate events witn reactor on (left) and reactor off (right) (Apollonio et al. 2003). Copyright (2003) Elsevier.*

Reactor ν flux	2%
Detector acceptance	1.5%
Total	2.7%

Table 2. *CHOOZ systematic errors.*

- *Add an identical detector close to the reactor.* This feature is the key improvement in all of the planned experiments. A near detector can be used to measure the reactor flux directly, eliminating the single largest uncertainty in the Chooz result. Also, only the relative acceptance of the near and far detectors is needed, rather than the absolute acceptance of a single far detector.

- *Optimize the baseline with the best information on Δm^2.* At first glance, choosing the optimum baseline seems obvious: site the far detector at the point where the oscillation probability is maximum. After consideration of statistical and systematic errors, however, the choice becomes more complicated. To minimize the statistical error in a rate measurement, one must consider the competition between the $1/R^2$ loss of statistics and the sinusoidal oscillation term. For the energy spectrum measurement, the distortion is different at different baselines (see Figure 15), resulting in a different optimal baseline.

  The choice of baseline also depends on the relative size of the statistical and systematic errors. With no systematic errors, both the rate and shape measurements have the best sensitivity per running time at a distance about 30% less than the oscillation maximum. If systematic errors are dominant, however, the rate measurement is best done close to the oscillation maximum, while the shape measurement is more sensitive at a shorter baseline. Therefore, the optimum baseline depends on the expected statistical and systematic errors. Figure 16 shows an example of the relative importance of rate and spectrum mea-

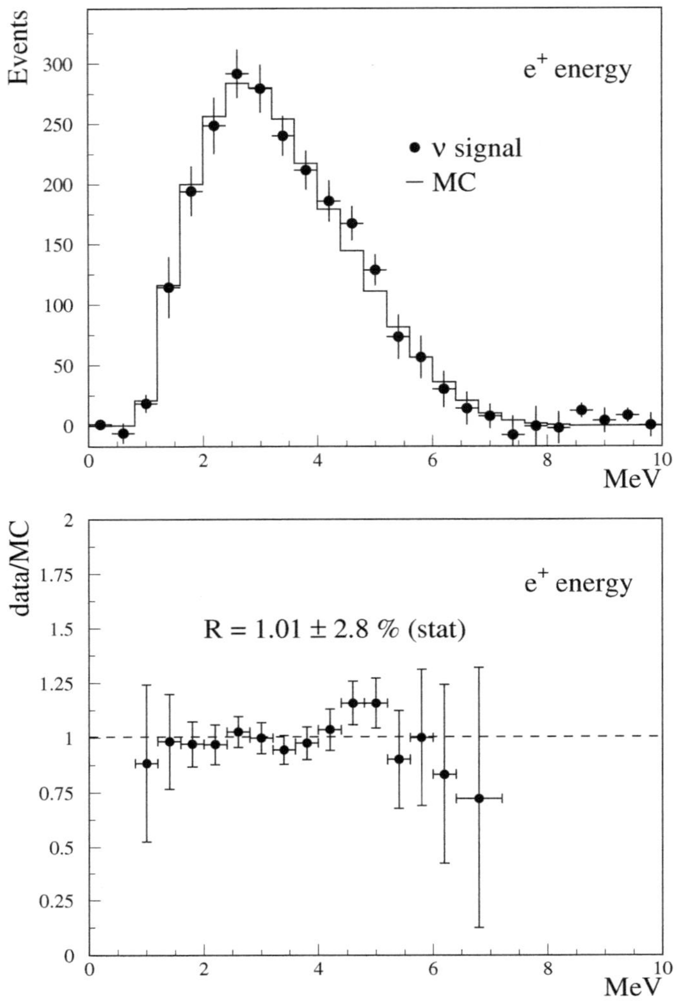

Figure 13. *(top) Chooz reconstructed positron energy spectrum for data and MC simulation assuming no oscillations. (b) Data/MC ratio. The errors shown are statistical only (Apollonio et al. 2003). Copyright (2003) Elsevier.*

surements as a function of "luminosity" (measured in reactor power times target mass).

- *Use larger detectors with improved designs.* To reduce statistical errors, new experiments will increase the target mass compared to previous experiments, and some will use higher power reactor installations. It is interesting to consider the most efficient way to increase the target mass. In particular, is it better to have many small detectors or fewer large detectors? Most acceptance systematic uncertainties are related to the surface area of the boundary between the Gd-loaded scintillator region and the surrounding regions. Larger, spherical detectors minimize the surface area to volume ratio, simplify reconstruction, and make it possible to study the radial dependence of signal and background. Of course, a larger number of small detectors offers the possibility of consistency checks, and reduced uncertainties if systematic errors in the different detectors are independent.

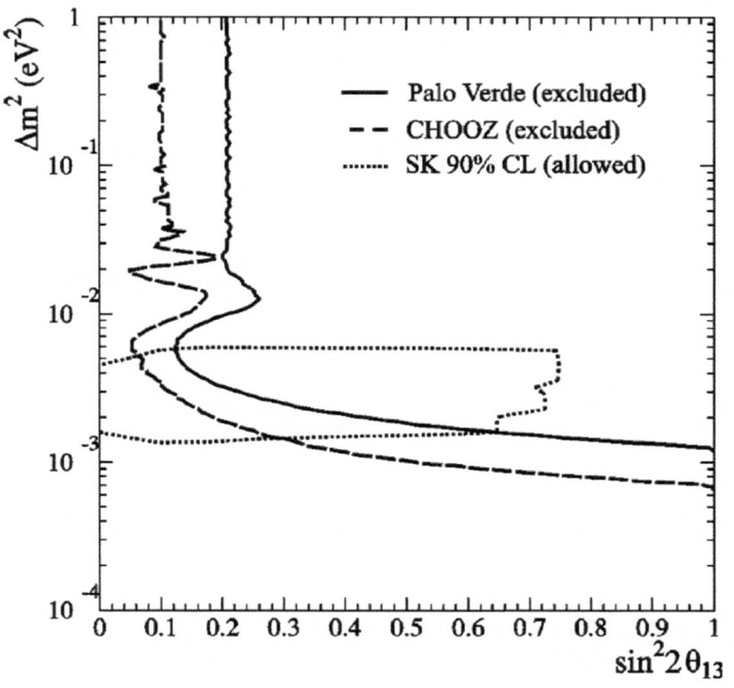

Figure 14. *Excluded regions for* Δm^2 *and* $\sin^2 2\theta_{13}$ *from Chooz and Palo Verde (to the right of the curves) and allowed region from Super-Kamiokande experiment (inside the dotted curve).*

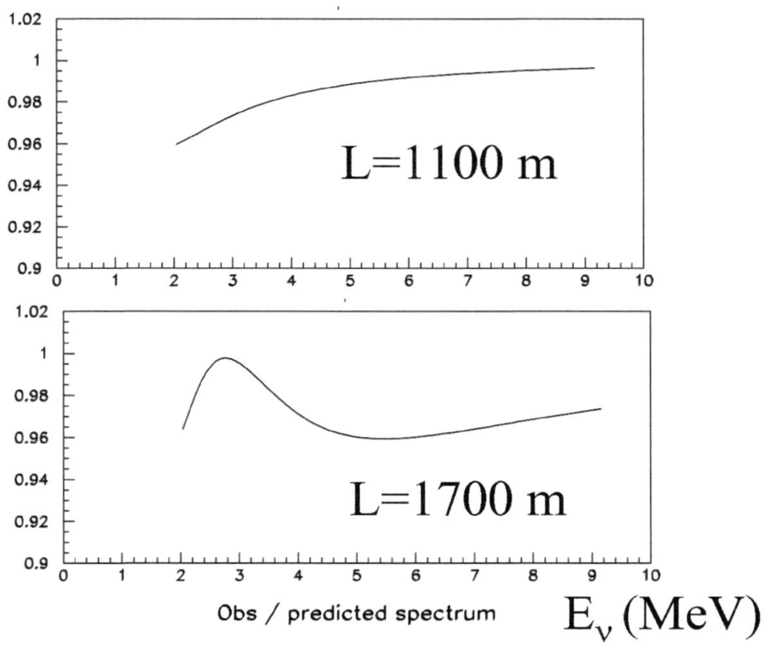

Figure 15. *Ratio of oscillated to unoscillated neutrino energy spectrum for two different baselines.*

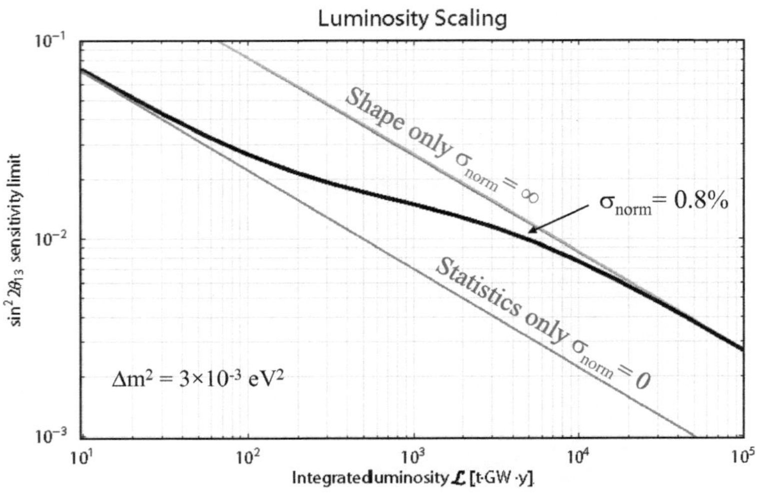

Figure 16. *The sensitivity to* $\sin^2 2\theta_{13}$ *as a function of the integrated luminosity for different values of the uncertainty in detector acceptance* (σ_{norm}) *(Huber et al. 2003)*

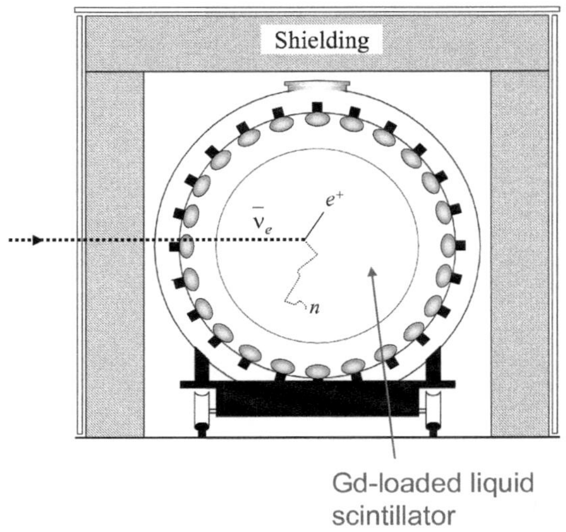

Figure 17. *Sketch of detector for* θ_{13} *measurement.*

For future experiments, the detectors and analysis strategy also will be designed to minimize relative acceptance differences between near and far detectors, reducing the second component of the systematic error in Table 2. All detectors will use a neutrino target of Gd-loaded LS, surrounded by a buffer region and phototubes, as shown schematically in Figure 17. To reduce systematic errors, the event selection will be based only on the coincidence of a positron signal, with $E >\sim 0.5$ MeV, and the gammas released from neutron capture on Gd, with $E >\sim 6$ MeV. Figure 18 shows a simulated distribution of the positron and neutron capture energy. There will be no explicit requirement on the reconstructed event position – the energy requirements will define the fiducial volume. It will be critical that the Gd-LS used in near and far detectors be the same, and that the volume of scintillator in each detector be known precisely.

- *Reduce backgrounds.* As discussed earlier, all of the background sources in reactor exper-

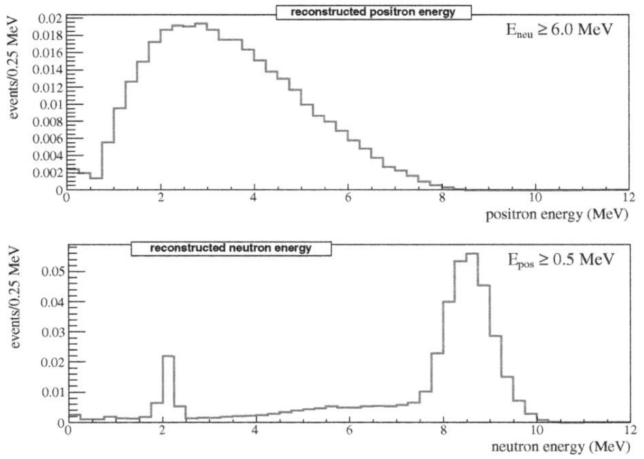

Figure 18. *Simulated reconstructed energy spectra for positrons (top) and neutrons (bottom). In the bottom plot, the peak at 2 MeV is from neutron capture on hydrogen and the peak near 8 MeV is from neutron capture on Gd.*

iments are associated with cosmic ray muons (even accidental backgrounds usually result from a coincidence between a radioactive decay and a muon-induced neutron). Therefore, experiments can reduce backgrounds using greater depth, shielding, and active veto systems. Note that since no future experiment is likely to have enough reactor-off time to measure backgrounds directly, treatment of backgrounds will be a greater challenge than for the original Chooz experiment.

5.2.3 Future experiments to measure θ_{13}

Currently, the Double Chooz experiment in France and the Daya Bay experiment in China are under construction. Plans also exist for reactor experiments in South Korea (RENO) and Brazil (ANGRA). Here, we will briefly summarize some general features of the Double Chooz (Ardellier et al. 2006) and Daya Bay (Guo et al. 2007) experiments.

Both the Double Chooz and Daya Bay experiments have chosen to use cylindrical detectors with a "3-zone" detector design (see Figure 19). Cylindrical detectors were chosen to simplify construction compared to "ideal" spherical detectors. For Double Chooz, the $\overline{\nu}_e$ target is contained in a central acrylic cylinder and consists of approximately 8.8 tonnes of LS loaded with 0.1% Gd. The region between the first cylinder and the next larger acrylic cylinder is filled with liquid scintillator without Gd. This so-called "gamma catcher" region allows detection of gammas resulting from neutron capture near the target boundary, reducing the low energy tail on the neutron capture energy distribution (see Figure 18). Finally, the third region is filled with mineral oil to shield the scintillator regions from backgrounds such as radioactivity of the PMT glass. The inner surface of this third region is instrumented with PMTs. The Double Chooz design employs a fourth cylindrical detector, outside the steel containment vessel, to veto cosmic ray muons.

The Daya Bay detector design also includes three regions with the same liquids and functions as Double Chooz. Each Daya Bay detector is larger, with a target mass of 20 tons, and phototubes are used only on the sides of the cylinder, rather than on all surfaces as in the Double

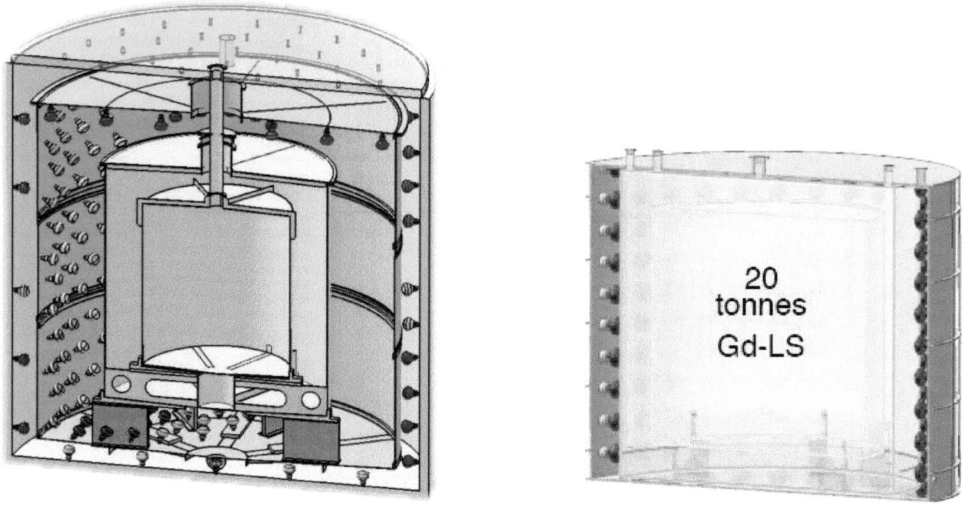

Figure 19. *(left) Diagram of the Double Chooz detectors. (right) Diagram of the Daya Bay detectors.*

Chooz design.

The layouts of the two experiments are shown in Figure 20. The Double Chooz experiment will use two "identical" detectors: a far detector will be installed in the old Chooz site, and a new experimental hall will be excavated for a near detector. The experiment will collect data for the first 18 months with only the far detector. The sensitivity as a function of running time is illustrated in Figure 21. For $\Delta m^2 \approx 2.5 \times 10^{-3}\text{eV}^2$, Double Chooz will be sensitive to $\sin^2 2\theta_{13} \sim 0.06$ after 1.5 years of data taking (beginning in 2008) with only the far detector, and to $\sin^2 2\theta_{13} \sim 0.03$ or better after 3 years of operation with two detectors.

As shown in Figure 20, the Daya Bay reactor complex has two clusters of reactors. Therefore, the Daya Bay experiment will use two different near sites to monitor the reactor power in addition to a far site. In the final configuration, there will be two detectors at each near site and 4 detectors at the far site. Figure 22 shows the projected sensitivity of the Daya Bay experiment. The greater detector mass and reactor power compared to the Double Chooz experiment will allow Daya Bay to reach a sensitivity of better than $\sin^2 \theta_{13} = 0.01$.

6 Conclusions

Reactors experiments have played an important role in investigating the properties of the neutrino. The worldwide program to understand neutrino oscillations and determine the mixing parameters, CP violating effects, and the mass hierarchy will require a broad range of measurements. A reactor experiment to measure θ_{13} will be a key part of this program.

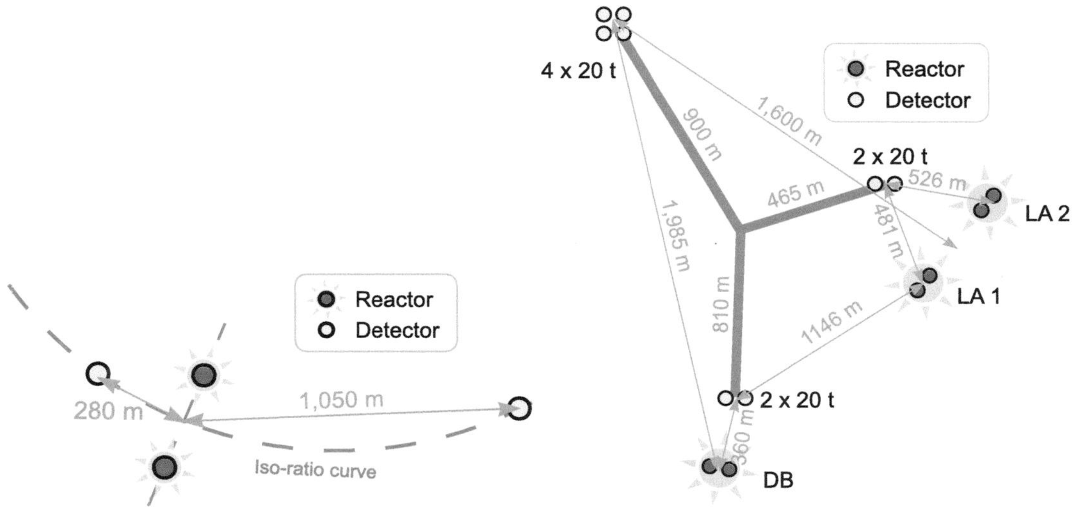

Figure 20. *(left) Layout of the Double Chooz experiment. (right) Layout of the Daya Bay experiment. Both figures are taken from (Mention et al. 2007).*

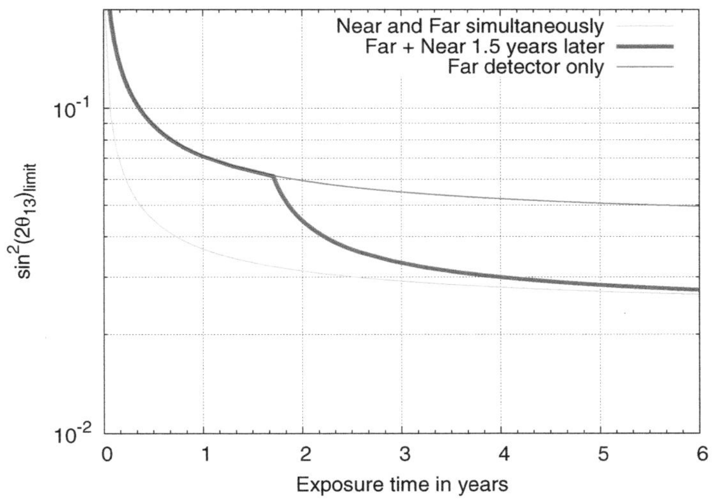

Figure 21. $\sin^2 2\theta_{13}$ *sensitivity for Double Chooz experiment as a function of time. The experiment will run for the first 18 months with only the far detector.*

Acknowledgments

I wish to thank the organizers of the school for arranging such a pleasant and interesting couple of weeks. I appreciate the comments and advice of Peter Fisher, Karsten Heeger, Jon Link, and Mike Shaevitz during the prepartion of these lectures.

References

Apollonio, M. et al. (2003), 'Search for neutrino oscillations on a long base-line at the Chooz nuclear power station', *Eur. Phys. J.* **C27**, 331–374.

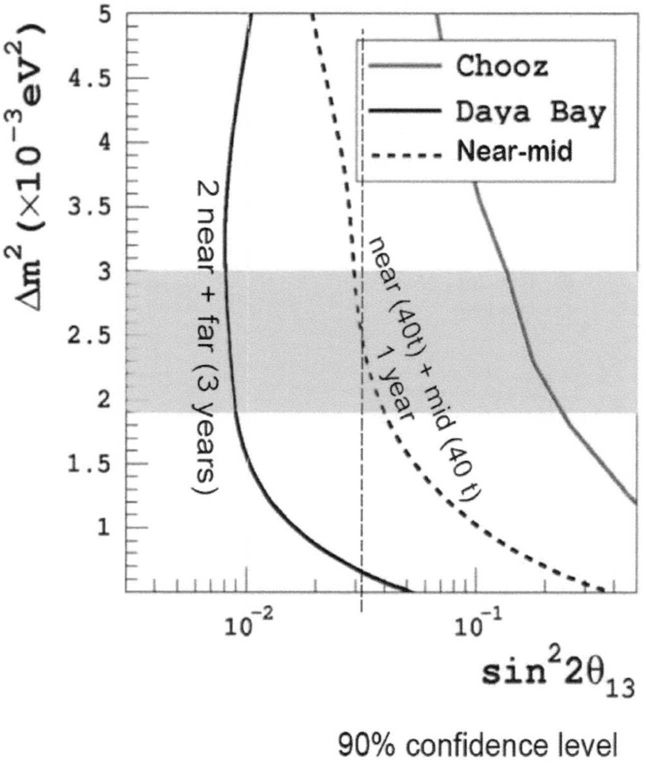

Figure 22. *Projected* $\sin^2 2\theta_{13}$ *sensitivity for Daya Bay experiment. The shaded region shows the allowed range for* Δm^2. *The rightmost curve is the current limit from the Chooz experiment.*

Araki, T. et al. (2005), 'Measurement of neutrino oscillation with Kamland: Evidence of spectral distortion', *Phys. Rev. Lett.* **94**, 081801.

Ardellier, F. et al. (2006), 'Double Chooz: A search for the neutrino mixing angle θ_{13}', hep-ex/0606025.

Arms, R.G. (2001), 'Detecting the neutrino', *Phys. perspect.* **3**, 314.

Bemporad, C., Gratta, G. & Vogel, P. (2002), 'Reactor-based neutrino oscillation experiments', *Rev. Mod. Phys.* **74**, 297.

Boehm, F. et al. (2001), 'Final results from the Palo Verde neutrino oscillation experiment', *Phys. Rev.* **D64**, 112001.

Declais, Y. et al. (1995), 'Search for neutrino oscillations at 15-meters, 40-meters, and 95-meters from a nuclear power reactor at Bugey', *Nucl. Phys.* **B434**, 503–534.

Eguchi, K. et al. (2003), 'First results from Kamland: Evidence for reactor anti- neutrino disappearance', *Phys. Rev. Lett.* **90**, 021802.

Guo, X. et al. (2007), 'A precision measurement of the neutrino mixing angle θ_{13} using reactor antineutrinos at daya bay', hep-ex/0701029.

Hahn, A. A. et al. (1989), 'Anti-neutrino spectra from Pu-241 and Pu-239 thermal neutron fission products', *Phys. Lett.* **B218**, 365–368.

Harris, D. (2007), this volume.

Huber, P., Lindner, M., Schwetz, T. & Winter, W. (2003), 'Reactor neutrino experiments compared to superbeams', *Nucl. Phys.* **B665**, 487–519.

Kayser, B. (2007), this volume.

Mention, G., Lasserre, T. & Motta, D. (2007), 'A unified analysis of the reactor neutrino program towards the measurement of the θ_{13} mixing angle', arXiv:0704.0498 [hep-ex].

Reines, F. & Cowan, C. L. (1956), 'The neutrino', *Nature* **178**, 446–449.

Reines, F. et al. (1960), 'Detection of the free anti- neutrino', *Phys. Rev.* **117**, 159.

Schreckenbach, K., Colvin, G., Gelletly, W. & Von Feilitzsch, F. (1985), 'Determination of the anti-neutrino spectrum from U-235 thermal neutron fission products up to 9.5-mev', *Phys. Lett.* **B160**, 325–330.

Zacek, G. et al. (1986), 'Neutrino oscillation experiments at the Gosgen nuclear power reactor', *Phys. Rev.* **D34**, 2621–2636.

Absolute Neutrino Mass Measurements

Beate Bornschein

Tritium Laboratory Karlsruhe, Forschungszentrum Karlsruhe, Germany

1 Introduction

The discovery of neutrino oscillation in the past decade proved that neutrinos have non-vanishing masses in contrast to their description within the Standard Model of particle physics. However, neutrino oscillation experiments are only measuring differences of neutrino mass squares and not absolute neutrino masses. Therefore, one of the most important tasks in neutrino physics, the determination of the neutrino mass scale to distinguish between hierarchical and degenerate neutrino mass models and the clarification of the role of neutrinos in the early universe, cannot be done by oscillation experiments. For that task other types of experiments are necessary.

The present lecture will give a survey of the methods used for the past five decades to measure absolute neutrino masses. The main focus will be on tritium β decay experiments, since for more than fifty years tritium has been the best isotope to search for a non-zero neutrino mass. The understanding of systematic effects is of utmost importance for each neutrino mass experiment. Mistakes in this field can lead to negative neutrino mass squares, something that happened in the nineties of the last century. Because of the importance of that topic a whole section is devoted to that question. Since the author[1] has been a member of the Mainz neutrino mass experiment for more than 15 years, the Mainz experiment has been chosen to demonstrate the influence of systematic effects. At the end of the lectures KATRIN, the next generation neutrino mass experiment with a sub-eV sensitivity, is introduced.

2 Absolute neutrino mass measurements – methods

The methods discussed in the following are belonging to the direct neutrino mass methods. Direct means that these methods do not require the non-conservation of total lepton family number or lepton number. These methods are essentially kinematic in nature and, with one exception, are investigating the kinematics of weak decays in laboratory experiments. The exception is the time-of-flight measurement of neutrinos emitted in a supernova (Section 2.3). All these

[1] with maiden name 'Beate Degen'.

experiments are making use of the relativistic energy momentum relation $E^2 = m^2c^4 + p^2c^2$ as well as of energy and momentum conservation. Because of this fact the standard unit being detected is not m_ν but m_ν^2.

In principle, a kinematical neutrino mass measurement yields information on the different mass eigenstates $m(\nu_\mathrm{i})$, but usually the different neutrino mass eigenstates cannot be resolved by the experiment. Therefore an average over neutrino mass eigenstates is obtained which is specific for the flavor of the weak decay and hence termed correspondingly. Since in a weak decay of flavor α the different neutrino mass eigenstates i are produced with the fraction $|U_{\alpha\mathrm{i}}{}^2|$ the following equation is valid:

$$m^2(\nu_\alpha) = \sum_i |U_{\alpha\mathrm{i}}{}^2| \cdot m^2(\nu_\mathrm{i}) \tag{1}$$

with $|U_{\alpha\mathrm{i}}^2|$ being an element of the neutrino mixing matrix. All experiments which are discussed in the following are highly complex and have to take into account a lot of systematic effects leading to years of additional studies and experiments. Within the present lecture notes it is not possible to go into detail in more than one experiment. For obvious reasons, the author has chosen the Mainz experiment (see Section 4) to go a little bit more into detail by discussing systematic effects.

2.1 Measurement of tau neutrino mass

The tau neutrino mass $m(\nu_\tau)$ is measured via the investigation of multi hadronic decays of τ pairs produced at electron-positron colliders. Because of the large mass of the τ, 1777 MeV/c^2 (Yao 2006 [PDG]), decays into five or six pions give the highest sensitivity on the tau neutrino mass. In this case the available phase space of the tau neutrino is restricted. The disadvantage is the low branching ratio for that kind of decay leading to low count rates. One example is the decay

$$\tau^- \rightarrow \pi^-\pi^-\pi^-\pi^+\pi^+(\pi^0) + \nu_\tau. \tag{2}$$

The bound of the tau neutrino mass is derived by describing the tau decay as a two-body decay (Barate 1998)

$$\tau^-(E_\tau, \vec{p}_\tau) \ \rightarrow \ h^-(E_\mathrm{h}, \vec{p}_\mathrm{h}) + \nu_\tau(E_\nu, \vec{p}_\nu) \tag{3}$$

where the hadronic system h^- is composed of three, five or six pions.

In the tau rest frame ($\vec{p}_\tau = 0$) the energy of the hadronic system can be derived in the following way (with c =1):

$$E_\nu = E_\tau - E_\mathrm{h} \tag{4}$$

Forming the square of that formula, substitute E with $m^2 + \vec{p}\ ^2$ and taking into account that $E_\tau = m_\tau$ and $\vec{p}_\mathrm{h}{}^2 = \vec{p}_\nu{}^2$ leads to

$$E_\mathrm{h} = \frac{m_\tau^2 + m_\mathrm{h}^2 - m_\nu^2}{2m_\tau} \tag{5}$$

The value of m_ν can in principle be derived by the measurement of E_h in the tau rest frame and m_h. Since these quantities can only be measured in the laboratory frame, they have to be transferred into this system. To do so one needs to know the angle between the direction of the tau and that of the hadronic system in the tau rest frame. Since the tau direction is not determined, the neutrino mass cannot be calculated directly. The way out is doing a two-dimensional analysis, as it is demonstrated in Figure 1. Shown is the $E_{h(lab)}$, m_h plane ($E_{h(lab)}$ = hadronic energy in laboratory frame) and the allowed kinematic regions for different tau neutrino masses. The allowed kinematic region is defined by the interval $\gamma(E_h \pm \beta p_h)$ in which $E_{h(lab)}$ has to fall. Here γ is the boost and β is the velocity of the tau which is known since the tau energy is assumed to be the beam energy. m_h and $E_{h(lab)}$ of one "‘event’" are determined from the measured momenta of the particles composing the hadronic system.

Two hypothetical events together with their error ellipses are plotted in the diagram to illustrate the principle. Event #1 lies on curve (2) indicating a neutrino mass of 30 MeV/c^2 and event #2 is located below curve (3) indicating a neutrino mass of more than 50 MeV/c^2.

From Figure 1 follows that a single observed event may be sufficient to constrain the tau neutrino mass. Because of that it is very important not to misinterpret background events from the tau decay. A much deeper discussion on systematic effects can be found in Barate (1998).

The actual upper limit for the tau neutrino mass has been achieved by the ALEPH experiment at LEP. Its two-dimensional analysis in the M_h, E_h plane leads to (Barate 1998):

$$m(\nu_\tau) < 18.2\ \mathrm{MeV/c^2} \quad (95\%\ \mathrm{C.L.}). \tag{6}$$

2.2 Measurement of muon neutrino mass

A first upper limit on the muon neutrino mass ('7 electron masses') was obtained in 1956 by Barkas and collaborators (Barkas, 1956) at the 184-inch cyclotron at the Lawrence Berkeley Laboratory. Their method, the study of the two-body decay of pions, is still the one that provides the best limits. Starting point is a pion decay at rest:

$$\pi^+ \to \mu^+ + \nu_\mu \ \text{ or } \ \pi^- \to \mu^- + \bar{\nu}_\mu. \tag{7}$$

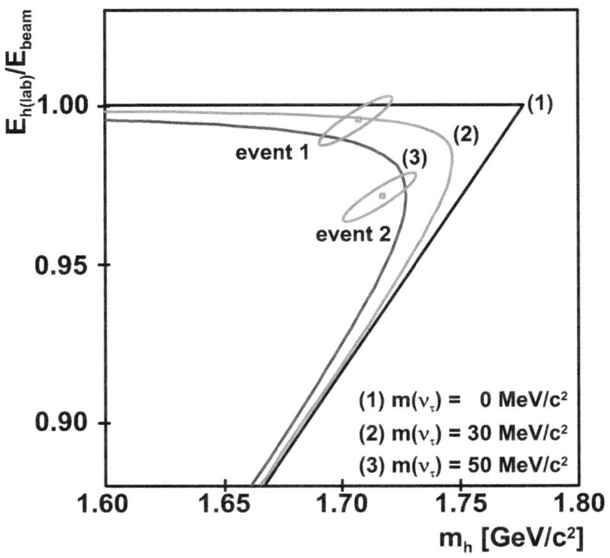

Figure 1. *Kinematic allowed regions for different values of the tau neutrino mass (after Barate 1998). Further discussion see text.*

Applying the standard formula $E^2 = m^2c^4 + p^2c^2$ and taking into account energy and momentum conservation results in the following formula for the mass of the muon neutrino:

$$m^2(\nu_\mu) = m^2(\pi) + m^2(\mu) - 2m(\pi)\sqrt{m^2(\mu) + p^2(\mu)} \tag{8}$$

with $E(\tau) = E(\mu) + E(\nu_\mu)$, $p(\pi) = 0$, $p(\nu_\mu) = p(\mu)$ and $c = 1$.

The intrinsic problem of this formula is that one has to calculate a difference of large numbers. To get the muon neutrino mass, the three different quantities $m(\pi)$, $m(\mu)$ and $p(\mu)$ therefore need to be measured with a very high precision. This has been done in the last decade with different experiments at different sites. The experiments listed below have been taken by the Particle Data Group (PDG) to evaluate the upper limit on the muon neutrino mass (Yao, 2006):

- Measurement of $m(\pi^-)$: The most accurate charged pion mass measurements are based upon x-ray wavelength measurements for transitions in π^--mesonic atoms (pion is stopped in matter and captured by an atom). The PDG averages the results of 2 experiments (Lenz 1998 and Jeckelmann 1994) and obtains $m(\pi^-) = 139.57018(35)$ MeV/c^2.

- Measurement of $m(\mu^+)$: The primary determination of the muon mass is done by measuring the ratio of muon mass to that of a nucleus, so that the result is obtained in u (atomic units). Since the conversion factor in MeV is more uncertain than the uncertainty of the mass measurement, PDG is giving the result in both units: $m(\mu) =$ 0.1134289264(39)u and $m(\mu) = 105.65833692(94)$ MeV/c^2, respectively. The actual value for the muon mass is based on the following formula:

$$\frac{m_\mu}{m_e} = \left(\frac{\mu_e}{\mu_p}\right)\left(\frac{\mu_\mu}{\mu_p}\right)^{-1}\left(\frac{g_\mu}{g_e}\right) \tag{9}$$

  where g stands for g-factor and μ for magnetic moment, e is the electron and p the proton. Measurements of the frequencies of transitions between Zeemann energy levels in muonium (μ^+e^- atom) can yield a value of μ_μ/μ_p. The muon mass is therefore linked to several other quantities which are known with different precision. The actual evaluation has been done by Mohr (2005).

- Measurement of $p(\mu^+)$: The actual value for $p(\mu^+)$ comes from Assamagan (1996) and has been determined from the decay $\pi^+ \rightarrow \mu^+\nu_\mu$ at rest by analyzing a surface muon beam in a magnetic spectrometer equipped with a silicon microstrip detector. The experiment has been performed at the Paul Scherrer Institute (PSI), the result is $p(\mu^+) =$ 29.79200(11) MeV/c.

The evaluation of the PDG for the muon neutrino mass is based on their average for the pion mass and the Assamagan value for the muon momentum and leads to

$$m(\nu_\mu) < 0.19 \text{ MeV/c}^2 \quad (90\% \text{ C.L.}). \tag{10}$$

2.3 Measurement of electron neutrino mass

The observation of neutrino oscillation has been an indirect proof for non-zero neutrino masses. The measured small values of Δm^2_{ij} (see Section 1) lead to the conclusion, that all neutrino

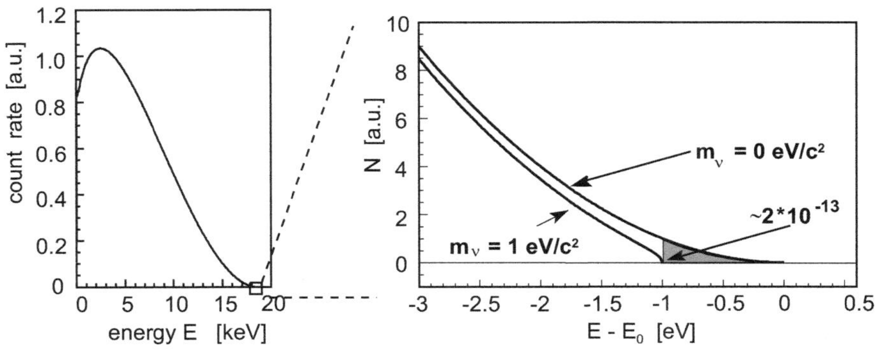

Figure 2. *The electron energy spectrum of tritium β decay: complete (left side) and zoom into the region around the endpoint* E_0.

masses are linked together: the measurement of a low limit of one neutrino species automatically also brings down the limits of the others. Since the lowest upper limit can experimentally be obtained for the electron neutrino, the definition of the neutrino mass scale is strongly connected to the measurement of the electron neutrino mass. Due to the great importance of this kind of experiments the remaining part of the present lecture notes only deals with electron neutrino mass measurement.

The most sensitive way to determine the neutrino mass scale without further assumptions is to measure the shape of a β spectrum near its endpoint. The idea that the neutrino mass can be deduced by the shape of the β spectrum was already recognized by Fermi in 1934 when he formulated the theory of beta decay. His statement was that the neutrino has a mass in the range of the electron mass or smaller (Fermi, 1934).

Figure 2 illustrates how a non-zero neutrino mass changes the shape of the β spectrum, exemplified shown for the tritium β decay. A discussion on the β formula is done in Section 3.1.

In the last seven decades the upper limit on the electron neutrino mass has been improved by more than five orders of magnitude. Table 1 gives a survey of the last 16 years indicating also that tritium is the key isotope for this approach. Tritium β decay experiments have been performed for more than 50 years, yielding in a large experience in that field. The method of tritium β decay experiments is discussed in detail in the next section including the question, why the tritium isotope is being the key isotope. The other two sources of neutrinos for direct neutrino mass measurements mentioned in Table 1 are rhenium (Re-187) and supernovae. The latter two methods are discussed very briefly in the following.

Rhenium cryogenic bolometer experiments

A comparatively new approach to directly measure the electron neutrino mass is the use of cryogenic bolometers. In this case, the β source can be identical to the β-electron spectrometer, which is de facto a calorimeter operated at very low temperatures (< 0.1 K). A so called 'source=detector' set-up is realized. Since calorimeters do measure the entire β spectrum at once, one needs to fulfill three basic requirements. First of all one has to use β decaying isotopes with a low endpoint to achieve enough statistic close to the endpoint. Second one needs to operate small calorimeters in large arrays to overcome the problem of pile up effects due to high count rate in a single detector, and third, the detector needs a very good energy resolution in region of 30 eV (or better) together with a short rise time.

Table 1. *Upper limits of the electron neutrino mass over the last 16 years (after Yao 2006.)*

Value [eV/c^2]	C.L. [%]	Reference	Neutrino source
<2.3	95	Kraus, 2005	^{3}H β decay
<2.5	95	Lobashev, 1999	^{3}H β decay
<21.7	90	Arnaboldi, 2003	^{187}Re β decay
<5.7	95	Loredo, 2002	SN1987A
<2.8	95	Weinheimer, 1999	^{3}H β decay
<4.35	95	Belesev, 1995	^{3}H β decay
<12.4	95	Ching, 1995	^{3}H β decay
<92	95	Hiddemann, 1995	^{3}H β decay
<19.6	95	Kernan, 1995	SN1987A
<7.0	95	Stoeffl, 1995	^{3}H β decay
<7.2	95	Weinheimer, 1993	^{3}H β decay
<11.7	95	Holzschuh, 1992	^{3}H β decay
<13.1	95	Kawakami, 1991	^{3}H β decay
<9.3	95	Robertson, 1991	^{3}H β decay
<14	95	Avignone, 1990	SN1987A

This new technique has been applied to the isotope Re-187, which has a 7 times lower endpoint energy than tritium (2.5 keV). Current rhenium microcalorimeters reach an energy resolution of $\Delta E = 20 - 30$ eV and yield an upper limit of $m_{\nu_e} < 15$ eV/c^2 (Sisti, 2004). To improve the limit leading groups in this field have proposed to start a new common experiment called MARE (Microcalorimeter Arrays for a Rhenium Experiment). The aim of MARE is to reach a sub-eV sensitivity on the electron neutrino mass in the next 10 to 20 years (Gatti, 2006, MARE proposal). A necessary requirement is the successful development of new detectors and their operation in an array of 10000 and more.

Time-of-flight (ToF) measurements of supernova neutrinos

The narrow time signal of a supernova (SN) neutrino burst of less than 10 s in combination with the very long-baseline between source (SN) and detector (e.g. underground water Cerenkov detectors as Kamiokande) of ten thousands of light years allows the investigation of small ToF effects resulting from non-zero neutrino masses (see e.g. Vogel, 2002). The flight time Δt for one neutrino can be expressed as

$$\Delta t = t_{\mathrm{E}} - t_{\mathrm{SN}} = \frac{L}{c} \cdot \left[1 + \frac{m^2 c^4}{2E^2} \right] \tag{11}$$

where t_{E} is the time arrival on earth, given in terms of the emission time from the supernova, t_{SN} and L the distance between SN and earth. E is the energy and m the mass of the neutrino. The equation can be derived from our basic formula $E^2 = m^2c^4 + p^2c^2$ by using $m^2 << E^2$ (and $1/\sqrt{1-x} \approx 1 + x/2$). For two neutrinos with different energies E_1 and E_2 (but same mass) the time difference of the detection is then given by

$$t_{E1} - t_{E2} = (t_{SN1} - t_{SN2}) + \frac{L}{c} \cdot \frac{m^2c^4}{2} \left[\frac{1}{E_1^2} - \frac{1}{E_2^2} \right] \tag{12}$$

From this follows that we need to measure at least 2 neutrinos with their energy and there exact time of arrival. In addition we have to make some assumptions on $t_{SN1} - t_{SN2}$ which depends on the modelling of supernovae.

This method provides an experimental sensitivity for the rest masses of ν_e, ν_μ and ν_τ of a few eV/c^2, if these additional assumptions concerning the time evolution of the neutrino burst are made. However, because on the model dependent information on $t_{SN1} - t_{SN2}$ the neutrino mass sensitivity also becomes model-dependent.

Since the main reaction with the highest cross section to detect a neutrino in a water detector is $\bar{\nu}_e$+ p $\rightarrow$ n + e$^+$, these detectors are essentially sensitive on $\bar{\nu}_e$. The PDG has taken several results on upper limits on the electron neutrino, see Table 1.

A small drawback of the ToF method is, that one has to wait for a supernova. Since the expectation for such an event is about 2 in 100 years in our galaxy, this method needs not only the existence of running big detectors but also some good luck.

3 Tritium β decay experiments – the key to neutrino mass

The most sensitive way to determine the electron neutrino mass is based on the investigation of the electron spectrum of tritium β decay

$$^3\mathrm{H} \rightarrow {}^3\mathrm{He}^+ + \mathrm{e}^- + \bar{\nu}_e. \tag{13}$$

The transition rate for a β decay can be derived from Fermi's Golden Rule (complete sequence in Altarelli, 2003) and is given by (velocity of light reintroduced):

$$\frac{d^2N}{dtdE} = C \cdot F(E, Z+1)\, p \left(E + m_e c^2\right) \left(E_0 - E\right) \sqrt{(E_0 - E)^2 - m_{\bar{\nu}_e}{}^2 c^4}\, \Theta(E_0 - E - m_{\bar{\nu}_e} c^2), \tag{14}$$

where E is the kinetic energy of the electron, m_e the electron mass, p the electron momentum and $m_{\bar{\nu}_e}$ the neutrino mass. E_0 corresponds to the total decay energy, $F(E, Z+1)$ is the Fermi function, taking into account the Coulomb interaction of the outgoing electron in the final state, the step function $\Theta(E_0 - E - m_{\bar{\nu}_e} c^2)$ ensures energy conservation and C is given by

$$C = \frac{G_F^2}{2\pi^3 \hbar^7 c^5} \cos^2(\Theta_c)\, |M_{had}|^2). \tag{15}$$

Here G_F is the Fermi constant, Θ_c the Cabibbo angle, and M_{had} the nuclear matrix element. Equation 14 holds only for the decay of a bare, infinitely heavy nucleus. In case of an atom or a molecule, the possible excitation of the electron shell due to the sudden change of the nuclear charge by one unit has to be taken into account. The atom or molecule will end up in a specific state of excitation energy V_i with a probability W_i. Therefore equation 14 has to be modified into a sum of β spectra with amplitude W_i and endpoint energies $E_{0,i} = E_0 - V_i$. Since both,

$|M_{\text{had}}|^2$ and F(E,Z+1) are independent of $m_{\bar{\nu}_e}$, the dependence of the spectral shape on $m_{\bar{\nu}_e}$ is given by the phase space only. The square-root term of equation 14 shows first that $m_{\bar{\nu}_e}{}^2$ is the experimental observable and second that the neutrino mass influences the β–spectrum only at the upper end just below E_0.

Figure 2 shows the signature of an electron neutrino with a mass of $m_{\bar{\nu}_e}$ = 1 eV/c^2 in comparison with the undistorted β spectrum. The influence of a non-zero neutrino mass is statistically significant only in an energy region close to the β endpoint. Therefore, only a very narrow region close to the endpoint need to be analyzed. Since the fraction of β decays in this region is proportional to a factor $(1/E_0)^3$, a very low endpoint energy is an important requirement for a β decay source.

Nevertheless, the requirements for a tritium β decay experiment with a sub-eV sensitivity on the neutrino mass are quite demanding. These experiments require a high β decay rate since the fraction of β particles within 1 eV below E_0 is 2×10^{-13}, a huge luminosity, i.e a large source area multiplied with a large accepted solid angle and a spectrometer with a very high energy resolution and a very low background rate.

3.1 Why tritium ?

Tritium has the following clear advantages as β emitter in neutrino mass investigations:

- Tritium has the second lowest endpoint energy of $E_0 = 18.6$ keV, maximizing the fraction of β–decays in this region.

- Tritium has a rather short half life of 12.3 y, which corresponds to a high specific activity. Only a small amount of source material is needed and the fraction of inelastic scattered decay electrons is low. Inelastic scattered decay electrons are a source of systematic uncertainties, their number should all the time be kept as small as possible.

- Tritium has a low nuclear charge, the inelastic scattering of outgoing β electrons within the β source is small. This goes in the same direction as the item above.

- The hydrogen isotope tritium and its daughter, the ^{3}He$^+$ ion, have a simple electronic shell configuration allowing precise calculations of the final state spectrum. Therefore, corrections due to the interaction of the outgoing β electron with the tritium source can be calculated in a simple and straightforward manner.

- The tritium β decay is a super-allowed nuclear transition, the nuclear matrix element is energy independent. Therefore, no corrections from the nuclear matrix element have to be taken into account.

The combination of all these features makes tritium an almost ideal β emitter for neutrino mass measurements.

A comparison with Re-187 shows, that rhenium has indeed an advantage with regard to its low endpoint energy of 2.47 keV. On the other hand, the disadvantages are the long half life of more than 10^{10} years, its high nuclear charge of Z=75 and the comparable complex electronic shell configuration. In addition the decay of Re-187 is unique forbidden leading to an energy dependent nuclear matrix element. Neutrino mass experiments with rhenium (see also Section 2.3) can only compensate for that by using the calorimeter method (detector = source), since here one needs not to account for energy losses inside the source as e.g. inelastic scattering or excited final states. Nevertheless, the challenge to calculate the nuclear matrix element with high precision still exists.

The almost ideal features of tritium as a β emitter have been the reason for a long series of tritium β decay experiments (see Figure 3, data from: Robertson (1991), Kawakami (1991), Holzschuh (1992), Sun (1993), Stoeffl (1995), Lobashev (1999), Weinheimer (1999)). It is remarkable that the error bars on m_ν^2 have decreased by nearly two orders of magnitude. Equally important is the fact that the problem of negative values of m_ν^2 of the early nineties has disappeared due to better understanding of systematic effects and improvements in the experimental set-ups.

3.2 Standard experimental set-up and MAC-E filter

The standard set-up of a tritium β decay experiment consists of four main components: the tritium source, the (magnetic) transport system, the spectrometer and the detector. The β decay electrons are guided into the spectrometer by a magnetic transport system. This transport system must not disturb the kinetic energy of the β decay electrons and has to pump residual tritium molecules originating from the source. Electrons, which have passed the spectrometer are then counted by the detector. Figure 4 gives a list of the main requirements for each component.

The high sensitivity of the Troitsk and the Mainz neutrino experiments (see Figure 3) is due to a new type of spectrometers, so-called MAC-E-Filters (Magnetic Adiabatic Collimation combined with an Electrostatic Filter). It combines high luminosity and low background with a high energy resolution, both essential to measure the neutrino mass from the endpoint region of a β decay spectrum.

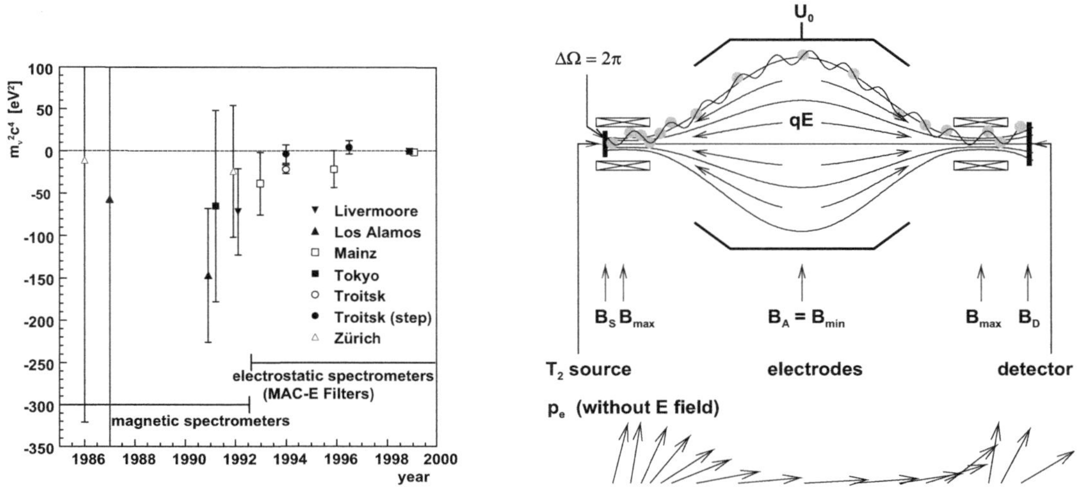

Figure 3. *Results of tritium β decay experiments (left) and the principle of an electrostatic spectrometer of the MAC-E-Filter type as it is used in Mainz and in Troitsk (right). Further discussion see Section 3.2.*

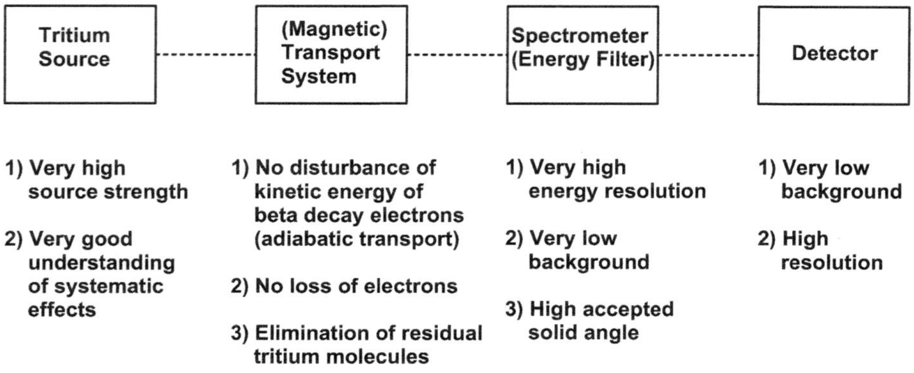

Figure 4. *Standard experimental set-up for a tritium β decay experiment.*

The main features of the MAC-E-Filter are illustrated in Figure 3. Two superconducting solenoids are providing a guiding magnetic field. The β electrons, which start from the tritium source in the left solenoid into the forward hemisphere, are guided magnetically on a cyclotron motion around the magnetic field lines into the spectrometer, thus resulting in an accepted solid angle of up to 2π. On their way into the center of the spectrometer the magnetic field B drops by many orders of magnitude. Therefore, the magnetic gradient force transforms most of the cyclotron energy $E_\perp$ into longitudinal motion. This is illustrated in Figure 3 by a momentum vector. Due to the slowly varying magnetic field the momentum transforms adiabatically and therefore the magnetic moment μ keeps constant (equation is given in non-relativistic approximation)

$$\mu = \frac{E_\perp}{B} = \text{const.} \tag{16}$$

This transformation can be summarized as follows: The β electrons, isotropically emitted at the source, are transformed into a broad beam of electrons flying almost parallel to the magnetic field lines. This parallel beam of electrons is energetically analyzed by applying an electrostatic potential generated by a system of cylindrical electrodes. All electrons which have enough energy to pass the electrostatic barrier are re-accelerated and collimated onto a detector, all others are reflected. The spectrometer, therefore, acts as an integrating high-energy pass filter.
The relative sharpness $\Delta E/E$ of this filter is given by the ratio of the minimum magnetic field B_{min} in the analyzing plane to the maximum magnetic field B_{max} between β electron source and spectrometer:

$$\frac{\Delta E}{E} = \frac{B_{\text{min}}}{B_{\text{max}}} \tag{17}$$

The β spectrum can be measured by scanning the electrostatic retarding potential.

In order to suppress electrons which have a very long path within the tritium source and therefore exhibit a high scattering probability, the electron source is placed in a magnetic field B_S (see Figure 3), which is lower than the maximum magnetic field B_{max}. This restricts the maximum accepted starting angle of the electrons Θ_{max} by the magnetic mirror effect to:

$$\sin\Theta_{\text{max}} = \sqrt{\frac{B_S}{B_{\text{max}}}}. \tag{18}$$

3.3 Recent tritium β decay experiments

Both recent tritium β decay experiments at Mainz (Kraus, 2005) and at Troitsk (Lobashev, 1999) use similar MAC-E-Filters with an energy resolution of 4.8 eV at Mainz and 3.5 eV at Troitsk. The sizes of the spectrometers are also similar: 1 m diameter and 4 m length at Mainz, and 1.5 m diameter and 7 m length at Troitsk, respectively. The major difference between these both experiments is the source principle used in their set-up (Figure 5): Mainz has used a thin film of molecular tritium, quench condensed on a 1.9 K cold graphite substrate and Troitsk uses a windowless gaseous molecular tritium source. This difference in the source principle leads to the fact that, to a large extent, both experiments have to take into account independent and complementary systematic effects. This fact is especially important with regard to the anomalous excess of count rate in the β spectrum which was reported by the Troitsk group (Belesev, 1995). Since this so-called Troitsk anomaly has not been observed at Mainz, also not during a simultaneous measurement phase over Christmas 2000 (Krause, 2005), it is very likely that the origin of the Troitsk anomaly has to be attributed to experimental effects.

The experiments at Troitsk and Mainz started around 1985 and have performed not only measurements of the tritium β spectrum but also dedicated experiments to investigate systematic effects. These efforts lasted over years, the systematic corrections became much better understood and their uncertainties were reduced significantly. A detailed discussion exemplified for the Mainz experiment is given in the next section.

The latest results of both experiments are the following:

- Mainz (Kraus, 2005):

$$m^2(\nu_e) = (-0.6 \pm 2.2_{\text{stat}} \pm 2.1_{\text{syst}})\ \text{eV}^2/\text{c}^4 \tag{19}$$

  which corresponds to an upper limit of

$$m(\nu_e) < 2.3\ \text{eV}/\text{c}^2\ \ (95\%\ \text{C.L.}) \tag{20}$$

- Troitsk (Lobashev, 2003):

$$m^2(\nu_e) = (-2.3 \pm 2.5_{\text{stat}} \pm 2.2_{\text{syst}})\ \text{eV}^2/\text{c}^4 \tag{21}$$

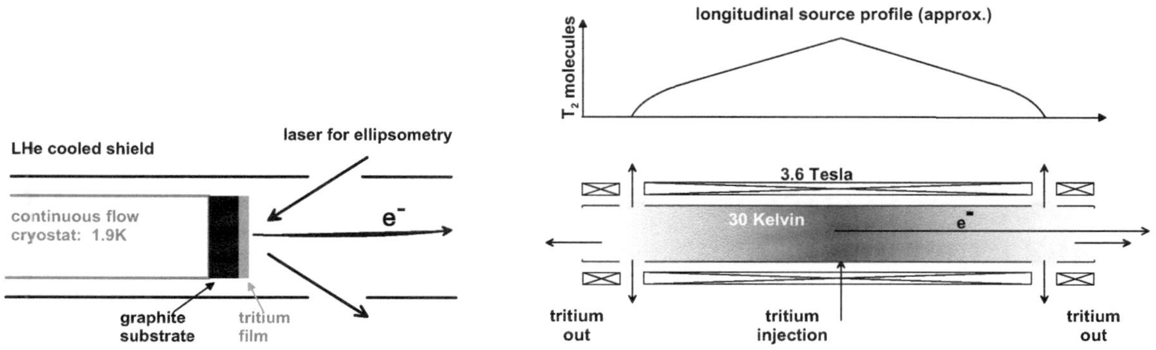

Figure 5. *Two source principles: Quench Condensed Tritium Source (QCTS, left) and Windowless Gaseous Tritium Source (WGTS, right). Ellipsometry is an optical method to measure the thickness of the tritium film.*

from which an upper limit on $m(\nu_e)$ is obtained of

$$m(\nu_e) < 2.05 \text{ eV}/c^2 \quad (95\% \text{ C.L.}) \tag{22}$$

under the assumption that the anomalous excess count rate near the β endpoint is described by an additional line.

Please notice, that the Particle Data Group (Yao, 2006) has chosen the published data from Kraus (2005) and Lobashev (1999) to derive a weighted upper limit for the electron neutrino mass of 2 eV/c^2. New data from both, the Mainz and the Troitsk experiment will not improve the sensitivity on m(ν_e) significantly, both experiments have reached their limit with regard to source strength, energy resolution of the spectrometer and the minimizing of systematic uncertainties. This fact clearly underlines the importance of a next generation tritium β decay experiment, which will be discussed in Section 5.

Short excursus on how to determine an upper limit from negative neutrino mass squares
This question is often asked by students, therefore, we will go a little bit into this method. Much more detailed information is available e.g. in Yao (2006, section about statistics) and the paper about Unified Approach (Feldman and Cousins, 1998). The reader is really encouraged to look into these papers.

First of all: A physical value which is in principle zero will lead to positive and negative measured values because of the Gaussian distribution of the measured values. The question is only how to proceed with the data. In the past, two possible solutions have been applied - the Bayesian approach and the method of confidence intervals. Both methods led to different upper limits of the neutrino mass. In 1998 Feldmann and Cousins published a 'Unified approach to the classical statistical analysis of small signals', in which they suggested a novel way to obtain a unified set of classical confidence intervals for setting upper limits and quoting two-sided confidence intervals. From this year on, de-facto all neutrino experiments publishing upper limits have used the tables from that publication.

How to get an upper limit by means of Unified Approach? Here is an example, using the data from Mainz: We have $m^2(\nu_e) = (-0.6 \pm 2.2_{\text{stat}} \pm 2.1_{\text{syst}})$ eV^2/c^4 which is equivalent to $m^2(\nu_e) = (-0.6 \pm 3.04_{\text{tot}})$ eV^2/c^4. The first step is to calculate m^2_ν in units of σ, here -0.6/3.04 = -0.176. The second step is to look into table 10 of the Unified Approach paper. The number -0.176 is equivalent to 'x_0' in the first column. The fourth column gives the upper limit with 95% Confidence Level in units of σ. In our case the result is $\approx 1.8\sigma = 1.8 \cdot 3.04$ eV^2/c^4 = 5.47 eV^2/c^4. The upper limit of the neutrino mass is than the square root: $\sqrt{5.47}$ eV^2/c^4 = 2.3 eV/c^2.

4 Importance of systematic effects – exemplified with the Mainz experiment

We start this section on systematic effects with a short description on how the standard data evaluation is done. We need this to understand the further discussion on systematic effects and their influence on the determination of the neutrino mass. After that follows a discussion on the possible origin of negative mass squares and, at the end, the motivation to use a gaseous tritium source for a next generation tritium β decay experiment.

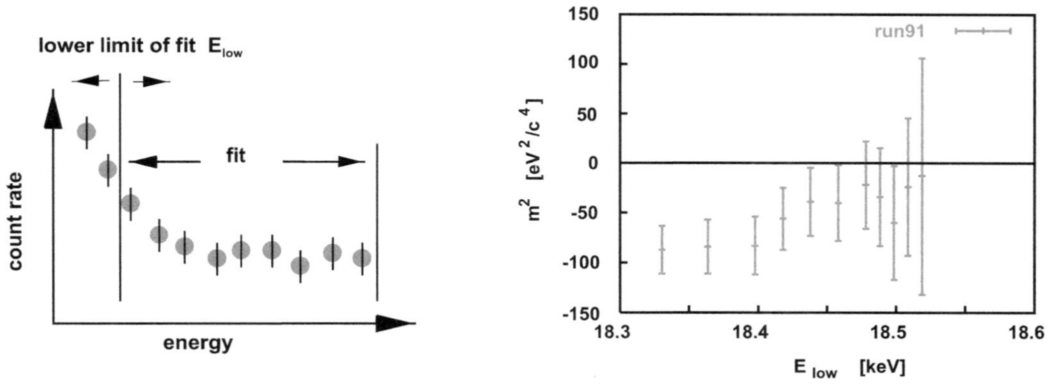

Figure 6. *Fit results of the 91-data obtained with a 4.2 K cold quench condensed tritium source. Displayed on the right is the neutrino mass squared m_ν^2 as a function of the lower limit of fit interval $E_{\rm low}$ (upper limit = 18.6 keV). An unphysical trend is visible towards negative values of m_ν^2 for larger fit intervals. Displayed on the left is a sketch of the beta spectrum to explain the variable $E_{\rm low}$.*

4.1 Former problem of negative m_ν^2 at Mainz – the dewetting effect

Figure 6 (left part) shows a sketch of a measured β-spectrum near its endpoint. Every fit to such a spectrum covers the pure background data (above the β endpoint E_0) and some data points within the spectrum. For a standard analysis several fits are done, varying the number of data points. For that the upper limit of the fit interval is kept constant and the lower limit of the fit interval is shifted into the spectrum. In other words: the neutrino mass square is calculated as a function of the lower limit of the fit interval. This procedure is done to have a consistency check. If the description of the data is correct, there should be no dependence of m_ν^2 on the fit interval.

Figure 6 (right part) shows as an example the result of such an analysis for an early data set from 1991 (Weinheimer, 1993), the error bars are representing the statistical uncertainties. Systematic uncertainties are not included. For small intervals below the endpoint the statistical uncertainty is relatively large and m_ν^2 is compatible with zero within the uncertainties. Adding more and more data points from further below the endpoint increases the data sample, the statistical uncertainties are getting smaller but from about 18.42 keV a significant trend towards negative fit results for m_ν^2 is visible. This unphysical behavior was observed for some years and could not be explained by the estimated and thus known systematic uncertainties.

To find an explanation, one has first to understand the 'behavior' of a fit: In the fit only the parameters A (source strength), E_0 (endpoint), BG (background rate) and m_ν^2 are free. All other inputs are determined by independent measurements or taken from literature. Examples are the function describing the inelastic scattering of the β decay electrons and the distribution of the final states. If one of these functions deviates more than statistically expected from the assumed value or if another systematic effect has not been included in the description of the fit, the the fit misuses one or more of its four parameters to compensate the discrepancies between the measured count rate and its description in the fit. The fit parameter m_ν^2 is very sensitive to such a mistake, because it is the only free parameter being able to change the shape of the spectrum. The challenge in the beginning of the nineties was to find the origin for the observed trend towards negative mass squares.

Simulations showed that the trend to negative mass squares can be stopped by adding an additional energy loss to the fit function, e.g. additional inelastic component or a second neutrino component. Figure 7 shows the influence of an energy loss on the β spectrum near its endpoint: β decay electrons which have undergone an energy loss are missing near the endpoint but are added in a region farer below the endpoint. As a result the shape of β spectrum is changed. A fit to such data without taking the energy loss into account automatically leads to a shifted endpoint E_0, thereby introducing additional count rate above the endpoint, and to negative neutrino mass squares. This scheme is valid for every energy loss process which has not been taken into account properly. This explanation was even more favored by the fact, that some previous experiments, which could only analyse larger intervals below the endpoint also obtained significant negative values for m_ν^2 as best fits (Figure 3).

The Mainz group has undertaken a large series of investigations to search for an undetected energy loss like process by checking the impact of the known systematic effects. In all these experiments no significant deviation from the expectations were found, except for one open question: Have the tritium films undergone a transition from a homogeneous quench-condensed film into a rough inhomogeneous one after preparation (see Figure 8)? This so-called dewetting transition, which is explained by a thermally activated surface diffusion process, had been investigated earlier for stable hydrogen isotopes. Dewetting would increase the fraction of inelastic scattering in the tritium film due to partly longer paths for the β decay electrons emitted from the tritium film which therefore results in a higher percentage of multi-scattering processes.

To answer the question, whether a dewetting transition of the tritium film took place and if yes, whether it can be sufficiently slowed down in the future to not disturb the β spectrum, an extended phase of investigation started in Mainz. H_2, HD and D_2 were used to study systematically the behavior of the hydrogen isotopes. In addition a few measurements with a hydrogen gas composition similar to that used in the earlier neutrino mass experiments have been done. Such a film consists of about 21% H_2, 38% HT and 41% T_2. In later neutrino experiments (Mainz phase II) the tritium purity was increased up to about 90%.

Figure 9 shows the principle method, the measurement of stray light at different source (film) temperatures. For this purpose thin films were quench-condensed on a substrate of graphite or aluminum at temperatures between 1.6 and 2.5 K. The thickness of the film was controlled during production via optical ellipsometry. Afterwards the ellipsometry laser was

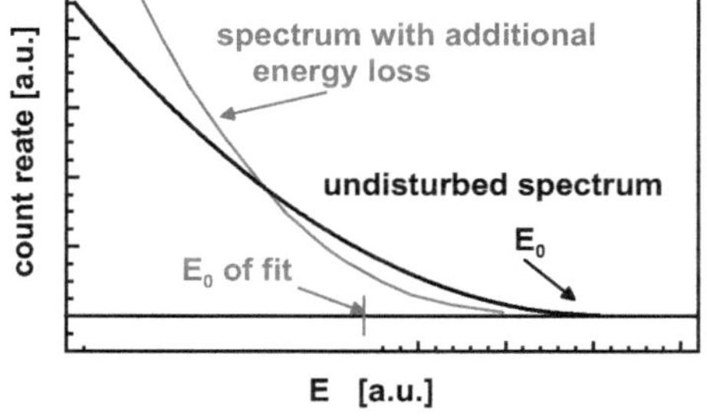

Figure 7. *Principle change of the β spectrum after introduction of an additional energy loss (sketch).*

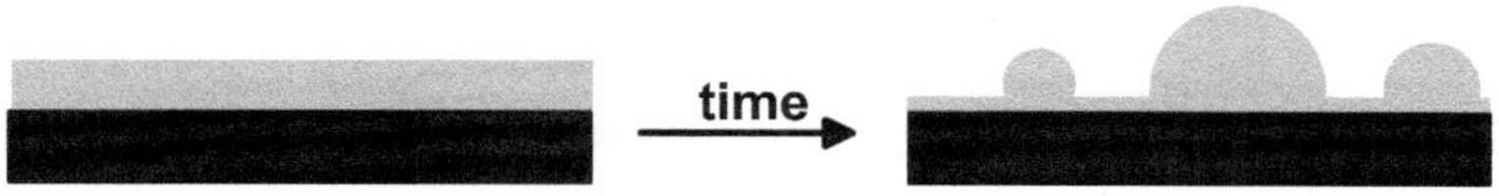

Figure 8. *Dewetting of a quench-condensed tritium film (sketch).*

kept beaming on the substrate and the stray light intensity in normal direction to the substrate was observed. In the next step the substrate temperature was increased to a value where the relaxation process and dewetting of the film should be studied. In case of a flat film one expects only light in the reflexion angle, a rough, i.e. a dewetted film would also cause stray light in the normal direction. The intensity of the scattered light was then registered as a function of time. This gives a measure of the growing film roughness upon annealing.

The diagram in Figure 9 shows a typical stray light signal. Displayed is the measured stray light intensity as a function of time. We start with a quench condensed film at 1.6 K. The film is flat and the stray light intensity is small. Then the temperature of the film is increased within 100 seconds from 1.6 K to 4.2 K. The stray light intensity increases with a certain time constant until it becomes more or less constant. After an additional raise of the substrate temperature to above 5 K the film is desorbed and the observed stray light of the flat substrate is reduced almost to zero.

The results of the investigations of the dewetting effect were (Fleischmann, 2000a, 2000b):

1. We did not find any substrate backing which suppresses the dewetting and clustering of quench condensed hydrogen films (H, D, T), but the transition speed is slowed down drastically for the heavier isotopes and by going to lower temperatures.

2. Dewetting time constants have been measured as a function of the film temperature, being in the range of a couple of minutes to several hours. The acceleration with temperature follows Arrhenius law: $\Delta t \propto \exp(\Delta E/kT)$ with E the activation energy and Δt the dewetting time. This yields activation energies for the different isotopes.

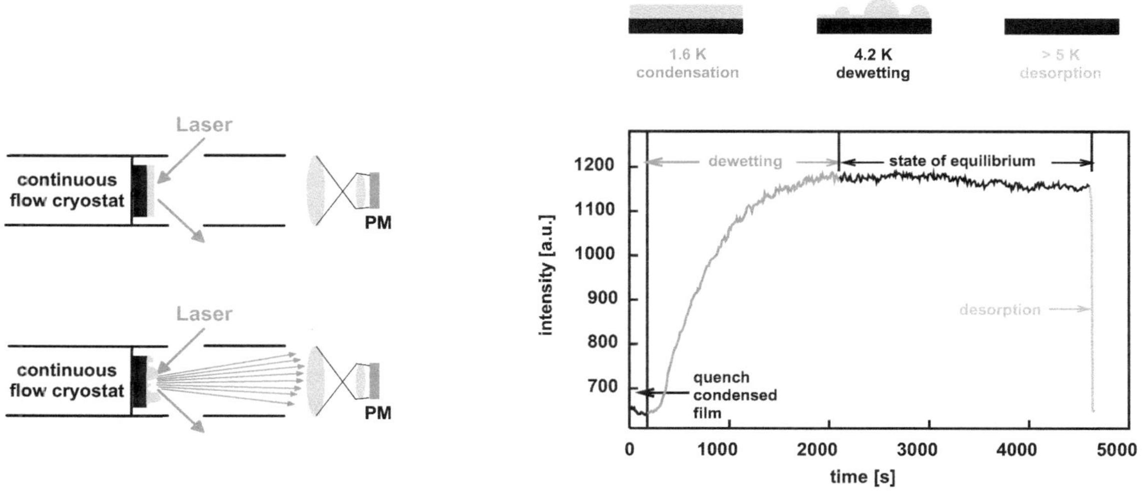

Figure 9. *Investigation of dewetting transition, principle set-up (left) and measured stray light signal (right).*

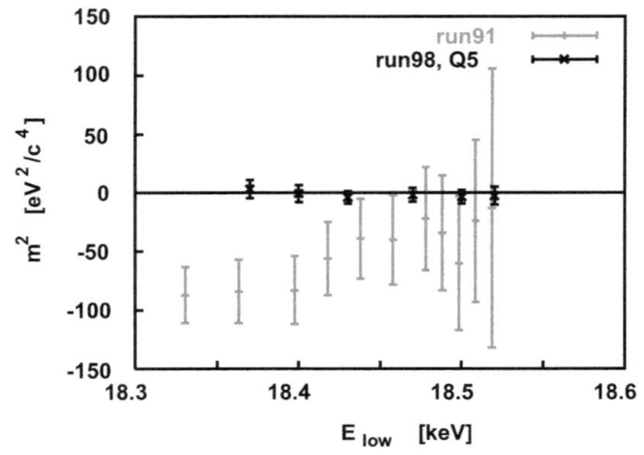

Figure 10. *Solving the former problem of negative neutrino mass squares in Mainz: avoid dewetting transition of the tritium film by going to lower temperatures.*

3. For tritium films a dewetting time of more than 1 year has been extrapolated for a source temperature below 1.9 K.

4. If the quench-condensed film is covered by just roughly a monolayer of heavier adsorbate like Ar, it withstands annealing up to the evaporation temperature without showing the characteristic increase in light scattering which is indicative of film roughening.

Since the dewetting transition can be avoided within a single measurement phase of about eight weeks by going to lower source temperature, the Mainz experiment did all further tritium β decay measurements with a source temperature of about 1.9 K. The solution discussed in 4) has been rejected since an additional layer causes additional energy loss leading to larger systematic uncertainties.

Figure 10 shows the fit results of the data obtained in 1998 with a 1.9 K cold quench condensed tritium source compared with those from 1991 with a source temperature 4.2 K. Displayed is m_ν^2 as a function of the lower limit of fit interval. There is no trend anymore towards negative values of m_ν^2 for larger fit intervals as is was the case for the Mainz data obtained with source temperatures of 4.2 K (1991) and 2.8 K (1994). This shows that a dewetting transition of the tritium film was indeed the reason for this behavior. Now this effect is safely suppressed by the much lower temperature of the tritium film. The former problem of negative mass squares in Mainz does not exist anymore.

4.2 Overview on systematic effects of the Mainz experiment

The main systematic uncertainties of the Mainz experiment originate from the physics and properties of the quench-condensed tritium film: the inelastic scattering of β electrons within the tritium film, the excitation of neighboring molecules due to the β decay, and the self-charging of the tritium film by radioactivity. In addition one has to take into account a continuously growing coverage of the source by 0.3 monolayers H_2 per day together with a certain evaporation of tritium from the source. Figure 11 illustrates the different effects. These effects and their influence on the neutrino mass were investigated by the Mainz group leading to the fact that the group for some time was more occupied with solid state physics and physics of low temperatures than with neutrino physics.

Figure 12 shows the influence of the different systematic effects on the uncertainty of the fit parameter m_ν^2 and as a comparison the behavior of the statistical uncertainty. Two items are typical for the experiment: the systematic uncertainties are decreasing with decreasing fit interval close to the endpoint and the statistical uncertainties are growing. This is due to the fact that most of the systematic effects are arising from processes connected to atomic and molecular physics, for example energy thresholds, see also discussion in Section 5.3. On the other hand the statistic is becoming worse in the region of low count rate leading to larger statistical uncertainties.

In the following subsection we will go a little bit deeper into the charging effect, since this is a novel effect and since it is the main drawback of a quench condensed tritium source for the use in a tritium β decay experiment.

4.3 The charging effect – the drawback of a thick quench condensed tritium source

The self-charging of a solid tritium film was discovered by chance during the tritium β decay experiment in Mainz as a shift of the endpoint energy of the measured β spectrum by several eV. This effect was too large to be explainable with instrumental uncertainties. It turned out that the shift is due to charging of the tritium film caused by its 1.5 GBq activity. This corresponds to an electron current of several hundred pA leaving the film, whereas the positively charged daughter molecules are mostly sticking on it. Without a sufficient compensation current from elsewhere the film will charge up positively.

For studying self-charging of tritium films quantitatively, a monochromatic electron source is suited much better than the continuous β spectrum. For this reason a submonolayer of radioactive Kr-83m was condensed on top of tritium films of controlled thickness. The isomer Kr-83m decays with a half life of 1.83 h and it has a 17.8 keV K32 conversion line not far from the endpoint of the tritium β spectrum at 18.6 keV. The source is quasi monoenergetic: the main peak at full energy has a Lorentzian width of only 2.8 eV. The surface potential of the tritium film is then derived by determining the exact energy of the K32 conversions electrons. A precision of about 1% of the linewidth has been feasible with the Mainz MACE-filter.

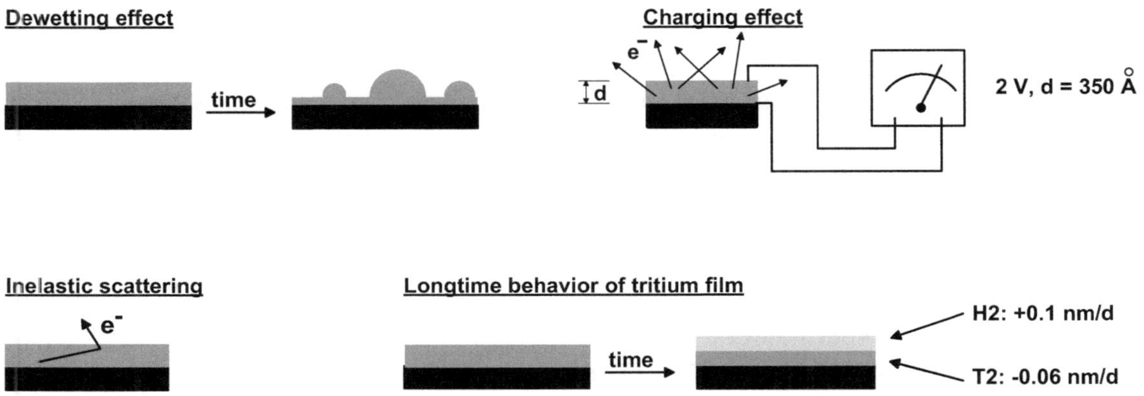

Figure 11. *Main systematic effects of the Mainz experiment, all being connected with the quench condensed tritium source.*

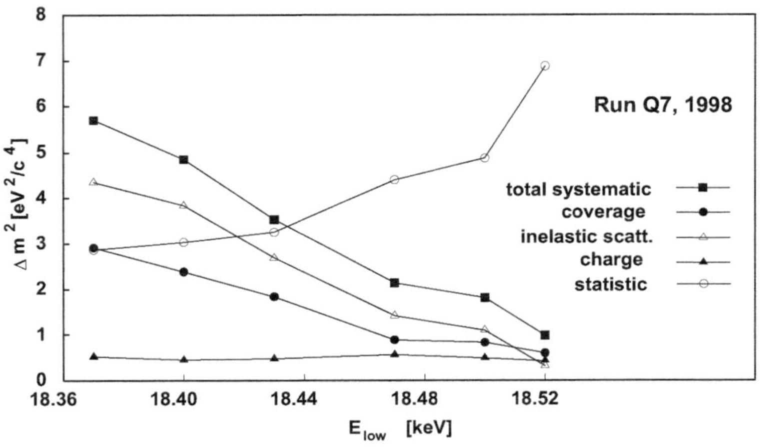

Figure 12. *Individual uncertainties of the observable m_ν^2 as a function of the lower limit of fit interval for Mainz run Q7. The upper limit is always 18.6 keV.*

It has been found that the film charges up within 30 min to a constant critical field strength of 62.6 MV/m (Bornschein, 2003) as it is shown in Figure 13. This results in a linearly increasing shift of the starting potential of tritium β decay electrons throughout the film, reaching about 2.5 V at the outer surface for a standard thickness of about 120 monolayer tritium which is equivalent to about 40 nm film thickness. The charging effect leads to a relatively slight decrease of the effective energy resolution of the Mainz experiment being in the order of 4 eV and therefore only to a small increase of systematic uncertainty for the fit parameter m_ν^2 (see Figure 12). However, in a next generation tritium β decay experiment with a required source strength of about 200 monolayer tritium film thickness equivalent (see next section) the charging effect would lead to an unacceptable electrical potential difference of 4 V between top and bottom of the film. In view of the aspired sub-eV sensitivity on the the next generation neutrino mass experiment which requires both, a high source strength, i.e high statistics and low systematic uncertainties, the usage of a quench condensed tritium source is disfavoured and a windowless gaseous tritium source is the first choice.

5 KATRIN – the next-generation neutrino mass experiment

The tritium β decay experiments at Troitsk and Mainz have reached their sensitivity limit of about 2 eV/c^2. To distinguish hierarchical from quasi-degenerate neutrino mass scenarios and to check the cosmological relevance of neutrino dark matter for the evolution of the universe require the improvement of the direct neutrino mass search by one order of magnitude at least. The step forward into the important sub-eV range of neutrino masses therefore calls for a new experimental effort - a next generation tritium β decay experiment.

The KATRIN collaboration has taken this challenge and has proposed to set-up an ultra-sensitive tritium β decay experiment based on the successful MAC-E-filter spectrometer technique and a very strong windowless gaseous tritium source (WGTS). A (not to) short survey of KATRIN will be given in the next subsections including a discussion on how and why one has to scale up the existing tritium β decay experiments to reach a sub-eV sensitivity. A complete description of KATRIN can be found in the KATRIN Design Report (Angrik, 2004).

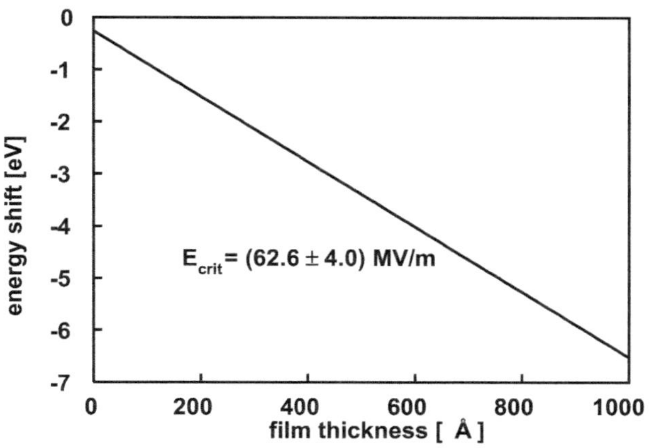

Figure 13. *Energy shift of electrons starting from the tritium film surface as a function of the film thickness. The charging effects leads to an energy shift of the tritium β decay electrons, which is between ≈ 0 eV (electrons starting near the substrate) and the maximal value (electrons starting from the film surface).*

5.1 Scaling factors for sub-eV tritium β decay experiment

The aim of KATRIN is to improve the sensitivity for m_{ν_e} by one order of magnitude (2 eV/c^2 $\rightarrow$ 0.2 eV/c^2). This corresponds to an necessary improvement of the fit parameter m_ν^2 by two orders of magnitude. The same holds for the statistical and systematic uncertainties, both have to be reduced by a factor of 100. This significant improvement on the neutrino mass requires an energy resolution of the spectrometer of $\Delta E \approx 1$ eV at the tritium β decay endpoint of 18.6 keV. This resolution corresponds to an improvement of about a factor of 4 compared to the experiments in Mainz and Troitsk. Since the energy interval of interest just below the β endpoint (= fit interval) rapidly decreases with a small neutrino mass, the signal rate has to be increased.

This can be achieved by a higher tritium source strength which is equivalent to a larger source area and a higher and optimized column density. Optimized means here a trade off between (unwanted) inelastic scattering and (wanted) high tritium β decay rate. Since the magnetic flux is conserved ($\Phi = B \cdot A =$ const), any increase of the source area requires the increase of the analyzing plane of the spectrometer. The same holds for a decrease of ΔE (because of the necessary reduction of the B-field in the analyzing plane).

To summarize: Compared with the experiments at Mainz and Troitsk the KATRIN experiment requires

- a larger signal rate ($\approx$ factor 100);
- a larger analyzing plane of the spectrometer ($\approx$ factor 80);
- a better energy resolution of 1 eV ($\approx$ factor 4);
- a reduction of systematic uncertainties ($\approx$ factor 100);
- an increase of measurement time ($\approx$ factor 10).

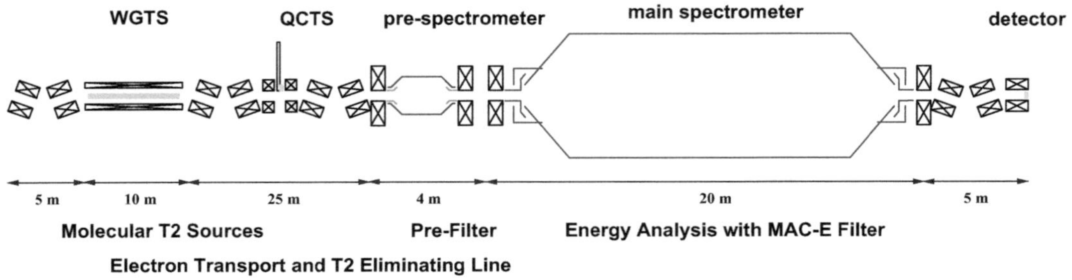

Figure 14. *Reference set-up of KATRIN (schematic view).*

5.2 The key components of KATRIN

The KATRIN experiment will be performed on the site of Forschungszentrum Karlsruhe (FZK). This allows to make use of the unique expertise of the on-site Tritium Laboratory Karlsruhe (TLK), which is the only scientific laboratory equipped with a closed tritium cycle and licensed to handle the required amount of tritium (license: 40 g tritium $\approx 1.5 \cdot 10^{16}$ Bq). A further unique advantage of choosing TLK as host laboratory is the possibility to operate the tritium related parts (and in particular the tritium source) of KATRIN within the existing TLK building close to the tritium handling facilities.

The reference set-up of KATRIN (schematic view) shown in Figure 14 corresponds to a nearly 70 m long linear configuration with about 20 superconducting solenoids, which adiabatically guide tritium β decay electrons from source to detector. The experimental configuration of KATRIN can be grouped into four major functional units:

- a high luminosity Windowless Gaseous Tritium Source (WGTS) delivering 10^{11} decay electrons per second during the standard operation mode of the experiment;
- an electron transport and tritium pumping section, comprising an active differential pumping section and a passive cryogenic pumping section;
- a system of two electrostatic retarding filters with a smaller pre-spectrometer for pre-filtering and a larger main spectrometer for energy analysis of β electrons;
- a semi-conductor based high-resolution low background detector to count the β electrons transmitted through the electrostatic filters.

In the following the features of the key components are briefly summarized. For detailed information the reader is referred to the KATRIN Design Report (Angrik, 2004).

Windowless gaseous tritium source
The WGTS will consist of a 10 m long cylindrical tube of 90 mm diameter filled with molecular tritium gas of high isotopic purity (>95%) and be kept on 30 K. The tritium gas will be injected through a capillary at the middle of the tube. It diffuses over a length of 5 m to both ends of the tube and is pumped out by a series of differential turbo molecular pumping stations. The result is a density profile over the source length of nearly triangular shape with a source column density of $(\rho d) \approx 5 \cdot 10^{17}$ molecules/cm^2 (see Figure 5). Superconducting solenoids will generate a homogeneous magnetic field of 3.6 T, which adiabatically guides the decay electrons to the tube ends. The requirements for the WGTS are quite demanding:

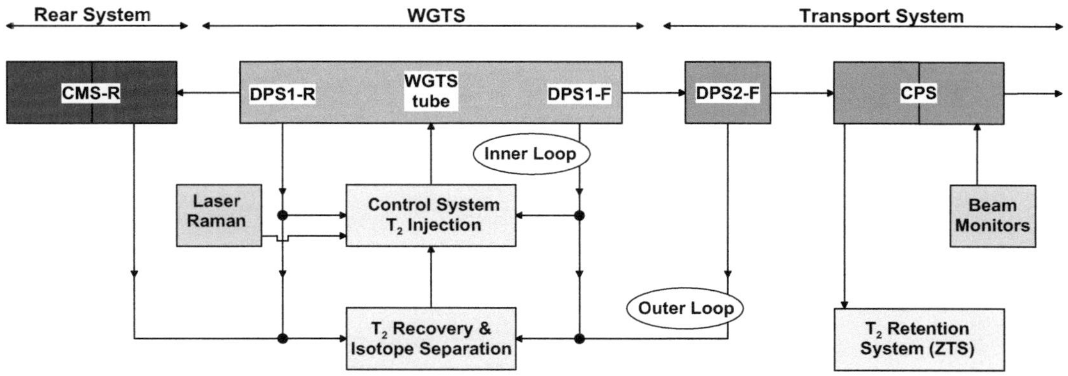

Figure 15. *Block diagram of the tritium related parts of KATRIN and their interfaces to the infrastructure of the Tritium Laboratory Karlsruhe.*

- To maintain the required column density of $(\rho d) \approx 5 \cdot 10^{17}$ molecules/cm^2 the tritium injection rate has to be in the order of 1.8 cm^3/s (standard conditions). This is equivalent to $1.7 \cdot 10^{11}$ Bq/s or 40 g tritium per day. The tritium purity should be higher than 95%.
- To minimize systematic uncertainties, the column density should be stable in the order of 0.1%. This requires an appropriate stability of the source parameters such as tube temperature, inlet pressure etc. over a whole measurement period of about 60 days per run, with 3 to 5 runs per year. This results in one of the most complex magnet-cryostat systems ever designed in the world.

Since the TLK tritium inventory is actually about 20 g, this task can only be solved by using a closed tritium loop and withdrawing only a small tritium fraction (1%) for clean-up.

The actual work on the WGTS and its connected tritium loops is focused a) on accompanying the manufacturing of the WGTS magnet&cryo system which presently is being manufactured at ACCEL company and b) on design and set-up of the closed tritium loops being done by TLK experts (see also Figure 15).

Electron transport and tritium pumping system

The electron transport system will guide the β decay electrons adiabatically from the source to the spectrometer ($B = 5.6$ T), while at the same time reducing the tritium flow rate towards the spectrometer. Since the background generated by tritium decay within the spectrometers must be less than 10^{-3} counts/s, the amount of tritium permissible in the main spectrometer is equivalent to a partial pressure of tritium of about 10^{-20} mbar. This leads to a maximal allowed tritium flow rate into the pre-spectrometer of the order of 10^{-14} mbar l/s. With a tritium injection rate of 1.8 mbar l/s and a retention factor of about 500 in the first differential pumping section DPS1 inside the WGTS the Transport System of KATRIN has to provide a tritium retention factor of about 10^{11}. This will be done by a combination of differential (DPS) and cryogenic (CPS) pumping sections. The cryo pump consists of a 4.5 K cold surface of the transport tube, covered with a thin layer of argon for better sorption. With this kind of argon frost pump one can suppress the tritium partial pressure to values below 10^{-14} mbar. To reduce the molecular beaming effect, the direct line-of-sight is prohibited by 20 or 15 degree bents between each pair of superconducting coils of about 1 m length. In the last 2 years the tritium test experiment TRAP (TRitium Argon frost Pump) has demonstrated that the required tritium

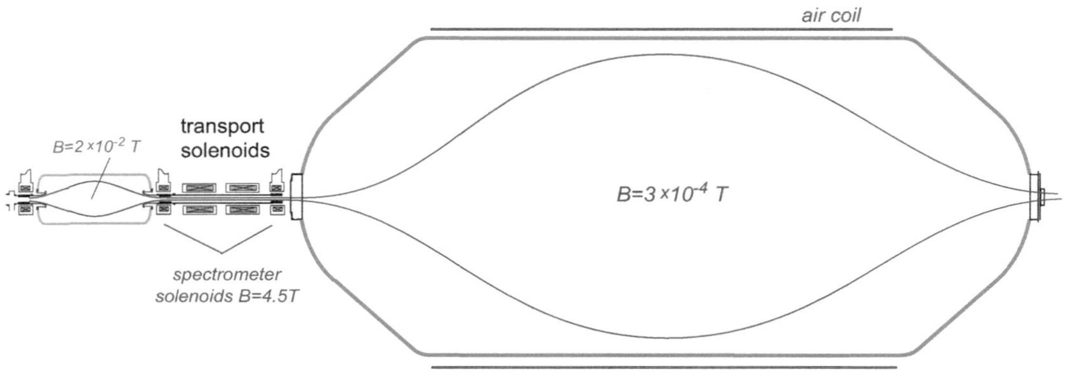

Figure 16. *Tandem spectrometer design: pre-spectrometer (left) and main spectrometer (right).*

reduction factor can be reached.

The actual work is focused on accompanying the manufacturing of the differential pumping system (DPS2-F) with delivery foreseen in 12/2007 and on the principle design of the cryo pump with tender action foreseen in 05/2007.

Electrostatic spectrometers

The energy of the β decay electrons are analyzed in a system of two electrostatic retarding spectrometers (MAC-E-Filter), a small pre-spectrometer and a large main spectrometer (see Figure 16). Since the principle of a MAC-E-Filter already has been explained in Section 3.2, only specific items of the KATRIN spectrometer will be discussed here.

The design of the KATRIN spectrometers is based on a novel electromagnetic concept: the retarding high voltage is directly connected to the hull of the spectrometer vessel itself. A nearly massless inner wire electrode at slightly more negative potential than the vessel itself suppresses low-energy electrons emanating from the inner surfaces of the spectrometer walls. These secondary electrons are mostly created by cosmic rays or environmental radioactivity from the vessel wall and are representing a potential source of background. This new method of strong background reduction has been developed and successfully tested at the Mainz spectrometer.

The tritium β decay electrons have to pass the smaller pre-spectrometer first, acting as a pre-filter. In normal tritium mode of the experiment it will reject low-energy electrons below 18.3 keV, which do not carry information on the neutrino mass. The remaining electrons enter the second, much larger spectrometer, where the energy spectrum close to the β decay endpoint is scanned with an energy resolution of 0.93 eV. The advantage of this tandem set-up is the reduction of the total flux of electrons from the tritium source into the pre-spectrometer by a factor of about 10^{-6}. This will minimize the background from ionization of residual gas molecules in the main spectrometer. For the same reason there is the requirement to reach and maintain extremely high vacuum conditions with a pressure of $< 10^{-11}$ mbar in both spectrometers constituting a major technology challenge for the KATRIN project.

The main emphasis of the pre-spectrometer hardware activities is actually given to the investigation of background systematic and the performance of the wire electrode. This information is a necessary input for the full set-up of the main spectrometer being planned for 2007/8. A major step was done in November 2006 with the successful delivery of the spectrometer

Figure 17. *Transport of main spectrometer vessel through village of Leopoldshafen near FZK. At the narrowest point there was only a few centimeter clearance to the houses.*

vessel at site of FZK (see Figure 17). Because of its size of 24 m length and 10 m diameter it has been shipped from its manufacturing site, Deggendorf at river Danube, to FZK via a more than 8000 km detour over Danube, Black See, Mediterranean, Gibraltar, Atlantic Ocean, River Rhine to Leopoldshafen and from there by transporter about 8 km over road to FZK. The direct way of less than 400 km was not possible because of the large dimension of the vessel.

Detector
All β electrons passing the retarding potential of the main spectrometer are re-accelerated to their initial energy and magnetically guided to the final plane detector (FPD) being located inside a separate superconducting solenoid with a large warm bore. The FPD will be a multi-pixel silicon semi-conductor detector with a ultra-high energy resolution and a very thin entrance window. The present concept is based on a large array of about 200 PIN photodiodes surrounded by low-level passive shielding and an active veto counter to reduce background.

Currently intensive R&D is devoted to the detector layout, background simulations are underway to optimize the shielding design and the pre-amplifier development. Test wafers are being manufactured.

5.3 The neutrino mass sensitivity of KATRIN

For a high sensitivity tritium β decay experiment like KATRIN, the relevant region of the β spectrum close to the endpoint E_0 is very narrow. A narrow energy interval is equivalent to a low count rate being also equivalent to a large statistical error. On the other hand, a narrow energy interval close to E_0 strongly reduces possible systematic uncertainties, since these uncertainties mainly arise from processes connected to atomic and molecular physics, such as inelastic scattering of tritium β electrons in the tritium source. Due to energy thresholds in such processes the fraction of β electrons not being disturbed is increasing with decreasing energy interval close to E_0 (see also Figure 11). Figure 18 gives an estimate of the statistical uncertainties in dependence of the fit interval. Displayed is here the expected statistical error after 3 years measurement time with a tritium column density of $5 \cdot 10^{17}$ cm^{-2}, 51 degree solid angle,

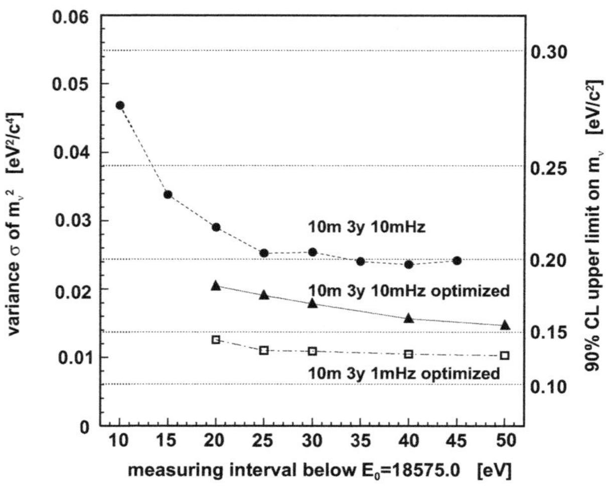

Figure 18. *Statistical uncertainty σ of the observable m_ν^2 and corresponding upper limit (90% C.L.) as a function of the analyzed interval for different configurations and background rates. The right axis indicates the upper limit on m_ν assuming the corresponding statistical uncertainty only for $m_\nu = 0$ eV/c^2 (Feldmann and Cousins, 1998).*

95% tritium purity and 0.93 eV energy resolution of the main spectrometer. '10m' stands for spectrometer with 10 m diameter and '10 mHz' or '1 mHz' is the background rate. It is clearly visible that of utmost importance are the reduction of background as well as the optimization of the measuring time distribution, i.e. how much time to spend at which retarding potential.

Detailed simulations of the KATRIN experiment including the investigation of possible systematic effects and their influence on the fit parameter m_ν^2 led to the following results:

- The total systematic uncertainty of the fit parameter m_ν^2 is expected to be $\sigma_{\text{sys}} \leq 0.017$ eV^2/c^4.

- For the KATRIN reference solution ('10m, 3y, 10 mHz, optimized', see discussion above) and a fit interval of [E_0-30 eV, E_0+5 eV], the statistical uncertainty of the fit parameter m_ν^2 amounts to $\sigma_{\text{stat}} \leq 0.018$ eV^2/c^4.

- Adding statistical and systematic uncertainties quadratically leads to a total uncertainty of $\sigma_{\text{tot}} \approx 0.025$ eV^2/c^4. Assuming a vanishing neutrino mass, this uncertainty translates into an upper limit L(90% C.L.) which is connected to the error of m_ν^2 via L(90% C.L.) = $\sqrt{1.64 \cdot \sigma_{\text{tot}}}$ (Feldmann and Cousins, 1998). After three years of measuring time this limit becomes $m_{\nu_e} < 0.2$ eV/c^2 with no finite neutrino mass being observed. On the other hand, a non-zero neutrino mass of 0.30 eV/c^2 would be detected with 3σ significance, a mass of 0.35 eV/c^2 even with 5σ (one has to divide the measured m_ν^2 by the total uncertainty, here: $0.35^2/0.025 = 4.9$).

This sensitivity improves the existing limits by one order of magnitude and also demonstrates the discovery potential of KATRIN for degenerate neutrino mass scenarios and for neutrino masses of cosmological relevance.

6 Summary and conclusions

Experiments for an absolute measurement of the neutrino mass have been performed since more than 5 decades, yielding in upper limits for ν_τ, ν_μ and ν_e. In all cases a sound knowledge of systematic effects is absolutely necessary. Its importance has been shown by way of example for the Mainz electron neutrino mass experiment.

The lowest direct upper limits on a neutrino mass of a few eV are obtained for the mass of the electron neutrino by means of the investigation of the tritium β decay. The recent tritium β decay experiments at Mainz and Troitsk have reached their sensitivity of about 2 eV/c^2. Since a mass determination with sub-eV sensitivity is needed to distinguish between hierarchical and degenerate neutrino mass models and to clarify the role of neutrinos in the early universe, a next generation direct neutrino mass experiment has to be set-up and operated. Discussing the different options shows that this experiment has to be a large tritium β decay experiment using a MAC-E-Filter. Such an experiment is being set-up by the KATRIN collaboration at the site of Forschungszentrum Karlsruhe.

Acknowledgments

I thank the organizers of the school for their invitation and their hospitality on site.

References

Altarelli G and Winter K (Editors), 2003, Neutrino Mass, p25-52 (Springer, Volume 190) .
Angrik J et al. (KATRIN Design Report 2004), 2005, *FZKA Scientific Report 7090* .
Arnaboldi C et al., 2003, *Physical Review Letters* **91** 161802 .
Assamagan K et al., 1996, *Physical Review* **D53** 6065 .
Avignone F T and Collar J I, 1990, *Physical Review* **D41** 682 .
Barkas W H et al., 1956, *Physical Review* **101** 778 .
Barate R et al., 1998, *European Physical Journal* **C2** 395 .
Belesev A I et al., 1995, *Physics Letters* **B350** 263 .
Ching C R et al., 1995 *International Journal of Modern Physics* **A10** 2841 .
Fermi E, 1934, *Zeitschrift für Physik* **88** 11 .
Feldmann G J and Cousins R D, 1998, *Physical Review* **D57** 3873 .
Gatti F, MARE proposal, 19/05/2006, (http://crio.mib.infn.it/wig/silicini/publications.html) .
Fleischmann L et al., 2000a, *Journal of Low Temperature Physics* **119** 615 .
Fleischmann L et al., 2000b, *European Physical Journal* **B16** 521 .
Hiddemann K H et all, 1995, *Journal of Physics* **G21** 639 .
Holzschuh E et al., 1992, *Physics Letters* **B287** 381 .
Jeckelmann B et al., 1994, *Physics Letters* **B335** 326 .
Kawakami H et al., 1991, *Physics Letters* **B256** 105 .
Kraus Ch et al., 2005, *European Physical Journal* **C40** 447 .
Kernan P J and Krauss L M, 1995, *Nuclear Physics* **B437** 243 .
Lenz S et al., 1998, *Physics Letters* **B416** 50 .
Lobashev V M, 2003, *Proceedings 17th Int. Conf. on Nuclear Physics in Astrophysics, Debrecen/Hungary, 2002*, in *Nucl. Phys. A* **719** 153 .

Lobashev V M, 1999, *Physics Letters* **B460** 227 .
Loredo T J and Lamb D Q, 2002, *Physical Review* **D65** 063002 .
Mohr P J and Taylor B N, 2005, *Review of Modern Physics* **77** 1 .
Robertson R G H et al., 1991, *Physical Review Letters***67** 957 .
Sisti M et al., 2004, *Nuclear Instruments and Methods* **A520** 125 .
Stoeffl W and Decman D J, 1995, *Physical Review Letters* **75** 3237 .
Weinheimer Ch et al., 1999, *Physics Letters* **B460** 219 .
Weinheimer Ch et al., 1993, *Physics Letters* **B300** 210 .
Vogel P, 2002, *Prog. Part. Nucl. Phys.* **48** 29 .
Yao W M et al. (Particle Data Group),2006, *J. Phys.* **G33** 1 .

Neutrinoless Double Beta Decay

Kai Zuber

Department of Physics and Astronomy, University of Sussex,
Falmer, Brighton BN1 9QH, UK

1 Introduction

Neutrino physics has gone through a revolution in the last ten years. Now it is beyond doubt that neutrinos have a non-vanishing rest mass. All the evidence stems from neutrino oscillation experiments, proving that neutrinos can change their flavour if travelling from a source to a detector. Oscillations violate the concept of single lepton number conservation but total lepton number is still conserved. Furthermore, the oscillation experiments are not able to measure absolute neutrino masses, because their results depend only on the differences of masses-squared, $\Delta m^2 = m_i^2 - m_j^2$, with m_i, m_j as the masses of two neutrino mass eigenstates. In the full three neutrino mixing framework the weak eigenstates ν_e, ν_μ and ν_τ can be expressed as superpositions of three neutrino mass eigenstates ν_1, ν_2 and ν_3 linked via a unitary matrix U:

$$\begin{pmatrix} \nu_e \\ \nu_\mu \\ \nu_\tau \end{pmatrix} = \begin{pmatrix} U_{e1} & U_{e2} & U_{e3} \\ U_{\mu 1} & U_{\mu 2} & U_{\mu 3} \\ U_{\tau 1} & U_{\tau 2} & U_{\tau 3} \end{pmatrix} \begin{pmatrix} \nu_1 \\ \nu_2 \\ \nu_3 \end{pmatrix} \tag{1}$$

This kind of mixing has been known in the quark sector for decades and the analogous matrix U is called Cabbibo-Kobayashi-Maskawa matrix. The corresponding mixing matrix in the lepton sector is named Pontecorvo-Maki-Nakagawa-Sato (PMNS)-matrix [**?**]. The unitary matrix U in eq. 1 can be parametrised in the following form

$$U = \begin{pmatrix} c_{12}c_{13} & s_{12}c_{13} & s_{13}e^{-i\delta} \\ -s_{12}c_{23} - c_{12}s_{23}s_{13}e^{i\delta} & c_{12}c_{23} - s_{12}s_{23}s_{13}e^{i\delta} & s_{23}c_{13} \\ s_{12}s_{23} - c_{12}s_{23}s_{13}e^{i\delta} & -c_{12}s_{23} - s_{12}c_{23}s_{13}e^{i\delta} & c_{23}c_{13} \end{pmatrix} \tag{2}$$

where $s_{ij} = \sin\theta_{ij}, c_{ij} = \cos\theta_{ij}\,(i, j = 1,2,3)$. The phase δ is a source for CP-violation and like in the quark sector cannot be removed by rephasing the neutrino fields. The Majorana case, ie. the requirement of particle and antiparticle to be identical, restricts the freedom to redefine the fundamental fields even further. The net effect is the appearance of a CP-violating phase even for two flavours. For three flavours two additional phases have to be introduced resulting in a mixing matrix of the form

$$U = U_{PMNS} diag(1, e^{i\alpha_2}, e^{i\alpha_3}) \tag{3}$$

with the two new Majorana phases α_2 and α_3. These phases again might only be accessible in double beta decay, they are not accessible in neutrino oscillation experiments. They are a further source of CP-violation.
Based on the observations from neutrino oscillations (see (Kayser 2006)), various neutrino mass models have been proposed. These can be categorized as normal hierarchy ($m_3 \gg m_2 \approx m_1$), inverted hierarchy ($m_2 \approx m_1 \gg m_3$) and almost degenerate ($m_3 \approx m_2 \approx m_1$) neutrinos (Fig. 1). A key result, based on the observed Δm^2 in atmospheric neutrinos, is the existence of a neutrino mass eigenstate in the region around 10-50 meV. This is the minimal value neccessary, because it corresponds to the square root of the measured Δm^2 in case one of the mass eigenstates is zero. Fixing the absolute mass scale is of outmost importance, because it will fix the mixing matrix and various other important quantities will then be determined, like the contribution of neutrinos to the mass density in the Universe.
Traditionally, laboratory experiments search for a finite neutrino rest mass by exploring the endpoint energy of the electron spectrum in tritium beta decay. Currently a limit for the electron neutrino mass of less than 2.2 eV has been achieved (Bornschein 2006). A similar limit is obtained by analysing recent cosmic microwave background measurements using the WMAP satellite combined with large scale galaxy surveys and Lyman-α systems, see e.g. (Lesgourges 2006). However, there are about two orders of magnitude difference with respect to the region below 50 meV and even the next generation beta decay experiment, called KATRIN, can at best lead to an improvement of a factor ten. However it should be noticed, that beta decay and double beta decay are measuring slightly different observables and are rather complementary than competitive. Therefore, very likely double beta decay is the only way to explore the region below 100 meV.

2 Double beta decay

Double beta decay is characterized by a nuclear process, changing the nuclear charge Z by two units while leaving the atomic mass A unchanged. It is a transition among isobaric isotopes. Therefore, it is a higher order process and can be seen as two simultaneous beta decays. This can only happen for even-even nuclei. All even-even nuclei have a ground state of spin 0 and a positive parity, hence the ground state transitions are characterised as $(0^+ \to 0^+)$ transitions. Thus, a neccessary requirement for double beta decay to occur is $m(Z,A) > m(Z+2,A)$ and for practical purposes β-decay has to be forbidden $m(Z,A) < m(Z+1,A)$ or at least strongly suppressed. The same ground state configurations and arguments might hold for isotopes on the right side of the even-even parabola. This would lead to the process of double positron decay or double electron capture, discussed later. In nature 35 isotopes are known, which show the specific ground state configuration necessary for double beta decay. Double beta decay was first discussed by M. Goeppert-Mayer (Goeppert-Mayer 1935) in the form of

$$(Z,A) \to (Z+2,A) + 2e^- + 2\bar{\nu}_e \quad (2\nu\beta\beta\text{-decay}) \tag{4}$$

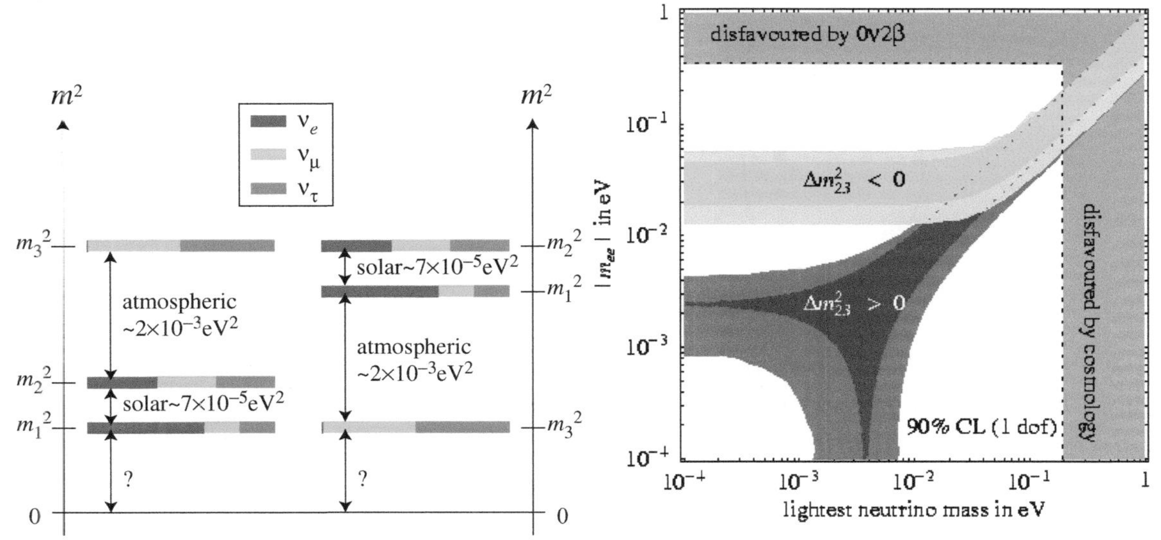

Figure 1. *Left: Possible configurations of neutrino mass states as suggested by oscillations. Currently a normal (left) and an inverted (right) hierarchy cannot be distinguished. The flavour composition is shown as well. Right: The effective Majorana mass $\langle m_{\nu_e} \rangle$ as a function of the lightest mass eigenstate m_1. Hierarchical mass patterns can be distinguished for $\langle m_{\nu_e} \rangle$ smaller than 50 meV, otherwise neutrinos can be considered as almost degenerate. Also shown in grey are the regions disfavoured by current $0\nu\beta\beta$-decay limits and a very optimistic limit (could be worse by an order of magnitude) from cosmology (from (Feruglia 2002)).*

This process can be seen as two simultaneous neutron decays (Fig. 2). Shortly after the classical papers of Majorana (Majorana 1937) discussing a 2 - component neutrino, Racah (Racah 1937) and Furry discussed another decay mode in the form of (Furry 1939)

$$(Z,A) \rightarrow (Z+2,A) + 2e^- \quad (0\nu\beta\beta\text{-decay}) \quad . \tag{5}$$

In contrast to neutrino oscillations which violate individual flavour lepton number, but keep total lepton number conserved, $0\nu\beta\beta$-decay violates total lepton number by two units. This process is forbidden in the Standard Model. It can be seen as two subsequent steps ("Racah - sequence") as shown in Fig. 2:

$$(Z,A) \rightarrow (Z+1,A) + e^- + \bar{\nu}_e$$
$$(Z+1,A) + \nu_e \rightarrow (Z+2,A) + e^-$$

First a neutron decays under the emission of a right-handed $\bar{\nu}_e$. This has to be absorbed at the second vertex as a left-handed ν_e. To fulfill these conditions neutrino and antineutrino have to be identical, requiring that neutrinos are Majorana particles, i.e. a 2-component object. This is different from all the other fundamental fermions where particles and antiparticles can be already distinguished by their charge. Majorana neutrinos are preferred by most Grand Unified Theories to explain the small magnitude of neutrino masses via the see-saw mechanism. Hence, double beta decay is generally considered to be the 'gold plated' channel to probe the fundamental character of neutrinos. Moreover, to allow for the helicity matching, a neutrino mass is required. The reason is that the wavefunction describing neutrino mass eigenstates for

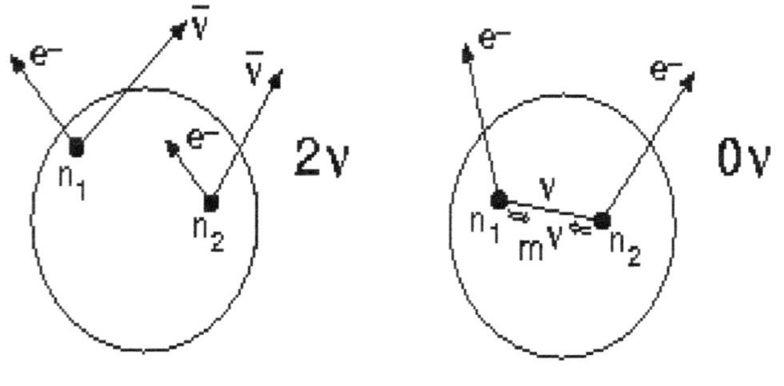

Figure 2. *Principle of double beta decay. Left: The simultaneous decay of two neutrons as an allowed higher order process ($2\nu\beta\beta$-decay). Right: The lepton-number violating mode ($0\nu\beta\beta$-decay) where the neutrino only occurs as a virtual particle. This process is not allowed in the Standard Model.*

$m_\nu > 0$ has no fixed helicity and therefore, besides the dominant left-handed contribution, has an admixture of a right-handed component (or vice versa for antineutrinos), which is proportional to m_ν/E_ν. Thus, for double beta decay to occur, massive Majorana particles are required. For recent reviews on double beta decay see (Elliott 2004, Zuber 2006).

The quantity measured in $0\nu\beta\beta$-decay is called the effective Majorana neutrino mass and is given for light neutrinos by

$$\langle m_{\nu_e} \rangle \;=\mid \sum_i U_{ei}^2 m_i \mid=\mid \sum_i \mid U_{ei} \mid^2 e^{i\alpha_i} m_i \mid \tag{6}$$

which can be written in case of CP-invariance ($e^{i\alpha} = 0, \pi$) as

$$\langle m_{\nu_e} \rangle \;=\mid m_1 \mid U_{e1}^2 \mid \pm m_2 \mid U_{e2}^2 \mid \pm m_3 \mid U_{e3}^2 \mid\mid \tag{7}$$

As can be seen, the different terms in the sum have a chance to interfere destructively, only the absolute value is measured at the end. On the other hand, beta decay measures

$$m_{\bar{\nu}_e} = \sum_i \mid U_{ei}^2 \mid m_i \tag{8}$$

which is independent of the fundamental character of the neutrino and does not allow destructive interference. As a result, a certain care should be taken if comparing neutrino masses obtained by β-decay and $0\nu\beta\beta$-decay , they should be seen as complementary measurements. A discussion of $\langle m_{\nu_e} \rangle$ within the context of the oscillation results is also useful. In terms of mixing angles it can be written as

$$\langle m_{\nu_e} \rangle \;= \cos^2\theta_{12}\cos^2\theta_{13} m_1 + \sin^2\theta_{12}\cos^2\theta_{13} e^{i\alpha_1} m_2 + \sin^2\theta_{13} e^{i\alpha_2} m_3 \tag{9}$$

Solar neutrino results combined with KamLAND reactor data are considered to be describe by θ_{12} and hence result in $\Delta m_{12}^2 = \Delta m_\odot^2 = 8\times 10^{-5}\ eV^2$ and a mixing angle of $\sin^2\theta_{12} = \sin^2\theta_\odot =$ 0.3, atmospheric and in the similar way long baseline neutrino oscillations can be described by $\Delta m_{23}^2 = \Delta m_{atm}^2 = 2.5\times 10^{-3}\ eV^2$ and $\sin^2\theta_{23} = \sin^2\theta_{atm} = 1$ and the remaining mixing angle is restricted by reactor data to be $\sin^2\theta_{13} = \sin^2\theta_R < 0.051$. Thus, equation 6 can be rewritten

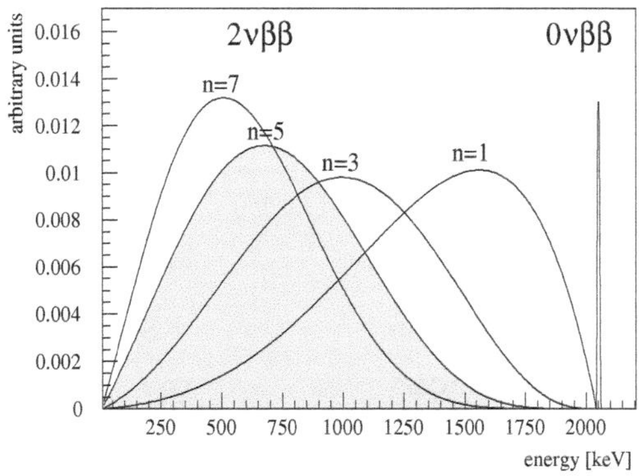

Figure 3. *Schematic drawing of the sum energy spectrum of electrons in double beta decay, here in case of ^{76}Ge.The $2\nu\beta\beta$-decay shows a continous spectrum (yellow), while $0\nu\beta\beta$-decay is a peak at the Q-value of the transition. The additional curves shown correspond to various Majoron emitting modes not discussed here.*

as

$$\langle m_{\nu_e}\rangle = \cos^2\theta_\odot \cos^2\theta_R m_1 + \sin^2\theta_\odot \cos^2\theta_R e^{i\alpha_1}\sqrt{m_1 + \Delta m_\odot^2} + \sin^2\theta_R e^{i\alpha_2}\sqrt{m_1 + \Delta m_\odot^2 + \Delta m_{atm}^2} \tag{10}$$

In case of CP invariance there are four possible ways how those terms can add up according to the sign in the second and third term. For a very small $\sin^2\theta_R$ there could be a vanishing $\langle m_{\nu_e}\rangle$ if $m_1 = \tan^2\theta_\odot m_2$. In the case of the inverted hierarchy and a negligible third term there is a lower bound on $\langle m_{\nu_e}\rangle$ due to the non-maximal mixing of solar neutrinos resulting in $\langle m_{\nu_e}\rangle = (\cos^2\theta_\odot - \sin^2\theta_\odot)\sqrt{\Delta m_{atm}^2} \approx 20$ meV (see Fig. 1).

3 General considerations

Being a nuclear decay, the actual experimental quantity measured is the half-life. As a higher order effect the expected half-lives for double beta decay are long, in the region of about 10^{20} years and beyond. The experimental signal of $0\nu\beta\beta$-decay is two electrons in the final state, whose energies add up to the Q-value of the nuclear transition, while for the $2\nu\beta\beta$-decay the sum energy spectrum of both electrons will be continuous (Fig. 3). The total decay rates, and hence the inverse half-lives, are a strong function of the available Q-value. The rate of $0\nu\beta\beta$-decay scales with Q^5 compared to a Q^{11}-dependence for $2\nu\beta\beta$-decay . Therefore isotopes with a high Q-value (above about 2 MeV) are normally considered for experiments. This restricts the candidates to those listed in Tab. 3. The measured half-life or its lower limit in case of non-observation of the process can be converted into a neutrino mass or an upper limit via

$$(T_{1/2}^{0\nu})^{-1} = G^{0\nu} \mid M^{0\nu} \mid^2 \left(\frac{\langle m_{\nu_e}\rangle}{m_e}\right)^2 \tag{11}$$

where $G^{0\nu}$ is the exactly calculable phase space integral (see (Boehm 1992) for numerical values) of the decay and $\mid M^{0\nu} \mid$ is the nuclear matrix element of the transition.

With reasonable assumptions on the nuclear matrix element it can be estimated that for a neutrino mass measurement of the order of 50 meV, half-lives in the region of $10^{26} - 10^{27}$

Table 1. *Compilation of $\beta^-\beta^-$-emitters with a Q-value of at least 2 MeV. Shown are the transition energies Q and the natural abundances.*

Transition	Q-value (keV)	nat. ab. (%)
$^{48}_{20}Ca \rightarrow ^{48}_{22}Ti$	4271	0.187
$^{76}_{32}Ge \rightarrow ^{76}_{34}Se$	2039	7.8
$^{82}_{34}Se \rightarrow ^{82}_{36}Kr$	2995	9.2
$^{96}_{40}Zr \rightarrow ^{96}_{42}Mo$	3350	2.8
$^{100}_{42}Mo \rightarrow ^{100}_{44}Ru$	3034	9.6
$^{110}_{46}Pd \rightarrow ^{110}_{48}Cd$	2013	11.8
$^{116}_{48}Cd \rightarrow ^{116}_{50}Sn$	2809	7.5
$^{124}_{50}Sn \rightarrow ^{124}_{52}Te$	2288	5.64
$^{130}_{52}Te \rightarrow ^{130}_{54}Xe$	2530	34.5
$^{136}_{54}Xe \rightarrow ^{136}_{56}Ba$	2458	8.9
$^{150}_{60}Nd \rightarrow ^{150}_{62}Sm$	3367	5.6

years must be explored, by no means an easy task. This can be shown by the following estimate. Assume the radioactive decay law in the approximation $T_{1/2} \gg t$

$$T^{0\nu}_{1/2} = \ln 2 a N_A m t / N_{\beta\beta} \tag{12}$$

with t the measuring time, m the sample mass, a is the natural abundance of the isotope of interest, N_A the Avogadro constant and $N_{\beta\beta}$ the number of double beta decays. Expecting a half-life of about 6×10^{26}yrs and to observe as little as one decay per year, the number of source atoms required is around 6×10^{26}. However, this corresponds to 1000 moles and using an average isotope of mass 100 like ^{100}Mo , would immediately imply using about 100 kg. Hence, even without any disturbing background, and full efficiency for detection, one needs about hundred kilogram of the isotope of interest, to observe one decay per year independent of the experimental approach! Even worse, in the background-limited case, the sensitivity on the half-life depends on experimental quantities according to

$$T^{0\nu}_{1/2} \propto a \cdot \varepsilon \cdot \sqrt{\frac{M \cdot t}{\Delta E \cdot B}} \tag{13}$$

with a the natural abundance of the isotope of interest, ε the detection efficiency, M the mass of source employed, t the measuring time, ΔE the energy resolution at the peak position in the sum energy spectrum of the electrons (Q-value) and B the background index, typically quoted in events/keV/kg/yr. In contrast to the background-free case, for a background-limited experiment the half-life sensitivity increases only with the square root of the measuring time and mass.

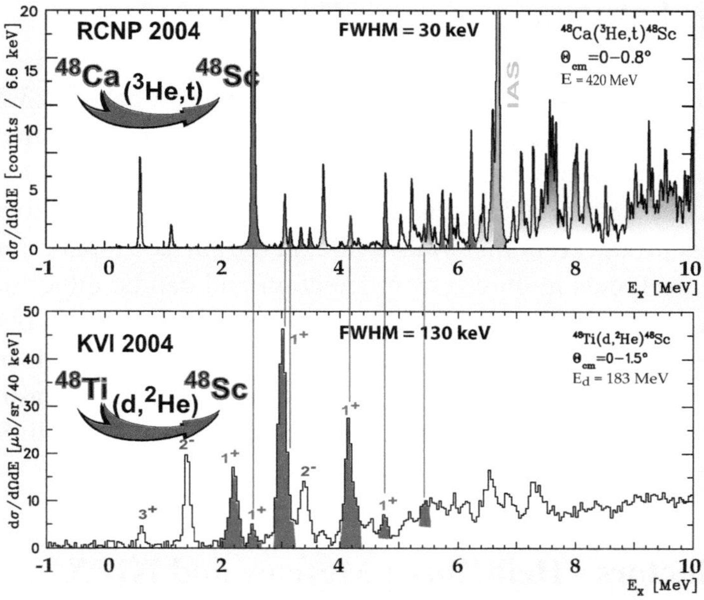

Figure 4. *Measured* $^{48}Ca(^{3}He,t)^{48}Sc$ *(RCNP Osaka) and* $^{48}Ti(d,^{2}He)^{48}Sc$ *(KVI Groningen) spectra in charge exchange reactions at* $0°$. *Intermediate states which are excited by both reactions are on top of each other (Frekers 2006a).*

3.1 Nuclear matrix elements

As can be seen in eq.11, the major ingredients in the conversion of measured half-lives into neutrino masses are the nuclear matrix elements involved. Those calculations are performed within the quasi random phase approximation (QRPA) or by using the nuclear shell model. While $2\nu\beta\beta$-decay matrix element are pure Gamow-Teller transitions as only 1^+-states in the intermediate nucleus are contributing, in $0\nu\beta\beta$-decay also higher multipoles contribute. A detailed discussion is beyond this article, for details see (Faessler 1998, Suhonen 1998, Ejiri 2000, Rodin 2006). There still seems to be an uncertainty of a factor 2-3 in the calculations, the treatment of short range-correlation of nucleons in the calculations is likely responsible for a significant part of the discrepancy (Kortelainen 2007). Hence, an initiative has recently been started to provide those calculations with more and better input from the experimental side to help as much as possible to settle the issue (Zuber 2005). Those measurements include charge exchange reactions measuring the beta transition strengths to 1^+-states. Some of the isotopes have already been measured at KVI Groningen with the (d,^{2}He) reactions complemented by (^{3}He,t) measurements performed at RCNP Osaka (Fig. 4). New ft-value measurements of electron capture for the intermediate nuclei are proposed using atomic traps (Frekers 2007). Those might help to solve the issue of how to fix the particle-particle coupling parameter g_{PP}, to which the 1^+-states calculations are very sensitive. Atomic traps will be used as well to determine the Q-values of some transitions more accurately by high precision mass spectrometry. In addition, ordinary muon capture and neutrino-nucleus scattering have been proposed to gain further information on the matrix elements involved. The hope is that all those measurements might allow to bring down the error on the matrix elements down to the level of 30 %.

4 Experimental status

The search for $0\nu\beta\beta$-decay relies on finding a peak in the region below 4.3 MeV, depending on the isotope (see Tab. 3). Common to all experimental approaches is the aim for a very low-background environment due to the long expected half-lives. Among the most common background sources are the natural decay chains of U and Th, ^{40}K, Rn, neutrons, atmospheric muons and radioisotopes produced in materials while on the surface (spallation).
All direct experiments are focusing on electron detection and can be either active or passive. Active detectors are such that source and detector are identical which is a big advantage, but often only measure the sum energy of both electrons. On the other hand, passive detectors (source and detector are different) allow to get more information like measuring energy and tracks of both electrons seperately, but usually have smaller source strength. Some running or past experiments will be described now in a little more detail.

4.1 Ge-semiconductors - Heidelberg Moscow and IGEX

Major progress has been achieved in the last decades by pushing half-life limits and increasing the sensitivity towards smaller and smaller neutrino masses using Ge-semiconductor devices. Source and detector are identical, the isotope under investigation is ^{76}Ge with a Q-value of 2039 keV. The big advantage is the excellent energy resolution of Ge-semiconductors (typically about 3-4 keV at 2 MeV). However, the technique only allows the measurement of the sum energy of the two electrons. A big step forward due to an increase in source strength was done by using enriched germanium (the natural abundance of ^{76}Ge is 7.8 %). Two experiments were performed recently, the Heidelberg-Moscow and the IGEX experiment. The Heidelberg-Moscow experiment in the Gran Sasso Laboratory took data from 1990-2003 using 11 kg of Ge enriched to about 86 % in ^{76}Ge in the form of five high purity Ge-detectors (HPGe). A background as low as 0.12 counts/year/kg/keV at the peak position has been achieved. After 53.9 $kg \times y$ of data taking the peak region reveals no signal and the obtained half-life limit is (Klapdor-Kleingrothaus 2001) $T_{1/2}^{0\nu} > 1.9 \times 10^{25}yrs(90\%CL)$ which can be converted using eq. 13 and the matrix elements given in (Staudt 1990) to an upper bound of $\langle m_{\nu_e} \rangle < 0.35$ eV. This is currently the best available bound coming from double beta decay . However, recently a subgroup of the collaboration found a small peak at the expected position (Klapdor-Kleingrothaus 2004) (Fig. 5). Taking the peak as real and based on 71.7 $kg \times y$ of data would point towards a half-life between $0.7 - 4.2 \times 10^{25}$yrs. Using the matrix elements calculated in (Staudt 1990) this would imply a range for the neutrino mass between 0.2-0.6 eV, which might be widened by using other matrix element calculations. If true, this would immediately result in the fact that neutrinos are almost degenerate. However, the discussion concerning the possible evidence is still quite controversial.

4.2 CdZnTe-semiconductors - COBRA

A new approach to take advantage of the good energy resolution of semiconductors is COBRA (Zuber 2001) located in the Gran Sasso Underground Laboratory (LNGS). In total, there are nine double beta emitters within the detector including those of $\beta^+\beta^+$ decay. The idea here is to use CdZnTe detectors, mainly to explore ^{116}Cd and ^{130}Te decay and ^{106}Cd for $\beta^+\beta^+$ de-

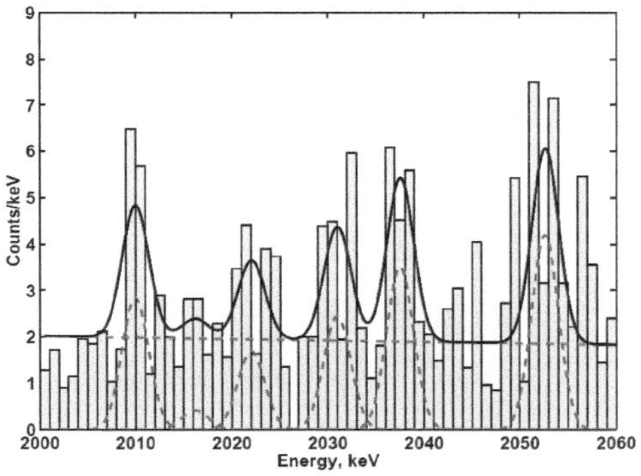

Figure 5. *Energy spectrum of the Heidelberg-Moscow experiment around the $0\nu\beta\beta$-decay region at 2040 keV (Klapdor-Kleingrothaus 2004).*

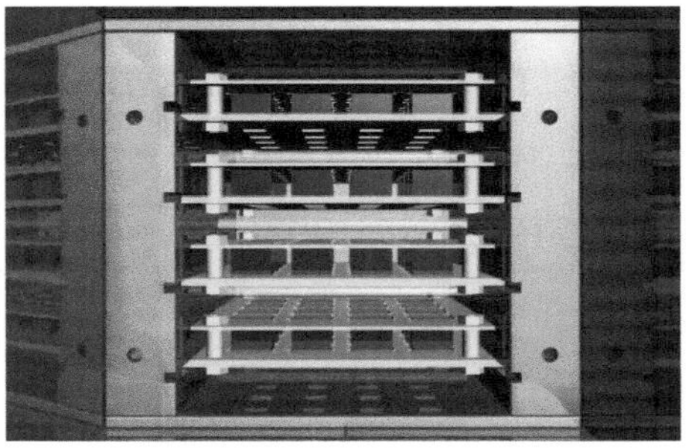

Figure 6. *Schematic layout of the COBRA 64 array in the form of a $4 \times 4 \times 4$ configuration. Each layer contains 16 CdZnTe semiconductor detectors.*

cay. The smallness of the detectors makes a search for coincidences a powerful way to reduce γ-background. The practical handling is simplified as these detectors are room temperature detectors. In case of pixelated detectors it offers tracking possibilities and even further background reduction by particle identification. Recent results obtained with four detectors can be found in (Zuber 2006a). Currently an upgrade to 64 detectors is ongoing, corresponding to about 0.42 kg CdZnTe (Fig. 6).

4.3 Cryogenic bolometers - CUORICINO

Currently two large scale experiments are running. The first technique uses cryogenic bolometers running at very low temperature (mK). The strong dependence of the specific heat as a function of temperature allows a suitable energy measurement. CUORICINO at the Gran Sasso Underground Laboratory in Italy, is operating 62 TeO_2 crystals, corresponding to about 40 kg, at 8 mK to search for the ^{130}Te decay with a Q-value of 2530 keV. The obtained half-life limit with 11.83 kg x yr statistics corresponds to (Arnaboldi 2005) $T^{0\nu}_{1/2}(^{130}Te\,) > 3 \times 10^{24} yr \quad (90\% CL)$ resulting in an upper bound on the neutrino mass of 0.2-1 eV, depending on the matrix elements used. An upgrade to CUORE, using 750 kg of TeO_2 is on its way.

Table 2. *Compilation of some obtained limits for $0\nu\beta\beta$-decay . However, notice the claimed evidence for ^{76}Ge. All results are 90 % CL, except ^{48}Ca (76 %) and ^{128}Te (68 %).* corresponds to a geochemical measurement.*

Isotope	Half-life limit (yrs)		ν mass limit (eV)
$^{48}_{20}Ca \rightarrow ^{48}_{22}Ti$	$> 9.5 \cdot 10^{21}$(76%)		< 8.3
$^{76}_{32}Ge \rightarrow ^{76}_{34}Se$	$> 1.9 \cdot 10^{25}$	(90%)	< 0.35
$^{76}_{32}Ge \rightarrow ^{76}_{34}Se$	$0.7 - 4.2 \cdot 10^{25}$	(90%)	$0.2 - 0.6$
$^{82}_{34}Se \rightarrow ^{82}_{36}Kr$	$> 2.1 \cdot 10^{23}$	(90%)	$< 1.2 - 2.3$
$^{100}_{42}Mo \rightarrow ^{100}_{44}Ru$	$> 5.8 \cdot 10^{23}$	(90%)	$< 0.6 - 1.2$
$^{116}_{48}Cd \rightarrow ^{116}_{50}Sn$	$> 1.7 \cdot 10^{23}$	(90%)	< 1.7
$^{128}_{52}Te \rightarrow ^{128}_{54}Xe$	$> 7.7 \cdot 10^{24}$	(68%)	$< 1.1^*$
$^{130}_{52}Te \rightarrow ^{130}_{54}Xe$	$> 3 \cdot 10^{24}$	(90%)	$< 0.2 - 1$
$^{136}_{54}Xe \rightarrow ^{136}_{56}Ba$	$> 4.4 \cdot 10^{23}$	(90%)	< 2.3
$^{150}_{60}Nd \rightarrow ^{150}_{62}Sm$	$> 2.1 \cdot 10^{21}$	(90%)	< 4.1

4.4 Time projection chambers - NEMO-3

The second experiment, NEMO-3 in the Frejus Underground Laboratory, is of the form of a passive experiment , which is mostly built in the form of time projection chambers (TPCs) where the double beta emitter is either the filling gas of the chamber (like ^{136}Xe) or is included in thin foils. The advantage is that energy measurements as well as tracking of the two electrons is possible. Disadvantages are the energy resolution and in case of using thin foils the limited source strength. NEMO-3 consists of a tracking (wire chambers) and a calorimetric (plastic scintillators) device put into a 25 G magnetic field. The total source strength is about 10 kg which in a first run is dominated by using enriched ^{100}Mo foils (about 7 kg). After 693 days limits of $T^{0\nu}_{1/2}(^{100}\text{Mo}) > 5.8 \times 10^{23} yr \quad (90\%CL)$ and $T^{0\nu}_{1/2}(^{82}Se) > 2.1 \times 10^{23} yr \quad (90\%CL)$, resulting in upper neutrino mass bound of 0.6 - 1.2 eV from ^{100}Mo , have been achieved. First results can be found in (Arnold 2005).
Observation of $2\nu\beta\beta$-decay has been quoted now for about a dozen isotopes. A complete listing of all experimental results obtained until the end of 2001 can be found in (Tretyak 2002), some newer values are in (Elliott 2004), with the most important results shown in Table 4.4.

It should be noted that double beta decay could also occur through other $\Delta L = 2$ processes besides light Majorana neutrino exchange. Whatever kind of new physics is providing this lepton number violation with two electrons in the final state will be restricted by the obtained experimental results. Among them are right-handed weak interactions, heavy Majorana neutrino exchange, double charged Higgs bosons, R-parity violating SUSY couplings (λ'_{111}), leptoquarks and other Beyond Standard Model physics.

Table 3. *Compilation of $\beta^+\beta^+$-emitters requiring a Q-value of at least 2048 MeV. Shown are the full transition energies Q-4m_ec^2 and the natural abundances.*

Transition	Q-4m_ec^2 (keV)	nat. ab. (%)
${}^{78}_{36}\mathrm{Kr} \rightarrow {}^{78}_{34}\mathrm{Se}$	838	0.35
${}^{96}_{44}\mathrm{Ru} \rightarrow {}^{96}_{42}\mathrm{Mo}$	676	5.5
${}^{106}_{48}\mathrm{Cd} \rightarrow {}^{106}_{46}\mathrm{Pd}$	738	1.25
${}^{124}_{54}\mathrm{Xe} \rightarrow {}^{124}_{52}\mathrm{Te}$	822	0.10
${}^{130}_{56}\mathrm{Ba} \rightarrow {}^{130}_{54}\mathrm{Xe}$	534	0.11
${}^{136}_{58}\mathrm{Ce} \rightarrow {}^{136}_{56}\mathrm{Ba}$	362	0.19

5 $\beta^+\beta^+$-decay

There is still more to investigate than the electron emitting double beta decay discussed. One example is the counterpart emitting two positrons. Three different decay channels can be considered for the latter

$$\begin{aligned} (Z,A) &\rightarrow (Z-2,A) + 2e^+ + (2\nu_e) \\ e^- + (Z,A) &\rightarrow (Z-2,A) + e^+ + (2\nu_e) \\ 2e^- + (Z,A) &\rightarrow (Z-2,A) + (2\nu_e) \end{aligned} \tag{14}$$

where the last two cases involve electron capture (EC). Especially the β^+/EC mode shows an enhanced sensitivity to right handed weak currents (Hirsch 1994). The experimental signatures of the decay modes involving positrons in the final state are promising because of two or four 511 keV photons. Despite this nice signature, they are less often discussed in the literature, because for each generated positron the available Q-value is reduced by 2 m_ec^2, which leads to much smaller decay rates than in comparable $0\nu\beta\beta$-decay . Hence, for $\beta^+\beta^+$ -decay to occur, the Q-values must be at least 2048 keV. Only six isotopes are known to have such a high Q-value, see Tab. 5. The full Q-value is only available in the EC/EC mode. Its detection is experimentally more challenging, basically requiring the concept of source equal to detector again. In the 0ν mode, because of energy and momentum conservation, additional particles must be emitted like an e^+e^- pair or an internal bremsstrahlung photon. There will be a resonant enhancement in the decay rate if the initial and final states are degenerate as has recently been explored in the context of radiative EC/EC (Sujkowski 2004). Such an enhancement is also expected if the ground state of the double beta emitter is degenerate with an excited state in the daughter, then the de-excitation gammas will serve as signal. Current half-life limits are of the order of 10^{20} yrs obtained with ^{106}Cd and ^{78}Kr for the modes involving positrons (Tretyak 2002). The ^{106}Cd system is currently explored by TGV2 (Stekl 2005) and COBRA. The COBRA experiment has the chance of simultaneously measuring 5 different isotopes for these decay channels (Zuber 2001). As the decay is intrinsic to the CdZnTe detectors one has a good chance to observe the EC/EC and, for the positron emitting modes, coincidences among the crystals can be used.

6 Future

The future activities are basically driven by three factors:

- Explore the claimed evidence observed for ^{76}Ge
- Increase sensitivity for neutrino masses down to 50 meV
- Explore further processes to disentangle the various underlying physics mechanisms discussed for neutrinoless double beta decay and to compensate for the nuclear matrix elements uncertainties.

To address the first topic, experiments have to come up with comparable experimental parameters to the Heidelberg-Moscow experiment, ie. about 10 yrs measuring time, 11 kg of high isotopical abundance (88 %), superb energy resolution and excellent low background in the peak region. Partly those parameters can be compensated by using an isotope with higher Q-value and more favourable matrix elements. As shown in the previous section, CUORICINO and NEMO-3 are starting to restrict the claimed region of neutrino masses. New large scale Ge-experiments are planned called MAJORANA and GERDA, the latter using the former Heidelberg-Moscow and IGEX Ge-semiconductor detectors. A novel technique using the idea of detecting the daughter ion as a possibility of background reduction is investigated by EXO. Independent of that, in the near future EXO will install 200 kg of enriched xenon in form of a LXe detector without ion tagging underground.
Concerning the second item various ideas and proposals are available which are listed in Table. 6. The last item requires the study of other processes like β^+/EC modes, transitions to excited 2^+-states (Doi 1985, Hirsch 1994) or LFV processes using charged leptons like $\mu \to e+\gamma$ (Vogel 2005). Furthermore, to account for the possible physics processes and matrix element uncertainties, the measurement of at least 3-4 different double beta isotopes might be necessary (Simkovic 2005).

7 Summary

While $2\nu\beta\beta$-decay decay is the rarest processes ever observed, there is an enormous physics potential in the lepton number violating process of $0\nu\beta\beta$-decay . In addition to the standard analysis, assuming the exchange of a light Majorana neutrino , various other kinds of $\Delta L = 2$ can severely be restricted. The evidence for a signal in agreement with neutrino masses between 0.2-0.6 eV, which would imply almost degenerate neutrinos, is hotly debated. If this turns out not to be real, the next benchmark number experiments are aiming for, is the 50 meV range, implying hundreds of kilograms of material. After identifying a positive signal, it will be necessary to figure out which lepton number violating physics process is dominating neutrinoless double decay and especially the contribution of light Majorana neutrino exchange. Covering also the nuclear matrix uncertainties it will be necessary to study several isotopes. Various experimental approaches are discussed to accomodate for this. New co-ordinated actions are on their way to provide the nuclear matrix element calculations with better experimental input parameters. Last, but not least, $0\nu\beta\beta$-decay might be the only opportunity to access two further possible CP-violating phases associated with the Majorana character of the neutrino. This

Table 4. *Compilation of proposals for future experiments. This table is a slightly modified version of the one given in (Zuber 2006) and does not claim to be complete.*

Experiment	Isotope	Experimental approach
CANDLES	^{48}Ca	Several tons of CaF_2 crystals in Liquid scintillator
CARVEL	^{48}Ca	100 kg $^{48}CaWO_4$ crystal scintillators
COBRA	^{116}Cd	420 kg CdZnTe semiconductors
CUORE	^{130}Te	750 kg TeO_2 cryogenic bolometers
DCBA	^{150}Nd	20 kg Nd layers between tracking chambers
EXO	^{136}Xe	1 ton Xe TPC (gas or liquid)
GERDA	^{76}Ge	$\sim$ 40 kg Ge diodes in LN_2, expand to larger masses
GSO	^{160}Gd	2t $Gd_2SiO_3 : Ce$ crystal scint. in liquid scintillator
MAJORANA	^{76}Ge	$\sim$ 180 kg Ge diodes, expand to larger masses
MOON	^{100}Mo	several tons of Mo sheets between scintillator
SNO++	^{150}Nd	1000 t of Nd-loaded liquid scintillator
SuperNEMO	^{82}Se	100 kg of Se foils between TPCs
Xe	^{136}Xe	1.56 t of Xe in liquid scintillator
XMASS	^{136}Xe	10 t of liquid Xe

might be important in the context of leptogenesis, explaining the observed baryon asymmetry in the Universe with the help of CP-violation in the lepton sector.

Acknowledgments

I thank the organizers of the school for their hospitality and creating a stimulating atmosphere.

References

Arnaboldi C et al, 2005, *Phys. Rev. Lett.* 95, **142501** , C Nones, Talk at DBD06 Workshop, ILIAS-WG1, Valencia, April 2006

Arnold R et al, 2005, *Phys. Rev. Lett.* 95, **182302** , R Saakyan, Talk at DBD06 Workshop, ILIAS-WG1, Valencia, April 2006

Bornschein B, *This volume*

Boehm F, Vogel P, *Physics of massive neutrinos*, Cambridge Univ Press 1992

Doi M, Kotani T, Takasugi, E, 1985, *Prog. Theo. Phys. Suppl.* 83, **1**

Ejiri H, 2000, *Phys. Rep.* 338, **265**

Elliott S, Vogel P, 2002, *Ann. Rev. Nucl. Part. Phys.* 52, **115**

Elliott S, Engel J 2004, *Journal of Physics G* 30, **R183**

Faessler A, Simkovic F, 1998, *Journal of Physics G* 24, **R2139**
Feruglio F, Strumia A, Vissani F, 2002, *Nucl. Phys.* **B** 637, **345**
Frekers D, Talk at DBD06 Workshop, ILIAS-WG1, Valencia, April 2006
Frekers D, Dilling J, Tanihata I, 2007, *Can. J. Phys.* 85, **57**
Furry W, 1939, *Phys. Rev.* 56, **1184**
Goeppert-Mayer M, 1935, *Phys. Rev.* 48, **512**
Hirsch M et al, 1994, *Z. Phys.* A 347, **151**
Kayser B, *This volume*
Klapdor-Kleingrothaus H V et al, 2001, *Europ. Phys. J.* A 12, **147**
Klapdor-Kleingrothaus H V et al, 2004, *Phys. Lett.* **B** 586, **198**
Kortelainen M et al, 2007, *Phys. Lett.* **B** 647, **128**
Lesgourgues J, Pastor S, 2006, *Phys. Rev.* 429, **307**
Majorana E, 1937, *Nuovo Cimento* 14, **171**
Racah G, 1937, *Nuovo Cimento* 14, **322**
Rodin V A et al, 2006, *Nucl. Phys.* A 766, **107** , err arXiv:07064304
Simkovic F, 2006, *Prog. Part. Nucl. Phys.* 57, **185**
Staudt A, Muto K, Klapdor - Kleingrothaus H V, 1990, *Europhys. Lett.* 13, **31**
Suhonen J, Civitarese O, 1998, *Phys. Rep.* 300, **123**
Sujkowski Z, Wycech S, 2004, *Phys. Rev.* C 70, **052501**
Stekl I, Proc. Medex 2005 conference
Tretyak V I, Zdesenko Y, 2002, At. Dat. and Nucl. Dat. Tab. 80,83
Vogel P, 2006, *Prog. Part. Nucl. Phys.* 57, **177**
Woicik M,2006, *Acta Pol.* B 37, **1911**
Zdesenko Y, 2002, *Rev. Mod. Phys.* 74, **663**
Zuber K 2001, *Phys. Lett.* **B** 519, **1**
Zuber K, Preprint nucl-ex/0511009, Preprint IPPP/05/56
Zuber K, 2006, *Prog. Part. Nucl. Phys.* 57, **235**

Superbeam, Beta Beam, and Neutrino Factory

Yoshitaka Kuno

Osaka University, Japan

1 Introduction

Since the discovery of neutrino oscillation, it has been known that neutrinos have masses, although their masses are very tiny. The neutrino masses are many orders of magnitude smaller than those of quarks and charged leptons. It suggests that neutrinos are fundamentally different from the other elementary particles and the neutrino masses are generated by distinct mechanism.

Particle physics on neutrinos is aiming to address the "Big Questions", which are

- What is the origin of neutrino mass ?
- Did neutrinos play a role in our existence ?
- Did neutrinos play a role in forming galaxies ?
- Did neutrinos play a role in birth of the Universe ?
- Are neutrinos telling us something about unification of matter and/or forces ?
- Will neutrinos give us more surprise ?

They are very difficult questions to answer. Before challenging the "Big Questions", intermediate neutrino questions can be asked. They are as follows.

- Are neutrinos a Dirac particle or Majorana particle ?
- What is the absolute mass scale of neutrinos ?
- How small is the neutrino mixing angle of θ_{13} ?
- Is CP violated in the neutrino sector ?
- Is the neutrino mixing angle of θ_{23} maximal ?
- Is the LSND result real, and Are there any sterile neutrinos ?

Future neutrino facilities should answer some of these intermediate questions. In this lecture, the future neutrino facilities considered and the prospects of neutrino physics which will be studied by these facilities are presented.

2 Neutrino Oscillation Physics

The three-generation neutrino mixing can be accommodated in the Standard Model. The flavor eigenstates of neutrinos can be presented by linear combinations of the mass eigenstates of neutrinos, as given by

$$\begin{pmatrix} \nu_e \\ \nu_\mu \\ \nu_\tau \end{pmatrix} = \begin{pmatrix} V_{e1} & V_{e2} & V_{e3} \\ V_{\mu 1} & V_{\mu 2} & V_{\mu 3} \\ V_{\tau 1} & V_{\tau 2} & V_{\tau 3} \end{pmatrix} \times \begin{pmatrix} \nu_1 \\ \nu_2 \\ \nu_3 \end{pmatrix}, \tag{1}$$

where ν_e, ν_μ, ν_τ are the flavor eigenstates of neutrinos, and ν_1, ν_2, ν_3 are the mass eigenstates of neutrinos, and V is the neutrino mixing matrix and is called the "Maki-Nakagawa-Sakata (MNS) Matrix". When the unitarity of the neutrino mixing matrix is assumed, V can be presented by

$$V = \begin{pmatrix} 1 & 0 & 0 \\ 0 & c_{23} & s_{23} \\ 0 & -s_{23} & c_{23} \end{pmatrix} \begin{pmatrix} c_{13} & 0 & s_{13}e^{-i\delta} \\ 0 & 1 & 0 \\ -s_{13}e^{i\delta} & 0 & c_{13} \end{pmatrix} \begin{pmatrix} c_{12} & s_{12} & 0 \\ -s_{12} & c_{12} & 0 \\ 0 & 0 & 1 \end{pmatrix} \begin{pmatrix} e^{-i\frac{\phi_1}{2}} & 0 & 0 \\ 0 & e^{-i\frac{\phi_2}{2}} & 0 \\ 0 & 0 & 1 \end{pmatrix} \tag{2}$$

$$= \begin{pmatrix} c_{12}c_{13} & s_{12}c_{13} & s_{13}e^{-i\delta} \\ -c_{23}s_{12} - s_{23}s_{13}c_{12}e^{i\delta} & c_{23}c_{12} - s_{23}s_{13}s_{12}e^{i\delta} & s_{23}c_{13} \\ s_{23}s_{12} - c_{23}s_{13}c_{12}e^{i\delta} & -s_{23}c_{12} - c_{23}s_{13}s_{12}e^{i\delta} & c_{23}c_{13} \end{pmatrix} \begin{pmatrix} e^{-i\frac{\phi_1}{2}} & 0 & 0 \\ 0 & e^{-i\frac{\phi_2}{2}} & 0 \\ 0 & 0 & 1 \end{pmatrix} \tag{3}$$

where, in Equation (2), the first matrix is responsible for the atmospheric neutrino oscillation, the second is for the reactor neutrino oscillation (plus a CP violating Dirac phase, δ), the third is for the solar neutrino oscillation, and the last is for CP violating Majorana phases, ϕ_1 and ϕ_2. At this moment, the mixing angles of θ_{12} and θ_{23}, and the mass squared difference of $\Delta m^2_{21} (= m^2_2 - m^2_1)$ and $|\Delta m^2_{32}| (= |m^2_3 - m^2_2|)$ are known. The known parameters, which are determined in the measurements of atmospheric neutrinos, reactor neutrinos and solar neutrinos, are summarized in Table 1.

Table 1. *Summary of the known MNS parameters*

Parameters	Δm^2_{32} (eV^2)	Δm^2_{21} (eV^2)	$\sin^2\theta_{23}$	$\sin^2\theta_{12}$	$\sin^2\theta_{13}$
Values	$(2.2^{+1.1}_{-0.8}) \times 10^{-3}$	$(8.1^{+1.0}_{-0.0}) \times 10^{-5}$	$0.5^{+0.18}_{-0.16}$	$0.3^{+0.08}_{-0.07}$	$0.0^{+0.047}_{-0}$

The other parameters such as θ_{13}, δ and the sign of Δm^2_{32} are not known, although the upper limit of θ_{13} is determined. Here, the sign of Δm^2_{32} determines the "neutrino mass hierarchy", as shown in Figure 1. And the non-zero imaginary Dirac phase, δ, would be responsible for

CP violation in the neutrino oscillation. The determination of these parameters are immediate topics in neutrino oscillation physics.

In addition, if neutrinos are a Majorana particle, they have Majorana phases. But it is not known whether neutrinos are a Dirac particle or a Majorana particle at this moment. Neutrino oscillation could not determine it. It should be studied by double beta decays. It is noted that the Majorana phases of the heavy right-handed neutrinos in the neutrino seesaw mechanism are responsible for leptogenesis (which is a scenario of creating the matter-dominated Universe by leptons). The Dirac phase, δ, which can be observed in neutrino oscillation, does not have any direct relation to leptogenesis.

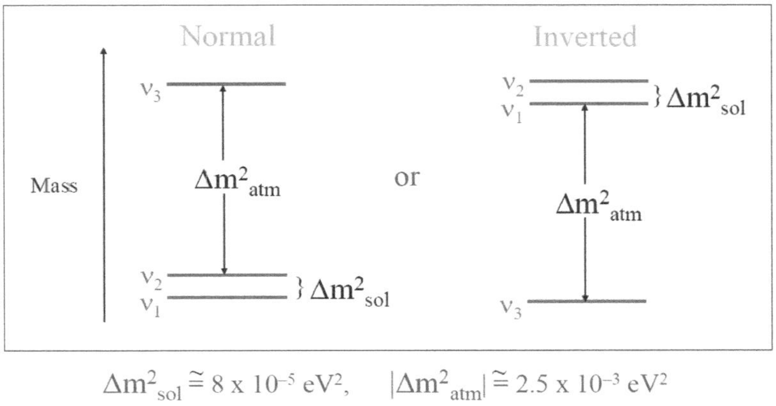

Figure 1. *Neutrino mass hierarchy. The normal hierarchy is $\Delta m^2_{32} (= m_3^2 - m_2^2) > 0$ and the inverted hierarchy is $\Delta m^2_{32} < 0$.*

2.1 Neutrino Oscillation Probability

The probabilities of oscillation from the neutrino flavors of ν_l to ν_m are given by

$$P(\nu_l \rightarrow \nu_m) = \left| \sum_j V_{mj} V_{ij}^* exp\left(-i\frac{m_j^2 L}{2E}\right)\right|^2 \tag{4}$$

$$P(\nu_l \rightarrow \nu_l) = 1 - \left| \sum_j V_{mj} V_{ij}^* exp\left(-i\frac{m_j^2 L}{2E}\right)\right|^2 \tag{5}$$

where V_{ik} is the MNS neutrino mixing matrix element, and m_j is the mass of the ν_j neutrino. L and E are a baseline distance from a neutrino source to neutrino detectors, and energy of neutrinos, respectively. The appearance oscillation channels are given in Equation (4) and the disappearance oscillation channels are given in Equation (5). In the two-neutrino flavor approximation, they are given by

$$P(\nu_l \rightarrow \nu_m) = \sin^2(2\theta)\sin^2\left(\frac{\Delta m^2 L}{4E}\right) \tag{6}$$

$$P(\nu_l \rightarrow \nu_l) = 1 - \sin^2(2\theta)\sin^2\left(\frac{\Delta m^2 L}{4E}\right) \tag{7}$$

where θ is the neutrino mixing angle in the two neutrino flavor approximation. Δm^2 is the mass-squared difference. Equation (6) can be also presented in the SI unit by

$$P(\nu_l \to \nu_m) = \sin^2(2\theta)\sin^2\left(\frac{1.27\Delta m^2(\text{eV}^2)L(\text{km})}{E(\text{GeV})}\right) \tag{8}$$

In the past, two neutrino flavor approximation was almost sufficient. However, in the era of precision studies, the three neutrino flavor framework should be used. Also, the approximation of neutrino oscillation in vacuum (vacuum oscillation) is not sufficient for experiments with long baseline distance. And the oscillation in matter, including the Mikheyev-Smirnov-Wolfenstein (MSW) effect, should be used. It is sometimes called "matter effect". The probability of neutrino oscillation in the three flavor framework will be presented in Section 6. The disappearance oscillation channels would have no information on θ_{13} and the CP-phase δ, but have capability to determine precisely $\sin^2 2\theta_{23}$ and Δm^2_{32}. On the other hand, the appearance oscillation channels would have sensitivities on θ_{13} and δ, but they have strong correlations between them. It is necessary to resolve the correlations. Details will be presented in Section 6.

The goals of neutrino oscillation experiments in the next rounds are given by

1. measurement of θ_{13} with high accuracy or improve the limit of θ_{13},

2. measurement of the CP violating phase δ, if $\theta_{13} \neq 0$ is determined.

3. determination of the neutrino mass hierarchy by measuring the sign of Δm^2_{32}, and

4. discrimination of whether θ_{23} is maximal (namely, $\theta_{23} = \frac{\pi}{4}$, or $\theta_{23} < \frac{\pi}{4}$, or $\theta_{23} > \frac{\pi}{4}$)

Table 2. *List of accelerator-based neutrino sources.*

Conventional beams (Superbeam)	based on pion decays $\pi^+ \to \mu^+ + \nu_\mu$ $\pi^- \to \mu^- + \bar{\nu}_\mu$	ν_μ and $\bar{\nu}_\mu$ available ν_e contamination >0.5% near detectors needed. $\nu_\mu \to \nu_e$ most interesting
Beta Beams	based on beta decays of nuclei ${}^Z_AX \to {}^{Z-1}_A Y + e^+ + \nu_e$ ${}^Z_AX \to {}^{Z+1}_A Y + e^- + \bar{\nu}_e$	ν_e and $\bar{\nu}_e$ available
Neutrino Factories	based on muon decays $\mu^+ \to e^+ + \nu_e + \bar{\nu}_\mu$ $\mu^- \to e^- + \bar{\nu}_e + \nu_\mu$	ν_e, $\bar{\nu}_e$ ν_μ, $\bar{\nu}_\mu$ available 10^{-4} backgrounds charged ID needed. $\nu_e \to \nu_\mu$ most interesting

3 Accelerator-based Neutrino Sources

Accelerator-based neutrino sources are definitely required to perform precise measurements of neutrino oscillation, since all the neutrino sources in nature are not strong enough. Three different kinds of the accelerator-based neutrino sources can be considered. They are pion decays, or beta decay of ions, or muon decays, as shown in Table 2. So far, many neutrino sources based on pion decays (called "conventional neutrino beams") have been constructed. Neutrino factories and beta beams are a future neutrino facility being planned and studied. In the following, we will discuss about superbeams in Section 4, neutrino factories in Section 5, and beta beams in Section 9.

4 Superbeams

The "superbeams" refer to future facilities of ν_μ or $\bar{\nu}_\mu$ beams, based on pion decays (namely conventional beams) in conjunction with a high proton intensity of the order of Mega Watt beam power and large neutrino detectors.

$$\pi^+ \to \mu^+ \nu_\mu \tag{9}$$

$$\pi^- \to \mu^- \bar{\nu}_\mu \tag{10}$$

The advantages of superbeams are as follows. First of all, technology and problems to construct superbeam facilities are understood, and secondary both wide-band beam (on beam axis) and narrow-band beam (off beam axis) can be built, where wide-band beam (narrow-band beam) imply a neutrino beam of wide (narrow) energy spread. The limitations are difficulty to achieve ν_e contamination of less than 0.5 %, which comes from muon decays and kaon decays.

Table 3. *List of current, next and future superbeam experiments*

Generation	Experiment	Location	Comment
Current	K2K	KEK	confirm atmospheric neutrino oscillation
	MINOS	Fermilab	measure $\sin^2 2\theta_{23}$ and $\Delta m^2_{23} \sim 10$ %
	MiniBooNE	Fermilab	confirm/refute the LSND result
	OPERA/ICARUS	CERN	measure $\nu_\mu \to \nu_\tau$
Next	T2K	J-PARC	with Super-Kamiokande
	NOνA	Fermilab	
Future	T2HK / T2KK	J-PARC	with Hyper-Kamiokande
	NOνA upgraded	Fermilab	
	CERN SPL-Frejus	CERN	

The list of current and future superbeam experiments are shown in Table 3. In the following, some of the future superbeam projects are presented.

4.1 The T2K Experiment

The T2K (Tokai to Kamioka) experiment is being prepared in Japan, after the K2K experiment was completed. The neutrino source is located at J-PARC, which is a high intensity proton accelerator facility under construction. The proton beam power of J-PARC is about 0.75 Mega Watts with 40 GeV beam energy. The neutrino detector is the Super-Kamiokande detector having a 22.5 kton fiducial volume. The Super-Kamiokande is a water cherenkov detector and has been fully restored from the accident by installing new photo-multipliers in year 2006. The baseline distance is 295 km.

The T2K neutrino beam is narrow-band, which is produced by taking neutrinos emitting off the proton beam axis. Due to the kinematics of two-body pion decay, the energy of neutrinos from decay of pions of different energy is almost similar, for the particular angle off from the beam axis, as shown in Figure 2 where an off-axis angle of 2.5 degrees would provide a best narrow-band beam. The rates of neutrino interaction events at Super-Kamiokande, with an off-axis angle of 2.5 degree, are about 1600 events for ν_μ charged current interaction and about 2200 events in total.

The goals of T2K after 5 year running are determinations of $\sin^2 2\theta_{23}$ and Δm^2_{32} with their errors of $\Delta(\sin^2 2\theta_{23}) < 0.01$ and $\Delta(\Delta m^2_{32}) < 1 \times 10^{-4}$ respectively, and a search for $\nu_\mu \to \nu_e$ appearance with $\sin^2 2\theta_{13} > 0.006$ in 90% confidence level (C.L.) exclusion sensitivity (or $\sin^2 2\theta_{13} \sim 0.018$ in a 3σ discovery sensitivity), where $\sin^2\theta_{23} = 0.5$, $\delta = 0$ and no matter effects are assumed.

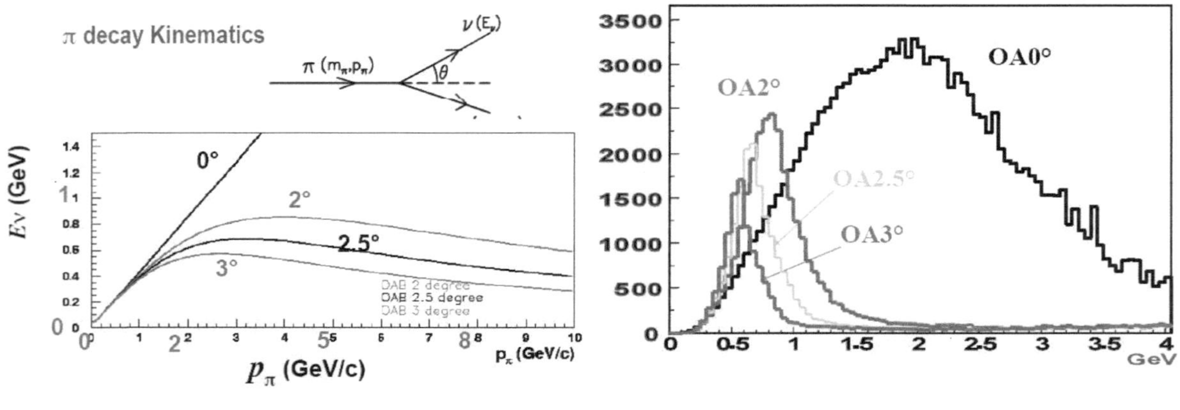

Figure 2. *Neutrino energy vs. pion energy as a function of off-axis angle in the two-body pion decay (left). Simulated neutrino energy spectra for different off-axis angles as a function of neutrino energy (right).*

4.2 The T2HK and T2KK Experiments

The T2HK experiment is a planned future extension of the T2K experiment. The neutrino source is the J-PARC proton accelerator with its upgrade to 4 Mega Watts. The upgrade is expected to be made by installing more rf cavities for a faster repetition rate and doubling a number of bunches with barrier buckets. The neutrino detector would be the Hyper-Kamiokande, which has 1 Mega-ton total mass (and a 0.5 Mega-ton fiducial volume). It is a future upgrade of the present Super-Kamiokande. The Hyper-Kamiokande consists of two detectors of 48 $\times$ 50 $\times$ 250 meters^3, as shown in Figure 3. The T2HK sensitivity on $\sin^2 2\theta_{13} > 0.001$ in 90% C.L. exclusion sensitivity with 10 % systematic and background after 5 year running. If $\sin^2 2\theta_{13} > 0.01$, a 3σ discovery of the CP phase δ could be made for $|\delta| > 20$ degrees with 2% systematic errors, after 2-year neutrino and 6-year anti-neutrino running. Here background and systematic errors are very important to improve sensitivity.

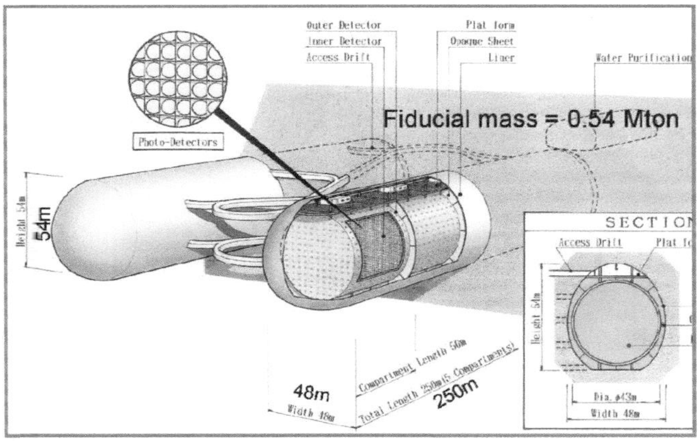

Figure 3. *Schematic layout of the proposed HyperKamiokande detector. It consists of two detectors. Total fiducial mass is 0,54 Mtons.*

The T2KK experiment is another future plan, where one of the Hyper-Kamiokande water tank is placed in Korea and the other one is at Kamioka in Japan. The motivations of the T2KK experiment are to achieve better CP sensitivity with reasonable systematic errors, say much more than 2%, and to determine the neutrino mass hierarchy with long baseline from Tokai to Korea. The advantages are as follows. The effect of CP violation is proportional to L/E and the longer baseline distance from Tokyo to Korea is better. The matter effect is larger for a longer baseline distance, and the neutrino mass hierarchy can be easily determined. Also the CP sensitivity can be kept reasonably high even with 10% systematic errors. The limitation is less event rates in Korea.

4.3 The NOνA Experiment

The NOνA experiment is a planned experiment using the NUMI beam from Fermilab (Harris, 2006). The NOνA detector is located 820 km away from Fermilab, in northern Minnesota, 12 km off axis from the NUMI beam. The NOνA detector is a totally active detector, consisting of 24 k ton liquid scintillator and 6 k ton PVC. The size of the NOνA detector is 15.7 $\times$ 15.7 $\times$ 132 meters^3. The detector consists of 1984 planes, each of which is formed by 12 extrusions. Each extrusion has 32 liquid scintillator cells, whose cell dimension is 3.9 cm $\times$ 6 cm $\times$ 15.7 m (of 0.15 radiation length). Each cell is readout by 0.8 mm wavelength shifting fibers into avalanche photo diodes (APD). With 30×10^{20} proton-on-target for 6 year running, the NOνA 3 σ discovery sensitivity on $\sin^2 2\theta_{13}$ is about $\sin^2 2\theta_{13} > 0.01$. The determination of Δm^2_{32} is expected to be $\Delta(\Delta m^2_{32}) < 1 \times 10^{-4}$. The mass hierarchy would be determined by comparing the neutrino and anti-neutrino runs. The experiment is planned to start in 2012. A future upgrade of a new proton driver with 8 GeV superconducting proton linac at Fermilab is being considered, and with it, the NOνA sensitivity would be improved by additional factor of two.

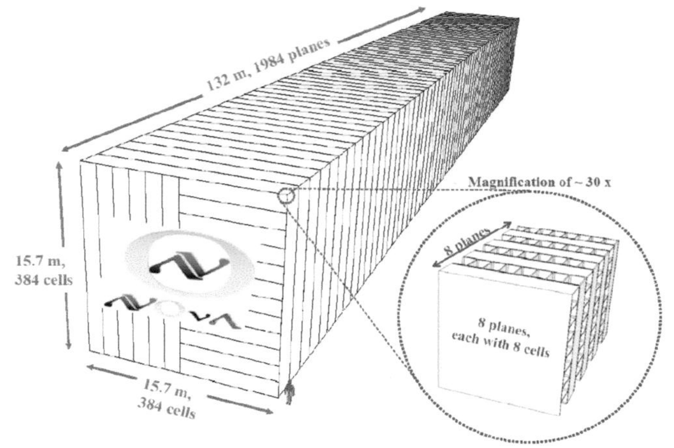

Figure 4. *Schematic layout of the NOvA detector. The detector has 24 kton liquid scintillator. It consists of 1984 planes, each of which has 12 extrusions. Each extrusion has 32 cells.*

4.4 The Very Long Baseline Experiment

The "very" long baseline experiment is another proposed neutrino oscillation experiment using a wide-band neutrino beam with a baseline distance of more than 2500 km. It is being studied at Brookhaven National Laboratory (BNL) and Fermilab. The motivation is to move the 2nd neutrino oscillation maximum to energy region where it can be resolved experimentally (namely above the detector threshold). It is shown in Figure 5. By doing this, stronger CP asymmetry can be observed at the 2nd oscillation maximum as well as the matter effect increases to resolve the neutrino mass hierarchy. By using a wide band neutrino beam it is possible to cover three energy regions, which are the first oscillation maximum that is sensitive to the mass hierarchy, the second oscillation maximum that is sensitive to CP asymmetry, and the higher maximum region that is sensitive to the solar neutrino oscillation. By comparing the three energy regions, potential correlations can be resolved.

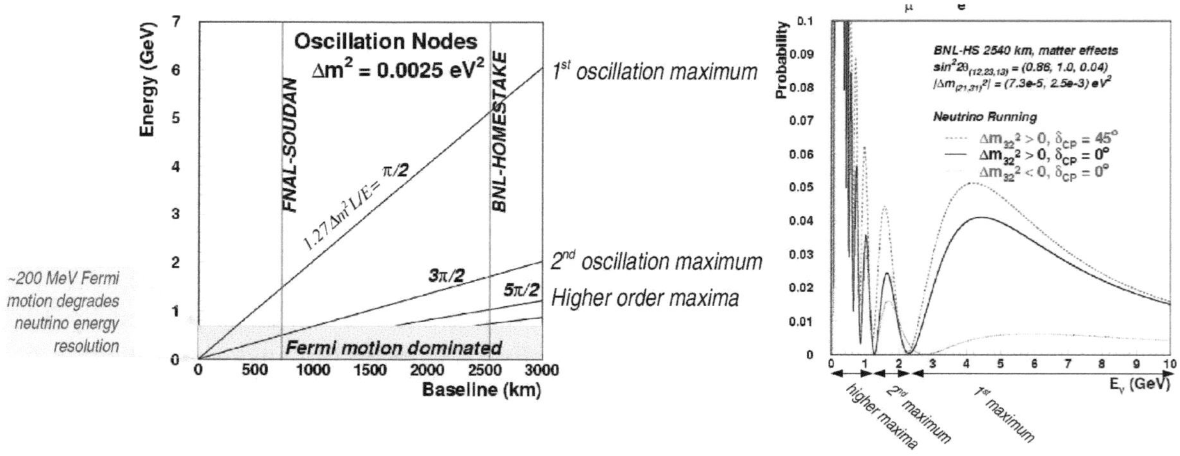

Figure 5. *Relation of Energy and a baseline distance for various oscillation maxima (left). The neutrino oscillation probability as a function of neutrino energy, estimated for the case from BNL to Homestake (2540 km) with the assumption of upgraded AGS beam power.*

4.5 The CERN to Frejus Experiment

The CERN to Frejus experiment is a future neutrino project under consideration at CERN. A neutrino beam from CERN is brought to the Frejus tunnel where a large water cherenkov detector, called MEMPHYS will be located. The baseline distance is about 130 km. The MEMPHYS is a water cherenkov detector consisting of three tanks of 145 ktons each (a total is 435 ktons). Its size is 65 meters in diameter, 65 meters in height. The neutrino beam is either ν_μ ($\bar{\nu}_\mu$) from a planned superconducting proton linac (SPL) or ν_e ($\bar{\nu}_e$) from the beta beam facility, as mentioned in Section 9. The SPL is a 2.2 GeV superconducting proton linear accelerator which is being planned at CERN, by recycling superconducting cavities used in the LEP ring. A total beam power of SPL would be 4 Mega Watts. An average neutrino energy is about 300 MeV (and its peak at 270 MeV). For the beta beam, ν_e from ^{18}Ne ions (for 5 years) and $\bar{\nu}_e$ from ^{6}He ions (for 5 years) with their energies of 400 MeV are considered. The sensitivity of the CERN-SPL to Frejus experiment has a 3σ discovery sensitivity of the CP-violating phase δ of more than 20 degrees when $\sin^2 2\theta_{13} > 0.01$.

5 Neutrino Factory

5.1 What Is a Neutrino Factory ?

As described in Section 3, there is another type of neutrino sources which is based on muon decays, $\mu^+ \to e^+\nu_e\bar{\nu}_\mu$ and $\mu^- \to e^-\bar{\nu}_e\nu_\mu$. This kind of neutrino source has several advantages. (1) Since muons mostly decay after pions and kaons decay, the contamination of wrong flavored neutrinos in a beam can be reduced when delayed neutrinos are taken, (2) four different neutrinos, ν_e, $\bar{\nu}_e$, ν_μ, $\bar{\nu}_\mu$, are available, and (3) the beam normalization can be precisely determined.

To obtain a higher intensity of a neutrino beam, one can increase proton beam power. In fact, proton accelerators with Mega Watt beam power is being constructed (for instance, J-PARC) or planned in the world. However, given the proton beam power, how can one obtain more neutrinos ? When the parent particles are accelerated to high energy (towards the detectors), more neutrinos are available by Lorentz boosting. The number of neutrinos at the very forward direction is proportional to the energy squared of the parent particles (E_μ), $N_\nu \propto E_\mu^2$. For the present accelerating technology, acceleration of pions is very difficult because of its too short lifetime of 26 nsec at its rest frame. Therefore, only muons, whose lifetime at its rest frame is 2.2 μsec, live long enough to accelerate.

The energy spectra of neutrinos from muon decay at rest are given by

$$\frac{dN(\nu_\mu)}{dx\,d\cos\theta} \propto 2x^2\left[(3-2x) \mp P_\mu(1-2x)\cos\theta\right], \quad \text{and} \tag{11}$$

$$\frac{dN(\nu_e)}{dx\,d\cos\theta} \propto 6x^2\left[(1-x) \mp P_\mu(1-x)\cos\theta\right]. \tag{12}$$

where $x = E_\nu/E_{max}$ and $E_{max} = m_\mu c^2/2$. The $\mp$ signs are for $\mu^\pm$. Their spectra are shown in Figure 6. As seen in Figure 6, the spectra of ν_μ and ν_e are quite different one another. Each neutrino energy spectrum from the accelerated muons would be similar to those from muon decay at rest in Figure 6, by scaling with the parent muon energy.

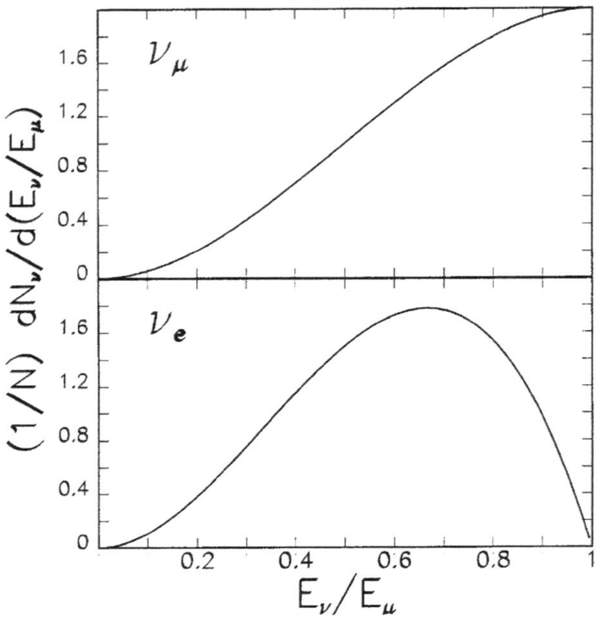

Figure 6. *Neutrino energy spectra from muon decays, $\mu^- \to e^- + \nu_\mu + \bar{\nu}_e$ ($\mu^+ \to e^+ + \bar{\nu}_\mu + \nu_e$). The top figure is for muon neutrinos and the bottom is for the electron neutrinos.*

The muons accelerated to high energy do not decay quickly. For instance, at 50 GeV, muon lifetime gets about 1 msec (γ=500). Therefore a muon storage ring is needed to store them until their decay. Two straight sections in the muon storage ring of a racetrack shape give automatically two neutrino beams provided to two different experiments. The angular divergence of the neutrino beam is about $1/\gamma$ where the muon energy is $E_\mu = \gamma m_\mu c^2$. This beam divergence corresponds about 20 cm at a 100 meter distance for $E_\mu = 50$ GeV.

The advantage of neutrino factories are summarized as follows.

- *Highly intense neutrino source* :
  It is a few orders of magnitude higher than the present at a few 10 GeV energy range.

- *Both muon (anti-)neutrinos and electron (anti-)neutrinos available* :
  It allows us to study various different modes.

- *Extremely low backgrounds* :
  Background levels for wrong-signed muons are very low, of the order of 10^{-4}.

- *Precise knowledge of neutrino flux* :
  Neutrino flux normalization can be measured at the level of 0.1 %.

5.2 Oscillation Physics at Neutrino Factory

Muon decays provide both ν_e ($\bar{\nu}_e$) and $\bar{\nu}_\mu$ (ν_μ) simultaneously. Therefore, the potential neutrino oscillation processes studied at a neutrino factory are 12 modes, as given in the following. There are many variety of oscillation modes to be studied. In particular, $\nu_e \to \nu_\mu$ oscillation, $\nu_e \to \nu_\tau$, and $\nu_\mu \to \nu_e$ oscillations are called "golden" channel, "silver" channel, and "platinum" channel, respectively.

$\mu^+ \to e^+\nu_e\bar{\nu}_\mu$	$\mu^- \to e^-\bar{\nu}_e\nu_\mu$		name
$\bar{\nu}_\mu \to \bar{\nu}_\mu$	$\nu_\mu \to \nu_\mu$	disappearance	
$\bar{\nu}_\mu \to \bar{\nu}_e$	$\nu_\mu \to \nu_e$	appearance	platinum channel
$\bar{\nu}_\mu \to \bar{\nu}_\tau$	$\nu_\mu \to \nu_\tau$	appearance (atm. osc.)	
$\nu_e \to \nu_\mu$	$\bar{\nu}_e \to \bar{\nu}_\mu$	appearance	golden channel
$\nu_e \to \nu_e$	$\bar{\nu}_e \to \bar{\nu}_e$	disappearance	
$\nu_e \to \nu_\tau$	$\bar{\nu}_e \to \bar{\nu}_\tau$	appearance	silver channel

Event rates at a neutrino factory is enormous. The charged-current event rates (N_{CC}) in the deep inelastic scattering (DIS) region can be obtained by

$$N_{CC}(\nu_l \to l) = N_\nu \cdot \sigma \propto \frac{E_\nu^2}{L^2} \cdot E_\nu = \frac{E_\nu^3}{L^2} \tag{13}$$

where the number of neutrinos is proportional to E_ν^2/L^2 and E_ν is the energy of neutrinos and L is a distance from the neutrino source and detectors. The DIS cross section (σ) is known to be proportional to E_ν. Typical numbers of events are about $3 \times 10^5(1 \times 10^6)$ events for $E_\nu = 20(30)$ GeV with a 10 kton detector for 10^{21} muon decays/year.

The number of the neutrino oscillation event rates (N_{OSC}) is given by

$$N_{OSC}(\nu_l \to l') = N_\nu \cdot \sigma \cdot P(\nu_l \to \nu_{l'}) \propto \frac{E_\nu^3}{L^2} \cdot \frac{L^2}{E_\nu^2} = E_\nu \tag{14}$$

where $P(\nu_l \to \nu_{l'})$ is the neutrino oscillation probability from ν_l to $\nu_{l'}$, and can be approximated to L^2/E_ν^2 from Equation (6) when L/E_ν is small. From Equation (14), the oscillation event rate is proportional to neutrino energy E_ν and independent of L, when L/E_ν is small.

5.3 Signature of Neutrino Oscillation at a Neutrino Factory

The muon decays have both an electron-flavored neutrino and a muon-flavored neutrino. How does one discriminate the flavor change ? Figure 7 shows the case when $\nu_e(\bar{\nu}_e)$ oscillates into $\nu_\mu(\bar{\nu}_\mu)$. As seen in Figure 7, the signature of neutrino oscillation is wrong-signed leptons. Therefore, the charge identification is needed. The wrong signed muons are clean signals, and therefore the background level would be about 10^{-4}.

To search for CP violation in neutrino oscillation, the oscillation probabilities of neutrinos and anti-neutrinos would be compared. However, the matter effect affects neutrinos and of anti-neutrinos differently. The matter effect enhances the oscillation probability of anti-neutrinos and decrease that of neutrinos, if $\Delta m_{32}^2 > 0$. On the other hand, it decreases that of anti-neutrinos and enhances that of neutrinos, if $\Delta m_{32}^2 < 0$. By using the matter effect, the sign of Δm_{32}^2 can be studied. However, to study CP violation in neutrino oscillation, the difference between neutrinos and anti-neutrinos caused by the matter effect should be removed. Figure 8 shows the ratio of $N(\bar{\nu}_e \to \bar{\nu}_\mu)/N(\nu_e \to \nu_\mu)$ as a function of baseline distance. This ratio is about 0.5 for small baseline distances. When the baseline distance is large, the ratio becomes different from

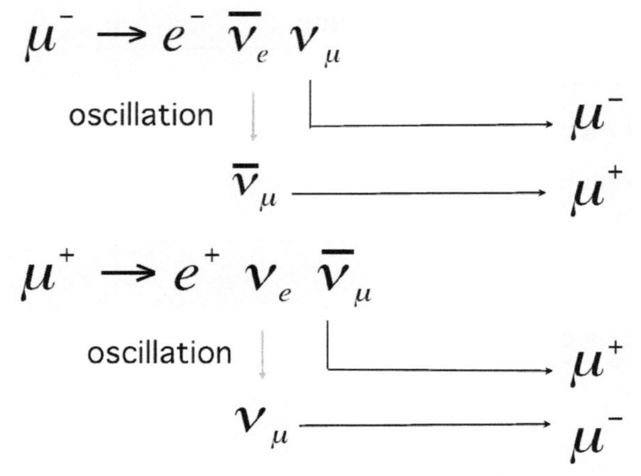

Figure 7. *Neutrino oscillation signature in a neutrino factory is wrong-signed leptons. Therefore, charge identification is needed. In general, charge identification of muons is easy, but that of electrons is difficult.*

0.5 due to the matter effect and CP violation. Whether the matter effect increases or decreases the ratio depends on the sign of Δm^2_{32}. And the bands show the effect of CP violation, where the CP violating phase δ changes from 0 to 2π.

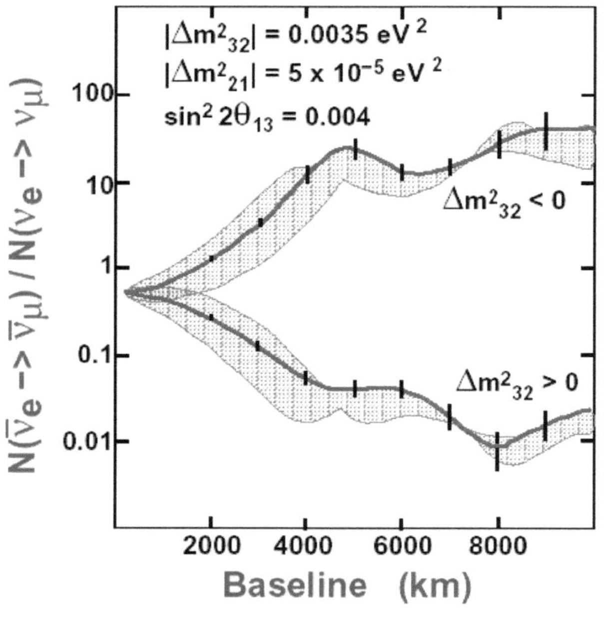

Figure 8. *Ratio of $N(\bar{\nu}_e \to \bar{\nu}_\mu)/N(\nu_e \to \nu_\mu)$ as a function of baseline distance. It is calculated for the case of 20 GeV neutrino factory with 4 MeV threshold. Two lines are for the different mass hierarchy. The statistical error represents the sample of 10^{21} muon decays with a 50 kton detector.*

6 Degeneracies

6.1 Three Flavor Oscillation Probability

The neutrino oscillation probability of the golden channel ($\nu_e \to \nu_\mu$ or $\bar{\nu}_e \to \bar{\nu}_\mu$) can be given by

$$P^{\pm}_{\nu_e\nu_\mu}(\theta_{13},\delta) \sim X_\pm \sin^2 2\theta_{13} + (Y^c_\pm \cos\delta \mp Y^s_\pm \sin\delta)\sin 2\theta_{13} + Z, \tag{15}$$

where $+$ ($-$) for neutrinos (anti-neutrinos), and the functions of $X_{\pm}, Y^c_{\pm}, Y^s_{\pm}$ and Z are given by

$$X_{\pm} = \sin^2\theta_{23}\left(\frac{\Delta_{23}}{B_{\mp}}\right)^2 \sin^2\left(\frac{B_{\mp}L}{2}\right) \quad (16)$$

$$Y^c_{\pm} = \sin 2\theta_{23}\sin 2\theta_{13}\frac{\Delta_{12}}{A}\frac{\Delta_{23}}{B_{\mp}}\sin\left(\frac{AL}{2}\right)\sin\left(\frac{B_{\mp}L}{2}\right)\cos\left(\frac{\Delta_{23}L}{2}\right) \quad (17)$$

$$Y^s_{\pm} = \sin 2\theta_{23}\sin 2\theta_{13}\frac{\Delta_{12}}{A}\frac{\Delta_{23}}{B_{\mp}}\sin\left(\frac{AL}{2}\right)\sin\left(\frac{B_{\mp}L}{2}\right)\sin\left(\frac{\Delta_{23}L}{2}\right) \quad (18)$$

$$Z = \cos^2\theta_{23}\sin^2 2\theta_{12}\left(\frac{\Delta_{12}}{A}\right)^2\sin^2\left(\frac{AL}{2}\right) \quad (19)$$

$$(20)$$

where $\Delta_{ij} = \Delta m^2_{ij}/2E$, and $B_{\pm} = |A \pm \Delta_{23}|$, and A is the matter parameters defined by $A = 2\sqrt{2}G_F n_e$. Here. G_F and n_e are the Fermi coupling constant, and a number of electrons in the earth matter.

6.2 Degeneracies of δ and θ_{13}

From Equation (15), it is clear that when only one counting measurement for a given L/E is performed, there are many sets of the θ_{13} and δ values which would give the same oscillation probability, in addition to the true set of θ_{13} and δ values (θ^*_{13}, δ^*). Namely $P(\theta_{13}, \delta) = P(\theta^*_{13}, \delta^*)$ (Rigolin, 2005). Therefore, there is no sensitivity to δ and a large uncertainty of θ_{13}. It is called "intrinsic degeneracy".

6.2.1 Combination of neutrinos and anti-neutrinos

To resolve the intrinsic degeneracy, additional measurements are needed. One can combine the oscillation probabilities of neutrinos (P_+) and anti-neutrinos (P_-). The oscillation probability of anti-neutrino oscillation can be obtained by changing δ to $-\delta$ and A to $-A$ in Equation (15). Namely

$$P_{\bar{\nu}_e\to\bar{\nu}_\mu} = P_{\nu_e\to\nu_\mu}(\delta \to -\delta, A \to -A) \quad (21)$$

The two counting measurements for neutrinos and anti-neutrinos would reduce the degeneracy, but there are still two solutions which would give the same neutrino oscillation probability. One is the true set of (θ^*_{13}, δ^*) and the other is a fake.

6.2.2 Combination of different L/E

When we perform two (or more than two) counting measurements for neutrinos (P_+) and anti-neutrinos for different L/E values, the intrinsic degeneracies of δ and θ_{13} can be in principle resolved. Alternatively, one can measure the energy spectra precisely and can bin them. Since each energy bin corresponds to different L/E, it would revolve the degeneracies. Thus, two (or more than two) counting measurements or binning of energy spectra are needed to resolve the intrinsic degeneracy.

6.3 Eight-fold Degeneracies

In addition to δ and θ_{13}, the following values are not well known;

- the sign of Δm^2_{32} (Δm^2_{32} or $-\Delta m^2_{32}$), and
- the octant of θ_{23} (θ_{23} or $\frac{\pi}{2} - \theta_{23}$)

The first one is often called "sign degeneracy" (Minakata and Nunokawa, 2001) and the second is called "octant degeneracy" (Fogli and Lisi 1996, and Barger et al, 2002). They are discrete degeneracies. With all together, the degeneracies of neutrino parameters become eight-fold. To resolve all the degeneracies, several measurements with high precisions are needed.

6.4 Degeneracy Resolution at a Neutrino Factory

6.4.1 Combination with the Silver and Platinum Channels

In a neutrino factory, there are other extra oscillation channels which would be useful to resolve the degeneracies if combined. One of them is the silver appearance channel, which is $\nu_e \to \nu_\tau$ (or $\bar{\nu}_e \to \bar{\nu}_\tau$) oscillation. The oscillation probability of the silver channel is given by changing $\sin^2\theta_{23} \to \cos^2\theta_{23}$ and $\sin^2\theta_{23} \to -\sin^2\theta_{23}$ in Equation (15). Namely,

$$P^{\pm}_{\nu_e\nu_\tau}(\theta_{13},\delta) \sim X^{\tau}_{\pm}\sin^2 2\theta_{13} + (Y^{\tau,c}_{\pm}\cos\delta \mp Y^{\tau,s}_{\pm}\sin\delta)\sin 2\theta_{13} + Z^{\tau}, \quad (22)$$

where + (−) for neutrinos (anti-neutrinos), and the functions of $X_{\pm}, Y^c_{\pm}, Y^s_{\pm}$ and Z are given by

$$X^{\tau}_{\pm} = \cos^2\theta_{23}\left(\frac{\Delta_{23}}{B_{\mp}}\right)^2 \sin^2\left(\frac{B_{\mp}L}{2}\right) \quad (23)$$

$$Y^{\tau,c}_{\pm} = -\sin 2\theta_{23}\sin 2\theta_{13}\frac{\Delta_{12}}{A}\frac{\Delta_{23}}{B_{\mp}}\sin\left(\frac{AL}{2}\right)\sin\left(\frac{B_{\mp}L}{2}\right)\cos\left(\frac{\Delta_{23}L}{2}\right) \quad (24)$$

$$Y^{\tau,s}_{\pm} = -\sin 2\theta_{23}\sin 2\theta_{13}\frac{\Delta_{12}}{A}\frac{\Delta_{23}}{B_{\mp}}\sin\left(\frac{AL}{2}\right)\sin\left(\frac{B_{\mp}L}{2}\right)\sin\left(\frac{\Delta_{23}L}{2}\right) \quad (25)$$

$$Z^{\tau} = \sin^2\theta_{23}\sin^2 2\theta_{12}\left(\frac{\Delta_{12}}{A}\right)^2 \sin^2\left(\frac{AL}{2}\right) \quad (26)$$

$$(27)$$

The combination of the golden and the silver channels (both neutrinos and anti-neutrinos) would resolve the degeneracies.

The other oscillation channel to be combined is the platinum channel, which is $\nu_\mu \to \nu_e$ (or $\bar{\nu}_\mu \to \bar{\nu}_e$). The oscillation probability of the platinum appearance channel is given by changing δ to $-\delta$ from that of the golden channel in Equation (15). To perform the silver oscillation channel, charge identification of electrons (or positrons), which is very challenging, is necessary.

6.4.2 Magic Baseline

Another consideration in neutrino factories is to locate the second detector at the magic baseline which is about $L = 7300 - 7600$ km. For $L = 7300 \sim 7600$ km, it is known that

$$\sqrt{2}G_F n_e L = 2\pi \quad \rightarrow \quad \sin\left(\frac{AL}{2}\right) = 0, \tag{28}$$

In this case, $Y^c_\pm$ and $Y^s_\pm$ and Z in Equation (15) are all zero, and only $X_\pm$ is not zero. Therefore, it allows one to make clean determinations of $\sin^2 2\theta_{13}$.

7 Sensitivity and Optimization at a Neutrino Factory

To maximize potential sensitivities to neutrino oscillation parameters in a neutrino factory, the neutrino energy and baseline distance have to be optimized. Recent intensive studies on optimization of a neutrino factory (Huber, Lindner, Rolinec, and Winter, 2006) show some basic guideline. Their standard setup for neutrino factory experiments is assumed to have about 5×10^{20} useful muon decays for each polarity (namely a total of 10^{21} muon decays) and detectors of 50 kton fiducial mass. They propose to have two detectors, one detector at about 4000 km (from 3000 to 5000 km) which is optimized for measurements of CP violation and the second detector to be at the magic baseline of 7500 km. The second detector serves precise determination of $\sin^2 2\theta_{13}$ as well as the neutrino mass hierarchy, as presented in Subsection 6.4.2. Figure 9 shows how the sensitivities on the neutrino mass hierarchy and CP violation could be improved by (1) adding the second detector at the magic baseline, (2) improvement of detectors having better energy resolution and lower thresholds, (3) utilizing the platinum channel with a 15 kton liquid argon detector. In Figure 9, the fraction of (true) δ_{CP} implies the coverage of CP phase δ in the full 2π range in which the required sensitivity can be achieved as a function of $\sin^2 2\theta_{13}$.

The comparison of the sensitivities to CP violation, mass hierarchy and $\sin^2 2\theta_{13}$ among various future neutrino facilities is shown in Figure 10. In this comparison, a neutrino factory which has two detectors, one at 4000 km and the other at 7500 km, with E_μ = 20 MeV, and a beta beam which has a water cherenkov detector of 400 kton fiducial volume with beam energy $\gamma = 300$ are assumed. From Figure 10, neutrino factories have sufficient sensitivity of $\sin^2 2\theta_{13}$ of down to 10^{-5} and that of the neutrino mass hierarchy and CP violation for $\sin^2 2\theta_{13} < 10^{-4}$. The neutrino factory overperforms for most of the cases, except for large $\sin^2 2\theta_{13}$ where beta beams are slightly better.

8 Accelerator Complex of Neutrino Factory

In this section, the basic accelerator concept of a neutrino factory is described. A neutrino factory complex consists of several components. They are listed in the following :

- *proton driver* : A proton driver provides protons of $1 - 4$ Mega Watts beam power on a target to produce pions. As a proton beam power is higher, the pion-production yield is higher.

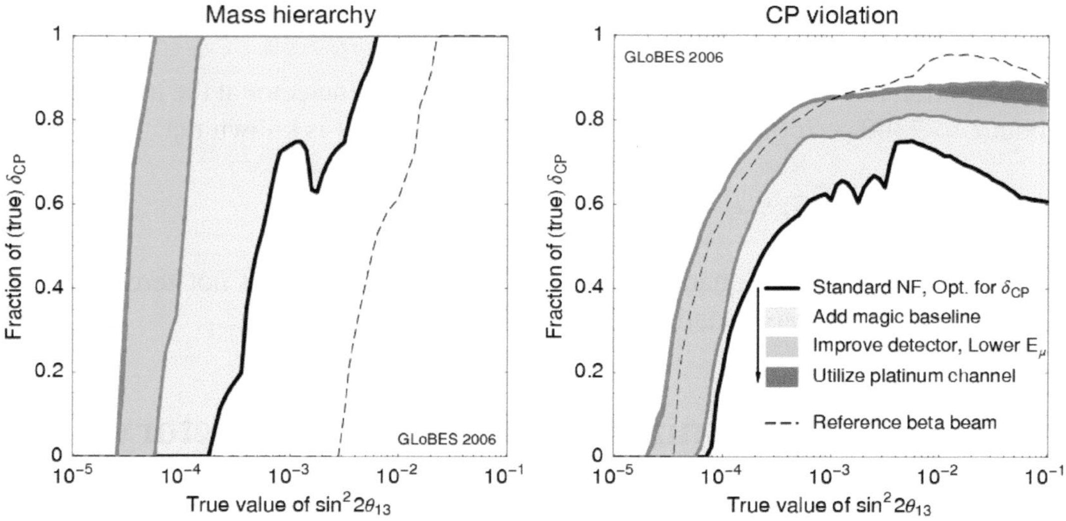

Figure 9. *Sensitivities to the neutrino mass hierarchy and CP violation as a function of* $\sin^2 2\theta_{13}$. *The sensitivities can be optimized and improved by (1) adding the second detector at the magic baseline, (2) improving the detectors, (3) utilizing the platinum channel (Huber et al. 2006). Copyright (2006) American Physical Society.*

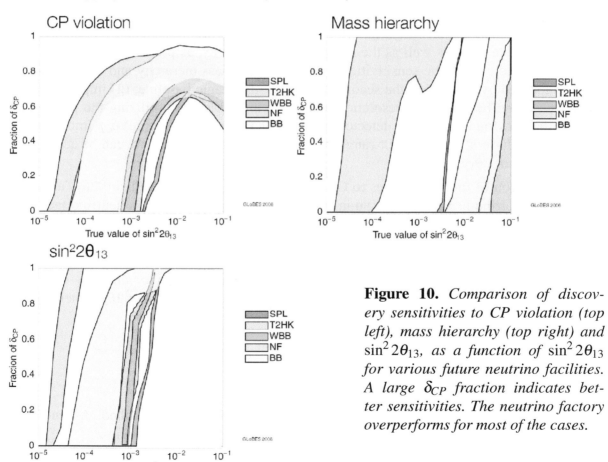

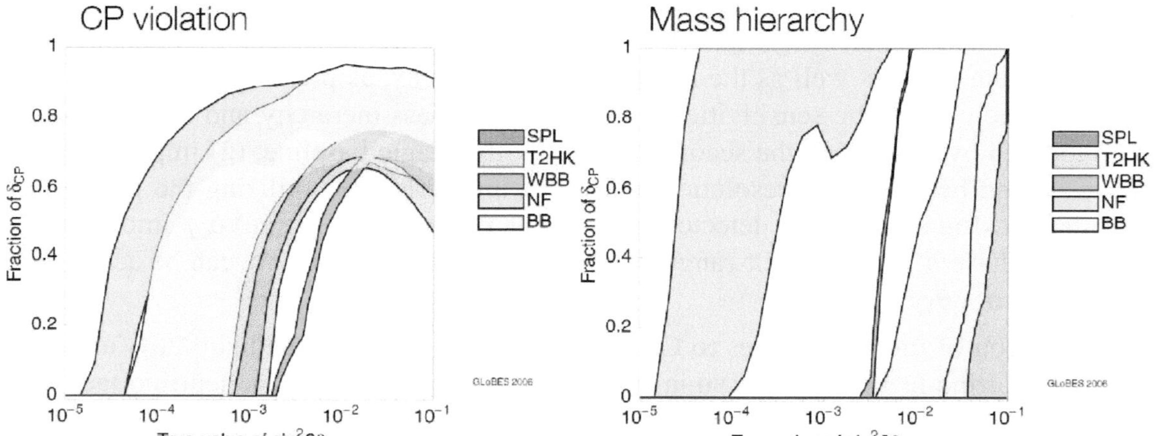

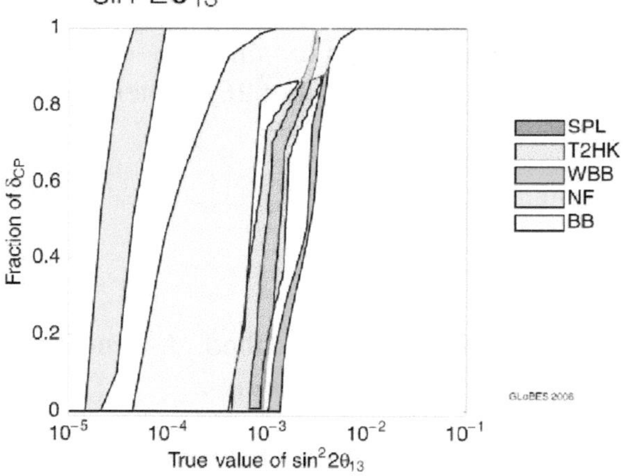

Figure 10. *Comparison of discovery sensitivities to CP violation (top left), mass hierarchy (top right) and* $\sin^2 2\theta_{13}$, *as a function of* $\sin^2 2\theta_{13}$ *for various future neutrino facilities. A large* δ_{CP} *fraction indicates better sensitivities. The neutrino factory overperforms for most of the cases.*

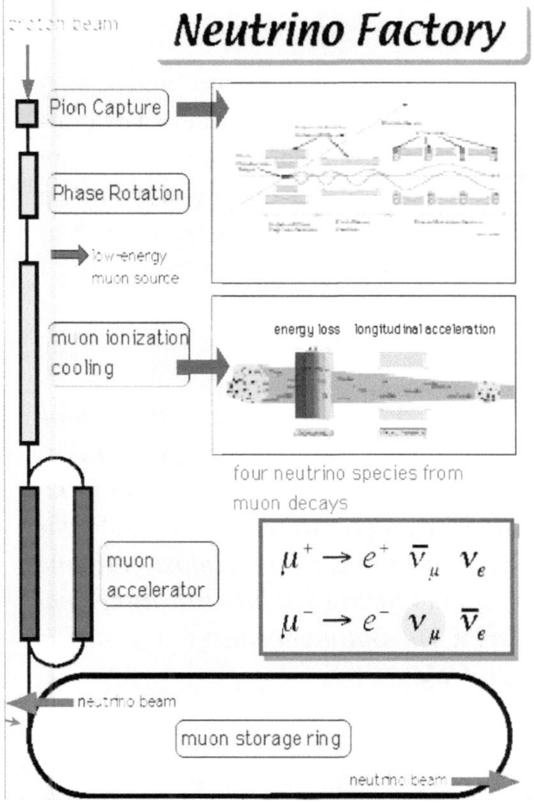

Figure 11. *A schematic layout of a neutrino factory. The facility consists of several components. They are a proton driver providing a proton beam of high beam power, pion capture, phase rotation by rf cavities after bunching, muon ionization cooling, muon acceleration, and a muon storage ring. From a muon storage ring, four different kinds of neutrinos will be available from* $\mu^+ \rightarrow e^+\bar{\nu}_\mu\nu_e$ *and* $\mu^- \rightarrow e^-\nu_\mu\bar{\nu}_e$.

- *Target and Pion Capture* : A target is immersed either in a high solenoidal magnetic field of 15 - 20 Tesla by surrounding superconducting solenoid magnets, or in a toroidal magnetic field by horn magnets. The idea is to collect as many pions as possible.

- *Bunching and Phase rotation* : A muon beam is bunched by rf cavities. After bunching the beam, another set of rf cavities with higher field gradients, is applied to make bunch rotation, which is to accelerate slow muons and decelerate fast muons to reduce the beam spread.

- *Ionization cooling* : A muon beam is then needed to cool its beam emittance (which is a measure of beam spread in space and directions) by using energy degraders and rf cavities to accelerate.

- *Acceleration* : A cooled muon beam is then accelerated to high energy such as 20 GeV or 50 GeV. There are several options of acceleration schemes. One of them is a recirculating linear accelerator (RLA) and the other is a fixed field alternating gradient (FFAG) synchrotron.

- *Storage Ring* : Muons thus accelerated are injected into a storage ring, where muons are stored until they decay.

In the following, each of the components will be described in detail.

8.1 Proton Driver

The yield of pion production is approximately proportional to proton beam energy above several GeV. It implies that the pion yield linearly increase as beam power. A proton driver considered in a neutrino factory needs beam power of about 1 - 4 Mega Watts to produce as many pions and muons as possible. Figure 12 shows pion yields per proton per energy (GeV) as a function of proton kinetic energy. From this, it is seen that the proton beam energy from 5 to 20 GeV is the best, and the pion-production yield drops at higher energy.

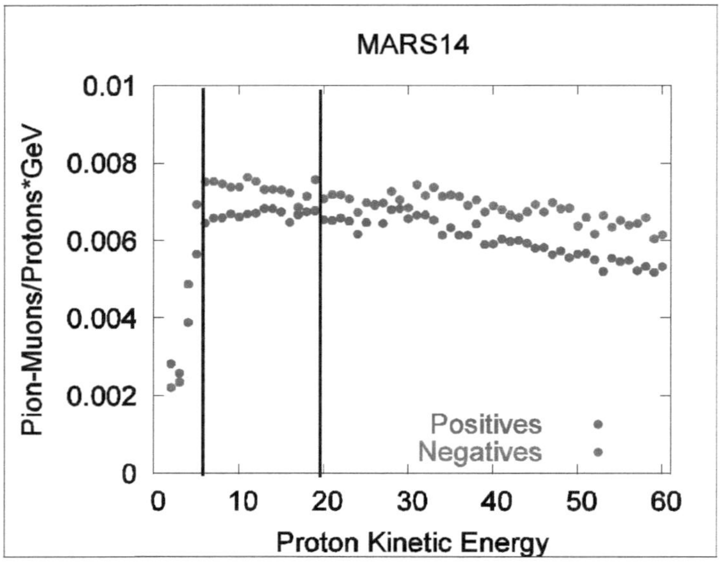

Figure 12. *Yields of a sum of pion and muon per proton per energy (GeV) as a function of proton kinetic energy. Above a few GeV, the yield/proton/GeV is almost flat and get decreasing slightly for higher energy. The optimum energy is from 5 to 20 GeV.*

8.2 Target and Pion Capture

To achieve a highly intense muon beam, it is important to maximize both the pion production by choosing appropriate target materials and the pion collection efficiency. For a target of pion production, heavy Z material is better to produce more pions. The target should sustain with high beam power of 1 to 4 Mega Watts. Normal solid metal targets wold be melt with such a high proton beam power. One of the choices is a liquid mercury jet target.

Pion collection of high efficiency is of critical importance for a neutrino factory. There are two options of the pion capture scheme. One is solenoid capture and the other is magnetic-horn capture, as shown in Figure 13.

8.2.1 Tests of Liquid Mercury Targets

So far, two experiments of testing a concept of a liquid mercury target were performed. One is the E951 at Brookhaven National Laboratory (BNL), in which liquid mercury jets of 2.5 meters/sec and 1 cm diameter were bombarded by 24 GeV protons of 4×10^{12}/sec under no magnetic field. Figure 14 shows photos of the bombardment. The second test was done at CERN, in which mercury jets of 12 meters/sec and 4 mm diameter were examined under magnetic fields of 10 tesla and 20 tesla, but without a proton beam.

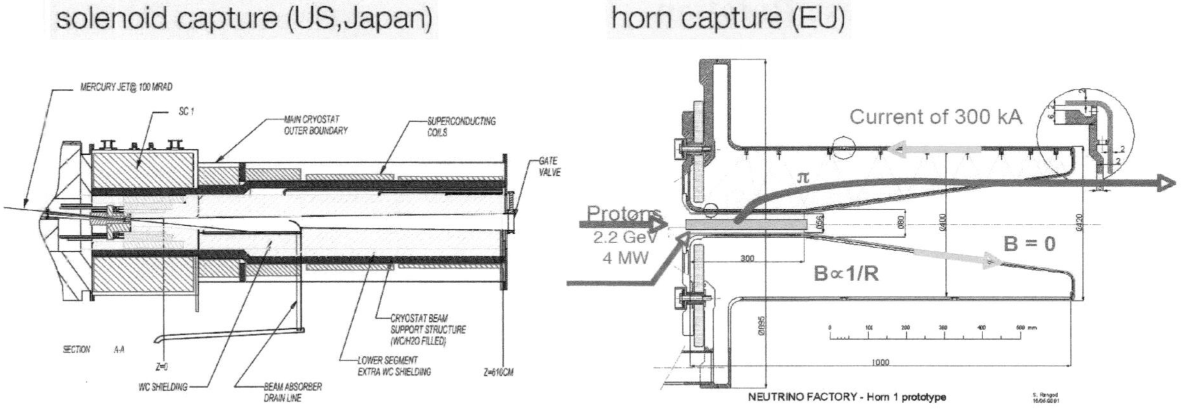

Figure 13. *Two pion capture schemes with large solid angle. One is solenoid pion capture (left) which is adopted in the US and Japanese designs. The other is magnetic-horn capture (right) which is considered in the CERN design.*

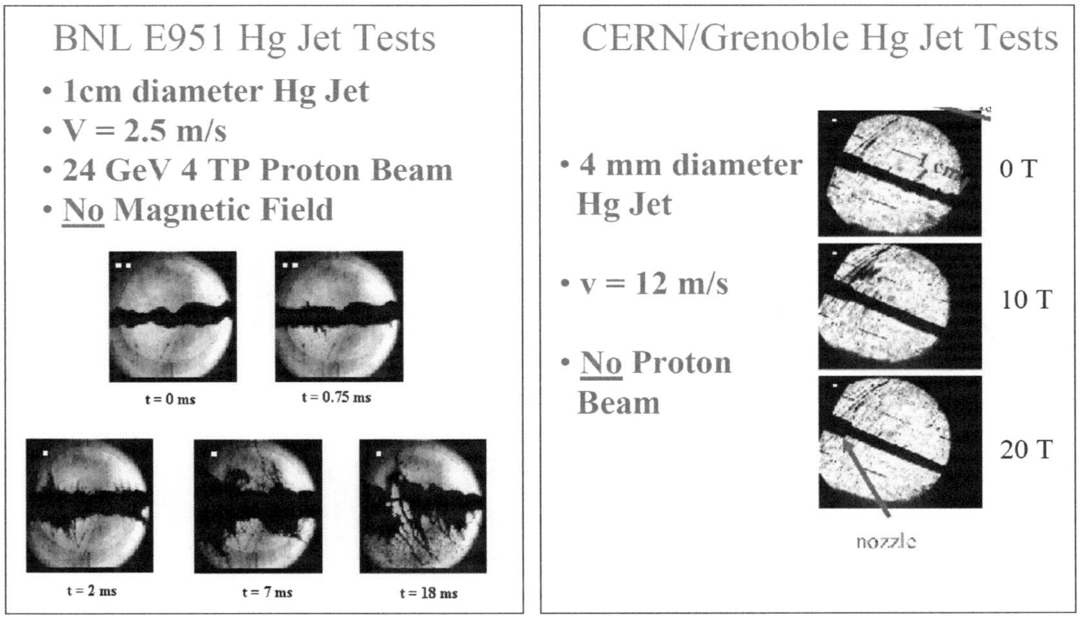

Figure 14. *Tests of a liquid mercury target. The left is a test at BNL with a 24 GeV proton beam. The right is a test at CERN with a magnetic field.*

A new experiment of testing mercury jets with both a proton beam and a high magnetic field was proposed and approved at CERN. In the new experiment, which is called "MERIT", a 24-GeV proton beam hit a target consisting of a free mercury jet inside a 15-T capture solenoid magnet. The experiment would be performed in the NTOF proton line upstream of the spallation target at CERN. The main piece of the apparatus is a 15-T copper magnet cooled with liquid nitrogen. The mercury jet is provided a closed loop. The MERIT experiment is supposed to run in year 2007.

8.3 Bunching and Phase Rotation

Pions thus produced decay into muons. The muons are generated over a very wide range of energies, and after a long drift, the bunch would widely spread longitudinally and has strong energy correlation in which higher energy particles are at the head and lower energy particles at the tail of the bunch. To make the long bunch to be captured and accelerated in a following rf system (of say 201 MHz), some beam treatment is needed. The idea is that the long bunch is separated into a train of a number of shorter bunches with a series of rf cavities applied ("bunching"). And then, for each of the separated bunch, phase rotation is performed, as shown in Figure 15. Phase rotation is to accelerate lower energy particles and accelerate higher energy particles by applying rf fields. The beam at the end of the buncher and phase rotation section has an average momentum of about 220 MeV/c. The present scheme is based on an rf cavity system. The other option which was discussed earlier was an induction linac based system. The present scheme of an rf system is known to be much cheaper than the latter. One of benefits on the rf-based system is ability to transport both signs of muons simultaneously.

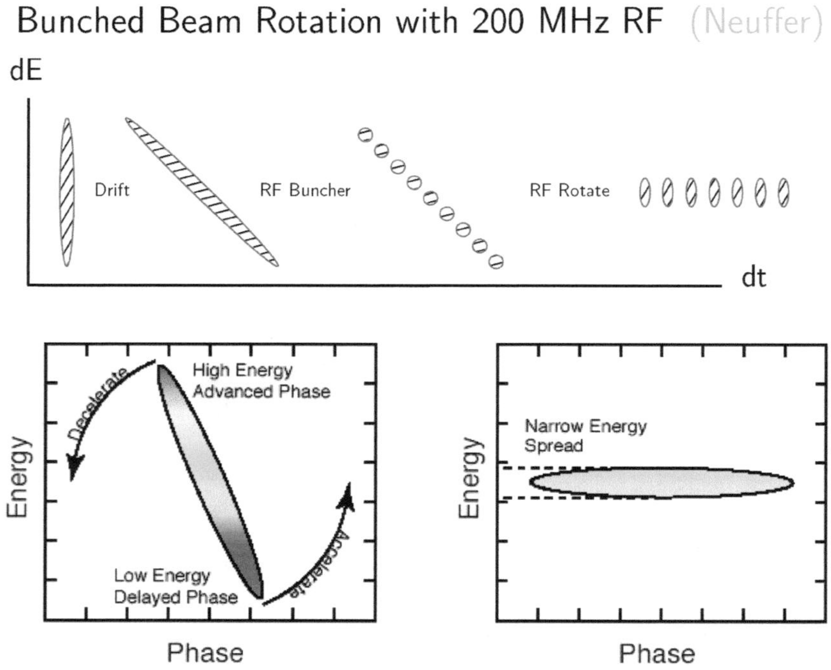

Figure 15. *Schematic figures on rf buncher and rf phase ration (top). The principle of phase rotation is shown (bottom), where higher energy particles are decelerated and lower energy particle are accelerated by rf cavities to make energy spread narrower.*

8.4 Cooling

The muon beam after bunching and phase rotation has still large transverse emittance. Reduction of the muon beam emittance is needed so that the following accelerating system can accept them. Normal beam cooling schemes such as electron beam cooling and statistical cooling do not work for the case of muons because of their short lifetime. The proposed cooling scheme is "ionization cooling", as shown in Figure 16. In the ionization cooling, the muon beam momenta in both longitudinal (parallel to the beam axis, p_l) and transverse (perpendicular to the beam axis, p_t) directions are reduced by passing through absorbers (such as liquid hydrogen), and then only the longitudinal momentum is restored by accelerating by rf cavities. As a result, the transverse momentum is reduced, whereas the longitudinal momentum does not change. Therefore the transverse beam emittance decreases. This process is repeated many times. The emittance change in the ionization cooling process is given by

$$\frac{d\varepsilon_t}{ds} = -\frac{1}{\beta^2}\left\langle\frac{dE_\mu}{ds}\right\rangle\frac{\varepsilon_t}{E_\mu} + \frac{1}{\beta^3}\frac{\beta_\perp(0.014)^2}{2E_\mu m_\mu X_0} \tag{29}$$

where ε_t, $\beta_\perp$ are the transverse beam emittance and the betatron amplitude at the absorber location respectively. E_μ, β and m_μ are the muon beam energy, the muon velocity, and the muon mass, respectively. s is a path length along the muon beam and $< dE_\mu/ds >$ is an average energy loss of the muons. X_0 is the radiation length of the absorber material. In Equation (29) the first term implies beam cooling and the second term does beam heating due to Coulomb multiple scattering. The emittance growth by Coulomb multiple scattering is controlled by focusing the beam at the location of absorbers so that the angular spread of the beam is reasonably large compared with the multiple scattering, namely making $\beta_\perp$ smaller. Also the absorber material should be chosen so as to reduce the multiple scattering by its long radiation length (X_0). And liquid hydrogen is the best.

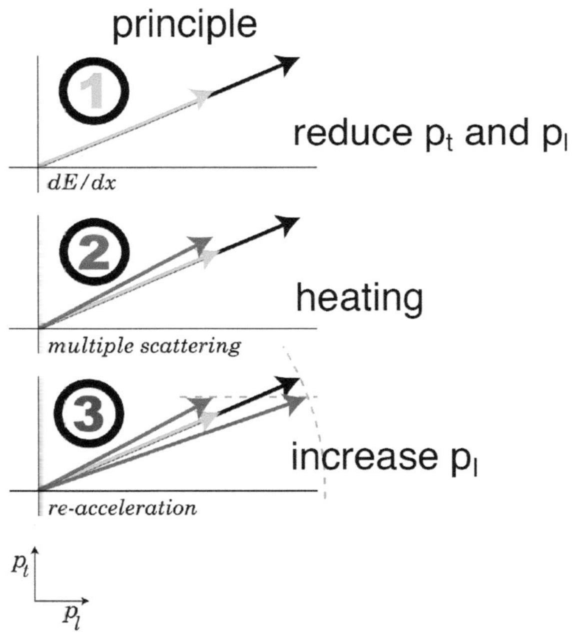

Figure 16. *Principle of ionization cooling. (1) Muons are passed through the energy absorber and lose both transverse momentum (p_t) and longitudinal momentum (p_l), as from the black momentum vector to a green momentum vector. (2) multiple scattering might contribute heating. (3) By restoring only p_l with rf fields, the beam emittance could be reduced.*

8.4.1 The MICE Experiment

The principle of ionization cooling has not been experimentally demonstrated yet. The Muon Ionization Cooling Experiment (MICE) was proposed and approved at Rutherford-Appleton Laboratory (RAL) in the UK to demonstrate the proof-in-principle of ionization cooling. In the MICE experiment, a section of cooling channel is built as shown in Figure 17. The muon beam from the internal target of the ISIS synchrotron in RAL is brought to the experimental setup which would measure the transverse emittance reduction of about 10 % with a relative accuracy of 1 %. Measurement of the beam emittance with absolute accuracy of 10^{-3} is challenging.

Since no ordinary methods of emittance measurement do have necessary precision, a single particle measurement is adopted so that the track parameters of each muon can be measured in an event-by-event base. The muon tracking chambers consisting of five stations of scintillating fibers are placed before and after the cooling channel. The cooling channel would consist of three sets of liquid hydrogen absorbers and two sets of rf cavities. The experiment is now in preparation and will start in the end of year 2007.

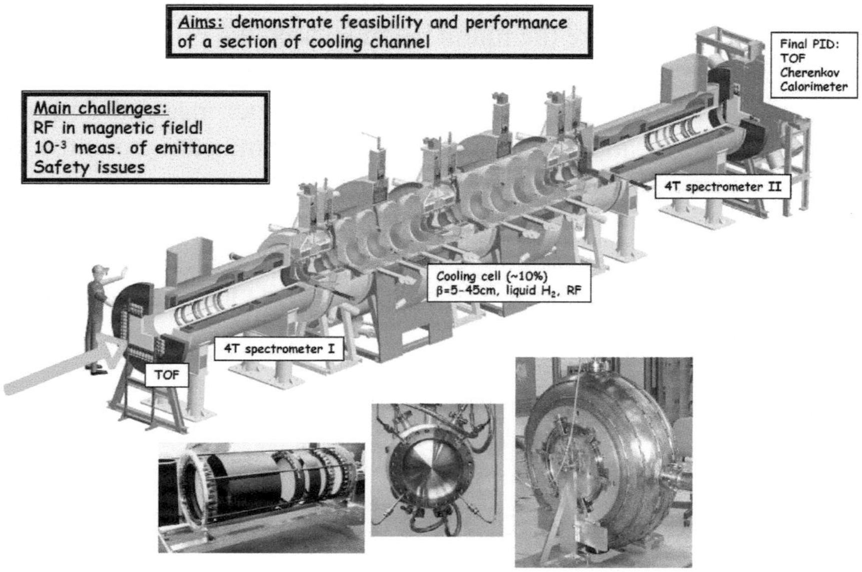

Figure 17. *Schematic layout of the MICE experimental setup.*

8.4.2 The MUCOOL Experiment

Although the proof-in-principle of ionization cooling is performed by the MICE experiment, there are many technical issues. One of them is whether the cooling channel can accept a high intensity particle beam. The MUCOOL experiment to examine capability of a high intensity beam is initiated at Fermi National Laboratory in the US, where a proton beam is planned to bring into the Muon Test Area (MTA). The experimental setup consists of 201 MHz rf cavities, liquid hydrogen absorbers with thin windows, and a 5 tesla cooling channel solenoid magnet.

8.5 Acceleration

Rapid acceleration from 200 MeV/c to 20–50 GeV is needed to reduce muon loss due to their lifetime. Therefore, ordinary synchrotrons do not work because of its slow repetition rates. There are three options of acceleration schemes considered, as shown in Figure 18. One is acceleration by a series of fixed field alternating gradient synchrotron (FFAG) of scaling type, and the second is that by FFAG of non-scaling type, and the third is a recirculating linear accelerator (RLA) of either racetrack-type or dog-bone-type. And combinations of some of the three acceleration schemes are also considered.

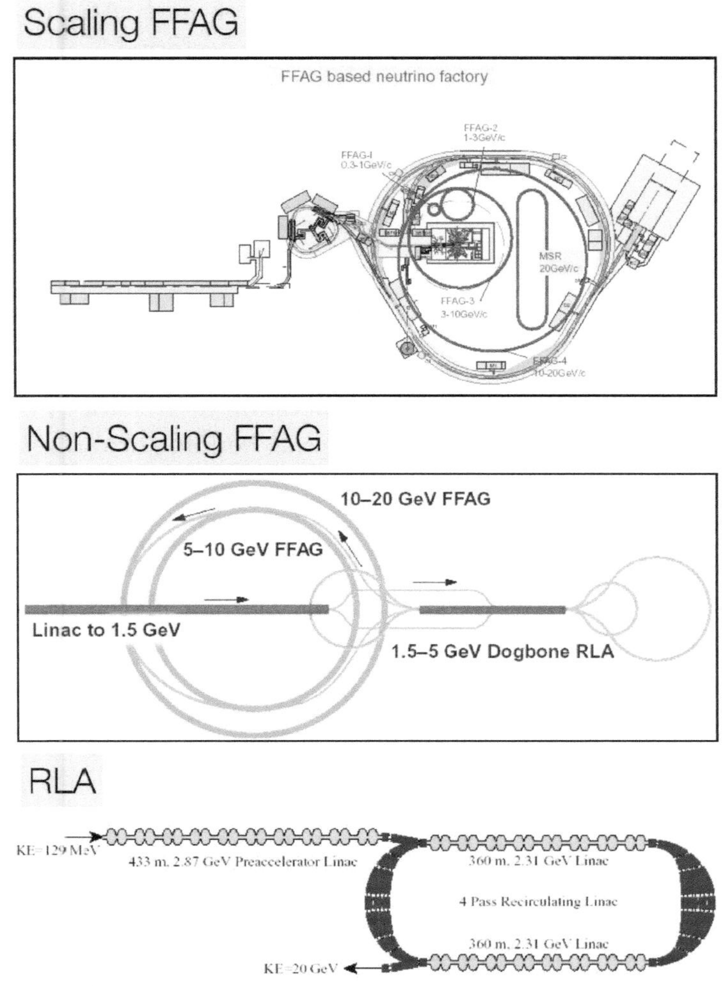

Figure 18. *Three acceleration schemes considered for a neutrino factory. They are based on (1) FFAG of scaling type (in the Japanese scheme), (2) FFAG of non-scaling type (in the US scheme), and (3) recirculating linear accelerators (RLA).*

8.5.1 What is FFAG ?

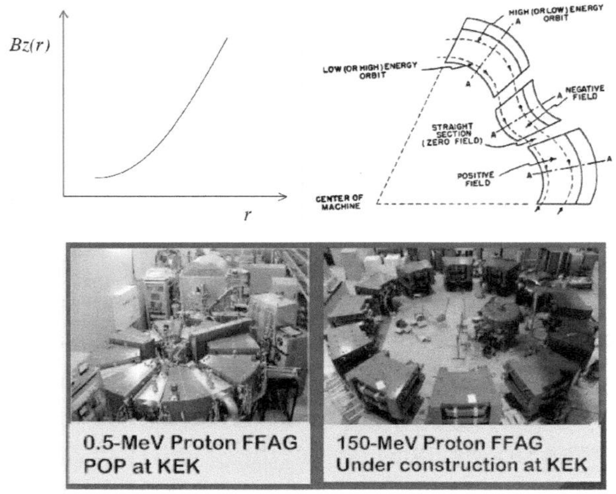

Figure 19. *(Top left) Magnetic field distribution in the FFAG ring of scaling type. (Top right) The magnets forming one cell of a FFAG ring of scaling type. (Bottom left) a photo of the proton FFAG of 0.5 MeV built at KEK. (Bottom right) a photo of the proton FFAG of 150 MeV at KEK.*

FFAG (fixed field alternating gradient synchrotron) is a new type of accelerators invented originally in Japan and recently being developed in the international framework. In FFAG, a magnetic field is fixed like a cyclotron, and therefore quick acceleration can be performed.

FFAG with scaling-type has zero chromaticity, which implies betatron tunes are the same for different momenta of beam particles. Therefore the trajectory of beam particles are similar, scaled by their momentum. The magnetic field at the location of radius, r, from the center of the FFAG ring is given by

$$B(r) = B_0 \left(\frac{r}{r_0}\right)^k \quad (k = 2 \sim 10) \tag{30}$$

where B_0 is a magnetic field at the radius r_0, and k is a field index. Since the magnetic field gets stronger for a larger radius, the average radius of beam trajectories does not increase very much as they are accelerated. Usually, one FFAG ring can achieve acceleration of a factor of about three or so from injection to extraction. Alternating focus-bending (*i.e.* bending towards the center of the ring) and defocus-bending (*i.e.* bending away from the center of the ring) in the FFAG ring, as shown in Figure 19, would provide strong beam focusing. From this, large transverse acceptance as well as large longitudinal acceptance can be achieved.

FFAG of non-scaling type has none-zero chromaticity, where the betatron tune changes as beam particles are accelerated. The advantage is to allow us to use linear beam elements such as dipole magnets, quadrupole magnets, sextapole magnets and so on, although the disadvantage is that beam particles pass across many beam resonances where a beam get unstable. It is being planned to construct a proof-of-principle machine of FFAG of non-scaling type.

8.6 Storage Ring

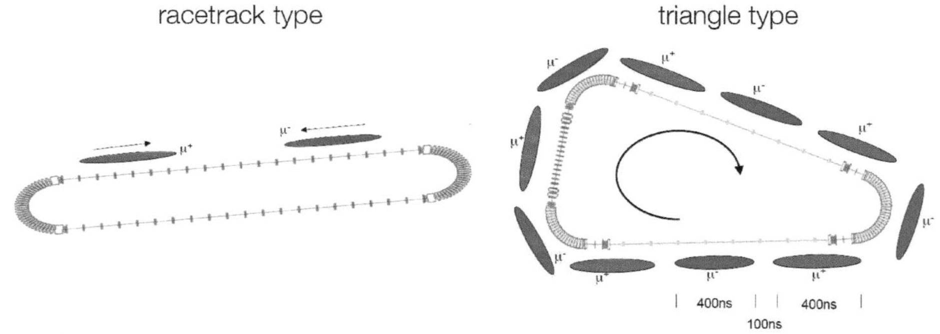

Figure 20. *Schematic layout of a racetrack ring (top) and a triangle ring (bottom) The bunches in red (blue) are for negative muons (positive muons).*

The accelerated muons are injected into the muon storage ring where muons are stored until they decay. Two types of the storage rings are considered as shown in Figure 20. One is a racetrack ring, and the other is a triangle ring. The racetrack ring has less fraction of the straight sections in which useful muon decays occur, but has more flexibility. The fraction could be up to 38 %. The triangle ring has more fraction of straight sections, but less flexibility. High-field superconducing arc magnets are used to minimize the arc length and maximize the fraction of the straight sections. One issue is how to avoid heat load from the muon-decay electrons hitting the superconducting magnets. Special coil design as well as collimators are considered. Both types of rings can have both signed muons being stored. Experimentally, they are discriminated by their decay timing. The detailed design of the muon storage ring depends on the locatiosn of detectors.

8.7 Neutrino Factory Studies in the World

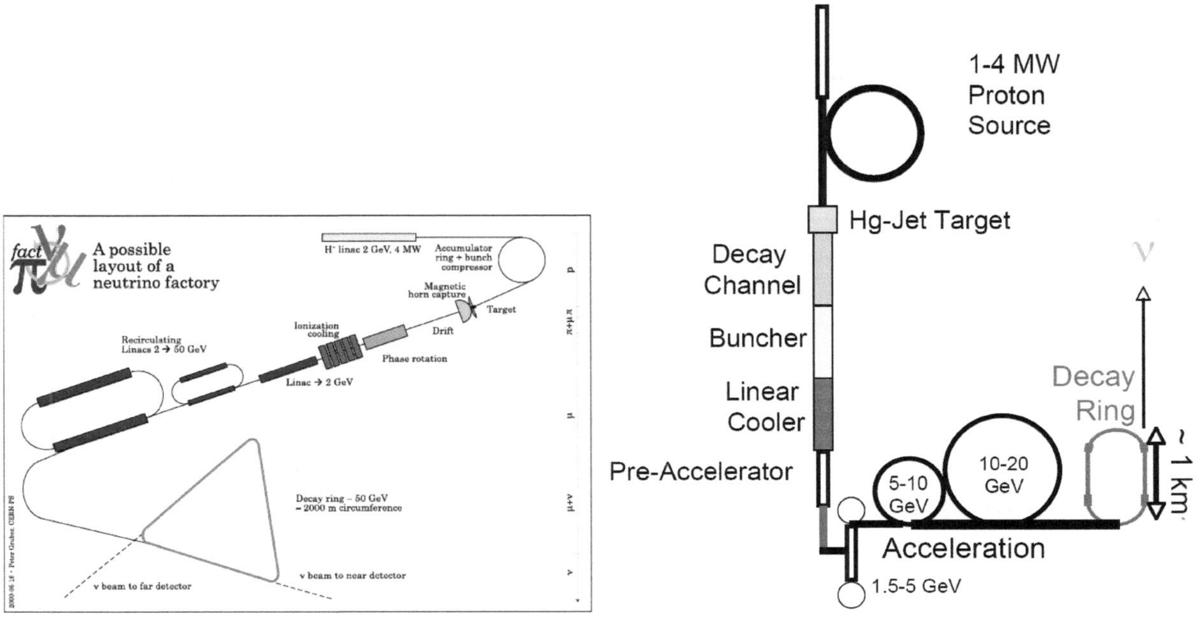

Figure 21. *Layouts of neutrino factories studied in the world. The layout of the CERN study (left) and the layout in the US study (right).*

Intensive studies and design works on neutrino factories have been performed in the world. They are such as the US, CERN, Japan, and the UK. The schematic layouts of the CERN design and the US are shown in Figure 21. The basic concept are almost the same, but they adopted different options in some systems. So far, those studies were made separately, but a new initiative to create an unified design in the world has been discussed and will start soon.

8.8 Options for Neutrino Factory Detectors

As described before, detectors for a neutrino factory should have charge identification and therefore detectors should be in a magnetic field. There are several options for neutrino factory detectors. They are (1) segmented magnetized detector, where many layers of sandwiches of magnetized steels and scintillator plates are placed, (2) a liquid argon detector, (3) a totally active scintillator detector (like the NOνA detector), and (4) an emulsion detector to detect τs in the silver channels. They are shown in Figure 22.

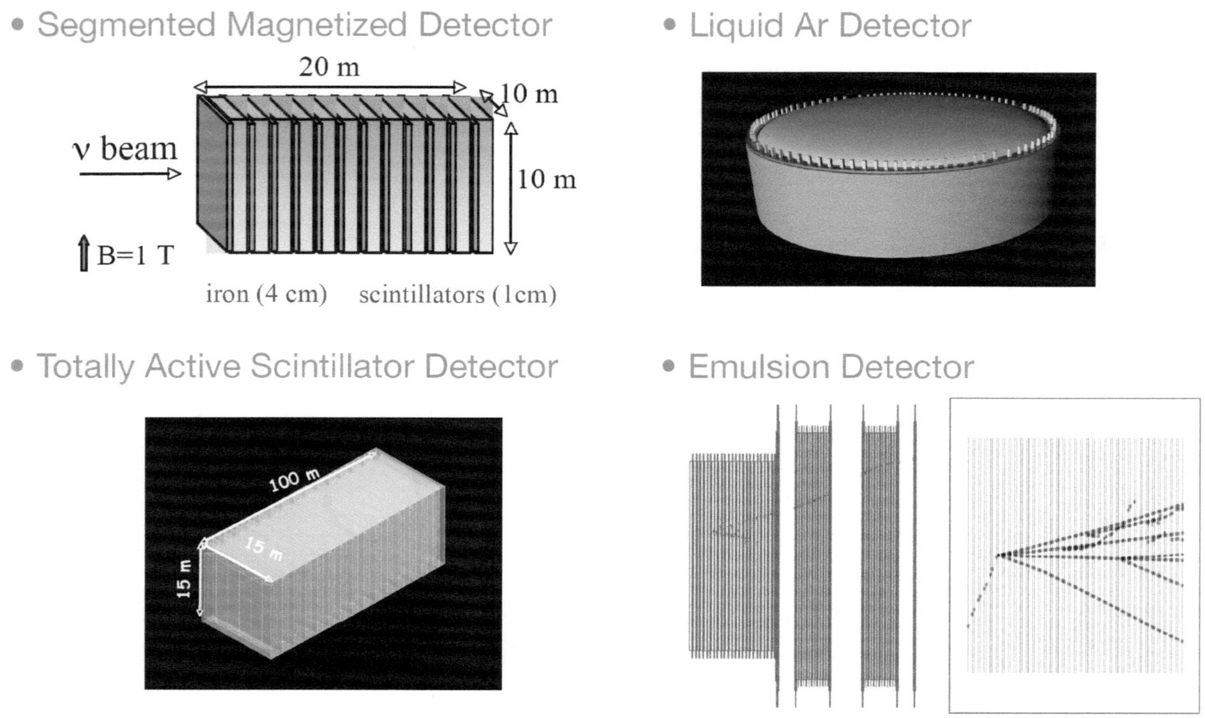

Figure 22. *Options for neutrino-factory detectors.*

9 Beta Beams

The "beta beam" is another future neutrino facility which produces pure and intense electron-neutrino (and anti electron-neutrino) beams (Zucchelli, 2002). The advantages are the following. First of all, the energy distribution and normalization of electron-neutrinos can be well known. Secondarily, when radioactive ions with a long lifetime ($\sim$sec) are selected, conventional acceleration, not rapid acceleration as in neutrino factories, can be used. In Europe, the EURISOL project to aim an next-generation facility for on-line production of radioactive isotopes has been initiated. And the studies of beam beams is organized as a part of the EURISOL design work.

9.1 Choice of Ions for Beta Beam

The choice of iron for beat beams has several considerations. They are that (1) ions can be produced at a reasonable amount, (2) ions should have not-too-short lifetime so as to be able to accelerate, and (3) ions should have not-too-long lifetime so that a reasonable amount of ions can decay even at high energy. The current choices are, for anti electron neutrinos ($\bar{\nu}_e$), ${}^6_2He$ which decays as

$$ {}^6_2He \rightarrow {}^6_3Li + e^- + \bar{\nu}_e, \quad \text{(average energy = 1.94 MeV)} \tag{31} $$

and, for electron neutrinos (ν_e), ${}^{18}_{10}Ne$ which decays as

$$ {}^{18}_{10}Ne \rightarrow {}^{18}_{9}F + e^+ + \nu_e. \quad \text{(average energy = 1.86 MeV)} \tag{32} $$

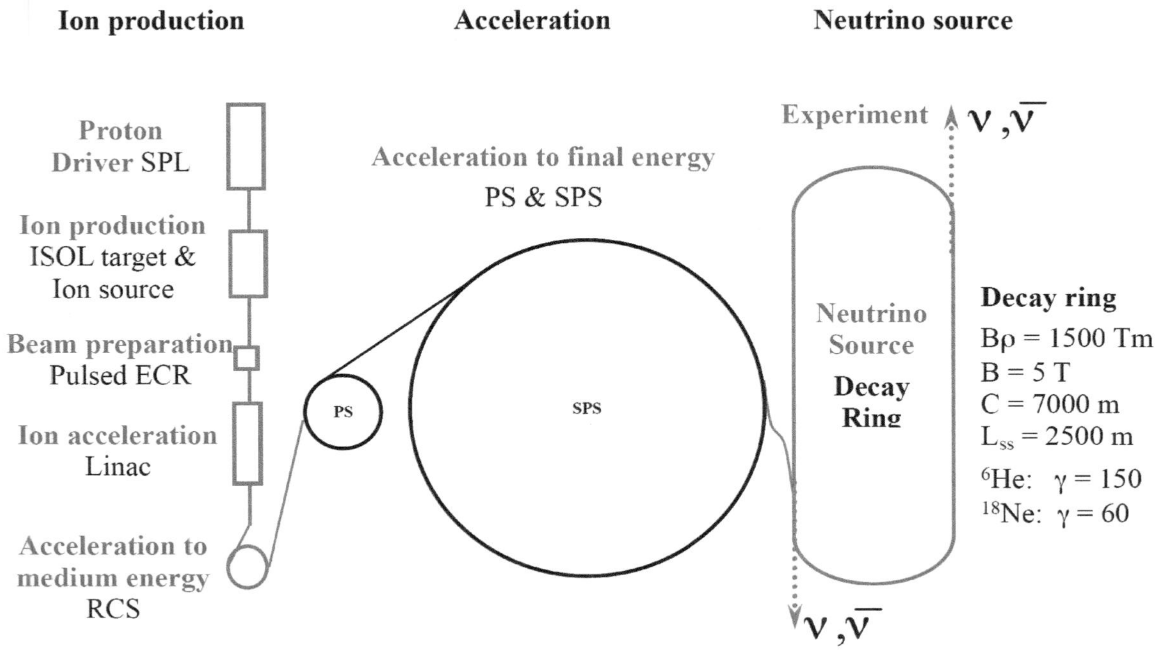

Figure 23. *Schematic layout of a beta beam facility proposed at CERN.*

To produce ${}^6_2He$ ions, a proton beam impinges on a water-cooled tungsten or liquid lead target core to produce neutrons. The neutrons hit surrounding BeO to produce ${}^6_2He$ by the reaction of ${}^9Be + n \rightarrow \alpha + {}^6He$. The expected rate of ion production is about 2×10^{13} ions/sec for 200 kW on target. On the other hand, to produce ${}^{18}Ne$ ions, a proton beam impinges on a MgO target. The production rate is 1×10^{12} ions/sec. Figure 23 shows a preliminary layout of a beta beam facility at CERN. In this scheme, ions are accelerated in the PS and SPS to γ=150 for ${}^6He$ and γ=60 for ${}^{18}Ne$ respectively.

Radioactive ions with electron capture would give mono-chromatic energy neutrinos. One of the candidate ions is ${}^{150}Dy$. The possible production rate is 1×10^{11} of ${}^{150}Dy$ ions/sec with 50 μA proton beam at TRIUMF. This rate is not sufficient, and an issue is whether 10^{18} ions can be achieved or not.

10 New Physics at Neutrino Oscillation Physics

It would be worth to ask yourself why high-precision determinations of the lepton mixing parameters are needed. It would be informative to look back the history of the quark mixing matrix. The Cabbibo-Kobayashi-Maskawa (CKM) quark mixing matrix has been studied extensively at *B* factories. In the case of the quark mixing matrix, the mixing angles were determined, chronologically from the largest to the smallest ones, and a non-zero CP-violating imaginary phase was finally measured and confirmed. After the completion of the precise measurements of the CKM matrix elements, searches for new physics has started by for instance investigating whether the quark unitarity triangle is closed or not. Will a similar history for the neutrino sector repeat ?

10.1 Is the Lepton Mixing Matrix Unitary ?

The studies presented so far for neutrino-oscillation physics assume that the neutrino mixing matrix is unitary. Under the unitarity assumption, the MNS neutrino mixing matrix could be presented by only four parameters which are 3 angles and 1 phase. If the unitary condition is removed, the (complex) neutrino mixing matrix can have 18 parameters, even for the 3 flavor generations, plus two Majorana phases.

$$V = \begin{pmatrix} V_{e1} & V_{e2} & V_{e3} \\ V_{\mu 1} & V_{\mu 2} & V_{\mu 3} \\ V_{\tau 1} & V_{\tau 2} & V_{\tau 3} \end{pmatrix} \times \begin{pmatrix} e^{-i\frac{\phi_1}{2}} & 0 & 0 \\ 0 & e^{-i\frac{\phi_2}{2}} & 0 \\ 0 & 0 & 1 \end{pmatrix} \tag{33}$$

In the comparison to the quark sector, we are still at the position to just complete the determination of the two large mixing angles, θ_{12} and θ_{23}. In coming future, we will measure the smallest mixing angle θ_{13}, and hopefully a CP-violating phase δ. It is likely that a history will repeat again and the unitarity tests of the MNS neutrino mixing matrix will be initiated after then. Therefore, it would be critical to consider such capabilities when future neutrino facilities are planned.

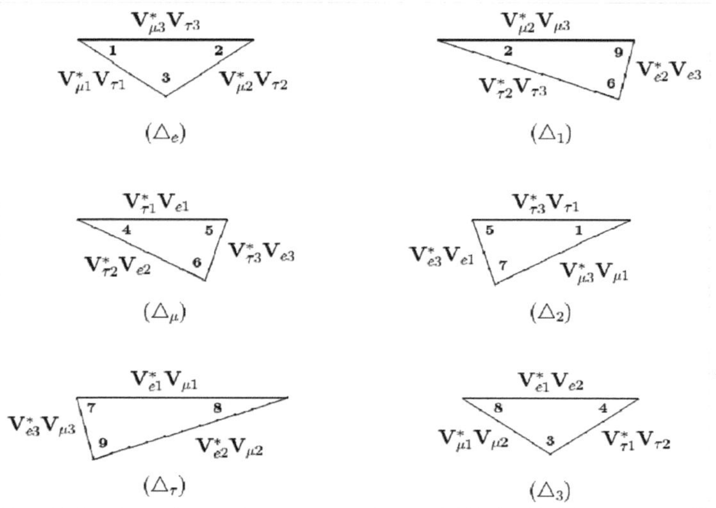

Figure 24. *Possible 6 unitarity triangles for neutrino oscillation. Since the contribution of new physics might be different for different channels, the triangles to be studied must be carefully chosen.*

There are two ways to test the unitarity of the MNS neutrino mixing matrix as follows.

- Normalization Tests :
  The normalization condition is given by, for instance,

$$|V_{e1}|^2 + |V_{e2}|^2 + |V_{e3}|^2 = 1. \tag{34}$$

To test this, the accurate determinations of each of the mixing matrix elements are needed. Instead of the above, a test of

$$\sum_n P(\nu_m \to \nu_n) = 1, \tag{35}$$

by using neutral current reaction rates is also proposed (Varger, Geer, Whisnant, 2004), where $P(\nu_m \to \nu_n)$ is the probability of neutrino oscillation from its flavor of ν_m to ν_n.

- Orthogonality Tests :
  In the orthogonality tests, the unitarity triangles are examined, as in the quark sector. In principle, 6 unitarity triangles can be formed, as seen in Figure 24. Since the contributions of new physics might be different for different oscillation channels, the triangles to be studied must be carefully chosen.

10.2 New Physics Interaction

What would be potential sources for non-unitarity ? One of the sources is that there are more than 3 generations, possibly including sterile neutrinos. The other potential sources are new physics interactions.

It has been discussed that three potential sources of new physics to neutrino oscillation can occur. They are new interaction at the neutrino production, namely $\mu \to e\nu\bar{\nu}$ decay, and that at the propagation in matter, and that at the neutrino detection ($\nu + N \to l + N'$). There have been extensive theoretical studies on new physics contributions. Also it should be noted that if the $SU(2)$ invariance in the lepton doublet is assumed, strong constraints from charged lepton flavor violation are applied. From that account, the neutrino oscillation channels involving the ν_τ appearance channels would be the best to search for new physics.

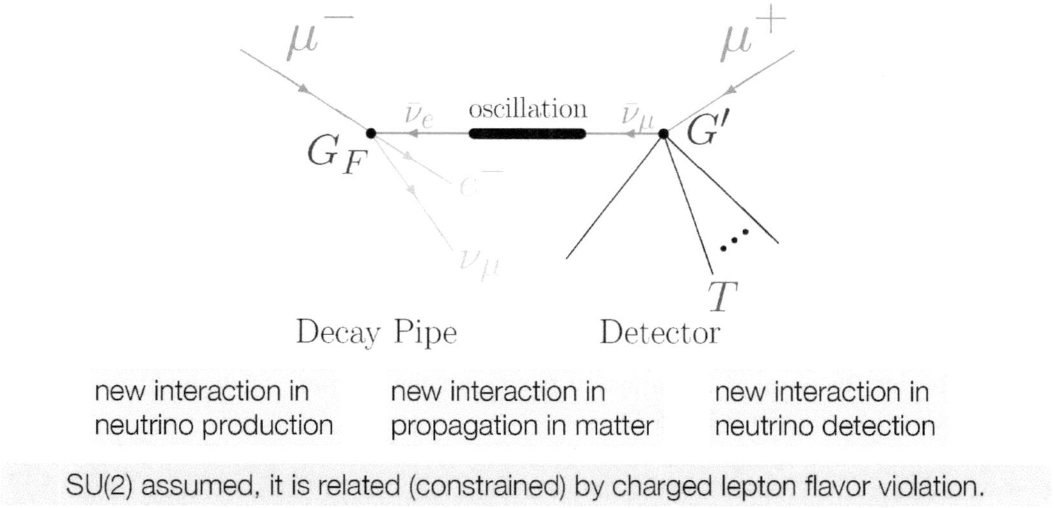

Figure 25. *Neutrino oscillation and new interaction. Potential contributions come from the neutrino production, the propagation in matter, and the neutrino detection.*

11 Summary

Neutrino oscillation physics is presented. The three major goals in neutrino oscillation physics are

- determination of θ_{13},

- determination of the neutrino mass hierarchy, and
- discovery of leptonic CP violation.

To disentangle the eight-fold degeneracies, several different kinds of oscillation measurements have to be performed. To accomplish these physics goals, future neutrino facilities such as superbeams, beta beams, and neutrino factories, are being studied extensively. In particular, studies and R&D of a neutrino factory are described in detail. These facilities might address some of the intermediate questions shown in Section 1. However, to address the "Big Questions", the bottom-up approaches are necessary, and potential sensitivities to new physics beyond the standard neutrino oscillation have to be envisaged when future neutrino facilities are planned.

Acknowledgments

I thank the organizers of the school for their hospitality and creating a stimulating environment.

References

Barger V, et al., 2002, Breaking eight fold degeneracies in neutrino CP violation, mixing, and mass hierarchy, *Physical Review* **D 65** 073023.

Barger V. Geer S, Whisnant K, 2004, Neutral currents and tests of three-neutrino unitarity in long-baseline experiments, *New Journal of Physics* **6** 135.

Fogli GL, and Lisi E, 1996, Tests of three flavor mixing in long baseline neutrino oscillation experiments,*Physical Review* **D 54** 3667.

Harris D, 2006, This volume.

Huber P, Lindner M, Rolinec M, and Winter W, 2006, Optimization of a neutrino factory oscillation experiment, *Physical Review* **D74** 073003.

MInakata H, and Nunokawa H, 2001, Exploring neutrino mixing with low-energy superbeams, *JHEP* **0110** 001.

Rigolin S, 2005, Physics Reach of β-beam and ν-factories : the problem of degeneracies, *Nuclear Physics (Proceedings Supplement)* **B155** 33.

Zucchelli P, 2002, A novel concept for a $\bar{\nu}_e/\nu_e$ neutrino factory: The beta beam, *Physics Letters* **B532** 168.

Section IV: Neutrinos in Cosmology

Leptogenesis: Standard Model and Alternatives

Wilfried Buchmüller

Deutsches Elektronen-Synchrotron DESY, Hamburg, Germany

1 Matter-Antimatter Asymmetry

One of the main successes of the standard early-universe cosmology is the prediction of the abundances of the light elements, D, ^{3}He, ^{4}He and ^{7}Li. Agreement between theory and observation is obtained for a certain range of the parameter η_B, the ratio of baryon density and photon density [1],

$$\eta_B^{BBN} = \frac{n_B}{n_\gamma} = (4.7 - 6.5) \times 10^{-10} , \tag{1}$$

where the present number density of photons is $n_\gamma \sim 400/\mathrm{cm}^3$. Since no significant amount of antimatter is observed in the universe, the ratio η_B coincides with the cosmological baryon asymmetry, $\eta_B = (n_B - n_{\bar{B}})/n_\gamma$.

The precision of measurements of the baryon asymmetry has dramatically improved with the observation of the acoustic peaks in the cosmic microwave background radiation (CMB). Most recently, the WMAP Collaboration has measured the baryon asymmetry with a (1σ) standard error of $\sim 5\%$ [2],

$$\eta_B^{CMB} = (6.1^{+0.3}_{-0.2}) \times 10^{-10} . \tag{2}$$

Such a matter-antimatter asymmetry can be dynamically generated in an expanding universe if the particle interactions and the cosmological evolution satisfy Sakharov's conditions [3],

- baryon number violation,
- *C* and *CP* violation,
- deviation from thermal equilibrium .

Although the baryon asymmetry is just a single number, it provides an important relationship between the standard model of cosmology, i.e., the expanding universe with Robertson-Walker metric, and the standard model of particle physics as well as its extensions.

At present there exist a number of viable scenarios for baryogenesis. They can be classified according to the different ways in which Sakharov's conditions are realized. In grand unified theories baryon number (B) and lepton number (L) are broken by the interactions of gauge bosons and leptoquarks. This is the basis of classical GUT baryogenesis [4]. In a similar way, the lepton number violating decays of heavy Majorana neutrinos lead to leptogenesis [5]. In the simplest version of leptogenesis the initial abundance of the heavy neutrinos is generated by thermal processes. Alternatively, heavy neutrinos may be produced in inflaton decays or in the reheating process after inflation. A further mechanism of baryogenesis can work in supersymmetric theories where the scalar potential has approximately flat directions. Coherent oscillations of scalar fields may then generate large asymmetries [6].

The crucial departure from thermal equilibrium can also be realized in various ways. One possibility is a sufficiently strong first-order electroweak phase transition at a critical temperature T_{EW}, which would make electroweak baryogenesis possible [7]. For the classical GUT baryogenesis and for thermal leptogenesis the departure from thermal equilibrium is due to the deviation of the number density of the decaying heavy particles from the equilibrium number density. How strong this departure from equilibrium is depends on the lifetime of the decaying heavy particles and the cosmological evolution.

A crucial ingredient of baryogenesis is the connection between baryon number and lepton number in the high-temperature, symmetric phase of the standard model. Due to the chiral nature of the electroweak interactions, baryon and lepton number are not conserved [8]. The divergence of the B and L currents,

$$J_\mu^B = \frac{1}{3} \sum_{generations} \left(\overline{q_L}\gamma_\mu q_L + \overline{u_R}\gamma_\mu u_R + \overline{d_R}\gamma_\mu d_R\right) , \tag{3}$$

$$J_\mu^L = \sum_{generations} \left(\overline{l_L}\gamma_\mu l_L + \overline{e_R}\gamma_\mu e_R\right) , \tag{4}$$

is given by the triangle anomaly,

$$\begin{aligned} \partial^\mu J_\mu^B &= \partial^\mu J_\mu^L \\ &= \frac{N_f}{32\pi^2}\left(-g^2 W_{\mu\nu}^I \widetilde{W}^{I\mu\nu} + g'^2 B_{\mu\nu}\widetilde{B}^{\mu\nu}\right) . \end{aligned} \tag{5}$$

Here N_f is the number of generations, and W_μ^I and B_μ are, respectively, the $SU(2)$ and $U(1)$ gauge fields with gauge couplings g and g'.

As a consequence of the anomaly, the change in baryon and lepton number is related to the change in the topological charge of the gauge field,

$$\begin{aligned} B(t_f) - B(t_i) &= \int_{t_i}^{t_f} dt \int d^3x \partial^\mu J_\mu^B \\ &= N_f[N_{cs}(t_f) - N_{cs}(t_i)] , \end{aligned} \tag{6}$$

where

$$N_{cs}(t) = \frac{g^3}{96\pi^2} \int d^3x \varepsilon_{ijk} \varepsilon^{IJK} W^{Ii} W^{Jj} W^{Kk} . \tag{7}$$

For vacuum to vacuum transitions W^{Ii} is a pure gauge configuration and the Chern-Simons numbers $N_{cs}(t_i)$ and $N_{cs}(t_f)$ are integers.

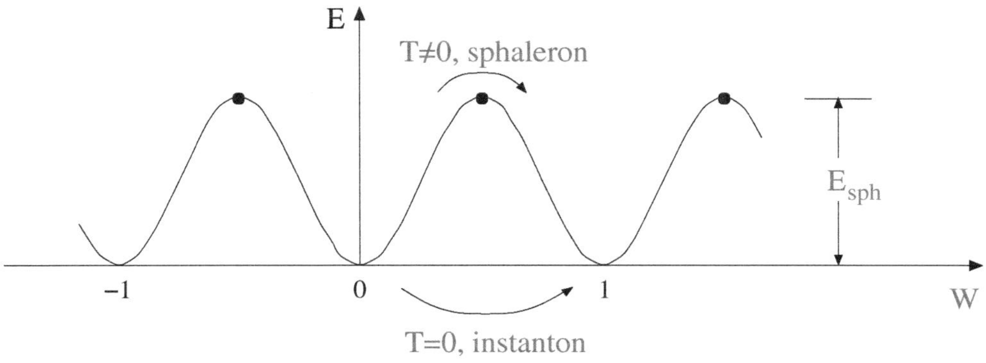

Figure 1. *Potential barrier in field space between vacua with different topological charge.*

In a non-abelian gauge theory there are infinitely many degenerate ground states, which differ in their value of the Chern-Simons number, $\Delta N_{cs} = \pm 1, \pm 2, \ldots$. The corresponding points in field space are separated by a potential barrier (cf. Fig. (1)) whose height is given by the so-called sphaleron energy E_{sph} [9]. Because of the anomaly, jumps in the Chern-Simons number are associated with changes of baryon and lepton number,

$$\Delta B = \Delta L = N_f \Delta N_{cs} \, . \tag{8}$$

Obviously, in the standard model the smallest jump is $\Delta B = \Delta L = \pm 3$.

In the semiclassical approximation, the probability of tunnelling between neighboring vacua is determined by instanton configurations. In the standard model, $SU(2)$ instantons lead to an effective 12-fermion interaction (cf. Fig. (2)),

$$O_{B+L} = \prod_{i=1\ldots3} (q_{Li} q_{Li} q_{Li} l_{Li}) \, , \tag{9}$$

which describes processes with $\Delta B = \Delta L = 3$, such as

$$u^c + d^c + c^c \to d + 2s + 2b + t + \nu_e + \nu_\mu + \nu_\tau \, . \tag{10}$$

The transition rate is determined by the instanton action and one finds [8]

$$\begin{aligned} \Gamma &\sim e^{-S_{\text{inst}}} \simeq e^{-\frac{4\pi}{\alpha}} \\ &= \mathscr{O}\left(10^{-165}\right) \, . \end{aligned} \tag{11}$$

Because this rate is extremely small, $(B+L)$-violating interactions appear to be completely negligible in the standard model.

This picture changes dramatically in a thermal bath at high temperatures. One can then have transitions between the gauge vacua not by tunnelling, but through thermal fluctuations over the barrier. The rate of these processes is related to the free energy of sphaleron field configurations [9] which carry topological charge. For temperatures larger than the height of the barrier, the exponential suppression in the rate provided by the Boltzmann factor disappears completely. Hence $(B+L)$-violating processes can occur at a significant rate and these processes can be in equilibrium in the expanding universe [7]. The sphaleron transition rate in the symmetric high-temperature phase has been evaluated by combining an analytical re-summation with numerical

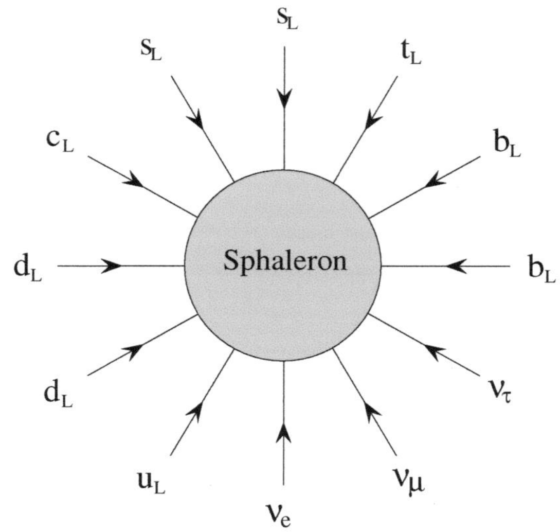

Figure 2. *One of the 12-fermion processes which are in thermal equilibrium in the high-temperature phase of the standard model.*

lattice techniques [10]. The result is, in accord with previous estimates, that B and L violating processes are in thermal equilibrium for temperatures in the range

$$T_{EW} \sim 100 \text{ GeV} < T < T_{SPH} \sim 10^{12} \text{ GeV} . \tag{12}$$

Sphaleron processes have a profound effect on the generation of the cosmological baryon asymmetry. Eq. (8) suggests that any $B+L$ asymmetry generated before the electroweak phase transition, i.e., at temperatures $T > T_{EW}$, will be washed out. However, since only left-handed fields couple to sphalerons, a non-zero value of $B+L$ can persist in the high-temperature, symmetric phase if there exists a non-vanishing $B-L$ asymmetry. An analysis of the chemical potentials of all particle species in the high-temperature phase yields the following relation between the baryon asymmetry and the corresponding L and $B-L$ asymmetries,

$$\langle B \rangle_T = c_S \langle B-L \rangle_T = \frac{c_S}{c_S - 1} \langle L \rangle_T . \tag{13}$$

Here c_S is a number $\mathcal{O}(1)$. In the standard model with three generations and one Higgs doublet one has $c_s = 28/79$.

An important ingredient in the theory of baryogenesis is also the nature of the electroweak transition. Since in the standard model baryon number, C and CP are not conserved, it is conceivable that the cosmological baryon asymmetry has been generated in a sufficiently strong first-order electroweak phase transition. Detailed studies during the past years have shown that for Higgs masses above the present LEP bound of 114 GeV electroweak baryogenesis is not viable, except for some supersymmetric extensions of the standard model [11]. In particular, the electroweak transition may have been just a smooth crossover, without any departure from thermal equilibrium. In this case it's sole effect in the cosmological evolution has been to switch off the $B+L$ changing sphaleron processes adiabatically.

Based on the relation (13) between baryon and lepton number we then conclude that $B-L$ violation is needed to explain the cosmological baryon asymmetry if baryogenesis took place before the electroweak transition, i.e., at temperatures $T > T_{EW} \sim 100$ GeV. In the standard model, as well as its supersymmetric version and its unified extensions based on the gauge group $SU(5)$, $B-L$ is a conserved quantity. Hence, no baryon asymmetry can be generated dynamically in these models and one has to consider extensions with lepton number violation[1]. On the other hand, lepton number violation can only be weak, since otherwise any baryon asymmetry would be washed out. The interplay of these two conflicting conditions leads to important constraints on neutrino properties and on extensions of the standard model in general.

2 Grand Unification and Thermal Leptogenesis

Lepton number is naturally violated in grand unified theories (GUTs) [13]. The unification of gauge couplings at high energies suggests that the standard model gauge group is part of a larger simple group,

$$G_{SM} = U(1) \times SU(2) \times SU(3) \subset SU(5) \subset SO(10) \ldots . \tag{14}$$

The simplest GUT is based on the gauge group $SU(5)$. Here quarks and leptons are grouped into the multiplets,

$$\mathbf{10} = (q, u^c, e^c)\,, \quad \mathbf{5}^* = (d^c, l)\,, \quad \mathbf{1} = N\,, \tag{15}$$

where $q = (u, d)$ and $l = (\nu, e)$. Unlike gauge fields, quarks and leptons are not unified in a single multiplet. In particular, the singlet neutrinos N, whose existence is crucial for the physics of massive neutrinos [14, 15], are not needed in $SU(5)$ models. Since the N's have no $SU(5)$ gauge interactions, they can have large Majorana masses M which are not controlled by the Higgs mechanism.

The spontaneous breaking of the electroweak symmetry $U(1) \times SU(2)$ leads to quark and lepton mass matrices. The part of the lagrangian relevant for neutrinos reads

$$\begin{aligned} \mathscr{L} &= \bar{\ell}_{Li} i\partial\!\!\!/ \ell_{Li} + \bar{N}_{Ri} i\partial\!\!\!/ N_{Ri} \\ &\quad + h_{ij} \bar{N}_{Ri} \ell_{Lj} H - \frac{1}{2} M_{ij} N_{Ri} N_{Rj} + \text{h.c.}\,, \end{aligned} \tag{16}$$

where $i, j = 1 \ldots 3$ are the family-number indices. We adopt, without losing generality, a basis where the matrix M_{ij} is diagonal. The Dirac neutrino mass matrix is $m_D = hv$, where $v = \langle H \rangle \sim 100$ GeV is the vacuum expectation value of the Higgs field H. The theory predicts six Majorana neutrinos as physical states, three heavy (N) and three light (ν), with masses

$$m_N \simeq M\,, \qquad m_\nu = -m_D^T \frac{1}{M} m_D\,. \tag{17}$$

The explanation of the smallness of the light neutrino masses in terms of the largeness of the heavy neutrino masses is the so-called seesaw mechanism.

All quarks and leptons of one generation are unified in a single multiplet in the GUT group $SO(10)$,

$$\mathbf{16} = \mathbf{10} + \mathbf{5}^* + \mathbf{1}\,. \tag{18}$$

[1]In the case of Dirac neutrinos, which have extremely small Yukawa couplings, one can construct leptogenesis models where an asymmetry of lepton doublets is accompanied by an asymmetry of right-handed neutrinos such that the total lepton number is conserved and $\langle B-L \rangle_T = 0$ [12].

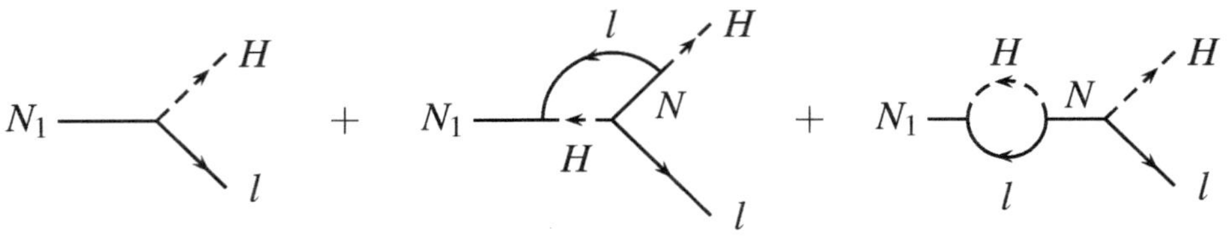

Figure 3. *Tree level and one-loop diagrams contributing to heavy neutrino decays whose interference leads to Leptogenesis.*

In the simplest pattern of symmetry breaking, $B-L$, a subgroup of $SO(10)$, is broken at the unification scale Λ_{GUT}. If Yukawa couplings of the third generation are $\mathcal{O}(1)$, as it is the case for the top-quark, one finds for the corresponding heavy and light neutrino masses:

$$M_3 \sim \Lambda_{GUT} \sim 10^{15}\ \text{GeV}\,, \quad m_3 \sim \frac{v^2}{M_3} \sim 0.01\ \text{eV}\,. \tag{19}$$

It is very remarkable that the light neutrino mass m_3 is of the same order as the mass differences $(\Delta m^2_{sol})^{1/2}$ and $(\Delta m^2_{atm})^{1/2}$ inferred from neutrino oscillations. This suggests that, via the seesaw mechanism, neutrino masses probe the grand unification scale! Like for quarks and charged leptons, one expects in GUTs a mass hierarchy also for the right-handed neutrinos. For instance, if their masses scale like the up-quark masses one has $M_1 \sim 10^{-5} M_3 \sim 10^{10}$ GeV.

The lightest of the heavy Majorana neutrinos, N_1, is ideally suited to generate the cosmological baryon asymmetry [5]. It can decay into final states with lepton or anti-lepton,

$$N_1 \to lH\,, \quad N_1 \to l^c H^c\,, \tag{20}$$

thereby violating lepton number conservation. N_1 decays to lepton-Higgs pairs then yield a lepton asymmetry $\langle L\rangle_T \neq 0$, which is partially converted to a baryon asymmetry $\langle B\rangle_T \neq 0$ by the sphaleron processes. The generated asymmetry is proportional to the *CP* asymmetry in N_1 decays, which arises from the interference of tree-level and one-loop vertex and self-energy contributions (cf. Fig. (3)). For hierarchical heavy neutrino masses, $M_1 \ll M_2, M_3$, one obtains [16],

$$\begin{aligned} \varepsilon_1 &= \frac{\Gamma(N_1 \to lH) - \Gamma(N_1 \to l^c H^c)}{\Gamma(N_1 \to lH) + \Gamma(N_1 \to l^c H^c)} \\ &\simeq \frac{3}{16\pi} \frac{M_1}{(h^\dagger h)_{11} v^2} \sum_{i=2,3} \text{Im}\left[\left(hh^\dagger\right)^2_{i1}\right] \frac{M_1}{M_i}\,. \end{aligned} \tag{21}$$

The CP asymmetry can be obtained in a very simple way by first integrating out the heavier neutrinos N_2 and N_3 in the lagrangian (16), which yields

$$\mathcal{L}_\nu^{eff} = h_{1j}\overline{N_{R1}}\ell_{Lj}H - \frac{1}{2}M_1\overline{N^c_{R1}}N_{R1} + \frac{1}{2}\eta_{ij}\ell_{Li}H\ell_{Lj}H + \text{h.c.}\,, \tag{22}$$

with

$$\eta_{ij} = \sum_{k=2}^{3} h^T_{ik}\frac{1}{M_k}h_{kj}\,. \tag{23}$$

The asymmetry ε_1 is then obtained from the interference of the Born graph and the one-loop graph involving the cubic and the quartic couplings. This includes automatically both, vertex and self-energy corrections [17] and yields an expression for ε_1 directly in terms of the light neutrino mass matrix:

$$\varepsilon_1 \simeq -\frac{3}{16\pi}\frac{M_1}{(hh^\dagger)_{11}v^2}\mathrm{Im}\left(h^* m_\nu h^\dagger\right)_{11} . \tag{24}$$

From this expression one easily obtains a rough estimate for ε_1 in terms of neutrino masses. Assuming dominance of the largest eigenvalue m_3 of the mass matrix m_ν, phases $\mathscr{O}(1)$ and an approximate cancellation of Yukawa couplings in numerator and denominator, one finds

$$\varepsilon_1 \sim \frac{3}{16\pi}\frac{M_1 m_3}{v^2} \sim 0.1\,\frac{M_1}{M_3} , \tag{25}$$

where we have again used the seesaw relation. In this example, the order of magnitude of the *CP* asymmetry is determined by the mass hierarchy of the heavy Majorana neutrinos. For a mass ratio as for up-type quarks, i.e., $M_1/M_3 \sim 10^{-5}$, one has $\varepsilon_1 \sim 10^{-6}$.

Given the *CP* asymmetry ε_1, one obtains for the baryon asymmetry,

$$\eta_B = \frac{n_B - n_{\bar{B}}}{n_\gamma} = -d\varepsilon_1\kappa_f \sim 10^{-10} . \tag{26}$$

Here the dilution factor $d \sim 10^{-2}$ accounts for the increase of the number of photons in a comoving volume element between baryogenesis and today, and the efficiency factor κ_f represents the effect of washout processes. In the estimate (26) we have assumed a typical value, $\kappa_f \sim 10^{-2}$. The correct value of the baryon asymmetry is then obtained as consequence of a large hierarchy of the heavy neutrino masses, which leads to a small *CP* asymmetry, and the kinematical factors d and κ_f [18]. The baryogenesis temperature

$$T_B \sim M_1 \sim 10^{10}\ \mathrm{GeV} , \tag{27}$$

corresponds to the time $t_B \sim 10^{-26}$ s, which characterizes the next relevant epoch before recombination, nucleosynthesis and the electroweak phase transition.

An important question concerns the relation between leptogenesis and neutrino mass matrices which can account for low-energy neutrino data. Many interesting models, some also very different from the example given above, have been discussed in the literature [15, 19]. Of particular interest is the connection with *CP* violation in other low energy processes [20]. Together with leptogenesis, improved measurements of neutrino parameters will have strong implications for the structure of grand unified theories.

3 Solving the Kinetic Equations

Leptogenesis takes place at temperatures $T \sim M_1$. For a decay width small compared to the Hubble parameter, $\Gamma_1(T) < H(T)$, heavy neutrinos are out of thermal equilibrium, otherwise they are in thermal equilibrium [4]. The borderline between the two regimes is given by $\Gamma_1 = H|_{T=M_1}$, which is equivalent to the condition that the effective neutrino mass

$$\widetilde{m}_1 = \frac{(m_D m_D^\dagger)_{11}}{M_1} \tag{28}$$

is equal to the 'equilibrium neutrino mass'

$$m_* = \frac{16\pi^{5/2}}{3\sqrt{5}} g_*^{1/2} \frac{v^2}{M_P} \simeq 10^{-3}\ \text{eV}\,. \tag{29}$$

Here we have used the Hubble parameter $H(T) \simeq 1.66\, g_* T^2/M_P$ where $g_* = g_{SM} = 106.75$ is the total number of degrees of freedom and $M_P = 1.22 \times 10^{19}\,\text{GeV}$ is the Planck mass.

It is quite remarkable that the equilibrium neutrino mass m_* is close to the neutrino masses suggested by neutrino oscillations, $\sqrt{\Delta m^2_{\text{sol}}} \simeq 8 \times 10^{-3}$ eV and $\sqrt{\Delta m^2_{\text{atm}}} \simeq 5 \times 10^{-2}$ eV. This encourages one to think that it may be possible to understand the cosmological baryon asymmetry via leptogenesis as a process close to thermal equilibrium. Ideally, $\Delta L = 1$ and $\Delta L = 2$ processes should be strong enough at temperatures above M_1 to keep the heavy neutrinos in thermal equilibrium and weak enough to allow the generation of an asymmetry at temperatures below M_1.

In general, the generated baryon asymmetry is the result of a competition between production processes and washout processes that tend to erase any generated asymmetry. Unless the heavy Majorana neutrinos are partially degenerate, $M_{2,3} - M_1 \ll M_1$, the dominant processes are decays and inverse decays of N_1 and the usual off-shell $\Delta L = 1$ and $\Delta L = 2$ scatterings [21, 22].

The Boltzmann equations for leptogenesis are[2]

$$\frac{dN_{N_1}}{dz} = -(D+S)\,(N_{N_1} - N_{N_1}^{\text{eq}})\,, \tag{30}$$

$$\frac{dN_{B-L}}{dz} = -\varepsilon_1\, D\,(N_{N_1} - N_{N_1}^{\text{eq}}) - W\, N_{B-L}\,, \tag{31}$$

where $z = M_1/T$. The number density N_{N_1} and the amount of $B-L$ asymmetry, N_{B-L}, are calculated in a portion of comoving volume that contains one photon at the onset of leptogenesis, so that the relativistic equilibrium N_1 number density is given by $N_{N_1}^{\text{eq}}(z \ll 1) = 3/4$. Alternatively, one may normalize the number density to the entropy density s and consider $Y_X = n_X/s$. If entropy is conserved, both normalizations are related by a constant.

There are four classes of processes that contribute to the different terms in the above equations: decays, inverse decays, $\Delta L = 1$ scatterings and $\Delta L = 2$ processes mediated by heavy neutrinos (cf. Fig. (4)). The first three processes all modify the N_1 abundance and try to push it towards its equilibrium value $N_{N_1}^{\text{eq}}$. Denoting by H the Hubble expansion rate, the term $D = \Gamma_D/(H z)$ accounts for decays and inverse decays, whereas the scattering term $S = \Gamma_S/(H z)$ represents the $\Delta L = 1$ scatterings. Decays also yield the source term for the generation of the $B-L$ asymmetry, the first term in Eq. (31), whereas all other processes contribute to the total washout term $W = \Gamma_W/(H z)$, which competes with the decay source term. The dynamical generation of the N_1 abundance and the $B-L$ asymmetry is shown in Fig. (5) for typical parameters.

[2]We use the conventions of [23]. We have also summed over the three lepton flavours neglecting the dependence on the lepton Yukawa couplings [24].

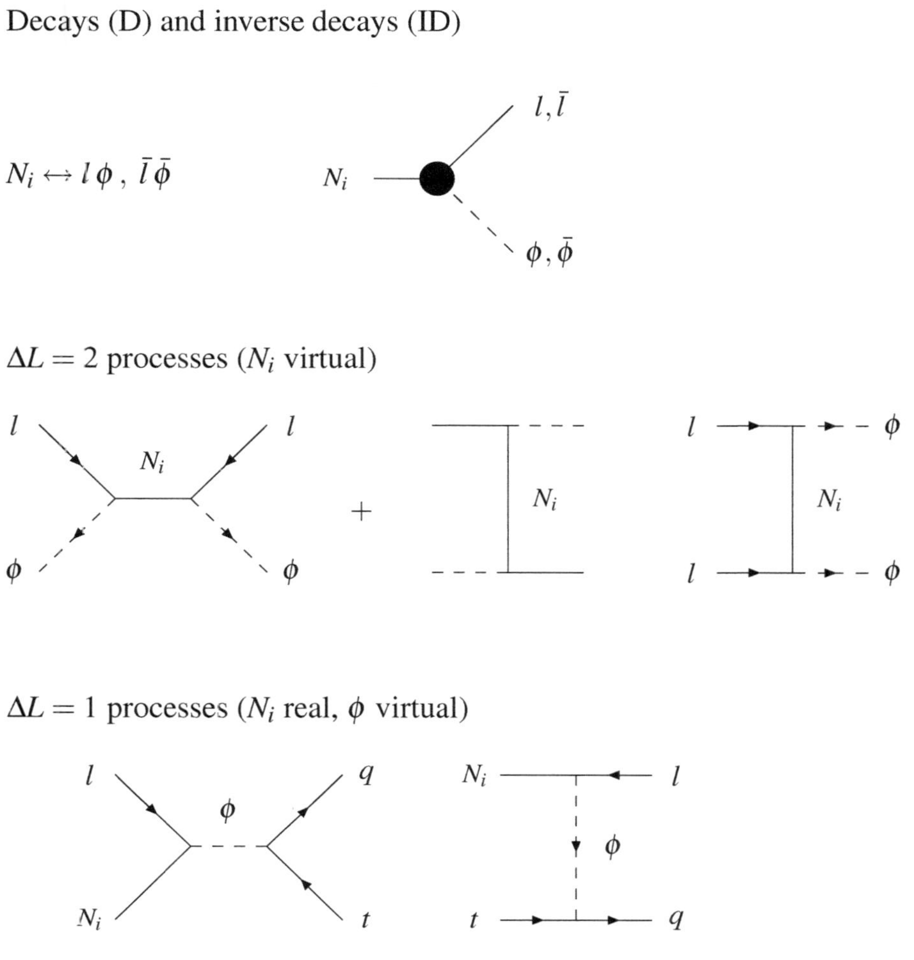

Figure 4. *Decays and inverse decays of heavy Majorana neutrinos, and lepton number violating scattering processes.*

Decays and inverse decays

It is very instructive to consider first a simplified picture in which decays and inverse decays are the only processes that are effective.[3] For consistency, in this approximation the real intermediate state contribution to the $2 \to 2$ processes has to be included. In the kinetic equations (30) and (31) one then has to replace $D+S$ by D and W by W_{ID}, respectively, where W_{ID} is the contribution of inverse decays to the washout term. The solution for N_{B-L} in this case is the sum of two terms [4],

$$N_{B-L}(z) = N_{B-L}^{\mathrm{i}}\, e^{-\int_{z_{\mathrm{i}}}^{z} dz'\, W_{ID}(z')} - \frac{3}{4}\,\varepsilon_1\,\kappa(z;\widetilde{m}_1)\,. \tag{32}$$

Here the first term accounts for an initial asymmetry which is partly reduced by washout, and the second term describes $B-L$ production from N_1 decays. It is expressed in terms of the

[3]This section follows closely [25].

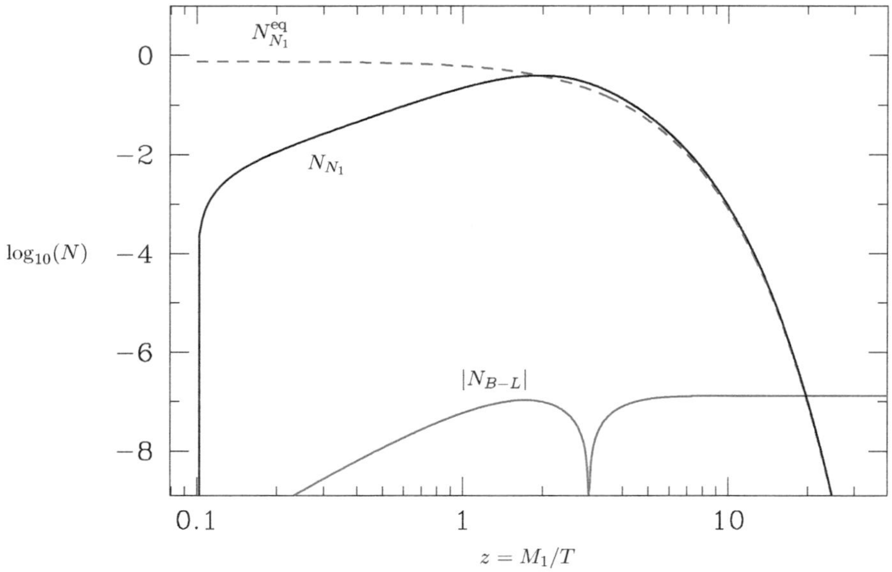

Figure 5. *The evolution of the N_1 abundance and the $B-L$ asymmetry for a typical choice of parameters, $M_1 = 10^{10}$ GeV, $\varepsilon_1 = 10^{-6}$, $\widetilde{m}_1 = 10^{-3}$ eV and $\overline{m} = 0.05$ eV. From [23].*

efficiency factor κ [24] which does not depend on the CP asymmetry ε_1,

$$\kappa(z) = \frac{4}{3}\int_{z_i}^{z} dz'\, D\left(N_{N_1} - N_{N_1}^{eq}\right) e^{-\int_{z'}^{z} dz''\, W_{ID}(z'')} \,. \tag{33}$$

As we shall see, decays and inverse decays are sufficient to describe qualitatively many properties of the full problem.

We will first study in detail the regimes of weak and strong washout. If just decays and inverse decays are taken into account, these regimes correspond, respectively, to the limits $K \ll 1$ and $K \gg 1$ of the decay parameter

$$K = \frac{\Gamma_D(z=\infty)}{H(z=1)} = \frac{\widetilde{m}_1}{m_*} \,, \tag{34}$$

introduced in the context of ordinary GUT baryogenesis [4]. Based on the insight into the dynamics of the non-equilibrium process gained from these limiting cases one can then obtain analytic interpolation formulas that describe rather accurately the entire parameter range.

To proceed, let us first recall some basic definitions and formulas. The decay rate is given by the formula [26],

$$\Gamma_D(z) = \Gamma_{D1} \left\langle \frac{1}{\gamma} \right\rangle \,, \tag{35}$$

where the thermally averaged dilation factor is given by the ratio of the modified Bessel functions K_1 and K_2,

$$\left\langle \frac{1}{\gamma} \right\rangle = \frac{K_1(z)}{K_2(z)} \,. \tag{36}$$

For the decay term D, one then obtains

$$D(z) = K\, z \left\langle \frac{1}{\gamma} \right\rangle . \tag{37}$$

The inverse decay rate is related to the decay rate by

$$\Gamma_{ID}(z) = \Gamma_D(z)\, \frac{N_{N_1}^{\rm eq}(z)}{N_l^{\rm eq}} , \tag{38}$$

where $N_l^{\rm eq}$ is the equilibrium density of lepton doublets. Because the number of degrees of freedom for heavy Majorana neutrinos and lepton doublets is the same, $g_{N_1} = g_l = 2$, one has

$$N_{N_1}^{\rm eq}(z) = \frac{3}{8} z^2 K_2(z) , \quad N_l^{\rm eq} = \frac{3}{4} . \tag{39}$$

This yields for the contribution of inverse decays to the washout term W:

$$W_{ID}(z) = \frac{1}{2} D(z)\, \frac{N_{N_1}^{\rm eq}(z)}{N_l^{\rm eq}} . \tag{40}$$

All relevant quantities are given in terms of the Bessel functions K_1 and K_2, which can be approximated by simple analytical expressions.

In the regime *far out of equilibrium*, $K \ll 1$, decays occur at very small temperatures, $z \gg 1$, and the produced $(B-L)$-asymmetry is not reduced by washout effects. In this case, using Eq. (30) with $S = 0$, the integral for the efficiency factor given in Eq. (33) becomes simply,

$$\kappa(z) \simeq \frac{4}{3} \left(N_{N_1}^{\rm i} - N_{N_1}(z) \right) . \tag{41}$$

The final value of the efficiency factor $\kappa_{\rm f} = \kappa(\infty)$ is proportional to the initial N_1 abundance. If $N_1^{\rm i} = N_1^{\rm eq} = 3/4$, then $\kappa_{\rm f} = 1$. But if the initial abundance is zero, then $\kappa_{\rm f} = 0$ as well. Therefore, in this region there is the well known problem that one has to invoke some external mechanism to produce the initial abundance of neutrinos. Moreover, an initial (B-L)-asymmetry is not washed out. Thus in the regime $K \ll 1$ the results strongly depend on the initial conditions and there is little predictivity.

In order to obtain the efficiency factor in the case of *vanishing initial N_1-abundance*, $N_{N_1}(z_{\rm i}) \equiv N_{N_1}^{\rm i} \simeq 0$, one has to calculate how heavy neutrinos are dynamically produced by inverse decays. This requires solving the kinetic equation Eq. (30) with the initial condition $N_{N_1}^{\rm i} = 0$.

Let us define a value $z_{\rm eq}$ by the condition

$$N_{N_1}(z_{\rm eq}) = N_{N_1}^{\rm eq}(z_{\rm eq}) . \tag{42}$$

Then Eq. (30) implies that the number density reaches its maximum at $z = z_{\rm eq}$. For $z > z_{\rm eq}$ the efficiency factor is always the sum of two contributions,

$$\kappa_{\rm f}(z) = \kappa^-(z) + \kappa^+(z) . \tag{43}$$

Here $\kappa^{-}(z)$ and $\kappa^{+}(z)$ correspond to the integration domains $[z_{\rm i}, z_{\rm eq})$ and $[z_{\rm eq}, z)$, respectively.

Consider first the case of *weak washout*, $K \ll 1$, which implies $z_{\rm eq} \gg 1$. One then finds,

$$N_{N_1}(z_{\rm eq}) \simeq \frac{9\pi}{16} K \,. \tag{44}$$

It turns out that to first order in K, there is a cancellation between κ^{+} and κ^{-}, yielding for the final efficiency factor

$$\kappa_{\rm f}(K) \simeq \frac{9\pi^2}{64} K^2 \,. \tag{45}$$

Note, that Eq. (45) does not hold for $K > 1$, because in this case $z_{\rm eq}$ becomes small, and washout effects change the result.

In the case of *strong washout*, $K \gg 1$, we can neglect the negative contribution κ^{-}, because the asymmetry generated at high temperatures is efficiently washed out. Now the neutrino abundance tracks closely the equilibrium behavior. Because $D \propto K$, one can solve Eq. (30) systematically in powers of $1/K$, which yields

$$D\left(N_{N_1}(z) - N_{N_1}^{\rm eq}(z)\right) = \frac{3}{2Kz} W_{ID}(z) + \mathscr{O}\left(\frac{1}{K}\right) , \tag{46}$$

where we have used properties of the Bessel functions. From Eqs. (33) and (46) one obtains for the efficiency factor

$$\kappa(z) = \frac{2}{K} \int_{z_{\rm i}}^{z} dz' \, \frac{1}{z'} W_{ID}(z') \, e^{-\int_{z'}^{z} dz'' \, W_{ID}(z'')} \,. \tag{47}$$

The integral is dominated by the contribution from a region around the value $z_{\rm B}$ where the integrand has a maximum, which is determined by the condition

$$W_{ID}(z_{\rm B}) = \left\langle \frac{1}{\gamma} \right\rangle^{-1} (z_{\rm B}) - \frac{3}{z_{\rm B}} \,. \tag{48}$$

For $K \gg 1$ one has $z_{\rm B} \gg 1$, and the condition (48) becomes approximately $W_{ID}(z_{\rm B}) \simeq 1$, with $W_{ID}(z) > 1$ for $z < z_{\rm B}$ and $W_{ID}(z) < 1$ for $z > z_{\rm B}$. This means that the asymmetry produced for $z < z_{\rm B}$ is essentially erased, whereas for $z > z_{\rm B}$, washout is negligible. Hence, the expression of Eq. (47) is a good approximation for the final efficiency factor.

One finds that a rather accurate expression for $z_B(K)$ is given by

$$z_B(K) \simeq 1 + \frac{1}{2} \ln\left(1 + \frac{\pi K^2}{1024}\left[\ln\left(\frac{3125\pi K^2}{1024}\right)\right]^5\right) . \tag{49}$$

The integral of Eq. (47), which gives the final efficiency factor in terms of $z_B(K)$, is well approximated by

$$\kappa_{\rm f}(K) \simeq \frac{2}{z_{\rm B}(K) K}\left(1 - e^{-\frac{1}{2} z_{\rm B}(K) K}\right) . \tag{50}$$

Both equations can also be extrapolated into the regime of weak washout, $K \ll 1$, where one obtains $\kappa_{\rm f} = 1$ corresponding to thermal initial abundance, $N_{N_1}^{\rm i} = N_{N_1}^{\rm eq} = 3/4$. At $K \simeq 3$ a rapid

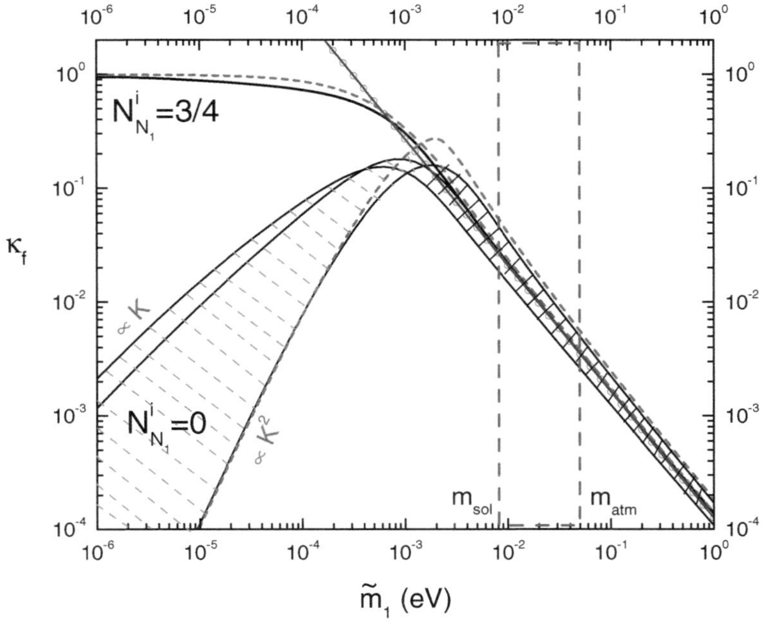

Figure 6. *Final efficiency factor when $\Delta L = 2$ washout terms are neglected. From [25].*

transition takes place from strong to weak washout. Even here analytical and numerical results agree within 30%. For the case of zero initial N_1 abundance one obtains an interpolation formula $\kappa_f(K)$ analogous to Eq. (50).

The above discussion of decays and inverse decays can be extended to include $\Delta L = 1$ and $\Delta L = 2$ scattering and washout processes. In the weak washout regime, $K \ll 1$, the main effect is that the efficiency factor of Eq. (45) is enhanced to $\kappa_f \propto K$. Relevant effects include scattering processes involving gauge bosons [27, 28] and thermal corrections to the decay and scattering rates [27, 29]. The range of different results is represented in Fig. (6) by the hatched region. An additional uncertainty in the weak washout regime is due to the dependence of the final results on the initial N_1 abundance and a possible initial asymmetry created before the onset of leptogenesis.

The situation is very different in the strong washout regime. Here the final efficiency factor is not sensitive to the neutrino production because a thermal neutrino distribution is always reached at high temperatures. For $\widetilde{m}_1 > m_* \simeq 10^{-3}\,\text{eV}$, the effect of $\Delta L = 1$ processes on the washout is not larger than about 50%, as indicated by the hatched region in Fig. (6). Within these uncertainties, the final efficiency factor is given by the simple power law:

$$\kappa_f = (2 \pm 1)\, 10^{-2} \left(\frac{0.01\,\text{eV}}{\widetilde{m}_1} \right)^{1.1 \pm 0.1} . \tag{51}$$

Both the scale of solar neutrino oscillations, $m_{\text{sol}} \equiv \sqrt{\Delta m^2_{\text{sol}}} \simeq 8 \times 10^{-3}\,\text{eV}$, and the scale of atmospheric neutrino oscillations, $m_{\text{atm}} \equiv \sqrt{\Delta m^2_{\text{atm}}} \simeq 0.05\,\text{eV}$, are larger than the equilibrium neutrino mass m_*. Hence, the range of neutrino masses, and therefore $\widetilde{m}_1$, indicated by neutrino oscillations lies entirely in the strong washout regime where theoretical uncertainties are small and the efficiency factor is still large enough to allow for successful leptogenesis.

4 Bounds on Neutrino Masses

Due to the simplicity of the leptogenesis process in the case of hierarchical heavy neutrinos, it is possible to obtain non-trivial constraints on light and heavy neutrino masses from the requirement of successful baryogenesis. The dependence of the scattering, decay and washout rates on the light and heavy neutrino mass matrices is very simple,

$$D,\, S,\, W-\Delta W \;\propto\; \frac{M_{\rm Pl}\widetilde{m}_1}{v^2}\,, \qquad \Delta W \;\propto\; \frac{M_{\rm Pl}M_1\,\overline{m}^2}{v^4}\,. \tag{52}$$

Here $\widetilde{m}_1$ is the effective neutrino mass (28), and $\overline{m}$ is the quadratic mean of the light neutrino masses $m_1 \leq m_2 \leq m_3$,

$$\overline{m}^2 = \mathrm{tr}\left(m_\nu^\dagger m_\nu\right) = m_1^2+m_2^2+m_3^2\,. \tag{53}$$

As long as the $\Delta L = 2$ washout term ΔW can be neglected, the efficiency factor κ_f is independent of the heavy neutrino mass M_1. However, for quasi-degenerate neutrinos, with increasing $\overline{m}$, the washout rate ΔW leads to an exponential suppression of the efficiency factor ($\omega \simeq 0.2$),

$$\kappa_f \simeq \kappa_f(\widetilde{m}_1)\exp\left[-\frac{\omega}{z_B}\left(\frac{M_1}{10^{10}\mathrm{GeV}}\right)\left(\frac{\overline{m}}{\mathrm{eV}}\right)^2\right]\,, \tag{54}$$

which eventually prevents successful leptogenesis. This leads to an upper bound on the light neutrino masses [23].

One can also obtain a lower bound on the heavy neutrino masses [30], because the second important factor for the baryon asymmetry (26), the *CP* asymmetry ε_1, satisfies an upper bound [30,31],

$$\varepsilon_1 \leq \varepsilon_1^{\max}(M_1,\widetilde{m}_1,\overline{m}) = \frac{3}{16\pi}\frac{M_1}{v^2}(m_3-m_1)(1+\ldots)\,, \tag{55}$$

with corrections which depend on $\widetilde{m}_1$.

Eqs. (54) and (55) for efficiency factor and *CP* asymmetry imply together the existence of a maximal baryon asymmetry which depends only on three neutrino mass parameters,

$$\begin{aligned}\eta_B \;&\leq\; \eta_B^{\max}(\widetilde{m}_1,M_1,\overline{m}) \\ &\simeq\; 0.01\;\varepsilon_1^{\max}(\widetilde{m}_1,M_1,\overline{m})\;\kappa(\widetilde{m}_1,M_1\overline{m}^2)\,.\end{aligned} \tag{56}$$

Requiring the maximal baryon asymmetry to be larger than the observed one,

$$\eta_B^{\max}(\widetilde{m}_1,M_1,\overline{m}) \geq \eta_B^{CMB}\,, \tag{57}$$

then yields a constraint on the neutrino mass parameters $\widetilde{m}_1$, M_1 and $\overline{m}$. For each value of $\overline{m}$ one obtains a domain in the ($\widetilde{m}_1$-M_1)-plane, which is allowed by successful baryogenesis. For $m_1 = 0$, corresponding to $\overline{m} \simeq m_{\rm atm}$, this domain is shown in Fig. (7). For $\overline{m} \geq 0.20$ eV this

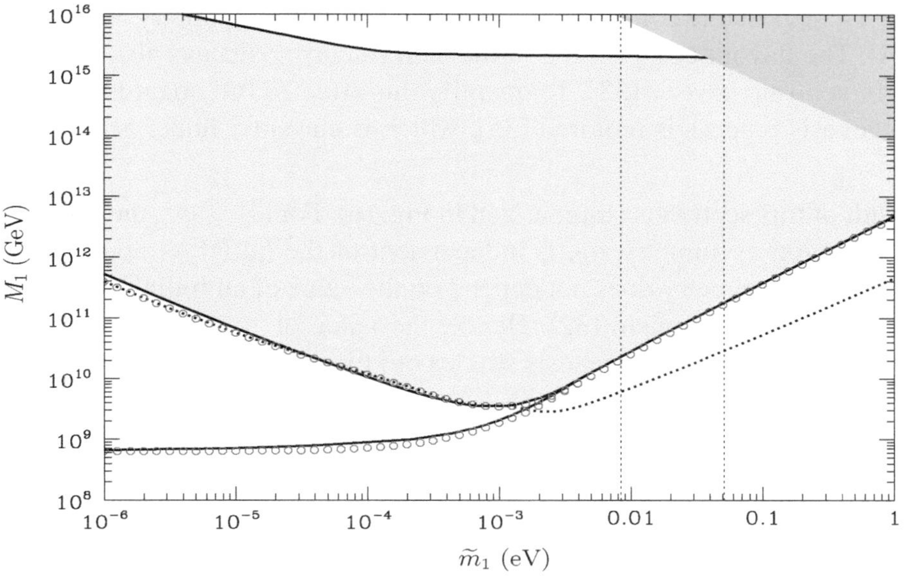

Figure 7. *Analytical lower bounds on M_1 (circles) and T_i (dotted line) for $m_1 = 0$, $\eta_B^{CMB} = 6 \times 10^{-10}$ and $m_{\text{atm}} = 0.05\,\text{eV}$. The analytical results for M_1 are compared with the numerical ones (solid lines). Upper and lower curves correspond to zero and thermal initial N_1 abundance, respectively. The vertical dashed lines indicate the range (m_{sol},m_{atm}). The gray triangle at large M_1 and large $\widetilde{m}_1$ is excluded by theoretical consistency. From [25]. Copyright (2005) Elsevier.*

domain shrinks to zero. One can easily translate this bound into upper limits on the individual neutrino masses. In a similar way, one finds a lower bound on M_1, the smallest mass of the heavy Majorana neutrinos. The resulting upper and lower bounds are [32]

$$m_i < 0.1\,\text{eV}\,, \quad M_1 > 4 \times 10^8\ \text{GeV}\,, \tag{58}$$

where we have assumed thermal initial N_1 abundance. The upper bound on the light neutrino masses holds for a normal as well as an inverted hierarchy of masses. For zero initial N_1 abundance one obtains the more restrictive lower bound $M_1 > 2 \times 10^9$ GeV. For $\widetilde{m}_1 > m_*$, the baryon asymmetry is generated at the temperature $T_B \simeq M_1/z_B < M_1$. Hence the lower bound on the reheating temperature T_i is less restrictive than the lower bound on M_1. The results of a detailed analytical and numerical calculation are summarized in Fig. (7). For the lower bound on the reheating temperature one finds $T_i > 2 \times 10^9$ GeV [25, 27].

What is the theoretical error on the upper bound for the light neutrino masses? In order to answer this question one needs a full quantum mechanical treatment of leptogenesis, a challenging problem! Conceptually interesting are thermal corrections at large temperatures, $T > M_1$, which correspond to loop corrections involving gauge bosons and the top quark [27]. Their effect is large if thermal masses are treated as kinematical masses in the evaluation of scattering matrix elements. At sufficiently high temperatures the process $N_1 \longrightarrow H\ell_L$ is then kinematically forbidden whereas the process $H \longrightarrow N_1\ell_L^\dagger$ is allowed by 'phase space'. It is important to clarify this issue for the treatment of non-equilibrium processes at high temperatures.

An important recent development in the theory of leptogenesis concerns the effect of the flavour composition of heavy neutrino decays on the generated lepton asymmetry [33]. Particularly interesting is the possible connection between the baryon asymmetry and *CP* violation at low energies [34]. The flavour composition of the heavy neutrino decays also affects the upper bound on the light neutrino masses [33]. To quantify this effect, a full quantum kinetic description of the leptogenesis process is required [35], which is currently under active investigation by several groups.

The main result of this section is summarized in Fig. (6). For $\widetilde{m}_1 > m_*$, the efficiency factor, and therefore the baryon asymmetry η_B, is independent of the initial N_1 abundance. Furthermore, the final baryon asymmetry does not depend on the value of an initial baryon asymmetry generated by some other mechanism [32]. Hence, the value of η_B is entirely determined by neutrino properties. In this way leptogenesis singles out the neutrino mass window

$$10^{-3}\ \mathrm{eV} < m_i < 0.1\ \mathrm{eV}\ , \tag{59}$$

where we have assumed that flavour effects can be neglected. Thermal leptogenesis opens a window into the physics of the early universe at temperatures $T_B = \mathscr{O}(10^{10}\ \mathrm{GeV})$, and we can ask what the implications are for dark matter, cosmology and particle physics.

Triplet models and resonant leptogenesis

Measurements in neutrino physics determine the parameters of the neutrino mass matrix,

$$m_\nu = -m_D^T \frac{1}{M} m_D + m_\nu^{\mathrm{triplet}}\ , \tag{60}$$

which in general contains a contribution from $SU(2)$ triplet fields [36] in addition to the seesaw term generated by $SU(2)$ singlet heavy Majorana neutrinos. So far, we have only considered the minimal case, $m_\nu^{\mathrm{triplet}} = 0$. Clearly, a dominant triplet contribution would destroy the connection between leptogenesis and low energy neutrino physics.

The discovery of quasi-degenerate neutrinos with masses above the bound 0.1 eV would require significant modifications of minimal leptogenesis and/or the seesaw mechanism. In this case $SU(2)$ triplet contributions to neutrino masses could be a possible way out [37, 38]. Clearly, one then has no upper bound on the light neutrino masses anymore. Yet leptogenesis with right-handed neutrino decays can still work yielding a slightly relaxed lower bound on the heavy neutrino masses. For instance, one may have $m_i \sim 0.35$ eV with $M_1 > 4 \times 10^8$ GeV [38].

Another way to reconcile quasi-degenerate light neutrinos with leptogenesis makes use of the enhancement of the CP asymmetry in the case of quasi-degenerate heavy neutrinos [16]. In the extreme case of 'resonant leptogenesis' [28], CP asymmetries $\varepsilon = \mathscr{O}(1)$ are reached for degeneracies $\Delta M/M = \mathscr{O}(10^{-10})$. In this case the right-handed neutrino masses may be as small as 1 TeV, which may lead to observable signatures at colliders. A number of models of this type have been constructed [39], some of which make use of the relative smallness of soft supersymmetry breaking terms [40].

5 Nonthermal Leptogenesis

Supersymmetry is an important ingredient of the unification of all interactions and all matter, and the supersymmetric standard model is a plausible scenario for producing new physics at the TeV scale. Thus, it is important to consider theories where supersymmetry is spontaneously broken in a hidden sector connected to ordinary matter by gravitational strength interactions – the supergravity framework. The seesaw mechanism is easily incorporated into this framework. However, the standard scenario with a neutralino lightest superparticle and an unstable heavy gravitino is in conflict with the constraints from primordial nucleosynthesis [41].

A possible solution to this problem may be provided by nonthermal leptogenesis [6, 42–44, 46], where one does not have a strong constraint on the reheating temperature. We will discuss here specifically nonthermal leptogenesis via inflaton decay [42, 43], which we consider an interesting scenario. In the next subsection, we present general arguments for this scenario and show that it suggests a lower bound on the mass of the heaviest light neutrino $m_3 > 0.01$ eV. In the subsequent subsection, we will also discuss the Affleck-Dine mechanism [6] for leptogenesis which, specifically in supersymmetric theories, is also an interesting mechanism to generate the matter-antimatter asymmetry.

Nonthermal leptogenesis via inflaton decay

Inflation early on in the history of the universe is one of the most attractive hypotheses in modern cosmology, because it not only solves long-standing problems in cosmology, like the horizon and the flatness problems, but also accounts for the origin of density fluctuations. In this subsection we discuss the hypothesis that the inflaton Φ decays dominantly into a pair of the lightest heavy Majorana neutrinos, $\Phi \longrightarrow N_1 + N_1$. We assume, for simplicity, that other decay modes including those into pairs of N_2 and N_3 are energetically forbidden. The produced N_1 neutrinos decay subsequently into $H + \ell_L$ or $H^\dagger + \ell_L^\dagger$. If the reheating temperature T_R is lower than the mass M_1 of the heavy neutrino N_1, then the out-of-equilibrium condition [3] is automatically satisfied.

The above two channels for N_1 decay have different branching ratios when *CP* conservation is violated. As discussed in Sect. 2, the asymmetry parameter ε can be expressed as [31][4]

$$\varepsilon = -\frac{3}{8\pi}\frac{M_1}{v^2} m_3 \delta_{\mathrm{eff}}\,, \tag{61}$$

where the effective CP-violating phase δ_{eff} is given by

$$\delta_{\mathrm{eff}} = \frac{\mathrm{Im}\left[h_{13}^2 + \frac{m_2}{m_3} h_{12}^2 + \frac{m_1}{m_3} h_{11}^2\right]}{|h_{13}|^2 + |h_{12}|^2 + |h_{11}|^2}\,. \tag{62}$$

Numerically, one obtains

$$\varepsilon \simeq -2 \times 10^{-6} \left(\frac{M_1}{10^{10}\mathrm{GeV}}\right) \left(\frac{m_3}{0.05\mathrm{eV}}\right) \delta_{\mathrm{eff}}\,. \tag{63}$$

[4] Because of supersymmetry, the expression for ε given below is a factor of 2 larger than that in Sect. 2.

The chain decays $\Phi \rightarrow N_1 + N_1$ and $N_1 \rightarrow H + \ell_L$ or $H^\dagger + \ell_L^\dagger$ reheat the universe producing not only the lepton-number asymmetry but also entropy for the thermal bath. The ratio of the lepton number to entropy density after reheating is estimated to be [43]

$$\begin{aligned} \frac{n_L}{s} &\simeq -\frac{3}{2}\varepsilon\frac{T_R}{m_\Phi} \\ &\simeq 3\times 10^{-10}\left(\frac{T_R}{10^6\text{GeV}}\right)\left(\frac{M_1}{m_\Phi}\right)\left(\frac{m_3}{0.05\text{eV}}\right), \end{aligned} \tag{64}$$

where m_Φ is the inflaton mass and we have taken $\delta_{\text{eff}} = 1$. This lepton-number asymmetry is converted into a baryon-number asymmetry through the sphaleron effects and one obtains

$$\frac{n_B}{s} \simeq -\frac{8}{23}\frac{n_L}{s}. \tag{65}$$

We should stress, here, an important merit of the inflaton-decay scenario: It does not require a reheating temperature $T_R \sim M_1$, but it requires only $m_\Phi > 2M_1$. On the other hand, for thermal leptogenesis to work it is necessary that $T_R \sim M_1$, which necessitates higher reheating temperature for leptogenesis to produce enough matter-antimatter asymmetry.

If one assumes that $T_R < 10^7$ GeV to satisfy the cosmological constraint on the gravitino abundance (cf. [41]) and uses $m_\Phi > 2M_1$, the observed baryon number to entropy ratio [2] gives the constraint on the heaviest light neutrino

$$m_3 > 0.01 \text{ eV}. \tag{66}$$

It is very interesting that the neutrino mass suggested by atmospheric neutrino oscillation experiments, $\sqrt{\Delta m^2_{\text{atm}}} \simeq 0.05$ eV, just satisfies the above constraint. However, to get this bound we assumed that the inflaton decays dominantly into a pair of N_1s. If this branching ratio is only 10 %, the lower bound on the neutrino mass exceeds the observed neutrino mass $\sqrt{\Delta m^2_{\text{atm}}} \simeq 0.05$ eV.

A variety of models have been considered to restore the bound of Eq. (66) by imposing a symmetry. It is perhaps most interesting to consider that the scalar partner of the heavy Majorana neutrino N_1 is the inflaton itself [44], and the inflaton decay into a lepton plus a Higgs boson gives an effective branching ration of 100%. In this model, one must assume that the initial value of the scalar field partner of N_1 is much larger than the Planck scale to cause inflation. This, however, is not easily realized in supergravity, because the minimal supergravity potential has an exponential factor, $\exp(\phi^*\phi/M_G^2)$, that prevents any scalar field ϕ from having a value larger than the reduced Planck scale $M_G \simeq 2.4 \times 10^{18}$ GeV.

Affleck-Dine leptogenesis

In the supersymmetric standard model, for unbroken supersymmetry, some combinations of scalar fields do not enter the potential, constituting so-called flat directions of the potential. Since the potential is almost independent of these fields, they may have large initial values in the early universe. Such flat directions receive soft masses in the supersymmetry breaking vacuum. When the expansion rate H_{exp} of the Universe becomes comparable to their masses, the flat directions begin to oscillate around the minimum of the potential. If the flat directions are made of scalar quarks and carry baryon number, the baryon-number asymmetry can be created during these coherent oscillations. This is the Affleck-Dine (AD) mechanism for baryogenesis [6].

QCD corrections, however, make the potential of the AD fields milder than $|\phi|^2$. This allows non-topological soliton solutions (Q-balls) to form in the early universe, as a result of the coherent oscillations in the flat directions. Because Q-balls have long lifetimes, their decays produce a huge amount of entropy at late times. To avoid this problem one must choose parameters in the SUSY theory so that the density of the lightest superparticle (LSP) does not exceed the dark matter density in the present universe (cf. [45]). Although this may not be a problem, it is much safer to consider flat directions without QCD interactions, because such directions most likely do not have Q-ball solutions.

The most interesting candidate [46] for such a flat direction is

$$\phi_i = (2H\ell_i)^{1/2}, \tag{67}$$

where ℓ_i is the lepton doublet field of the *i-th* family. Here, H and ℓ_i represent the scalar components of the corresponding chiral multiplets. The Yukawa interactions of H make the potential of ϕ_i steeper than the mass term and hence there is no instability of the coherent oscillation (i.e. there are no Q-ball solutions). Because this flat direction carries lepton number, a lepton asymmetry will be created during the coherent oscillation (AD leptogenesis) [46]. Sphaleron processes then transmute, in the usual fashion, this lepton asymmetry into a baryon asymmetry.

The seesaw mechanism induces a dimension-five operator in the superpotential for the theory,[5]

$$W = \frac{m_\nu}{2v^2}(\ell H)^2, \tag{68}$$

where we have used a basis in which the neutrino mass matrix is diagonal. With this superpotential we have a SUSY-invariant potential for the flat direction ϕ given by

$$V_{\text{SUSY}} = \frac{m_\nu^2}{4v^4}|\phi|^6. \tag{69}$$

In addition to the SUSY-invariant potential we have a SUSY-breaking potential,

$$\delta V = m_\phi^2|\phi|^2 + \frac{m_{\text{SUSY}}m_\nu}{8v^2}(a_m\phi^4 + \text{h.c.}). \tag{70}$$

Here, a_m is a complex number. We take $m_\phi \simeq m_{\text{SUSY}} \simeq 1$ TeV and $|a_m| \sim 1$. The second term in δV is very important, because it gives rise to the lepton-number generation.

We assume that the flat direction ϕ acquires a negative $(\text{mass})^2$ induced by the inflaton potential and rolls down to the point balanced by the SUSY-invariant potential V_{SUSY} during inflation. Thus, the AD field ϕ has an initial value of $\sqrt{H_{\text{inf}}v^2/m_\nu}$, where H_{inf} is the Hubble constant (the expansion rate) during inflation. ϕ decreases in amplitude gradually after inflation, and begins to oscillate around the potential minimum when the Hubble constant H_{exp} of the Universe becomes comparable to the SUSY-breaking mass m_ϕ. At the beginning of the oscillation, the AD field has the value $|\phi_0| \simeq \sqrt{m_\phi v^2/m_\nu}$ which, as shown below, is an effective initial value for leptogenesis.

[5]For ease of notation we have dropped the subscript *i* below.

Let us consider now lepton-number generation in this scenario. The evolution of the AD field ϕ is described by

$$\frac{\partial^2 \phi}{\partial t^2} + 3H_{\rm exp}\frac{\partial \phi}{\partial t} + \frac{\partial V}{\partial \phi^*} = 0\,, \tag{71}$$

where $V = V_{\rm SUSY} + \delta V$. Because the lepton number is given by

$$n_{\rm L} = i\left(\frac{\partial \phi^*}{\partial t}\phi - \phi^*\frac{\partial \phi}{\partial t}\right)\,, \tag{72}$$

the evolution of $n_{\rm L}$ is given by

$$\frac{\partial n_{\rm L}}{\partial t} + 3H_{\rm exp}n_{\rm L} = \frac{m_{\rm SUSY}m_\nu}{2v^2}{\rm Im}(a_m^*\phi^{*4})\,. \tag{73}$$

The motion of ϕ in the phase direction generates the lepton number. This is predominantly created just after the AD field ϕ starts its coherent oscillation, at a time $t_{\rm osc} \simeq 1/H_{\rm osc} \simeq 1/m_\phi$, because the amplitude $|\phi|$ damps as t^{-1} during the oscillation. Thus, we obtain for the lepton number

$$n_{\rm L} \simeq \frac{m_{\rm SUSY}m_\nu}{2v^2}\delta_{\rm eff}|a_m\phi_0^4| \times t_{\rm osc}\,, \tag{74}$$

where $\delta_{\rm eff} = \sin(4\arg\phi + \arg a_m)$ represents an effective *CP*-violating phase. Using $m_{\rm SUSY} \simeq m_\phi$, $|\phi_0| \simeq \sqrt{m_\phi v^2/m_\nu}$ and $t_{\rm osc} \simeq 1/m_\phi$, we find

$$n_{\rm L} \simeq \delta_{\rm eff}m_\phi^2\frac{v^2}{2m_\nu}\,. \tag{75}$$

After the end of inflation, the inflaton begins to oscillate around the potential minimum and $n_{\rm L}/\rho_{\rm inf}$ stays constant until the inflaton decays. Here $\rho_{\rm inf}$ is the energy density of the inflaton. The inflaton decay reheats the Universe producing entropy s. Because $\rho/s = 2T_R/4$, we find for the lepton-number asymmetry the expression

$$\frac{n_{\rm L}}{s} \simeq \left(\frac{\rho_{\rm inf}}{s}\right)\left(\frac{n_{\rm L}}{\rho_{\rm inf}}\right) \simeq \delta_{\rm eff}\frac{3T_R}{4M_G}\frac{v^2}{6m_\nu M_G}\,. \tag{76}$$

Here we have used $\rho_{\rm inf} \simeq 3m_\phi^2 M_G^2$ at the beginning of the AD field oscillation (when most of the lepton number is generated). This lepton-number asymmetry is converted to a baryon-number asymmetry by sphaleron processes. In this way one obtains for the baryon-number asymmetry

$$\frac{n_{\rm B}}{s} \simeq \frac{1}{23}\frac{v^2 T_R}{m_\nu M_G^2}\,. \tag{77}$$

The observed ratio $n_{\rm B}/s \simeq 0.9 \times 10^{-10}$ implies $m_\nu \simeq 10^{-9}$ eV for $T_R \simeq 10^6$ GeV. This small mass corresponds to the mass of the lightest neutrino.

6 Outlook

In the previous sections we have reviewed the basic ideas of leptogenesis, with emphasis on its simplest version, thermal leptogenesis with hierarchical heavy neutrinos. In the past years a detailed understanding of this nonequilibrium process at high temperature in the early universe has been achieved, and it is very remarkable that the experimental evidence for small neutrino masses fits very well together with the observed value of the matter-antimatter asymmetry.

Current research on leptogenesis mostly focusses on the following themes: (1) It is important to understand the dependence of the generated baryon asymmetry on the flavour composition of the heavy Majorana neutrino decays; this is part of the necessary development of a full quantum kinetical description of the nonequilibrium process. (2) Leptogenesis has to be considered within the framework of unified theories beyond the standard model, which should eventually lead to a prediction of the observed baryon asymmetry in terms of other observables. (3) Thermal leptogenesis is not compatible with the most popular supersymmetric extension of the standard model, with a neutralino LSP and a gravitino with a mass of a few TeV. An attractive alternative is low energy supersymmetry with a gravitino LSP. It is very exciting that these different possibilities will soon be tested at the LHC.

Acknowledgments

These lectures are strongly based on work with P. Di Bari and M. Plümacher and on the review article with R. D. Peccei and T. Yanagida. I am grateful to my collaborators, and I would also like to thank the organizers for their splendid hospitality and for creating a stimulating atmosphere in St. Andrews.

References

[1] W. M. Yao *et al.* [Particle Data Group], J. Phys. G **33** (2006) 220.

[2] WMAP Collaboration, C. L. Bennett, *et al*, Astrophys. J. Suppl. **148**, 1 (2003); D. N. Spergel, *et al*, Astrophys. J. Suppl. **148**, 175 (2003).

[3] A. D. Sakharov, JETP Lett. **5**, 24 (1967).

[4] E. W. Kolb and M. S. Turner, *The Early Universe* (Addison Wesley, New York, 1990).

[5] M. Fukugita and T. Yanagida, Phys. Lett. **B174**, 45 (1986).

[6] I. Affleck and M. Dine, Nucl. Phys. **B249**, 361 (1985).

[7] V. A. Kuzmin, V. A. Rubakov and M. A. Shaposhnikov, Phys. Lett. **B155**, 36 (1985).

[8] G. t' Hooft, Phys. Rev. Lett. **37**, 8 (1976).

[9] F. R. Klinkhammer and N. S. Manton, Phys. Rev. **D30**, 2212 (1984).

[10] D. Bödeker, G. D. Moore and K. Rummukainen, Phys. Rev. **D61**, 056003 (2000).

[11] J. M. Cline, arXiv:hep-ph/0609145.

[12] K. Dick, M. Lindner, M. Ratz and D. Wright, Phys. Rev. Lett. **84**, 4039 (2000).

[13] G. G. Ross, *Grand Unified Theories* (Reading, Benjamin/Cummings, 1984).

[14] B. Kayser, these proceedings.

[15] G. Altarelli, these proceedings.

[16] M. Flanz, E. A. Paschos and U. Sarkar, Phys. Lett. **B345**, 248 (1995); L. Covi, E. Roulet and F. Vissani, Phys. Lett. **B384**, 169 (1996); W. Buchmüller and M. Plümacher, Phys. Lett. **B431**, 354 (1998).

[17] W. Buchmüller and S. Fredenhagen, Phys. Lett. **B483**, 217 (2000).

[18] W. Buchmüller and M. Plümacher, Phys. Lett. **B389**, 73 (1996).

[19] For recent discussions and references, see: R. N. Mohapatra, S. Nasri and H. Yu, Phys. Lett. **B615**, 231 (2005); Z.-Z. Xing, Phys. Rev. **D70**, 071302 (2004); N. Cosme, JHEP **0408**, 027 (2004); W. Rodejohann, Eur. Phys. J. **C32**, 235 (2004); W. Grimus and L. Lavoura, J. Phys. **G30**, 1073 (2004); V. Barger, D. A. Dicus, H.-J. He and T. Li, Phys. Lett. **B583**, 173 (2004); P. H. Chankowski and K. Turzyński, Phys. Lett. **B570**, 198 (2003); L. Velasco-Sevilla, JHEP **0310**, 035 (2003); E. Kh. Akhmedov, M. Frigerio and A. Yu. Smirnov, JHEP **0309**, 021 (2003); G. C. Branco, R. González Felipe, F. R. Joaquim, I. Masina, M. N. Rebelo and C. A. Savoy, Phys. Rev. **D67**, 07025 (2003).

[20] G. C. Branco and M. N. Rebelo, New J. Phys. **7**, 86 (2005).

[21] M. A. Luty, Phys. Rev. **D45**, 455 (1992).

[22] M. Plümacher, Z. Phys. **C74**, 549 (1997).

[23] W. Buchmüller, P. Di Bari and M. Plümacher, Nucl. Phys. **B643**, 367 (2002).

[24] R. Barbieri, P. Creminelli, A. Strumia and N. Tetradis, Nucl. Phys. **B575**, 61 (2000).

[25] W. Buchmüller, P. Di Bari and M. Plümacher, Ann. Phys. **315**, 305 (2005) [hep-ph/0401240].

[26] E. W. Kolb and S. Wolfram, Nucl. Phys. **B172**, 224 (1980); Nucl. Phys. **B195**, 542(E) (1982).

[27] G. F. Giudice, A. Notari, M. Raidal, A. Riotto and A. Strumia, Nucl. Phys. **B685**, 89 (2004).

[28] A. Pilaftsis and T. E. J. Underwood, Nucl. Phys. **B692**, 303 (2004).

[29] L. Covi, N. Rius, E. Roulet and F. Vissani, Phys. Rev. **D57**, 93 (1998).

[30] S. Davidson and A. Ibarra, Phys. Lett. **B535**, 25 (2002).

[31] K. Hamaguchi, H. Murayama and T. Yanagida, Phys. Rev. **D65**, 043512 (2002).

[32] W. Buchmüller, P. Di Bari and M. Plümacher, Nucl. Phys. **B665**, 445 (2003).

[33] E. Nardi, Y. Nir, E. Roulet and J. Racker, JHEP **0601** (2006) 164; A. Abada, S. Davidson, F. X. Josse-Michaux, M. Losada and A. Riotto, JCAP **0604** (2006) 004.

[34] A. Abada, S. Davidson, A. Ibarra, F. X. Josse-Michaux, M. Losada and A. Riotto, JHEP **0609** (2006) 010; S. Blanchet and P. Di Bari, arXiv:hep-ph/0607330; S. Pascoli, S. T. Petcov and A. Riotto, arXiv:hep-ph/0611338; G. C. Branco, R. Gonzalez Felipe and F. R. Joaquim, Phys. Lett. B **645** (2007) 432; S. Antusch and A. M. Teixeira, JCAP **0702** (2007) 024.

[35] S. Blanchet, P. Di Bari and G. G. Raffelt, arXiv:hep-ph/0611337.

[36] G. Lazarides, Q. Shafi and C. Wetterich, Nucl. Phys. **B181**, 287 (1981); R. N. Mohapatra and G. Senjanović, Phys. Rev. **D23**, 165 (1981); C. Wetterich, Nucl. Phys. **B187**, 343 (1981).

[37] T. Hambye, Y. Lin, A. Notari, M. Papucci and A. Strumia, Nucl. Phys. **B695**, 169 (2004); T. Hambye and G. Senjanović, Nucl. Phys. **B582**, 73 (2004); W. Rodejohann, Phys. Rev. **D70**, 073010 (2004); P.-H. Gu and X.-J. Bi, Phys. Rev. **D70**, 063511 (2004); G. D'Ambrosio, T. Hambye, A. Hektor, M. Raidal and A. Rossi, Phys. Lett. **B604** 199 (2004).

[38] S. Antusch and S. F. King, Phys. Lett **B597**, 199 (2004).

[39] S. Dar, S. Huber, V. N. Senoguz and Q. Shafi, Phys. Rev **D69**, 077701 (2004); C. H. Albright and S. M. Barr, Phys. Rev. **D70**, 033013 (2004); A. Pilaftsis, Phys. Rev. Lett. **95** (2005) 081602.

[40] Y. Grossman, T. Kashti, Y. Nir and E. Roulet, Phys. Rev. Lett. **91**, 251801 (2003); G. D'Ambrosio, G. F. Giudice and M. Raidal, Phys. Lett. **B575**, 75 (2003); T. Hambye, J. March-Russell and S. M. West, JHEP **0407**, 070 (2004); L. Boubekeur, T. Hambye and G. Senjanović, Phys. Rev. Lett. **93**, 111601 (2004); Y. Grossman, T. Kashti, Y. Nir and E. Roulet, JHEP **0411**, 080 (2004); M. C. Chen and K. T. Mahanthappa, Phys. Rev. **D70**, 113013 (2004); R. Allahverdi and M. Drees, Phys. Rev. **D79**, 123522 (2004).

[41] For a discussion and references, see W. Buchmuller, R. D. Peccei and T. Yanagida, Ann. Rev. Nucl. Part. Sci. **55** (2005) 311.

[42] G. Lazarides and Q. Shafi, Phys. Lett. **B258**, 305 (1991); G. Lazarides, C. Panagiotakopoulos and Q. Shafi, Phys. Lett. **B315**, 325 (1993).

[43] K. Kumekawa, T. Moroi and T. Yanagida, Progr. Theor. Phys. **92**, 437 (1994); T. Asaka, K. Hamaguchi, M. Kawasaki and T. Yanagida, Phys. Lett. **B464**, 12 (1999); Phys. Rev. **D61**, 083512 (2000); G. F. Giudice, M. Peloso, A. Riotto and I. Tkachev, JHEP **08**, 014 (1999).

[44] H. Murayama, H. Suzuki, T. Yanagida and J. Yokoyama, Phys. Rev. Lett. **70**, 1912 (1993); J. R. Ellis, M. Raidal and T. Yanagida, Phys. Lett. **B581**, 9 (2004).

[45] M. Dine and A. Kusenko, Rev. Mod. Phys. **76** (2004) 1.

[46] H. Murayama and T. Yanagida, Phys. Lett. **B322**, 349 (1994); M. Dine, L. Randall and S. Thomas, Nucl. Phys. **B458**, 291 (1996).

Cosmological Aspects of Neutrino Physics

Sergio Pastor

Instituto de Física Corpuscular (CSIC-Universitat de Valéncia), Valencia, Spain

1 Introduction

In this contribution I summarize the topics discussed in my three lectures at the School on Neutrino Cosmology. This is one of the best examples of the very close ties that have developed between nuclear physics, particle physics, astrophysics and cosmology. I tried to present the most interesting aspects, but many others that were left out can be found in the review by Dolgov (2002).

We begin with a description of the properties and evolution of the background of relic neutrinos that fill the Universe. Then we review the influence of neutrinos on Primordial Nucleosynthesis and the possible effects of neutrino oscillations on Cosmology. The largest part of this contribution is devoted to the impact of massive neutrinos on cosmological observables, that can be used to extract bounds on neutrino masses from present data. Finally we discuss the sensitivities on neutrino masses from future cosmological experiments.

Note that light massive neutrinos could also play a role in the generation of the baryon asymmetry of the Universe from a previously created lepton asymmetry. In these leptogenesis scenarios, one can also obtain quite restrictive bounds on light neutrino masses, which are however model-dependent. We do not discuss this subject here, as it was covered by another lecturer at the School (Buchmüller (2006)).

For further details, the reader is referred to the short reviews on neutrino cosmology by Hannestad (2004) or Elgaroy and Lahav (2005), while more information can be found in Hannestad (2006) and in particular in Lesgourgues and Pastor (2006). A more general review on the connection between particle physics and cosmology can be found in Kamionkowski and Kosowsky (1999).

2 The cosmic neutrino background

The existence of a relic sea of neutrinos is a generic feature of the standard hot big bang model. The number of neutrinos is only slightly below that of relic photons that constitute the cosmic microwave background (CMB). This cosmic neutrino background (CNB) has not been detected yet, but its presence is indirectly established by the accurate agreement between the calculated and observed primordial abundances of light elements, as well as from the analysis of the power spectrum of CMB anisotropies. In this section we will summarize the evolution and main properties of the CNB.

2.1 Relic neutrino production and decoupling

Produced at large temperatures by frequent weak interactions, cosmic neutrinos of any flavour $(\nu_e, \nu_\mu, \nu_\tau)$ were kept in equilibrium until these processes became ineffective in the course of the expansion of the early Universe. While coupled to the rest of the primeval plasma of relativistic particles such as electrons, positrons and photons, neutrinos had a momentum spectrum with an equilibrium Fermi-Dirac distribution with temperature T:

$$f_{\rm eq}(p,T) = \left[\exp\left(\frac{p-\mu_\nu}{T}\right)+1\right]^{-1} . \tag{1}$$

This is just one example of the general case of particles in equilibrium and applies to fermions or bosons, relativistic or non-relativistic, as shown e.g. in Kolb and Turner (1990). In the previous equation we have included a neutrino chemical potential μ_ν that would exist in the presence of a neutrino-antineutrino asymmetry, but we will see later in section 5.1 that its contribution can be safely ignored.

As the Universe cools, the weak interaction rate Γ_ν falls below the expansion rate. One says that neutrinos decouple from the rest of the plasma. An estimate of the decoupling temperature $T_{\rm dec}$ can be found by equating the thermally averaged value of the weak interaction rate Γ_ν with the expansion rate:

$$\Gamma_\nu = \langle \sigma_\nu n_\nu \rangle \,, \tag{2}$$

where $\sigma_\nu \propto G_F^2$ is the cross section of the electron-neutrino scattering processes with G_F being the Fermi constant and n_ν the neutrino number density. The expansion rate is given by the Hubble parameter H

$$H^2 = \frac{8\pi\rho}{3M_P^2} \,, \tag{3}$$

where the total energy density $\rho \propto T^4$ is dominated by relativistic particles. $M_P = 1/G^{1/2}$ is the Planck mass with G being the Gravitational constant. If we approximate the numerical factors to unity, with $\Gamma_\nu \approx G_F^2 T^5$ and $H \approx T^2/M_P$, we obtain the rough estimate of $T_{\rm dec} \approx 1$ MeV. More accurate calculations give slightly higher values of $T_{\rm dec}$ which are flavour dependent since electron neutrinos and antineutrinos are in closer contact with electrons and positrons, as shown e.g. in Dolgov (2002).

Although neutrino decoupling is not described by a unique $T_{\rm dec}$, it can be approximated as an instantaneous process. The standard picture of *instantaneous neutrino decoupling* is very simple (see e.g. Kolb and Turner (1990) or Dodelson (2003)) and reasonably accurate. In this approximation, the spectrum in Equation (1) is preserved after decoupling, since both neutrino momenta and temperature redshift identically with the expansion of the Universe. In other words, the number density of non-interacting neutrinos remains constant in a comoving volume since the decoupling epoch. We will see later that active neutrinos cannot possess masses much larger than 1 eV, so they were ultra-relativistic at decoupling. This is the reason why the momentum distribution in Equation (1) does not depend on the neutrino masses, even after decoupling, i.e. there is no neutrino energy in the exponential of $f_{\rm eq}(p)$.

When calculating quantities related to relic neutrinos, one must consider the various possible degrees of freedom per flavour. If neutrinos are massless or Majorana particles, there are two degrees of freedom for each flavour, one for neutrinos in a negative helicity state and one for antineutrinos in a positive helicity state, respectively. Instead, for Dirac neutrinos there are in principle two more degrees of freedom, corresponding to the two additional helicity states. However, the extra degrees of freedom should be included in the computation only if they are populated and brought into equilibrium before the time of neutrino decoupling. In practice, the Dirac neutrinos with the "wrong-helicity" states do not interact with the plasma at temperatures of the order of MeVs. These have a vanishingly small density with respect to the usual left-handed neutrinos unless these have masses close to the keV range, as is explained in section 6.4 of Dolgov (2002). Such a large mass is excluded for active neutrinos. Thus the relic density of active neutrinos does not depend on their nature, either Dirac or Majorana particles.

Shortly after neutrino decoupling the temperature drops below the electron mass, favouring $e^{\pm}$ annihilations that heat the photons. If one assumes that this entropy transfer did not affect the neutrinos because they were already completely decoupled, it is easy to calculate the change in the photon temperature before any $e^{\pm}$ annihilation and after the electron-positron pairs disappear by assuming entropy conservation of the electromagnetic plasma. The result is

$$\frac{T_\gamma^{\rm after}}{T_\gamma^{\rm before}} = \left(\frac{11}{4}\right)^{1/3} \simeq 1.40102\,, \tag{4}$$

which is also the ratio between the temperatures of relic photons and neutrinos $T_\gamma/T_\nu = (11/4)^{1/3}$. The evolution of this ratio during the process of $e^{\pm}$ annihilations is shown in the left panel of Figure 1, while one can see in the other plot how in this epoch the photon temperature decreases with the expansion less than the inverse of the scale factor a. Instead the temperature of the decoupled neutrinos always falls as $1/a$.

It turns out that the standard picture of neutrino decoupling described above is slightly modified: the processes of neutrino decoupling and $e^{\pm}$ annihilations are sufficiently close in time so that some relic interactions between $e^{\pm}$ and neutrinos exist. These relic processes are more efficient for larger neutrino energies, leading to non-thermal distortions in the neutrino spectra at the per cent level and a slightly smaller increase of the comoving photon temperature, as noted in a series of works (see the full list given in the review by Dolgov (2002)). A proper calculation of the process of non-instantaneous neutrino decoupling demands solving the momentum-dependent Boltzmann equations for the neutrino spectra, a set of integro-differential kinetic equations that are difficult to solve numerically. The most recent analysis of this problem has included the effect of flavour neutrino oscillations on the neutrino decoupling process (Mangano et al. (2005)). One finds an increase in the neutrino energy densities with respect to the instantaneous decoupling approximation of 0.73% and 0.52% for ν_e's and $\nu_{\mu,\tau}$'s, respectively. The value of the comoving photon temperature after $e^{\pm}$ annihilations is now a factor 1.3978 larger, instead of 1.40102. These changes modify the contribution of relativistic relic neutrinos to the total energy density which is taken into account using $N_{\rm eff} \simeq 3.046$, as defined later in Equation (13). In practice, the distortions calculated in Mangano et al. (2005) only have small consequences on the evolution of cosmological perturbations, and for many

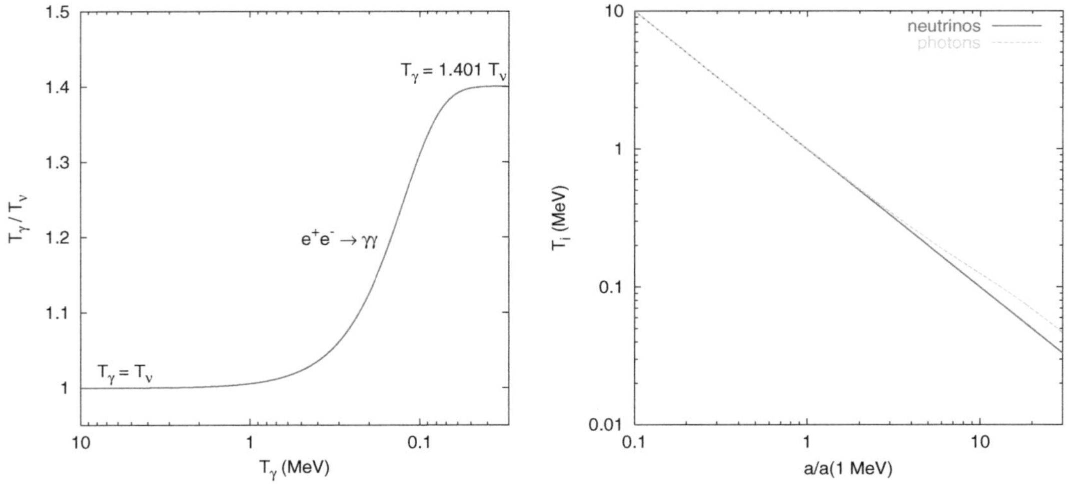

Figure 1. *Photon and neutrino temperatures during the process of $e^\pm$ annihilations: evolution of their ratio (left) and their decrease with the expansion of the Universe (right).*

purposes they can be safely neglected.

Any quantity related to relic neutrinos can be calculated after decoupling with the spectrum in Equation (1) and T_ν. For instance, the number density per flavour n_ν is closely related to the number of relic photons n_γ and fixed by the temperature,

$$n_\nu = \frac{3}{11}\, n_\gamma = \frac{6\zeta(3)}{11\pi^2}\, T_\gamma^3 \,, \tag{5}$$

where ζ is the Riemann zeta function. This leads to a present value of 113 neutrinos and antineutrinos of each flavour per cm^3. Instead, the energy density for massive neutrinos should in principle be calculated numerically, with two well-defined analytical limits,

$$\rho_\nu(m_\nu \ll T_\nu) = \frac{7\pi^2}{120}\left(\frac{4}{11}\right)^{4/3} T_\gamma^4 \,, \tag{6}$$

$$\rho_\nu(m_\nu \gg T_\nu) = m_\nu n_\nu \,. \tag{7}$$

2.2 Background evolution

Let us discuss the evolution of the CNB after decoupling in the expanding Universe, which is described by the Friedmann-Robertson-Walker metric (Dodelson (2003))

$$ds^2 = dt^2 - a(t)^2\, \delta_{ij} dx^i dx^j \,, \tag{8}$$

where we assumed negligible spatial curvature. Here $a(t)$ is the scale factor usually normalized to unity now ($a(t_0) = 1$) and related to the redshift z as $a = 1/(1+z)$. General relativity tells us the relation between the metric and the matter and energy in the Universe via the Einstein equations, whose time-time component is the Friedmann equation

$$\left(\frac{\dot{a}}{a}\right)^2 = H^2 = \frac{8\pi G}{3}\rho = H_0^2 \frac{\rho}{\rho_c^0} \,, \tag{9}$$

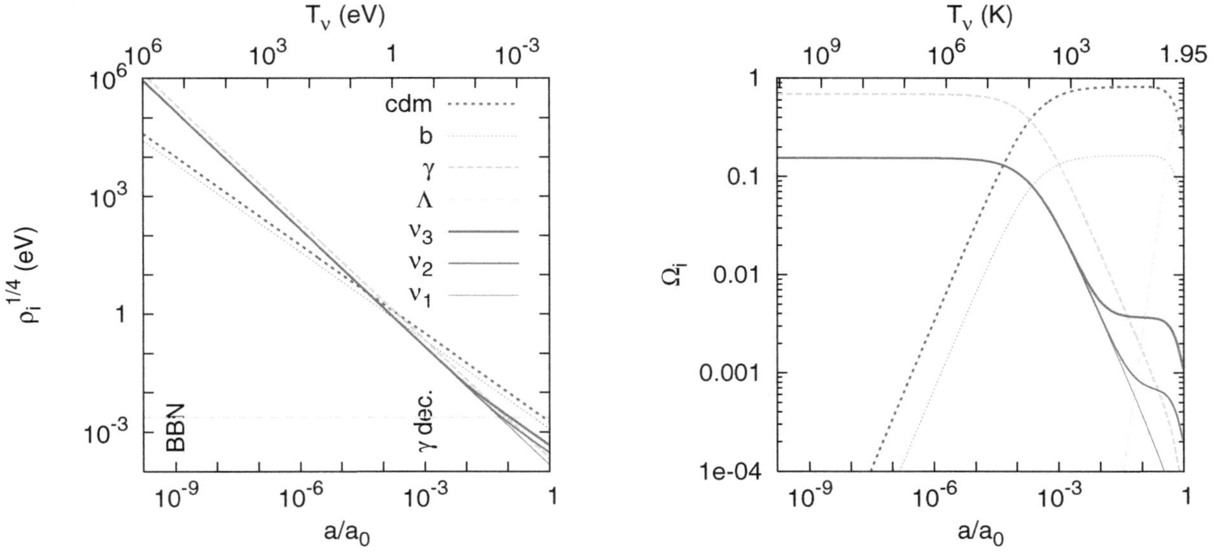

Figure 2. *Evolution of the background energy densities (left) and density fractions Ω_i (right) from the time when $T_\nu = 1$ MeV until now, for each component of a flat ΛMDM model with $h = 0.7$ and current density fractions $\Omega_\Lambda = 0.70$, $\Omega_b = 0.05$, $\Omega_\nu = 0.0013$ and $\Omega_{cdm} = 1 - \Omega_\Lambda - \Omega_b - \Omega_\nu$. The three neutrino masses are $m_1 = 0$, $m_2 = 0.009$ eV and $m_3 = 0.05$ eV.*

that gives the Hubble rate in terms of the total energy density ρ. At any time, the critical density ρ_c is defined as $\rho_c = 3H^2/8\pi G$, and the current value H_0 of the Hubble parameter gives the critical density today

$$\rho_c^0 = 1.8788 \times 10^{-29} h^2 \text{ g cm}^{-3} . \tag{10}$$

where $h \equiv H_0/(100 \text{ km s}^{-1} \text{Mpc}^{-1})$. The different contributions to the total energy density are

$$\rho = \rho_\gamma + \rho_{cdm} + \rho_b + \rho_\nu + \rho_\Lambda , \tag{11}$$

and the evolution of each component is given by the energy conservation law in an expanding Universe $\dot{\rho} = -3H(\rho + p)$, where p is the pressure. Thus the homogeneous density of photons ρ_γ scales like a^{-4}, that of non-relativistic matter (ρ_{cdm} for cold dark matter and ρ_b for baryons) like a^{-3}, and the cosmological constant density ρ_Λ is of course time-independent. Instead, the energy density of neutrinos contribute to the radiation density at early times but behave as matter after the non-relativistic transition.

The evolution of all densities is shown on the left plot of Figure 2, starting at MeV temperatures until now. We also display the characteristic times for the end of Primordial Nucleosynthesis and for photon decoupling or recombination. The evolution of the density fractions $\Omega_i \equiv \rho_i/\rho_c$ is shown on the right panel, where it is easier to see which of the Universe components is dominant. First radiation in the form of photons and neutrinos (Radiation Domination or RD) dominates, then matter which can be CDM, baryons and massive neutrinos at late times (Matter Domination or MD) and finally the cosmological constant density takes over at low redshift (typically $z < 0.5$).

Massive neutrinos are the only particles that present the transition from radiation to matter, when their density is clearly enhanced (upper solid lines in Figure 2). Obviously the contribution of massive neutrinos to the energy density in the non-relativistic limit is a function of the mass (or the sum of all masses for which $m_i \gg T_\nu$), and the present value Ω_ν could be of order unity unity for eV masses (see section 6).

3 Neutrinos and Primordial Nucleosynthesis

In the course of its expansion, when the early Universe was only less than a second old, the conditions of temperature and density of its nucleon component were such that light nuclei could be created via nuclear reactions. For a recent review, see Steigman (2006). During this epoch, known as Primordial or Big-Bang Nucleosynthesis (BBN), the primordial abundances of light elements were produced: mostly ^{4}He but also smaller quantities of less stable nuclei such as D, ^{3}He and ^{7}Li. Heavier elements could not be produced because of the rapid evolution of the Universe and its small nucleon content. This is related to the small value of the baryon asymmetry which, normalized to the photon density, $\eta_b \equiv (n_b - n_{\bar{b}})/n_\gamma$, is about a few times 10^{-10}. Measuring these primordial abundances today is a very difficult task, because stellar process may have altered the chemical compositions. Still, data on the primordial abundances of ^{4}He, D and ^{7}Li exist and can be compared with the theoretical predictions to learn about the conditions of the Universe at such an early period. Thus BBN can be used as a cosmological test of any non-standard physics or cosmology (Sarkar (1996)).

The physics of BBN is well understood, since in principle it only involves the Standard Model of particle physics and the time evolution of the expansion rate as given by the Friedmann equation (3). In the first phase of BBN the weak processes that keep the neutrons and protons in equilibrium,

$$n + \nu_e \leftrightarrow p + e^- \qquad n + e^+ \leftrightarrow p + \bar{\nu}_e \tag{12}$$

freeze and the neutron-to-proton ratio becomes a constant. This is later diminished due to neutron decays, $n \to p + e^- + \bar{\nu}_e$. This ratio largely fixes the produced ^{4}He abundance. Later all the primordial abundances of light elements are produced and their value depend on the competition between the nuclear reaction rates and the expansion rate. These values can be quite precisely calculated with a BBN numerical code (see e.g. Cuoco et al. (2004)). At present there exist a nice agreement with the observed abundance of D for a value of the baryon asymmetry $\eta = 6.1 \pm 0.6 \times 10^{-10}$ (Steigman (2006)). This also agrees with the asymmetry region determined by CMB and large-scale structure data (LSS). Instead, the predicted primordial abundance of ^{4}He tends to be a bit larger than the observed value. However, it is difficult to consider this as a serious discrepancy, because the accuracy of the observations of ^{4}He is limited by systematic uncertainties.

There are two main effects of relic neutrinos at BBN. The first one is that they contribute to the relativistic energy density of the universe (if $m_\nu \ll T_\nu$), thus fixing the expansion rate. This is why BBN gave the first allowed range of the number of neutrino species before this was measured with accelerators (see next section). On the other hand, BBN is the last period of the Universe sensitive to neutrino flavour, since electron neutrinos and antineutrinos play a direct role in the processes in Equation (12). We will see some examples of BBN bounds on the effective number or neutrino oscillations in the following sections.

4 Extra radiation and the effective number of neutrinos

In the standard case, neutrinos together with photons fix the expansion rate during the cosmological era when the Universe is dominated by radiation. Their contribution to the total radiation content can be parametrized in terms of the effective number of neutrinos $N_{\rm eff}$, through the relation

$$\rho_{\rm r} = \rho_\gamma + \rho_\nu = \left[1 + \frac{7}{8}\left(\frac{4}{11}\right)^{4/3} N_{\rm eff}\right] \rho_\gamma \,, \tag{13}$$

where we have normalized to the photon energy density because its value today is known from the measurement of the CMB temperature. This equation is valid when neutrino decoupling is complete and holds as long as all neutrinos are relativistic.

We know that the number of light neutrinos sensitive to weak interactions (flavour or active neutrinos) equals three from the analysis of the invisible Z-boson width at LEP, $N_\nu = 2.984 \pm 0.008$ (Yao et al. (2006)), and we saw in a previous section from the analysis of neutrino decoupling that they contribute as $N_{\rm eff} \simeq 3.046$. Any departure of $N_{\rm eff}$ from this last value would be due to non-standard neutrino features or to the contribution of other relativistic relics. For instance, the energy density of a hypothetical scalar particle ϕ in equilibrium with the same temperature as neutrinos would be $\rho_\phi = (\pi^2/30)\, T_\nu^4$, leading to a departure of $N_{\rm eff}$ from the standard value of $4/7$. A detailed discussion of cosmological scenarios where $N_{\rm eff}$ is not fixed to three can be found in Sarkar (1996) or Dolgov (2002).

In the previous section we saw that the expansion rate during BBN fixes the produced abundances of light elements, and in particular that of ^{4}He. Thus the value of $N_{\rm eff}$ can be constrained at the BBN epoch from the comparison of theoretical predictions and experimental data on the primordial abundances of light elements. In addition, a value of $N_{\rm eff}$ different from the standard one would modify the transition epoch from a radiation-dominated to a matter-dominated Universe, which has some consequences on some cosmological observables such as the power spectrum of CMB anisotropies, leading to independent bounds on the radiation content. These are two complementary ways of constraining $N_{\rm eff}$ at very different epochs.

Here, as an example, we will only describe the results of a very recent analysis by Mangano et al. (2006) (see the references therein for a list of recent works), who considered both BBN and CMB/LSS data. The allowed regions on the plane defined by $N_{\rm eff}$ and the baryon contribution to the present energy density $\Omega_{\rm b}h^2$ are shown in Figure 3. The BBN contours were calculated with the ^{4}He and D abundances, leading to the allowed range $N_{\rm eff} = 3.1^{+1.4}_{-1.2}$ (95% C.L.). Instead, the filled contours correspond to the regions found from the analysis of the most recent cosmological data on CMB temperature anisotropies and polarization, Large Scale galaxy clustering from SDSS and 2dF and luminosity distances of type Ia Supernova. We will describe these measurements later in section 8. The bounds on the effective number of neutrinos for this case are $N_{\rm eff} = 5.2^{+2.7}_{-2.2}$, while in a less conservative analysis data on the Lyman-α absorption clouds (Ly-α) and the Baryonic Acoustic Oscillations (BAO) from SDSS were added, leading to $N_{\rm eff} = 4.6^{+1.6}_{-1.5}$.

These ranges are in reasonable agreement with the standard prediction of $N_{\rm eff} \simeq 3.046$, shown as a horizontal line on the plots in Figure 3. Moreover, they show that there exists an allowed region of $N_{\rm eff}$ values that is common at early (BBN) and more recent epochs, although some tension remains particularly when adding Lyman-α and BAO data. However, the reader

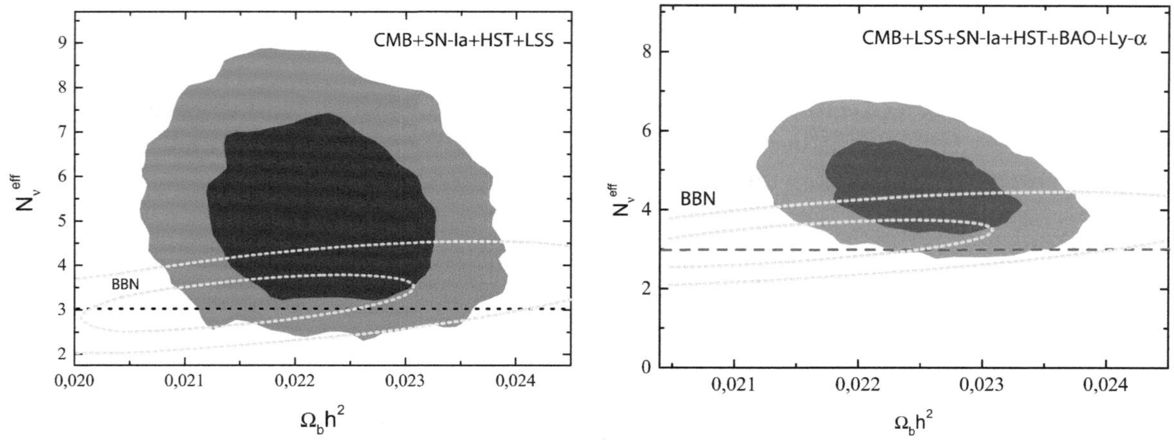

Figure 3. *Allowed regions on the ω_b-N_ν^{eff} plane at 68% and 95% C.L. from the analysis of BBN (dotted lines) and cosmological data from other observables as shown (filled contours). From Mangano et al. (2006). Copyright (2007) IOP Publishing.*

should be cautious in the interpretation of any of these allowed regions as an indication for relativistic degrees of freedom beyond the contribution of flavour neutrinos.

5 Neutrino oscillations in the Early Universe

Nowadays there exist compelling evidences for flavour neutrino oscillations from a variety of experimental data on solar, atmospheric, reactor and accelerator neutrinos. These are very important results, because the existence of flavour change implies that neutrinos mix and have non-zero masses, which in turn requires particle physics beyond the Standard Model. Thus it is interesting to check whether neutrino oscillations can modify any of the cosmological observables. More on neutrino oscillations and their implications can be found in Kayser (2006) or any of the existing reviews such as Maltoni et al. (2004) or Fogli et al. (2006), to which we refer the reader for more details.

It turns out that in the standard cosmological picture all flavour neutrinos were produced with the same energy spectrum. As we saw in section 2.1, we do not expect any effect from the oscillations among these three states. This is true up to the small spectral distortion caused by the heating of neutrinos from e^+e^- annihilations, described before (Mangano et al. (2005)). In this section we will briefly consider two cases where neutrino oscillations could have cosmological consequences: flavour oscillations with non-zero relic neutrino asymmetries and active-sterile neutrino oscillations.

5.1 Active-active: relic neutrino asymmetries

A non-zero relic neutrino asymmetry exists when the number densities of neutrinos and antineutrinos of a given flavour are different. Such a putative asymmetry could have been produced by some mechanism in the very early Universe well before the thermal decoupling epoch. This can be quantified assuming that a given flavour is characterized by a Fermi-Dirac distribution as in Equation (1) with a non-zero chemical potential μ_ν or equivalently with the dimensionless degeneracy parameter $\xi_\nu \equiv \mu_\nu/T$ (for antineutrinos, $\xi_{\bar{\nu}} = -\xi_\nu$). In such a case, sometimes one says that the relic neutrinos are degenerate, but not in the sense of equal masses.

Degenerate electron neutrinos have a direct effect on BBN: we saw in section 3 that any change in the $\nu_e/\bar{\nu}_e$ spectra modifies the primordial neutron-to-proton ratio, which in this case is $n/p \propto \exp(-\xi_{\nu_e})$. Therefore, a positive ξ_{ν_e} decreases the primordial ^{4}He mass fraction, while a negative ξ_{ν_e} increases it, leading to an allowed range

$$-0.01 < \xi_{\nu_e} < 0.07\,, \tag{14}$$

compatible with $\xi_{\nu_e} = 0$ and very restrictive for negative values. In addition a non-zero relic neutrino asymmetry always enhances the contribution of the CNB to the relativistic energy density, since for any ξ_ν one has a departure from the standard value of the effective number of neutrinos N_{eff} given by

$$\Delta N_{\text{eff}} = \frac{15}{7}\left[2\left(\frac{\xi_\nu}{\pi}\right)^2 + \left(\frac{\xi_\nu}{\pi}\right)^4\right] \tag{15}$$

We have seen that this increased radiation modifies the outcome of BBN and that bounds on N_{eff} can be obtained. In addition, another consequence of the extra radiation density is that it postpones the epoch of matter-radiation equality, producing observable effects on the spectrum of CMB anisotropies and the distribution of cosmic large-scale structures (LSS). Both independent bounds on the radiation content can be translated into flavour-independent limits on ξ_ν.

Altogether these cosmological limits on the neutrino chemical potentials or relic neutrino asymmetries are not very restrictive, since at least for BBN their effect in the ν_μ or ν_τ sector can be compensated by a positive ξ_{ν_e}. For example, an analysis of the combined effect of a non-zero neutrino asymmetry on BBN and CMB/LSS yields the allowed regions (Hansen et al. (2002))

$$-0.01 < \xi_{\nu_e} < 0.22, \qquad |\xi_{\nu_{\mu,\tau}}| < 2.6, \tag{16}$$

in agreement with similar but more updated bounds as cited in Lesgourgues and Pastor (2006). These limits allow for a very significant radiation contribution of degenerate neutrinos, leading many authors to discuss the implications of a large neutrino asymmetry in different physical situations (see e.g. Dolgov (2002)).

It is obvious that the limits in Equation (16) would be modified if neutrino flavour oscillations were effective before BBN, equalizing the neutrino chemical potentials. Actually, it was shown by Dolgov et al. (2002) that this is the case for the neutrino mixing parameters in the region favoured by present data. This result is obtained only after the proper inclusion of the refractive terms produced by the background neutrinos, which synchronize the oscillations of neutrinos with different momenta (which would evolve independently without them). In summary, since flavour equilibrium is reached before BBN, the restrictive limits on ξ_{ν_e} in Equation (14) apply to all flavours. The current bounds on the common value of the neutrino degeneracy parameter $\xi_\nu \equiv \mu_\nu/T$ are $-0.05 < \xi_\nu < 0.07$ at 2σ (Serpico and Raffelt (2005)). Thus the contribution of a relic neutrino asymmetry can be safely ignored, in turn implying that the cosmic neutrino radiation density is close to its standard value.

5.2 Active-sterile

In addition to the three species or flavour or active neutrinos as we saw from accelerator data, there could also exist extra massive neutrino states that are sterile, i.e. singlets of the Standard Model gauge group and thus insensitive to weak interactions. These sterile states ν_s were either not present in the early Universe or severely suppressed with respect to the active ones, but they could be populated through the effect of active-sterile oscillations if non-zero mixing exists, additional to that among the flavour states. This "thermalization" of the sterile neutrinos is a well-known phenomenon that is very difficult to avoid unless the cosmological scenario is drastically modified. For instance, the suppression of active-sterile oscillations may require very large pre-existing neutrino asymmetries.

There are two possible regimes for the active-sterile oscillations in cosmology, depending if they are effective before or after the decoupling of the flavour states. If oscillations occur before thermal decoupling, the abundance of sterile neutrinos grows via the conversion of the active states, whose spectrum is kept in equilibrium by the frequent weak interactions. In such a case, the contribution of sterile neutrinos to the radiation energy density is additional to that of the active states, leading to an extra N_{eff} between 0 and 1 for only one sterile species. This would be larger for more states depending on how effective the thermalization of the ν_s's was. Instead, for active-sterile oscillations effective after $T < 1$ MeV, i.e. after decoupling, the total number of neutrinos is conserved. The main feature is that large distortions may appear on the spectra of both the sterile and the active states. This has consequences on BBN when the active neutrinos are ν_e's (see e.g. Kirilova and Panayotova (2006), also for the case of non-zero initial ν_s abundance).

The detailed evolution of active-sterile oscillations in the early Universe is a difficult task, that requires solving the corresponding kinetic equations taking into account the medium effects. More details can be found for instance in the works by Dolgov and Villante (2004) and Cirelli et al. (2005). Detailed numerical calculations were carried out also including the unavoidable mixing between the flavour neutrinos (a 4×4 neutrino case). In general, one can obtain bounds on the region of active-sterile mixing parameters from the comparison with BBN and CMB/LSS data, as can be seen in the figures of the previous references or in the summary in Dolgov (2002).

Albeit the current data on neutrino oscillations does not seem to favour the existence of mixing with sterile states, the cosmological bounds provide complementary information valid also for some regions of parameters beyond the sensitivity of the laboratory experiments. Note, however, that if the results of the Liquid Scintillator Neutrino Detector (LSND) (Aguilar et al. (2001)), an experiment that has measured the appearance of electron antineutrinos in a muon antineutrino beam, are confirmed by the ongoing MiniBoone experiment, a fourth sterile neutrino would be required with mass of $\mathcal{O}$(eV). Its existence would have profound consequences in cosmology, first of all because it is very difficult to avoid its full thermalization via active-sterile oscillations.

6 Massive neutrinos as Dark Matter

Nowadays the existence of Dark Matter (DM), the dominant non-baryonic component of the matter density in the Universe, is well established. A priori, massive neutrinos are excellent DM candidates, in particular because we are certain that they exist, in contrast with other candidate particles. Together with CMB photons, relic neutrinos can be found anywhere in the Universe with a number density given by the present value of Equation (5) of 339 neutrinos and antineutrinos per cm^3. Their energy density in units of the critical value of the energy density (see Equation (10)) is

$$\Omega_\nu = \frac{\rho_\nu}{\rho_c^0} = \frac{\sum_i m_i}{93.14 h^2 \text{ eV}} \ . \tag{17}$$

Here $\sum_i m_i$ includes all masses of the neutrino states which are non-relativistic today. It is also useful to define the neutrino density fraction f_ν with respect to the total matter density

$$f_\nu \equiv \frac{\rho_\nu}{(\rho_{\text{cdm}} + \rho_{\text{b}} + \rho_\nu)} = \frac{\Omega_\nu}{\Omega_{\text{m}}} \tag{18}$$

In order to check whether relic neutrinos can have a contribution of order unity to the present values of Ω_ν or f_ν, we should consider which neutrino masses are allowed by non-cosmological data. Oscillation experiments measure the differences of squared neutrino masses $\Delta m^2_{21} = m^2_2 - m^2_1$ and $\Delta m^2_{31} = m^2_3 - m^2_1$, the relevant ones for solar and atmospheric neutrinos, respectively (Kayser (2006)). As a reference, we take the following 3σ ranges of mixing parameters from an update of Maltoni et al. (2004),

$$\begin{aligned} \Delta m^2_{21} &= (7.9^{+1.0}_{-0.8}) \times 10^{-5} \text{ eV}^2 \qquad |\Delta m^2_{31}| = (2.6 \pm 0.6) \times 10^{-3} \text{ eV}^2 \\ s^2_{12} &= 0.30^{+0.10}_{-0.06} \qquad s^2_{23} = 0.50^{+0.18}_{-0.16} \qquad s^2_{13} \leq 0.04 \end{aligned} \tag{19}$$

Here $s_{ij} = \sin\theta_{ij}$, where θ_{ij} ($ij = 12, 23$ or 13) are the three mixing angles. Unfortunately oscillation experiments are insensitive to the absolute scale of neutrino masses, since the knowledge of $\Delta m^2_{21} > 0$ and $|\Delta m^2_{31}|$ leads to the two possible schemes shown in Figure 1 of Lesgourgues and Pastor (2006), but leaves one neutrino mass unconstrained. These two schemes are known as normal (NH) and inverted (IH) hierarchies, characterized by the sign of Δm^2_{31}, positive and negative, respectively. For small values of the lightest neutrino mass m_0, i.e. m_1 (m_3) for NH (IH), the mass states follow a hierarchical scenario, while for masses much larger than the differences all neutrinos share in practice the same mass and then we say that they are degenerate. In general, the relation between the individual masses and the total neutrino mass can be found numerically, as shown in Figure 4.

There are two types of laboratory experiments searching for the absolute scale of neutrino masses, a crucial piece of information for constructing models of neutrino masses and mixings (Altarelli (2006)). The neutrinoless double beta decay $(Z,A) \to (Z+2,A) + 2e^-$ (in short $0\nu2\beta$) is a rare nuclear processes where lepton number is violated and whose observation would mean that neutrinos are Majorana particles. If the $0\nu2\beta$ process is mediated by a light neutrino, the results from neutrinoless double beta decay experiments are converted into an upper bound or a measurement of the effective mass $m_{\beta\beta}$

$$m_{\beta\beta} = |c^2_{12} c^2_{13} m_1 + s^2_{12} c^2_{13} m_2 e^{i\phi_2} + s^2_{13} m_3 e^{i\phi_3}| \ , \tag{20}$$

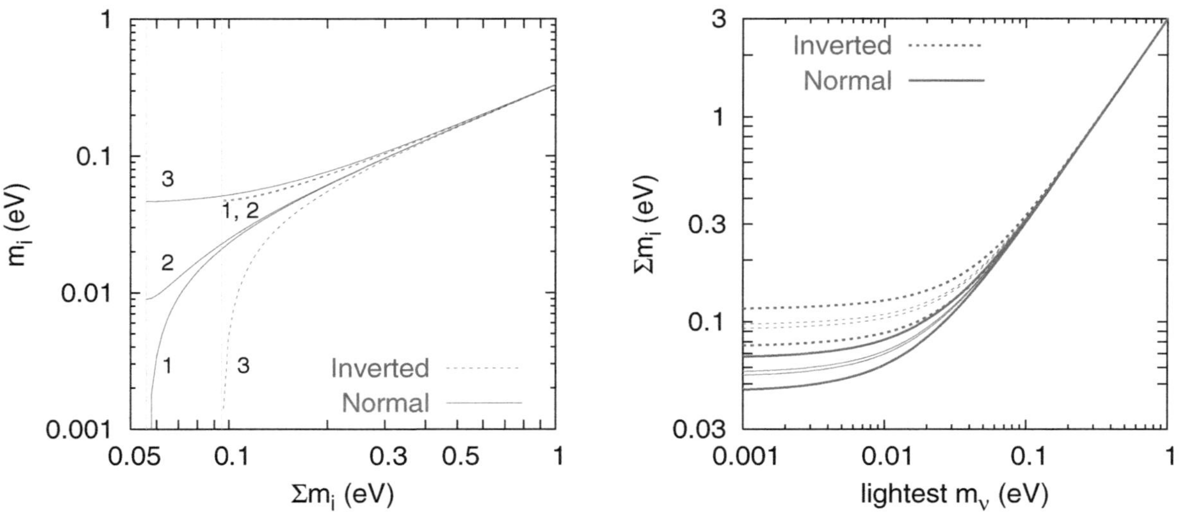

Figure 4. *Expected values of neutrino masses according to the values in Equation (19). Left: individual neutrino masses as a function of the total mass for the best-fit values of the Δm^2. Right: ranges of total neutrino mass as a function of the lightest state within the 3σ regions (thick lines) and for a future determination at the 5% level (thin lines). From (Lesgourgues and Pastor 2006). Copyright (2006) Elsevier.*

where $\phi_{2,3}$ are the two Majorana phases that appear in lepton-number-violating processes. See Zuber (2006) for more details and the current experimental results.

Beta decay experiments, which involve only the kinematics of electrons, are in principle the best strategy for measuring directly the neutrino mass (Bornschein (2006)). The current limits from tritium beta decay apply only to the range of degenerate neutrino masses, so that $m_\beta \simeq m_0$, where

$$m_\beta = (c_{12}^2 c_{13}^2 m_1^2 + s_{12}^2 c_{13}^2 m_2^2 + s_{13}^2 m_3^2)^{1/2}, \tag{21}$$

is the relevant parameter for beta decay experiments. The bound at 95% CL is $m_0 < 2.05 - 2.3$ eV from the Troitsk and Mainz experiments, respectively. This value is expected to be improved by the KATRIN project to reach a discovery potential for $0.3 - 0.35$ eV masses or a sensitivity of 0.2 eV at 90% CL. Taking into account this upper bound and the minimal values of the total neutrino mass in the normal (inverted) hierarchy, the sum of neutrino masses is restricted to the approximate range

$$0.06\,(0.1)\ \text{eV} \lesssim \sum_i m_i \lesssim 6\ \text{eV} \tag{22}$$

As we will discuss in the next sections, cosmology is at first order sensitive to the total neutrino mass $\sum_i m_i$ if all states have the same number density, providing information on m_0 but blind to neutrino mixing angles or possible CP violating phases. Thus cosmological results are complementary to terrestrial experiments. The interested reader can find the allowed regions in the parameter space defined by any pair of parameters $(\sum_i m_i, m_{\beta\beta}, m_\beta)$ in Fogli et al. (2006).

Now we can find the possible present values of Ω_ν in agreement with the three neutrino masses shown in Figure 4 and the approximate bounds of Equation (22). Note that even if the three neutrinos are non-degenerate in mass, Equation (17) can be safely applied, because we know from neutrino oscillation data that at least two of the neutrino states are non-relativistic today, since both $(\Delta m^2_{31})^{1/2} \simeq 0.05$ eV and $(\Delta m^2_{21})^{1/2} \simeq 0.009$ eV are larger than the temperature $T_\nu \simeq 1.96$ K $\simeq 1.7 \times 10^{-4}$ eV. If the third neutrino state is very light and still relativistic, its relative contribution to Ω_ν is negligible and Equation (17) remains an excellent approximation of the total density. One finds that Ω_ν is restricted to the approximate range

$$0.0013\,(0.0022) \lesssim \Omega_\nu \lesssim 0.13 \tag{23}$$

where we already included that $h \approx 0.7$. This applies only to the standard case of three light active neutrinos. In general a cosmological upper bound on Ω_ν has been used since the 1970s to constrain the possible values of neutrino masses. For instance, if we demand that neutrinos should not be heavy enough to overclose the Universe ($\Omega_\nu < 1$), we obtain an upper bound $\sum_i m_i \lesssim 45$ eV (again fixing $h = 0.7$). Moreover, since from present analyses of cosmological data we know that the approximate contribution of matter is $\Omega_{\rm m} \simeq 0.3$, the neutrino masses should obey the stronger bound $\sum_i m_i \lesssim 15$ eV. We see that with this simple argument one obtains a bound which is roughly only a factor 2 worse than the bound from tritium beta decay, but of course with the caveats that apply to any cosmological analysis. In the three-neutrino case, these bounds should be understood in terms of $m_0 = \sum_i m_i/3$.

Dark matter particles with a large velocity dispersion such as neutrinos are called hot dark matter (HDM). The role of neutrinos as HDM particles has been widely discussed since the 1970s, and the reader can find a historical review in Primack (2001). It was realized in the mid-1980s that HDM affects the evolution of cosmological perturbations in a particular way: it erases the density contrasts on wavelengths smaller than a mass-dependent free-streaming scale. In a universe dominated by HDM, this suppression is in contradiction with various observations. For instance, large objects such as superclusters of galaxies form first, while smaller structures like clusters and galaxies form via a fragmentation process. This top-down scenario is at odds with the fact that galaxies seem older than clusters.

Given the failure of HDM-dominated scenarios, the attention then turned to cold dark matter (CDM) candidates, i.e. particles which were non-relativistic at the epoch when the universe became matter-dominated, which provided a better agreement with observations. Still in the mid-1990s it appeared that a small mixture of HDM in a universe dominated by CDM fitted better the observational data on density fluctuations at small scales than a pure CDM model. However, within the presently favoured cosmological ΛCDM model which contains CDM and a cosmological constant Λ (or some form of dark energy) which is dominant at late times there is no need for a significant contribution of HDM. Instead, one can use the available cosmological data to find how large the neutrino contribution can be, as we will see later.

Before concluding this section, we would like to mention the case of a sterile neutrino with a mass of the order of a few keV's and a very small mixing with the flavour neutrinos. Such "heavy" neutrinos could be produced by active-sterile oscillations but not fully thermalized, so that they could play the role of dark matter and replace the usual CDM component. But due to their large thermal velocity (slightly smaller than that of active neutrinos), they would behave as Warm Dark Matter and erase small-scale cosmological structures. Their mass can be bounded from below using Lyman-α forest data from quasar spectra, and from above using

X-ray observations. The viability of this scenario is currently under careful examination, see e.g. Viel et al. (2006) for a recent discussion and a list of references.

7 Effects of neutrino masses on cosmology

In this section we will briefly describe the main cosmological observables and the effects that neutrino masses cause on them. A more detailed discussion of the effects of massive neutrinos on the evolution of cosmological perturbations can be found in Secs. 4.5 and 4.6 of Lesgourgues and Pastor (2006).

7.1 Brief description of cosmological observables

Although there exist many different types of cosmological measurements, here we will restrict the discussion to those that are more important for obtaining an upper bound or eventually a measurement of neutrino masses.

First of all, we have the CMB temperature anisotropy power spectrum, defined as the angular two-point correlation function of CMB maps $\delta T/\bar{T}(\hat{n})$ with $\hat{n}$ being a direction in the sky. This function is usually expanded in Legendre multipoles

$$\left\langle \frac{\delta T}{\bar{T}}(\hat{n}) \frac{\delta T}{\bar{T}}(\hat{n}') \right\rangle = \sum_{l=0}^{\infty} \frac{(2l+1)}{4\pi} C_l P_l(\hat{n}\cdot\hat{n}') \, , \quad (24)$$

where $P_l(x)$ are the Legendre polynomials. For Gaussian fluctuations, all the information is encoded in the multipoles C_l which probe correlations on angular scales $\theta = \pi/l$. We have seen that each neutrino family can only have a mass of the order of 1 eV, so that the transition of relic neutrinos to the non-relativistic regime is expected to take place after the time of recombination between electrons and nucleons, i.e. after photon decoupling. Since the shape of the CMB spectrum is related mainly to the physical evolution *before* recombination, it will be only marginally affected by the neutrino mass, except for an indirect effect through the modified background evolution. There exists interesting complementary information to the temperature power spectrum if the CMB polarization is measured, and currently we have some less precise data on the temperature $\times$ E-polarization (TE) correlation function and the E-polarization self-correlation spectrum (EE).

The current Large Scale Structure (LSS) of the Universe is probed by the matter power spectrum, observed with various techniques described in the next section (directly or indirectly, today or in the near past at redshift z). It is defined as the two-point correlation function of non-relativistic matter fluctuations in Fourier space

$$P(k,z) = \langle |\delta_{\rm m}(k,z)|^2 \rangle \, , \quad (25)$$

where $\delta_{\rm m} = \delta\rho_{\rm m}/\bar{\rho}_{\rm m}$, and $\bar{\rho}_{\rm m}$ is the average energy density of matter. Usually $P(k)$ refers to the matter power spectrum evaluated today (at $z=0$). In the case of several fluids, e.g. CDM, baryons and non-relativistic neutrinos, the total matter perturbation can be expanded as

$$\delta_{\rm m} = \frac{\sum_i \bar{\rho}_i \, \delta_i}{\sum_i \bar{\rho}_i} \, . \quad (26)$$

Since the energy density is related to the mass density of non-relativistic matter through $E = mc^2$, $\delta_{\rm m}$ represents indifferently the energy or mass power spectrum. The shape of the matter power spectrum is affected by the free-streaming caused by small neutrino masses of $\mathcal{O}$(eV). Thus it is the key observable for constraining m_ν with cosmological methods. We will show later in Figure 5 the typical shape of both the CMB temperature anisotropy spectrum C_l and the matter power spectrum $P(k)$.

7.2 Neutrino free-streaming

After thermal decoupling, relic neutrinos constitute a collisionless fluid, where the individual particles free-stream with a characteristic velocity that, on average, is the thermal velocity $v_{\rm th}$. It is possible to define a horizon as the typical distance on which particles travel between time t_i and t. When the Universe was dominated by radiation or matter $t \gg t_i$, this horizon is, as usual, asymptotically equal to $v_{\rm th}/H$, up to a numerical factor of order one. Similar to the definition of the Jeans length (see section 4.4 in Lesgourgues and Pastor (2006)), we can define the neutrino free-streaming wavenumber and length as

$$k_{FS}(t) = \left(\frac{4\pi G\bar{\rho}(t)a^2(t)}{v_{\rm th}^2(t)}\right)^{1/2}, \qquad \lambda_{FS}(t) = 2\pi\frac{a(t)}{k_{FS}(t)} = 2\pi\sqrt{\frac{2}{3}}\frac{v_{\rm th}(t)}{H(t)}\,. \tag{27}$$

where $a(t)$ is the scale factor. As long as neutrinos are relativistic, they travel at the speed of light and their free-streaming length is simply equal to the Hubble radius. When they become non-relativistic, their thermal velocity decays like

$$v_{\rm th} \equiv \frac{\langle p\rangle}{m} \simeq \frac{3T_\nu}{m} = \frac{3T_\nu^0}{m}\left(\frac{a_0}{a}\right) \simeq 150(1+z)\left(\frac{1\,{\rm eV}}{m}\right){\rm km\,s^{-1}}\,, \tag{28}$$

where we used for the present neutrino temperature $T_\nu^0 \simeq (4/11)^{1/3}T_\gamma^0$ and $T_\gamma^0 \simeq 2.726$ K. This gives for the free-streaming wavelength and wavenumber during matter or Λ domination

$$\lambda_{FS}(t) = 7.7\frac{1+z}{\sqrt{\Omega_\Lambda+\Omega_m(1+z)^3}}\left(\frac{1\,{\rm eV}}{m}\right)h^{-1}{\rm Mpc}\,, \tag{29}$$

$$k_{FS}(t) = 0.82\frac{\sqrt{\Omega_\Lambda+\Omega_m(1+z)^3}}{(1+z)^2}\left(\frac{m}{1\,{\rm eV}}\right)h{\rm Mpc}^{-1}, \tag{30}$$

where Ω_Λ and Ω_m are the cosmological constant and matter density fractions, respectively, evaluated today. So, after the non-relativistic transition and during matter domination, the free-streaming length continues to increase, but only like $(aH)^{-1} \propto t^{1/3}$, i.e. more slowly than the scale factor $a \propto t^{2/3}$. Therefore, the comoving free-streaming length λ_{FS}/a actually decreases like $(a^2H)^{-1} \propto t^{-1/3}$. As a consequence, for neutrinos becoming non-relativistic during matter domination, the comoving free-streaming wavenumber passes through a minimum $k_{\rm nr}$ at the time of the transition, i.e. when $m = \langle p\rangle = 3T_\nu$ and $a_0/a = (1+z) = 2.0\times10^3(m/1\,{\rm eV})$. This minimum value is found to be

$$k_{\rm nr} \simeq 0.018\,\Omega_{\rm m}^{1/2}\left(\frac{m}{1\,{\rm eV}}\right)^{1/2}h{\rm Mpc}^{-1}\,. \tag{31}$$

The physical effect of free-streaming is to damp small-scale neutrino density fluctuations: neutrinos cannot be confined into (or kept outside of) regions smaller than the free-streaming length, for obvious kinematic reasons. There exists a gravitational back-reaction effect that also damps the metric perturbations on those scales. Instead, on scales much larger than the free-streaming scale the neutrino velocity can be effectively considered as vanishing and after the non-relativistic transition the neutrino perturbations behave like CDM perturbations. In particular, modes with $k < k_{\rm nr}$ are never affected by free-streaming and evolve like in a pure ΛCDM model.

7.3 Impact of massive neutrinos on the matter power spectrum

The small initial cosmological perturbations in the early Universe evolve, under the linear regime at any scale at early times and on the largest scales more recently, and produce the structures we see today. We can not review here all the details (see Lesgourgues and Pastor (2006) and references therein), but we will emphasize the main effects caused by massive neutrinos in the framework of the standard cosmological scenario: a Λ Mixed Dark Matter (ΛMDM) model, where Mixed refers to the inclusion of some HDM component.

First let us describe the changes in the background evolution of the Universe. We have seen that massless neutrinos are always part of the radiation content, so in this case the present value of the matter contribution $\Omega^0_{\rm m}$ is equal to the contribution of CDM and baryons. Instead, massive neutrinos contribute to radiation at early times but to matter after becoming non-relativistic. Thus with respect to the massless neutrino case, massive neutrinos also contribute to $\Omega^0_{\rm m}$, reducing the values of $\Omega^0_{\rm CDM}$ and $\Omega^0_{\rm b}$. As a result, this transition is delayed if these massive neutrinos have not yet become non-relativistic at the time of radiation/matter equality which is the epoch of the Universe when it starts to be dominated by matter and the contribution of radiation becomes subdominant. The consequence of a late equality for the LSS matter power spectrum is the following: since on sub-Hubble scales the matter density contrast $\delta_{\rm m}$ grows more efficiently during MD than during RD, the matter power spectrum is suppressed on small scales relatively to large scales.

At the perturbation level, we also saw that free-streaming damps small-scale neutrino density fluctuations. This produces a direct effect on the matter power spectrum (see section 4.5 of Lesgourgues and Pastor (2006)) that depends on the value k with respect to $k_{\rm nr}$ in Equation (31),

$$
\begin{aligned}
P(k) &= \left\langle \left(\frac{\delta\rho_{\rm cdm} + \delta\rho_{\rm b} + \delta\rho_\nu}{\rho_{\rm cdm} + \rho_{\rm b} + \rho_\nu} \right)^2 \right\rangle \\
&= \left\langle \left(\frac{\Omega_{\rm cdm}\,\delta_{\rm cdm} + \Omega_{\rm b}\,\delta_{\rm b} + \Omega_\nu\,\delta_\nu}{\Omega_{\rm cdm} + \Omega_{\rm b} + \Omega_\nu} \right)^2 \right\rangle \\
&= \begin{cases} \langle \delta^2_{\rm cdm} \rangle & \text{for} \quad k < k_{\rm nr}, \\ [1 - \Omega_\nu/\Omega_{\rm m}]^2 \, \langle \delta^2_{\rm cdm} \rangle & \text{for} \quad k \gg k_{\rm nr}, \end{cases}
\end{aligned} \tag{32}
$$

with $\Omega_{\rm m} \equiv \Omega_{\rm cdm} + \Omega_{\rm b} + \Omega_\nu$. Thus the role of the neutrino masses would be simply to cut the power spectrum by a factor $[1 - \Omega_\nu/\Omega_{\rm m}]^2$ for $k \gg k_{\rm nr}$. However, it turns out that the presence

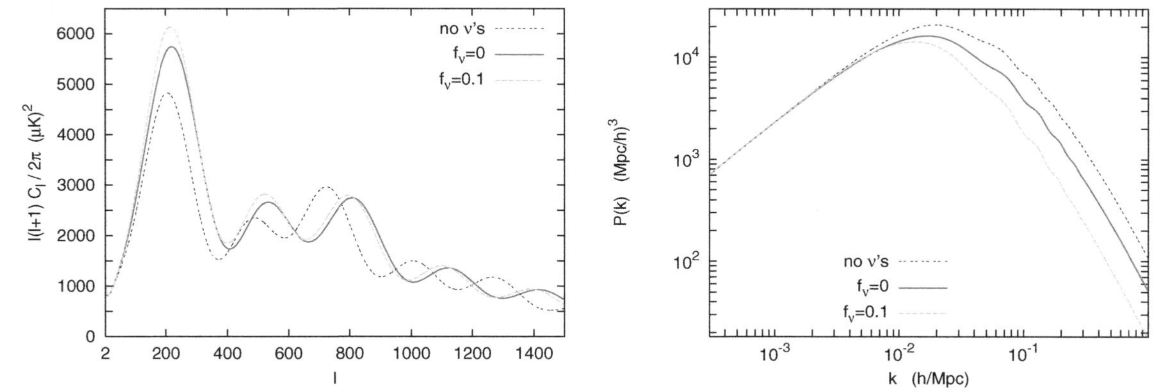

Figure 5. *CMB temperature anisotropy spectrum C_l^T and matter power spectrum $P(k)$ for three models: the neutrinoless ΛCDM model, a more realistic ΛCDM model with three massless neutrinos ($f_\nu \simeq 0$), and finally a ΛMDM model with three massive degenerate neutrinos and a total density fraction $f_\nu = 0.1$. In all models, the values of the cosmological parameters ($\omega_b = \Omega_b h^2$, $\omega_m = \Omega_m h^2$, Ω_Λ, A_s, n, τ) have been kept fixed. From (Lesgourgues and Pastor 2006). Copyright (2006) Elsevier.*

of neutrinos actually modifies the evolution of the CDM and baryon density contrasts in such way that the suppression factor is greatly enhanced, more or less by a factor four.

In conclusion, the combined effect of the shift in the time of equality and of the reduced CDM fluctuation growth during matter domination produces an attenuation of small-scale perturbations for $k > k_{nr}$. It can be shown that for small values of f_ν this effect can be approximated in the large values of k limit by the well-known linear expression (Hu et al. (1998))

$$\frac{P(k)^{f_\nu}}{P(k)^{f_\nu=0}} \simeq 1 - 8 f_\nu \, . \tag{33}$$

For the comparison with the data, one could use instead some better analytical approximations to the full MDM or ΛMDM matter power spectrum, valid for arbitrary scales and redshifts, as listed in Lesgourgues and Pastor (2006). However, nowadays the analyses are performed using the matter power spectra calculated by Boltzmann codes such as CMBFAST (Seljak and Zaldarriaga (1996)) or CAMB (Lewis et al. (2000)), that solve numerically the evolution of the cosmological perturbations.

An example of $P(k)$ with and without massive neutrinos is shown in Figure 5, where the effect of m_ν at large k's can be clearly visible. Such suppression is probably better seen in Figure 6, where we plot the ratio of the matter power spectrum for ΛMDM over that of ΛCDM, for different values of f_ν and three degenerate massive neutrinos, but for fixed parameters $(\omega_m, \Omega_\Lambda)$. For large k's, Equation (33) is a reasonable first-order approximation for $0 < f_\nu < 0.07$.

Is it possible to mimic the effect of massive neutrinos on the matter power spectrum with some combination of other cosmological parameters? If so, one would say that a parameter degeneracy exists, reducing the sensitivity to neutrino masses. This possibility depends on the interval $[k_{min}, k_{max}]$ in which the $P(k)$ can be accurately measured. Ideally, if we could have

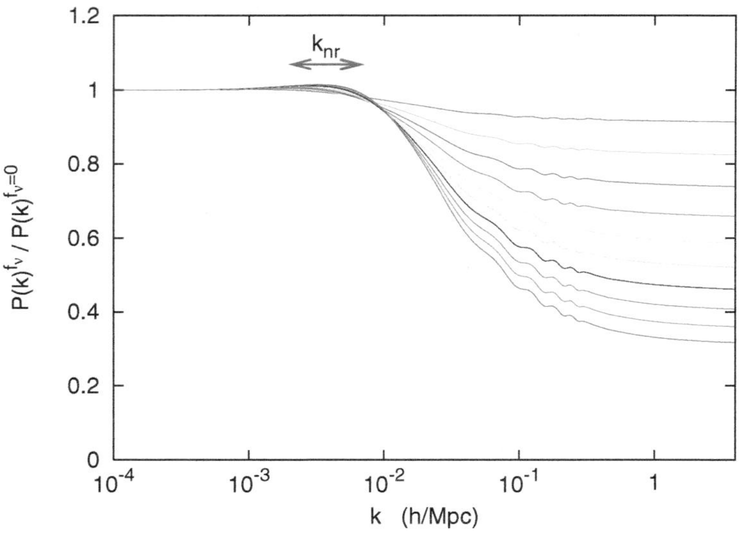

Figure 6. *Ratio of the matter power spectrum including three degenerate massive neutrinos with density fraction f_ν to that with three massless neutrinos. The parameters $(\omega_m, \Omega_\Lambda) = (0.147, 0.70)$ are kept fixed, and from top to bottom the curves correspond to $f_\nu = 0.01, 0.02, 0.03, \ldots, 0.10$. The individual masses m_ν range from 0.046 to 0.46 eV, and the scale k_{nr} from $2.1 \times 10^{-3} h\,Mpc^{-1}$ to $6.7 \times 10^{-3} h\,Mpc^{-1}$ as shown on the top of the figure. From (Lesgourgues and Pastor 2006). Copyright (2006) Elsevier.*

$k_{\rm min} \leq 10^{-2} h\,{\rm Mpc}^{-1}$ and $k_{\rm max} \geq 1\,h\,{\rm Mpc}^{-1}$, the effect of the neutrino mass would be non-degenerate, because of its very characteristic step-like effect. In contrast, other cosmological parameters like the scalar tilt or the tilt running change the spectrum slope on all scales. The problem is that usually the matter power spectrum can only be accurately measured in the intermediate region where the mass effect is neither null nor maximal: in other words, many experiments only have access to the transition region in the step-like transfer function. In this region, the neutrino mass affects the slope of the matter power spectrum in a way which can be easily confused with the effect of other cosmological parameters. Because of these parameter degeneracies, the LSS data alone cannot provide significant constraints on the neutrino mass, and it is necessary to combine them with other cosmological data, in particular the CMB anisotropy spectrum, which could lift most of the degeneracies. Still, for exotic models with e.g. extra relativistic degrees of freedom, a constant equation-of-state parameter of the dark energy different from -1 or a non-power-law primordial spectrum, the neutrino mass bound can become significantly weaker.

7.4 Impact of massive neutrinos on the CMB anisotropy spectrum

For neutrino masses of the order of 1 eV (about $f_\nu \leq 0.1$) the three neutrino species are still relativistic at the time of photon decoupling, and the direct effect of free-streaming neutrinos on the evolution of the baryon-photon acoustic oscillations is the same in the ΛCDM and ΛMDM cases. Therefore, the effect of the mass is indirect, appearing only at the level of the background evolution: the fact that today the neutrinos account for a fraction Ω_ν of the critical density implies some change either in the present value of the spatial curvature, or in the relative density of other species. If neutrinos were heavier than a few eV, they would already be non-relativistic at decoupling. This case would have more complicated consequences for the CMB, as described in (Dodelson et al. (1996)). However, we will see later that this situation is disfavoured by current upper bounds on the neutrino mass.

Let us describe one example: we choose to maintain a flat Universe where the sum of all Ω_i is one, with fixed $\omega_b = \Omega_b h^2$ and $\omega_m = \Omega_m h^2$, and Ω_Λ. Thus, while Ω_b and Ω_Λ are constant, Ω_{cdm} is constrained to decrease as Ω_ν increases. The main effect on the CMB anisotropy spectrum results from a change in the time of equality. Since neutrinos are still relativistic at decoupling, they should be counted as radiation instead of matter around the time of equality, which is found by solving $\rho_b + \rho_{cdm} = \rho_\gamma + \rho_\nu$. This gives $a_{eq} = \Omega_r/(\Omega_b + \Omega_{cdm})$, where Ω_r stands for the radiation density extrapolated until today *assuming that all neutrinos would remain massless*, given by Equation (13) with $N_{eff} \simeq 3.04$. So, when f_ν increases, a_{eq} increases proportionally to $[1 - f_\nu]^{-1}$: equality is postponed. This produces an enhancement of small-scale perturbations, especially near the first acoustic peak. Also, postponing the time of equality increases slightly the size of the sound horizon at recombination. These two features explain why in Figure 5 the acoustic peaks are slightly enhanced and shifted to the left in the ΛMDM case.

Since the effect of the neutrino mass on CMB fluctuations is indirect and appears only at the background level, one could think that it would be possible to cancel this effect exactly by changing the value of other cosmological parameters (i.e. a parameter degeneracy). It can be actually shown that in the simplest ΛMDM model, with only seven cosmological parameters, one cannot vary the neutrino mass while keeping fixed a_{eq} and all other quantities governing the CMB spectrum. Therefore, it is possible to constrain the neutrino mass using CMB experiments alone (**?**), although neutrinos are still relativistic at decoupling. This conclusion can be altered in more complicated models with extra cosmological parameters. For instance, allowing for an open Universe or varying the number of relativistic degrees of freedom. In such extended models the CMB alone is not sufficient for constraining the mass, but fortunately the LSS power spectrum can lift the degeneracy.

8 Current bounds on neutrino masses

In this section we review how the available cosmological data is used to get information on the absolute scale of neutrino masses, complementary to laboratory experiments. Note that the bounds in the next subsections are all based on the Bayesian inference method, and the upper bounds on the sum of neutrino masses are given at 95% C.L. after marginalization over all free cosmological parameters. We refer the reader to section 5.1 of Lesgourgues and Pastor (2006) for a detailed discussion on this statistical method, as well as for most of the references for the experimental data or parameter analysis.

8.1 CMB anisotropies

The experimental situation of the measurement of the CMB anisotropies is dominated by the third-year release of WMAP data (WMAP3), which improved the already precise TT and TE angular power spectra of the first-year release (WMAP1), and adds a detection of the E-polarization self-correlation spectrum (EE). On similar or smaller angular scales than WMAP, we have results from experiments that are either ground-based (ACBAR, VSA, CBI, DASI, ...) or balloon-borne (ARCHEOPS, BOOMERANG, MAXIMA, ...). When using data from different CMB experiments, one should take into account that they overlap in some multipole region, and that not all data are uncorrelated. We saw in the previous section that the signature on the CMB spectrum of a neutrino mass smaller than 0.5 eV is small but does not vanish due to a background effect, proportional to Ω_ν, which changes some characteristic times and scales in the evolution of the Universe, and affects mainly the amplitude of the first acoustic peak as well as the location of all the peaks. Therefore, it is possible to constrain neutrino masses using CMB experiments only, down to the level at which this background effect is masked by instrumental noise, or by cosmic variance, or by parameter degeneracies in the case of some cosmological models beyond the minimal Λ Mixed Dark matter framework.

Here it is assumed that the total neutrino mass was the only additional parameter with respect to a flat ΛCDM cosmological model characterized by 6 parameters. This will be the case for the bounds reviewed in this section, unless specified otherwise. In this framework, many analyses support the conclusion that a sensible bound on neutrino masses exists using CMB data only, namely of order of $2-3$ eV for the total mass $M_\nu \equiv \sum_i m_i$ depending on the data included in addition to WMAP. This is an important result, since it does not depend on the uncertainties from LSS data discussed next.

8.2 Galaxy redshift surveys

We have seen that free-streaming of massive neutrinos produces a direct effect on the formation of cosmological structures. As shown in Figure 6, the presence of neutrino masses leads to an attenuation of the linear matter power spectrum on small scales. In a seminal paper, Hu et al. (1998) showed that an efficient way to probe neutrino masses of order eV was to use data from large redshift surveys, which measure the distance to a large number of galaxies, giving us a three-dimensional picture of the universe. At present, we have data from two large projects: the 2 degree Field (2dF) galaxy redshift survey, whose final results were obtained from more than 220,000 galaxy redshifts, and the Sloan Digital Sky Survey (SDSS), which will be completed soon with data from one million galaxies.

One of the main goals of galaxy redshift surveys is to reconstruct the power spectrum of matter fluctuations on very large scales, whose cosmological evolution is described entirely by linear perturbation theory. However, the linear power spectrum must be reconstructed from individual galaxies which underwent a strongly non-linear evolution. A simple analytic model of structure formation suggests that on large scales, the galaxy-galaxy correlation function should be, not equal, but proportional to the linear matter density power spectrum, up to a constant factor that is called the light-to-mass bias (b). This parameter can be obtained from independent methods, which tend to confirm that the linear biasing assumption is correct, at least in first approximation.

A conservative way to use the measurements of galaxy-galaxy correlations in an analysis of cosmological data is to take the bias as a free parameter, i.e. to consider only the shape of the matter power spectrum at the corresponding scales and not its amplitude (denoted as galaxy clustering data). An upper limit on M_ν between 0.8 and 1.7 eV is found from the analysis of galaxy clustering data (SDSS and/or 2dF, leaving the bias as a free parameter) added to CMB data. These values improve those found with CMB data only. In general, one obtains weaker bounds on neutrino masses using preliminary SDSS results instead of 2dF data, but this conclusion could change after the next SDSS releases. The bounds on neutrino masses are more stringent when the amplitude of the matter power spectrum is fixed with a measurement of the bias, instead of leaving it as a free parameter. The upper limits on M_ν are reduced to values of order $0.5-0.9$ eV although some analyses also add Lyman-α data (see next subsection).

Finally, a galaxy redshift survey performed in a large volume can also be sensitive to the imprint created by the baryon acoustic oscillations (BAO) on the power spectrum of non-relativistic matter at large scales. Since baryons are only a subdominant component of the non-relativistic matter, the BAO feature is manifested as a small single peak in the galaxy correlation function in real space that was recently detected from the analysis of the SDSS luminous red galaxy (LRG) sample. The observed position of this baryon oscillation peak provides a way to measure the angular diameter distance out to the typical LRG redshift of $z = 0.35$, which in turn can be used to constrain the parameters of the underlying cosmological model. The SDSS measurement was included by Goobar et al. (2006) to get a bound of 0.44 eV on the total neutrino mass M_ν.

8.3 Lyman-α forest

The matter power spectrum on small scales can also be inferred from data on the so-called Lyman-α forest. This corresponds to the Lyman-α absorption by the neutral hydrogen in the intergalactic medium of photons traveling from distant quasars ($z \sim 2-3$). As an effect of the Universe expansion, photons are continuously red-shifted along the line of sight, and can be absorbed when they reach a wavelength of 1216 Angstroms in the rest-frame of the intervening medium. Therefore, the quasar spectrum contains a series of absorption lines, whose amplitude as a function of wavelength traces back the density and temperature fluctuations of neutral hydrogen along the line of sight. It is then possible to infer the matter density fluctuations in the linear or quasi-linear regime.

In order to use the Lyman-α forest data, one needs to recover the matter power spectrum from the spectrum of the transmitted flux, a task that requires the use of hydro-dynamical simulations for the corresponding cosmological model. This is a difficult procedure, and given the various systematics involved in the analysis the robustness of Lyman-α forest data is still a subject of intense discussion between experts.

In any case, the recovered matter power spectrum is again sensitive to the suppression of growth of mass fluctuations caused by massive neutrinos, and in many cosmological analyses the Lyman-α data is added to CMB and other LSS data. For a free bias, one finds that Lyman-α data help to reduce the upper bounds on the total neutrino mass to the level $M_\nu < 0.5-0.7$ eV. But those analyses that include Lyman-α data and a measurement of the bias do not always lead to a lower limit, ranging from 0.4 to 0.7 eV. Finally, Goobar et al. (2006) found the upper bound $M_\nu < 0.30$ eV adding simultaneously Lyα and BAO data, both from SDSS.

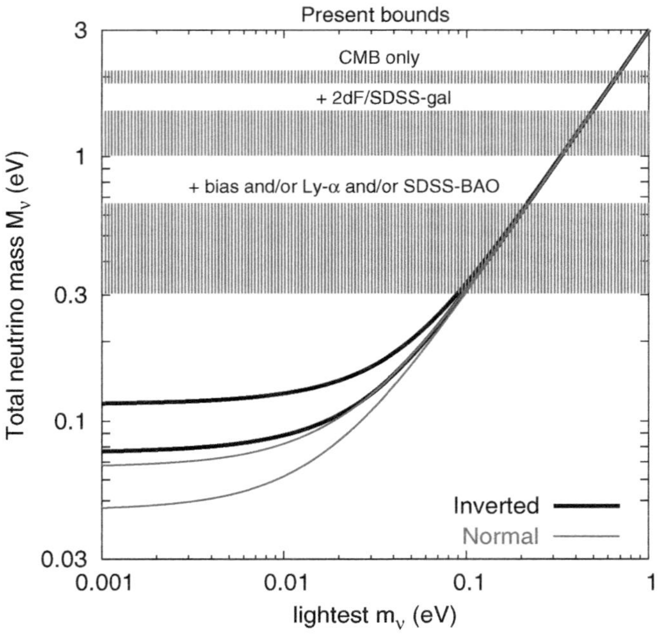

Figure 7. *Current upper bounds (95%CL) from cosmological data on the sum of neutrino masses, compared to the values in agreement at a 3σ level with neutrino oscillation data in Equation (19). From (Lesgourgues and Pastor 2006). Copyright (2006) Elsevier.*

8.4 Summary and discussion of current bounds

The upper bounds on M_ν from the previous subsections are representative of an important fact: a single cosmological bound on neutrino masses does not exist. A graphical summary is presented in Figure 7, where the cosmological bounds correspond to the bands, which were grouped according to the included set of data and whose thickness roughly describe the spread of values obtained from similar cosmological data : $2-3$ eV for CMB only, $0.9-1.7$ eV for CMB and 2dF/SDSS-gal or $0.3-0.9$ eV with the inclusion of a measurement of the bias and/or Lyman-α forest data and/or the SDSS measurement of the baryon oscillation peak.

Note that it is usual to add other measurements of cosmological parameters to CMB and LSS data. Probably the most important case is the measurement of the present value of the Hubble parameter by the Key Project of the Hubble Space Telescope, giving $h = 0.72 \pm 0.08\,(1\sigma)$, which excludes low values of h and leads to a stronger upper bound on the total neutrino mass. In addition, one can include the constraints on the current density of the dark energy component deduced from the redshift dependence of type Ia supernovae (SNIa) luminosity, which measures the late evolution of the expansion rate of the Universe. For a flat Universe with a cosmological constant, these constraints can be translated into bounds for the matter density $\Omega_{\rm m}$.

One can see from Figure 7 that current cosmological data probe the region of neutrino masses where the 3 neutrino states are degenerate, with a mass $M_\nu/3$. This mass region is conservatively bounded to values below approximately 1 eV from CMB results combined only with galaxy clustering data from 2dF and/or SDSS, i.e. the shape of the matter power spectrum for the relevant scales. The addition of further data leads to an improvement of the bounds, which reach the lowest values when data from Lyman-α and/or the SDSS measurement of the baryon oscillation peak are included or the bias is fixed. In such cases the contribution of a total neutrino mass of the order $0.3-0.6$ eV seems already disfavoured. It is interesting to compare these bounds with those coming from tritium beta decay and neutrinoless double beta decay, as recently done in Fogli et al. (2006).

Finally, we remind the reader that the impressive cosmological bounds on neutrino masses shown in Figure 7 may change if additional cosmological parameters, beyond those included in the minimal ΛCDM, are allowed. This could be the case whenever a new parameter degeneracy with the neutrino masses arises. For instance, in the presence of extra radiation parametrized by a larger $N_{\rm eff}$ the bound on M_ν gets less stringent (Crotty et al. (2004)). These results are interesting for the 4-neutrino mass schemes that also incorporate the results of the LSND experiment (see e.g. Dodelson et al. (2006) for a recent analysis). Another possible parameter degeneracy exists between neutrino masses and the parameter w, that characterizes the equation of state of the dark energy component X ($p_X = w\rho_X$). This degeneracy can be broken by adding data on baryon acoustic oscillations (Goobar et al. (2006)). Finally, the cosmological implications of neutrino masses could be very different if the spectrum or evolution of the cosmic neutrino background was non-standard (see the discussion in section 5.7 of Lesgourgues and Pastor (2006)).

9 Future sensitivities on neutrino masses from cosmology

In the near future we will have more precise data on cosmological observables from various experimental techniques and experiments. If the characteristics of these future experiments are known with some precision, it is possible to assume a "fiducial model", i.e. a cosmological model that would yield the best fit to future data, and to estimate the error bar on a particular parameter that will be obtained after marginalizing the hypothetical likelihood distribution over all the other free parameters. Technically, the simplest way to forecast this error is to compute a Fisher matrix, a technique has been widely used in the literature, for many different models and hypothetical datasets, now complemented by Monte Carlo methods. Here we will focus on the results for $\sigma(M_\nu)$, the forecast 68% CL error on the total neutrino mass, assuming various combinations of future observations: CMB anisotropies measured with ground-based experiments or satellites such as PLANCK, galaxy redshift surveys and galaxy cluster surveys. In particular, it has recently been emphasized the potentiality for measuring small neutrino masses in weak lensing experiments, which will look for the lensing effect caused by the large scale structure of the neighboring universe, either on the CMB signal (Lesgourgues et al. (2006)) or on the apparent shape of galaxies (measured by cosmic shear surveys, see e.g. Hannestad et al. (2006)). We refer the reader to section 6 of Lesgourgues and Pastor (2006) for further details.

We give a graphical summary of the sensitivities to neutrino masses forecast for different cosmological data in Figure 8, compared to the allowed values of neutrino masses in the two possible 3-neutrino schemes. One can see from this figure that there are very good prospects for testing neutrino masses in the degenerate and quasi-degenerate mass regions above 0.2 eV or so. A detection at a significant level of the minimal value of the total neutrino mass in the inverted hierarchy scheme will demand the combination of future data from CMB lensing

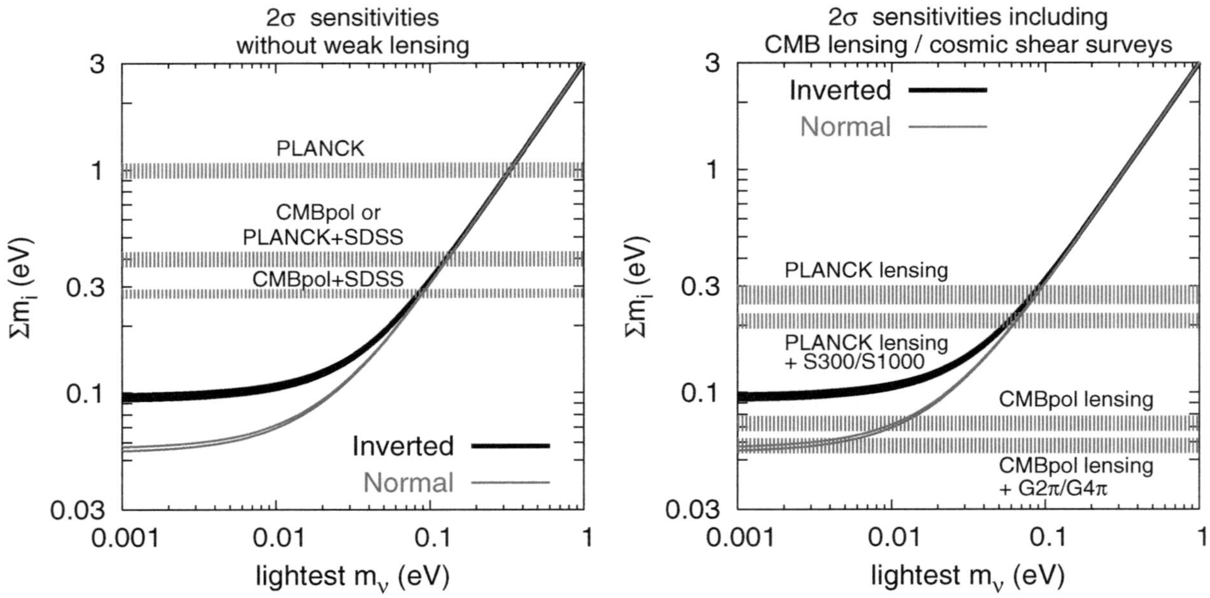

Figure 8. *Forecast 2σ sensitivities to the total neutrino mass from future cosmological experiments compared to the values in agreement with present neutrino oscillation data in Equation (19) (assuming a future determination at the 5% level). Left: sensitivities expected for future CMB experiments (without lensing extraction), alone and combined with the completed SDSS galaxy redshift survey. Right: sensitivities expected for future CMB experiments including lensing information, alone and combined with future cosmic shear surveys. Here CMBpol refers to a hypothetical CMB experiment roughly corresponding to the* INFLATION PROBE *mission. From (Lesgourgues and Pastor 2006). Copyright (2006) Elsevier.*

and cosmic shear surveys, whose more ambitious projects will provide a 2σ sensitivity to the minimal value in the case of normal hierarchy (of order 0.05 eV). The combination of CMB observations with future galaxy cluster surveys (Wang et al. (2005)), derived from the same weak lensing observations, as well as X-ray and Sunyaev–Zel'dovich surveys, should yield a similar sensitivity.

10 Conclusions

Neutrinos, despite the weakness of their interactions and their small masses, can play an important role in Cosmology, which we have reviewed in this contribution. In addition, cosmological data can be used to constrain neutrino properties, providing information on these elusive particles that complements the efforts of laboratory experiments. In particular, the data on cosmological observables have been used to bound the effective number of neutrinos (including a potential extra contribution from other relativistic particles).

But probably the most important contribution of Cosmology to our knowledge of neutrino properties is the information it can provide on the absolute scale of neutrino masses. We have seen that the analysis of cosmological data can lead to either a bound or a measurement of the sum of neutrino masses, an important result complementary to terrestrial experiments such as tritium beta decay and neutrinoless double beta decay experiments. In the near future, thanks to the data from new cosmological experiments we could even hope to test the minimal values of neutrino masses guaranteed by the present evidences for flavour neutrino oscillations. For this and many other reasons, we expect that neutrino cosmology will remain an active research field in the next years.

Acknowledgments

I thank the organizers of the School for their invitation and hospitality. Many of the topics discussed here were developed in an enjoyable collaboration with Julien Lesgourgues. This work was supported by the European Network of Theoretical Astroparticle Physics ILIAS/N6 under contract number RII3-CT-2004-506222 and the Spanish grants FPA2005-01269 and GV/05/017 of Generalitat Valenciana, as well as by a Ramón y Cajal contract of MEC.

References

Aguilar A et al. [LSND Collaboration] (2001), *Physical Review D* **64**, 112007.

Altarelli G (2006), This volume.

Bornschein B (2006), This volume.

Buchmüller W (2006), This volume.

Cirelli M et al. (2005), *Nuclear Physics B* **708**, 215.

Crotty P et al. (2004), *Physical Review D* **69**, 123007.

Cuoco A et al. (2004), *International Journal of Modern Physics A* **19**, 4431.

Dodelson S et al. (1996), *Astrophysical Journal* **467**, 10.

Dodelson S (2003), *Modern Cosmology* (Academic Press).

Dodelson S et al. (2006), *Physical Review Letters* **97**, 041301.

Dolgov A D (2002), *Physics Reports* **370**, 333.

Dolgov A D et al. (2002), *Nuclear Physics B* **632**, 363.

Dolgov A D and Villante F L (2004), *Nuclear Physics B* **679**, 261.

Elgaroy and Lahav O (2005), *New Journal of Physics* **7**, 61.

Fogli G L et al. (2006), *Progress in Particle and Nuclear Physics* **57**, 742.

Goobar A et al. (2006), *Journal of Cosmology and Astroparticle Physics* **0606**, 019.

Hannestad S (2004), *New Journal of Physics* **6**, 108.

Hannestad S (2006), *Annual Review of Nuclear and Particle Science* **56**, 137.

Hannestad S et al. (2006), *Journal of Cosmology and Astroparticle Physics* **0606**, 025.

Hansen S H et al. (2002), *Physical Review D* **65**, 023511.

Hu W et al. (1998), *Physical Review Letters* **80**, 5255.

Ichikawa K et al. (2005), *Physical Review D* **71**, 043001.

Kamionkowski M and Kosowsky A (1999), *Annual Review of Nuclear and Particle Science* **49**, 77.

Kayser B (2006), This volume.

Kirilova D P and Panayotova M P (2006), astro-ph/0608103.

Kolb E W and Turner M S (1990), *The Early Universe* (Addison-Wesley).

Lesgourgues J et al. (2006), *Physical Review D* **73**, 045021.

Lesgourgues J and Pastor S (2006), *Physics Reports* **429**, 307.

Lewis A et al. (2000), *Astrophysical Journal* **538**, 473. See also the webpage `http://camb.info`

Maltoni M et al. (2004), *New Journal of Physics* **6**, 122.

Mangano G et al. (2005), *Nuclear Physics B* **729**, 221.

Mangano G et al. (2006), *Journal of Cosmology and Astroparticle Physics* **03** (2007) 006; astro-ph/0612150.

Primack J R (2001), astro-ph/0112336.

Sarkar S (1996), *Reports of Progress in Physics* **59** (1996) 1493

Seljak U and Zaldarriaga M (1996), *Astrophysical Journal* **469**, 437. See also the webpage `http://cmbfast.org`

Serpico P D and Raffelt G G (2005), *Physical Review D* **71**, 127301.

Steigman G (2006), *International Journal of Modern Physics E* **15**, 1.

Viel M et al. (2006), *Physical Review Letters* **97**, 071301.

Wang S et al. (2005), *Physical Review Letters* **95**, 011302.

Yao W M et al. [Particle Data Group] (2006), *Journal of Physics G* **33**, 1.

Zuber K (2006), This volume.

Index

* 9 7 8 0 3 6 7 3 8 6 4 9 8 *

T - #0559 - 071024 - C400 - 279/216/19 - PB - 9780367386498 - Gloss Lamination